Brock/Springer Series in Contemporary Bioscience

The Rumen Protozoa

Brock/Springer Series in Contemporary Bioscience

Series Editor: Thomas D. Brock
University of Wisconsin-Madison

Tom Fenchel
ECOLOGY OF PROTOZOA:
The Biology of Free-living Phagotrophic Protists

Johanna Döbereiner and Fábio O. Pedrosa
NITROGEN-FIXING BACTERIA IN NONLEGUMINOUS CROP PLANTS

Tsutomu Hattori
THE VIABLE COUNT: Quantitative and Environmental Aspects

Roman Saliwanchik
PROTECTING BIOTECHNOLOGY INVENTIONS:
A Guide for Scientists

Hans G. Schlegel and Botho Bowien (Editors)
AUTOTROPHIC BACTERIA

Barbara Javor
HYPERSALINE ENVIRONMENTS: Microbiology and Biogeochemistry

Ulrich Sommer (Editor)
PLANKTON ECOLOGY: Succession in Plankton Communities

Stephen R. Rayburn
THE FOUNDATIONS OF LABORATORY SAFETY:
A Guide for the Biomedical Laboratory

Gordon A. McFeters (Editor)
DRINKING WATER MICROBIOLOGY:
Progress and Recent Developments

Mary Helen Briscoe
A RESEARCHER'S GUIDE TO SCIENTIFIC AND MEDICAL ILLUSTRATIONS

Max M. Tilzer and Colette Serruya (Editors)
LARGE LAKES: Ecological Structure and Function

Jürgen Overbeck and Ryszard J. Chróst (Editors)
AQUATIC MICROBIAL ECOLOGY:
Biochemical and Molecular Approaches

(Continued on page 442)

Alan G. Williams Geoffrey S. Coleman

The Rumen Protozoa

With 113 Figures and 115 Tables

Springer-Verlag
New York Berlin Heidelberg London Paris
Tokyo Hong Kong Barcelona Budapest

Alan G. Williams
Rumen Microbiology Section
Department of Biological Science and Technology
Hannah Research Institute
Ayr, Scotland KA6 5HL
United Kingdom

Geoffrey S. Coleman
Institute of Animal Physiology and Genetic Research
Babraham, Cambridge
United Kingdom

Cover illustration: Two *Entodinium* spp. feeding on bacteria that are colonizing plant material in the rumen of a cow.

Library of Congress Cataloging-in-Publication Data
Williams, Alan G. (Alan George)
The rumen protozoa/Alan G. Williams, Geoffrey S. Coleman.
p. cm.—(Brock/Springer series in contemporary bioscience)
Includes bibliographical references and index.

ISBN-13: 978-1-4612-7664-7 e-ISBN-13: 978-1-4612-2776-2
DOI: 10.1007/978-1-4612-2776-2

1. Ciliata. 2. Rumen—Microbiology. I. Coleman, Geoffrey S. (Geoffrey Stafford) II. Title. III. Series.
QL368.A22W55 1991
599.73′504132—dc20 91-11655

Printed on acid-free paper.

Softcover reprint of the hardcover 1st edition 1992

Production managed by Linda H. Hwang.
Typeset by Asco Trade Typesetting Ltd., Hong Kong.

9 8 7 6 5 4 3 2 1

Dedicated to those who appreciate the true value of scientific endeavor

Preface

For the past 150 years microbiologists and protozoologists have looked at rumen contents and pondered on the role and classification of the mass of protozoa visible under the light microscope. Although many different types of rumen bacteria, including some of the strict anaerobes, have been grown in the laboratory, the cultivation of the rumen protozoa has remained difficult. Protozoologists have, therefore, always had to classify the ciliates into different genera and species without being able to culture them. This has led, particularly among the entodiniomorphid ciliates, to a proliferation of species, some of which have been described only once. Reliance on morphological characteristics alone for species identification is problematic as we know that the morphology of some ciliates can be changed dramatically by alterations in the environmental conditions. However, for the sake of completeness, we have tried to include all previously described species. The literature still continues to contain descriptions of new species and we suspect that we shall have to wait for a long time before the resources are available to resolve, by cultural and DNA studies, the recurrent problems of speciation.

At the time of the writing of this preface the study of rumen protozoa is threatened by the withdrawal of support for agricultural research in developed countries. This is a matter of concern not only as the protozoa are fascinating microorganisms that we are only now beginning to understand, but also because of their impact on the health, nutrition, and productivity of the host ruminant. The current controversy as to whether the protozoa are beneficial or detrimental to the host is explored in this book. Although it seems likely that the effects of the protozoa depend upon the host, the type of feed, and other less clearly defined factors, we suspect that the answers will not be easy to find and that the debate will continue for many years to come. Animal performance may be improved by manipulating the size or generic composition of the protozoal population. However, such manipulations cannot be made randomly and must be based on a thorough understanding of the activities and capabilities of the protozoa in the rumen ecosystem. These answers will only be obtained by further detailed studies of the metabolism and role of the ruminal protozoa.

This book attempts, for the first time, to collate the available information on the rumen protozoa. The book contains details of the important anatomical features of the protozoa that are of value for their recognition and identification, and includes sections on the techniques for protozoal isolation and cultivation. Other chapters are dedicated to the ecology, occurrence, biochemistry, and role of the protozoa in the ruminal environment. The potential that exists to improve ruminant productivity by the manipulation of the protozoal population or its activities is fully discussed. We hope that this compilation of knowledge about the rumen protozoa will stimulate further research effort. We hope also that it will persuade all those working on rumen metabolism of the importance of the protozoa so that in future they will cease to regard the protozoa as a "black box" but will seek information about the numbers and types of protozoa present, their role in the ecosystem, and contribution to the health and nutritional well-being of the ruminant host.

Acknowledgment. The micrograph used for the cover illustration was kindly supplied by Dr. K.N. Joblin, DSIR Grasslands, Palmerston North, New Zealand.

Alan G. Williams
Geoffrey S. Coleman

Contents

1

Introduction

The rumen protozoa were first observed by Gruby and Delafond (1843) and from their dramatic appearance were assumed to be of importance in the metabolism and nutrition of the host. These protozoa are principally ciliates and are of two types: the Entodiniomorphid protozoa (the Oligotrich protoza of the older literature) and the Holotrichs (Fig. 1.1). The former are characterized by the presence of a firm pellicle and of cilia situated only on the peristome and sometimes elsewhere. In contrast the holotrich protozoa have more flexible pellicles which, in the most commonly occurring species, are almost completely covered in cilia. The structure and classification of these ciliates will be considered in more detail in Chapters 2 and 3 respectively. Flagellate protozoa are also present in most rumens, although some of the organisms that were earlier described as flagellates are now known to be phycomycete fungi (see Chapter 13). In the next few chapters the entodiniomorphid and holotrich protozoa will be considered separately, since they are both structurally and metabolically different. However, where the effect of the protozoa on the host animal is considered, there is usually insufficient evidence for this separation and "the protozoa" will be considered as though only one organism is involved. This is clearly unsatisfactory but it is the best that can be done at present.

The rumen is an unusual environment for microorganisms. It is warm (39°C), has a low redox potential (although it may have free oxygen in the gas phase), and is rich (as far as particle-ingesting protozoa are concerned) in particles such as starch grains (especially if starch is fed), as well as plant fibres and bacteria. During feeding there are also injections of soluble compounds such as sugars and proteins.

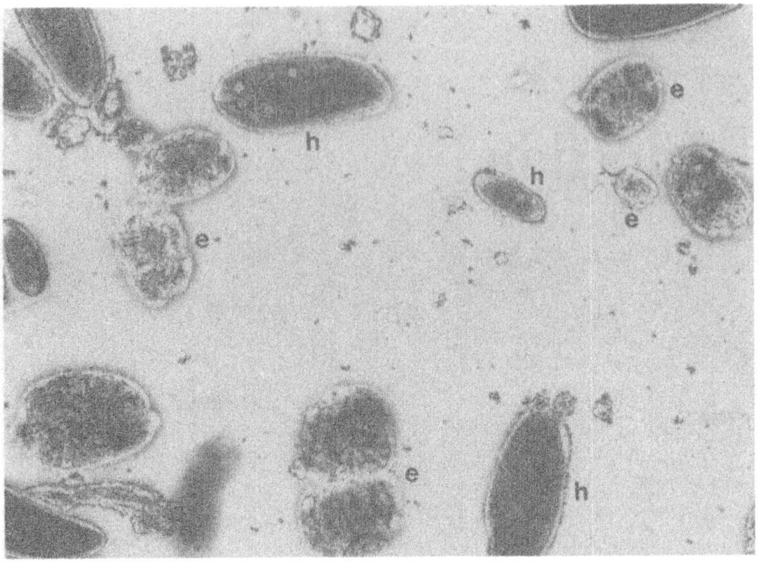

Fig. 1.1 Low power photomicrograph of ciliate protozoa isolated from the rumen of a sheep. h = holotrich protozoa; e = entodiniomorphid protozoa.

The entodiniomorphid protozoa are well adapted to the environment and utilize particulate rather than soluble food materials. In contrast, the holotrichs can use soluble food materials and are more aerotolerant.

The ciliate protozoa (entodiniomorphs and holotrichs) are found in the rumens of all wild and domesticated ruminants and also in camelids (camel, llama, alpaca, vicuna, guanaco) which are pseudoruminants. They are also found in the gastrointestinal contents of some nonruminants such as the hippopotamus and capybara. Many different species of protozoa are usually found in any one rumen, but occasionally animals with only one or a few species are found. This is particularly true where an animal has been subjected to any kind of feed-related stress such as starvation or a rumen acidosis. Under these conditions some or all of the rumen protozoa may be lost. In the authors' experience, ruminants kept for laboratory experiments may have only entodinia, for example, in their rumens because of the effects of previous experiments and therefore may not be typical of other animals of the same species and breed in the area. The protozoa have no resistant phase in their life cycle and can only be transferred to a young or adult animal by close contact with an infected ruminant. A young ruminant reared without such contact remains ciliate-free for life. Similarly an adult ruminant that has had its rumen protozoa killed deliberately or that has lost them accidently, as from an acidosis, will remain protozoa-free until infected. However, as will be discussed in Chapter 9, not all protozoal species behave in the same way when transmission between animals is considered.

The importance of the rumen protozoa to the health and well-being of ruminants and to the growth of the young animal has been debated at length ever since Becker (1929) first showed that adult animals survived normally after their rumen ciliates had been removed by chemical means. For many years the evidence suggested that the growth of lambs and calves was unaffected or slightly increased by the presence of protozoa, but it has now been shown that in ruminants given poor-quality rations, protozoa may be detrimental to the growth of the young animal. It seems likely that the debate will continue for some years to come.

However, despite this debate there is general agreement on the effect of the presence or absence of rumen ciliates on various rumen parameters. For example, defaunation of the rumen decreases the rate of cellulose and protein digestion and the rate of methane production and ammonia concentration while increasing the bacterial population density, the efficiency of bacterial protein synthesis and the rate of nitrogen flow to the duodenum. All these effects are the direct or indirect result of the engulfment of bacteria by the ciliates during their normal feeding processes. The engulfed bacteria are then killed and digested. Some of the digestion products are used for protozoal growth while the remainder, principally amino acids, are released back into the rumen where they are fermented by the bacteria. Some of the fermentation products are used for bacterial growth but the remainder appear free in the rumen fluid. This recycling of bacterial carbon and nitrogen represents an energy and nitrogen loss as far as the host animal is concerned and may be of crucial importance in animals fed a low-nitrogen ration.

The host animal obtains many of the amino acids it requires by digesting, lower down the alimentary tract, the bodies of rumen microorganisms. These pass from the reticulo-rumen into the omasum and then the abomasum, which is the true stomach. The evidence available suggests that protozoa are more valuable to the host than the bacteria because the former are more digestible than the latter. However, the protozoa pass out of the rumen relatively more slowly than the bacteria and therefore the relative proportions of protozoa and bacteria in the rumen give a false impression of the importance of the protozoa. The loss of bacterial nitrogen due to protozoal activity is therefore compensated for by the increased digestibility of protozoal bodies. The exact result as far as, for example, host growth is concerned depends the degree of compensation and other unquantifiable factors. At present we are unable to predict the exact effect of defaunation on host performance. This is particularly true with animals fed a ration rich in starch where there are massive populations of rumen ciliate protozoa.

2

Identification and Classification of Entodiniomorphid Protozoa

2.1 Taxonomy of Entodiniomorphid Protozoa

Most of the ciliates to be described below belong to the order Entodiniomorphida (Doflein and Reichenow, 1929) and the family Ophryoscolecidae (Stein, 1858), as shown in Table 2.1. The exceptions are a *Parentodinium* sp., which belongs to the family Cycloposthiidae, and *Rhinozeta* spp., which belong to the family Rhinozetidae.

The identification of these protozoa is not easy for a number of reasons. If the starting material is rumen contents and if the host has recently been fed, the protozoa are so full of storage polysaccharide that no internal detail is visible and it is impossible to see the skeletal plates (if any). A description of the structure of entodiniomorphid protozoa is found in Chapter 7. With increase in time after feeding, the amount of general body polysaccharide decreases until either on staining with iodine, or by the use of phase contrast optics, it is often possible to see the plate(s) lying just under the right (in the classical sense) surface of the organisms. The plates are storage organelles, so they also progressively lose their amylopectin on starvation of the protozoa and, in ciliates just about to die, it is very difficult to see the plates. The shape of the macronucleus can also only be seen in starved organisms.

There are no culture collections of rumen ciliates, so the investigator has to resort to photomicrographs for identification Ogimoto and Imai, 1981; Coleman, 1979b) or line drawings (Dogiel, 1927; Kofoid and MacLennan, 1930, 1932, 1933; Latteur, 1968, 1969, 1970). These are all general references and give information on many of the ciliates described up to the time

Table 2.1 The genera of the subfamilies of the family Ophryoscolecidae

Subfamily					
Entodiniiae	Diplodiniinae	Epidiniinae	Opisthotrichinae	Ophryoscolecinae	Caloscolecinae
Entodinium	*Eodinium* *Diplodinium* *Eudiplodinium* *Eremoplastron* *Ostracodinium* *Metadinium* *Diploplastron* *Elytroplastron* *Enoploplastron* *Polyplastron*	*Epidinium* *Epiplastron*	*Opisthotrichum*	*Ophryoscolex*	*Caloscolex*

Based on the classifications proposed by Latteur (1966d); Kofoid and MacLennan (1930, 1932, 1933); and Hungate (1978).

of their writing. There are also many, usually more recent, papers describing individual species, and these will be referred to at the appropriate place in the text. The terms "dorsal," "ventral," "right," and "left" will be used in their classical sense with the macronucleus assumed to be on the dorsal side. A schematic diagram showing this orientation is shown in Fig. 7.1. In the orientation described by Lubinsky (1958a, b) this is different between the entodinia and the more complex species and can give rise to confusion.

A list of references giving information on the ciliate types found in different hosts is given in Table 9.2.

2.2 The Genus *Entodinium*

The entodinia are the smallest, simplest, and most common of the rumen ciliates and are found in almost all ruminants without exception. They are also the most difficult to classify into species because of the large number of closely related organisms and the difficulty in deciding between intraspecific and interspecific variation. As none of the protozoa had been cultured when the major descriptive works were published, the authors had to decide arbitrarily which features were taxonomically important and which slightly different protozoa belonged to the same species. For example, *Entodinium caudatum* only has its characteristic caudal spines when grown in the presence of *E. bursa*, and a range of intermediary forms between the normal and spineless forms can be seen in clone cultures in vivo and in vitro (Poljansky and Strelkow, 1938; Coleman et al., 1977). Poljansky and Strelkow (1938) also showed that the distribution of the different forms of a clone culture in vivo can be affected by the ration fed to the host. They also suggested that *E. loboso-spinosum* could be a form of *E. caudatum*. Lubinsky (1957a) extended this idea and showed that *E. caudatum*, *E. simulans*, and *E. rectangulatum* all had similarly variable caudal spination although he only used slaughter house material and could not be certain which forms were variations of which protozoon. He believed that the position of the contractile vacuole and of the lateral groove were important taxonomic characters. Latteur (1968, 1969) extended this further and described seven groups of entodinia based on the position of the contractile vacuole in relation to the macronucleus. Latteur, like Hungate (1966) and the present authors, believed that caudal spination was variable within a species. Letteur's (1968, 1969) seven types are as follows:

1. Dorsopulsatum: vacuole applied to the dorsal face of the extreme anterior end of the macronucleus.
2. Troncopulsatum: vacuole is located in the median dorsoventral plane adjacent to the anterior face of the macronucleus.
3. Anteropulsatum: vacuole is situated anterior to the macronucleus on a line ventral to its axis.

4. Lateropulsatum: vacuole is applied to the left face of the anterior extremity of the macronucleus.
5. Cephalopulsatum: vacuole is directly anterior to the macronucleus in a line with its longitudinal axis.
6. Proximopulsatum: vacuole is under the left dorsal surface, three-quarters of the way up the cell from the posterior end and near the left surface of the macronucleus but not immediately adjacent to it.
7. Sinistropulsatum: vacuole is on the lateral median plane, three-quarters of the way up the cell from the posterior end, under the left surface of the cell.

However Poljansky and Strelkow (1938) found some variability in the distance between the contractile vacuole and the macronucleus in *E. caudatum*, and Hungate (1978) pointed out that the position of the contractile vacuole can vary within a species.

General Description of the Entodinia The entodinia are characterized by the presence of an adoral ciliary band found in all Ophryoscolecid protozoa and the absence of the dorsal band characteristic of the higher organisms. The entodinia also lack a skeletal plate and have only one contractile vacuole. Most of these protozoa are ellipsoidal in shape but are distinctly flattened laterally. They have an extrudable peristome, carrying the ciliary band, at the anterior end. At the posterior end they have a cytoproct (anus) that opens between the caudal lobes or spines or onto the smoothly rounded end of the organism. The macronucleus, the long axis of which runs parallel to the main axis of the organism, is often of characteristic size and shape and is situated in the ectoplasm on (by definition) the dorsal surface. The micronucleus, which is often difficult to see, is a small spherical body (1–5 μm in diameter, depending on species) situated in a depression on the ventral surface of the macronucleus. A detailed description of the structures of all these organelles is given in Chapter 7.

Over 120 species and forms (some of which are undoubtedly the same) have been described and are listed in Table 2.2. This table also contains information about their physical characteristics, the reference publication where it was first described and where it received its present name (if different), and references to where the reader can find photomicrographs and line drawings. Photomicrographs of some species are given in Figs. 2.1–2.6. More detailed information about 100 species of entodinia is given below.

Detailed Descriptions of the *Entodinium* Species

1. *Entodinium abruptum* is almost spherical in shape and is found in deer. It was originally described from mule deer by MacLennan (1935).

Table 2.2 Characteristics of the *Entodinium* species

Species (Fig. no.)[a]	Original description/ name[b]	Latteur's classification (1968, 1969)	Length (μm)	Width (μm)	Spines	Lobes
1. *E. abruptum*	42	Sinistro-	28–32	23–26	0	2
2. *E. aculeatum*	36	Latero-	34–40	26–28	3	0
3. *E. acutonucleatum*	36	Latero-	25–39	22–28	3	0
4. *E. acutum*	36	Latero-	30–42	23–27	3	0
5. *E. alces*	17		29–45	21–33	0	0
6. *E. alpinum*	8		32–48	25–32	0	1
7. *E. anteronucleatum*	22					
f. *dilobum*	22	Latero-	52–82	39–50	0	2
f. *laeve*	22	Latero-	51–80	39–49	0	0
f. *monolobum*	22	Latero-	52–82	39–50	0	1
8. *E. attenuatum*	26		62	42	0	0
9. *E. augustatum*	14/38	Latero-	48–62	20–30	2	0
10. *E. babici*	53	Cephalo-	27–42	22–32	0	0
11. *E. bengalensis*	1		35–50	30–40	3	0
12. *E. bicarinatum*	14/15	Cephalo-	60	45	0	0
13. *E. bicaudatum*	5	Cephalo-	22–34	21–34	2	0
14. *E. biconcavum*	36	Latero-	28–41	22–28	0	1
15. *E. bicornutum*	22	Antero-	26–37	21–28	2	0
16. *E. bidentatum*	45					
17. *E. bifidum*	24/36	Latero-	31–40	20–22	2	0
18. *E. bimastus* (2.1)	24	Latero-	30–40	28–37	0	1
19. *E. birostratum*	55					
20. *E. bovis*	54	Antero-	26–38	23–34	0	0
21. *E. brevispinum*	35	Latero-	24–30	15–18	1	0
22. *E. bubalum*	28		25–45	20–25	2	0
23. *E. bursa* (2.2)	51	Proximo-	80–115	70–90	0	0
24. *E. caudatum*	51					
Spined (2.2)		Latero-	49–69	48–56	3	0
Spineless (2.3, 2.4)		Latero-	48–74	34–41	0	0
25. *E. cervi*	37	Latero-	53–70	35–53	0	1
26. *E. chatterjeei*	16	Cephalo-	26–35	15–18	0	0
27. *E. ciculum*	18		36–53	23–31	2	0
28. *E. constrictum*	17		31–43	20–29	0	0
29. *E. contractum*	35	Latero-	39–46	27–34	0	0
30. *E. convexum*	42	Latero-	21–40	15–30	0	0
31. *E. costatum*	42	Latero-	30–42	21–29	0	1
32. *E. couturieri*	9		67–99	48–70	0	0
33. *E. curtum*	35	Proximo-	40–53	26–38	1	0
34. *E. dalli*	17		28–38	23–41	1	0
35. *E. damae*	50		28–36	15–22	0	0
36. *E. dentatum*	51		60–90	30–50	6	0
37. *E. dilobum*	16	Latero-	42–55	28–36	0	2
38. *E. dubardi*	3					
f. *aethiopicum*	43	Latero-	30–40	20–25	1	1
f. *crassicaudatum*	24	Latero-	30–40	20–25	1	0
f. *dubardi*	3/21	Latero-	23–57	18–36	0	0
f. *gracilicaudatum*	3/24	Latero-	30–40	20–25	1	0
39. *E. ekendrae*	16	Cephalo-	30–35	30–33	2	1
40. *E. ellipsoideum*	36	Proximo-	65–120	50–81	0	0
41. *E. elongatum*	24	Latero-	60–78	40–52	0	0
42. *E. exiguum*	22	Cephalo-	21–29	14–18	0	0
43. *E. fujitai*	28		23–32	18–25	2	0

	Macronucleus			Illustrated in Ogimoto and Imai (1981)	Other reference[b]
	Body shape	Size (μm)	Shape		
1.	Subspherical	7 × 21	Wedge	–	39
2.	Oval	4 × 17	Curved rod	+	39
3.	Subspherical	4 × 29	Curved rod	+	39
4.	Oval	5 × 25	Rod	–	
5.	Ellipsoidal	4 × 21	Bent rod	–	56
6.	Ellipsoidal	3 × 20	Rod	–	
7.					
	Ellipsoidal	8 × 25	Short rod	+	24
	Ellipsoidal	9 × 27	Short rod	–	24, 56
	Ellipsoidal	8 × 25	Short rod	–	24
8.	Elongated		Curved rod	–	41
9.	Long, thin	6 × 47	Long rod	–	24
10.	Rectangular		Short rod	–	39
11.	Ellipsoidal	5 × 37	Curved rod	–	
12.	Ellipsoidal	8 × 22	Short rod	–	24
13.	Spherical	7 × 8	Spherical	–	
14.	Oval	5 × 17	Variable	+	
15.	Spherical	4 × 9	Irregular	–	
16.				–	
17.	Oval	6 × 11–22	Rod	+	
18.	Subspherical	6 × 30	Curved wedge	+	36
19.				–	
20.	Spherical	4 × 17	Club-shaped	+	56
21.	Small, slender	3 × 13	Wedge	–	38
22.	Ellipsoidal	4 × 26	Curved rod	–	
23.	Subspherical	11 × 63	Curved rod	+	13, 39, 44
24.					
	Ellipsoidal	8 × 29	Rod	+	11, 12, 13, 38
	Ellipsoidal		Rod		44, 46, 48
25.				–	
26.	Slender	6 × 9	Ellipsoidal	+	
27.	Ellipsoidal	5 × 24	Rod	–	
28.	Ellipsoidal	6 × 8	Ellipsoidal	–	
29.	Pear-shaped	5 × 31	Club-shaped	–	
30.	Oval	4 × 20	Curved rod	–	39
31.	Oval	4 × 16	Rod	+	39
32.	Rectangular	5 × 28	Curved wedge	–	
33.	Ellipsoidal	9 × 32	Curved rod	–	
34.	Subspherical	6 × 8	Subspherical	–	
35.	Ellipsoidal	4 × 17	Rod bent at 90°	–	41
36.	Rectangular	5 × 42	Rod	–	
37.	Ellipsoidal	8 × 30	Rod	–	38
38.					
				–	38
	Ellipsoidal	4 × 15	Curved rod	–	38
	Ellipsoidal	5 × 24	Curved rod	+	38, 57
	Ellipsoidal	3 × 9	Rod	–	38
39.	Subspherical	9 × 14	Oval	+	39
40.	Subspherical	10 × 60	Triangular	–	
41.	Slender	4 × 22	Curved rod	–	
42.	Rectangular	4 × 8	Short rod	+	
43.	Curved	3 × 9	Short rod	+	

Table 2.2 *Continued.*

Species (Fig. no.)[a]	Original description/ name[b]	Latteur's classification (1968, 1969)	Length (μm)	Width (μm)	Spines	Lobes
44. *E. furca*	14/15					
f. *dilobum*	24	Latero-	42–55	28–36	0	2
f. *furca* (2.5)	24	Latero-	60	35	2	0
f. *monolobum*	24	Latero-	38–44	20–24	0	1
45. *E. fyferi*	52		20–30	17–24	0	2
46. *E. gibberosum*	36	Latero-	38–51	22–35	2	0
47. *E. hamatum*	2					
48. *E. ibicis*	9		32–51	18–29	0	0
49. *E. incurvatum*	25					
50. *E. indicum*	36	Tronco-	25–40	22–28	3	0
51. *E. javanicum*	30		29–42	19–21	0	2
52. *E. kalaharicus*	56		26–47	17–24	0	0
53. *E. laterale*	36	Sinistro-	19–28	18–21	3	0
54. *E. laterospinum*	36	Latero-	25–32	14–19	1	0
55. *E. loboso-spinosum* (2.1)	21	Latero-	30–46	20–23	1	1
56. *E. longicorpum*	34		31–52	9–19	0	0
57. *E. longinucleatum*	22					
f. *acutonucleatum*	36	Latero-	25–39	22–28	3	0
f. *longinucleatum*	22	Latero-	44–64	29–36	0	0
f. *spinonucleatum* (2.6)	18	Latero-	26–41	24–33	1–2	0–1
f. *spinolobum*	29	Latero-	45–63	31–40	2	1
58. *E. longispinosus*	6	Latero-	26–31	21–26	4	0
59. *E. lucii*	56		27–40	16–24	0	0
60. *E. mamillatum*	14/15		48–70	35–40	0	1
61. *E. medium*	42	Latero-	28–50	19–34	0	0
62. *E. minimum*	49		40–42	17–18	0	0
63. *E. montanum*	5	Cephalo-	36–53	22–32	0	0
64. *E. muntiacum*	1	Cephalo-	25–40	20–25	0	0
65. *E. nanellum*	20	Latero-	20–25	10–13	0	0
66. *E. nanum*	5	Cephalo-	18–26	12–17	0	0
67. *E. ogimotoi*	28		30–47	22–35	0	0
68. *E. orbicularis*	5	Cephalo-	22–33	18–32	0	0
69. *E. ovale*	24	Sinistro-	20–40	12–20	0	0
70. *E. ovalis*	33	Sinistro-	20–40	12–20	0	0
71. *E. ovibos*	19		24–44	18–33	0	0
72. *E. ovinum*	24	Latero-	53–69	32–41	0	0
73. *E. ovoideum*	36	Latero-	30–50	21–29	0	0
74. *E. ovoido-nucleatum*	16	Dorso-	25–30	22–25	3	0
75. *E. ovum-rajae*	23					
f. *deficiens*	23	Cephalo-	46–55	32–48	1	0
f. *ovum-rajae*	23	Cephalo-	46–55	32–48	2	0
76. *E. parvum*	4					
f. *monospinosum*	32	Latero-	35–45	15–19	1	0
f. *parvum*	4	Latero-	30–35	20–24	0	0
77. *E. pisciculum*	36	Antero-	38–58	22–29	2	0
78. *E. pristinum*	42	Latero-	26–34	18–23	0	0
79. *E. protuberans*	5	Cephalo-	20–45	11–27	0	0
80. *E. quadricuspis*	22					
f. *quadricuspis*	22	Latero-	24–34	21–32	4	0
f. *stauracanthum*	22	Latero-	32–40	27–31	2	2
81. *E. rectangulatum*	36					
f. *caudatum*	40	Sinistro-	23–47	23–39	1	2
f. *dubardi*	40	Sinistro-	26	17	0	0

Table 2.2 *Continued.*

	Body shape	Macronucleus Size (μm)	Macronucleus Shape	Illustrated in Ogimoto and Imai (1981)	Other reference[b]
44.					
	Ellipsoidal	8 × 30	Rod	–	
	Ellipsoidal	8 × 46	Curved rod	–	
	Ellipsoidal	4 × 26	Curved rod	–	
45.	Subspherical	3–6 × 10	Irregular	–	
46.	Humpbacked	4–8 × 30–45	Long notched	–	
47.	Elongated		Wedge	–	
48.	Rectangular	3 × 17	Curved wedge	–	
49.	S-shaped		Curved rod	–	
50.	Ellipsoidal	3–10 × 19–29	Wedge	+	38
51.	Ellipsoidal	4 × 14	Rod	–	
52.	Ellipsoidal	5 × 20	Notched wedge	–	
53.	Ellipsoidal	3–8 × 8–18	Wedge	–	
54.	Wedge-shaped	3 × 15	Curved rod	–	
55.	Rectangular	4 × 6–20	Club-shaped	–	24, 55
56.	Elongated	1–5 × 19–31	Long, narrow	–	
57.					
	Ellipsoidal	4–29	Curved rod	–	
	Oval	4–8 × 48	Curved rod	+	44, 47
	Ellipsoidal	3 × 31	Curved rod	–	
	Oval	4 × 48	Curved rod	–	
58.	Ellipsoidal	3 × 16	Rod, thick middle	–	39
59.	Rectangular	3 × 28	Rod	–	
60.	Ellipsoidal	9 × 34	Rod	–	24
61.	Ellipsoidal	4–7 × 16–28	Rod	–	38
62.	Triangular	4 × 15	Rod	+	39
63.	Ellipsoidal	4 × 19	Rod	–	39
64.	Oval	5 × 8	Oval	–	39
65.	Ellipsoidal	2–5 × 11–21	Curved rod	+	24
66.	Ellipsoidal	2–3 × 6	Club-shaped	–	39
67.	Rectangular	4 × 19	Rod	+	
68.	Spherical	4 × 6	Oval	–	39
69.	Ellipsoidal	5 × 20	Rod bent at 90°	–	39
70.	Ellipsoidal	5 × 20	Rod bent at 90°	–	39
71.	Ellipsoidal	2–5 × 9–29	Club-shaped	–	
72.	Ellipsoidal	5 × 38–50	Rod	–	28, 35
73.	Ellipsoidal	4–5 × 23–40	Rod	–	28, 35
74.	Rectangular	5 × 11	Oval	–	38
75.					
	Ellipsoidal	8 × 21	Ellipsoidal	–	39
	Ellipsoidal	8 × 23	Curved rod	–	39
76.					
	Ellipsoidal	3 × 23	Rod	–	24, 34, 54, 56
	Rectangular	3 × 20	Rod	+	38
77.	Elongated	3 × 22–33	Rod	–	38
78.	Oval	4 × 13	Rod	–	39
79.	Elongated	6 × 7	Oval	–	
80.					38
	Subspherical	4 × 13	Rod	–	38
	Ellipsoidal	4 × 17	Rod	–	
81.					39
	Rectangular	7 × 21	Club-shaped	–	39
	Rectangular	4 × 31	Club-shaped	+	39

Table 2.2 *Continued.*

Species (Fig. no.)[a]	Original description/ name[b]	Latteur's classification (1968, 1969)	Length (μm)	Width (μm)	Spines	Lobes
f. *loboso-spinosum*	40	Sinistro-	18–33	13–25	1	1
f. *rectangulatum*	36	Sinistro-	29–45	23–35	3	0
82. *E. rhomboideum*	36	Latero-	30–47	25–30	0	0
83. *E. rostratum* (2.1, 2.5)	27	Antero-	28–41	18–21	1	0
84. *E. rotundum*	42	Latero-	27–40	21–31	0	0
85. *E. rupicaprae*	7		32–48	19–39	0	0
86. *E. setnai*	16	Cephalo-	50–60	38–42	0	3
87. *E. sierrae*	5	Cephalo-	45–65	30–40	0	0
88. *E. simplex* (2.5)	21/24	Latero-	38–50	21–29	0	0
89. *E. simulans*	40					
f. *caudatum*	40	Sinistro-	27–44	21–34	1	2
f. *dubardi*	40	Sinistro-	21–47	16–34	0	1
f. *loboso-spinosum*	40	Sinistro-	25–46	22–34	1	1
90. *E. spino-nucleatum*	3	Proximo-	50–70	29–40	1	2
91. *E. spinonucleatum*	18		26–41	24–33	1–2	0
92. *E. triacum*	4					
f. *dextrum*	24	Latero-	25–48	25–36	3	0
f. *triacum*	4/24	Latero-	25–45	18–30	3	0
93. *E. triangulatum*	18		25–41	22–29	2	0
94. *E. tricostatum*	36	Tronco-	22–33	25–31	3	0
95. *E. tricuspis*	6				0	3
96. *E. tridentatum*	45					
97. *E. truncatum*	5	Cephalo-	40–50	28–40	0	0
98. *E. tsunodai*	28		28–40	23–28	4	0
99. *E. vorax*	22					
f. *bispinosum*	22/24	Proximo-	60–78	40–52	2	0
f. *vorax* (2.2)	22/24	Proximo-	80–121	52–83	0	0
100. *E. wertheimi*	9		42–58	26–38	0	0

[a] A number in parentheses indicates that this species is illustrated in the figure corresponding to that number in this chapter.
[b] References are as follows: 1, Banerjee (1955); 2, Bretschneider (1934); 3, Buisson (1923a); 4, Buisson (1923b); 5, Bush and Kofoid (1948); 6, Chavarria (1933); 7, Christl (1955); 8, Christl (1958a); 9, Christl (1958b); 10, Coleman (1969a); 11, Coleman (1979b); 12, Coleman and Sandford (1980); 13, Coleman et al. (1977); 14, Cunha (1914a); 15, Cunha (1914b); 16, Das Gupta (1935); 17, Dehority (1974); 18, Dehority (1979); 19, Dehority (1985); 20, Dogiel (1923a); 21, Dogiel (1925a); 22, Dogiel (1925d); 23, Dogiel (1926b); 24, Dogiel (1927); 25, Dogiel (1932); 26, Dogiel (1935); 27, Fiorentini (1889); 28, Imai (1981); 29, Imai (1984); 30, Imai

2. *Entodinium aculeatum* was first described from *Bos indicus* in India and Sri Lanka by Kofoid and MacLennan (1930). It has also been found in cattle in Thailand and Indonesia (Imai, 1985; Imai and Ogimoto, 1984). This species has three caudal spines, two of which can be larger and heavier than the other and one of which can be bifurcate. This protozoon is believed by Latteur (1968) to be *E. triacum* (Buisson, 1923a, 1923b) forma *triacum* (Buisson, 1923a, 1923b).

3. *Entodinium acutonucleatum* was first described in *Bos indicus* in India and Sri Lanka by Kofoid and MacLennan (1930). It also occurs in small numbers in Brazilian cattle (Dehority, 1979), although in the latter hosts the protozoa are wider than originally described. This organism is

Table 2.2 *Continued.*

	Body shape	Macronucleus Size (μm)	Macronucleus Shape	Illustrated in Ogimoto and Imai (1981)	Other reference[b]
	Rectangular	4 × 13	Club-shaped	+	39
	Rectangular	7 × 22	Wedge	+	38
82.	Subspherical	3–8 × 27	Wedge	+	
83.	Curved	3 × 21	Rod	+	24, 36, 38, 44
84.	Spherical	7 × 8	Spherical	−	39
85.	Ellipsoidal	1–4 × 23	Curved wedge	−	
86.	Ellipsoidal	11 × 11	Spherical	−	39
87.	Rectangular	7 × 12	Oval	−	39
88.	Ellipsoidal	5 × 16	Wedge	+	10, 38, 44, 48
89.					
	Rectangular	14 × 23	Club-shaped	+	39
	Rectangular	4 × 18	Club-shaped	−	39, 48
	Rectangular	4 × 17	Club-shaped	+	39
90.	Rectangular	5 × 67	Bent rod	−	39
91.	Ellipsoidal	3 × 30	Curved rod	−	
92.					
	Ellipsoidal	6 × 24	Rod	−	38
	Ellipsoidal	3 × 25	Rod	−	38, 55
93.	Curved	2–5 × 21	Triangular	−	
94.	Ellipsoidal + ribs	7–10 × 21	Club-shaped	−	38
95.					
96.					
97.	Ellipsoidal	5 × 14	Rod	−	39
98.	Ellipsoidal	3 × 24	Rod	+	
99.					
	Ellipsoidal	10 × 46	Curved rod	−	
	Subspherical	11 × 67	Curved rod	−	13
100.	Ellipsoidal	3 × 32	Curved rod	−	

(1985); 31, Imai (1988); 32, Imai and Ogimoto (1983); 33, Jameson (1925b); 34, Kleynhans and Van Hoven (1976); 35, Kofoid and Christenson (1933); 36, Kofoid and MacLennan (1930); 37, Kubikova (1935); 38, Latteur (1968); 39, Latteur (1969); 40, Lubinsky (1957a); 41, Lubinsky (1958a); 42, MacLennan (1935); 43, Moriggi (1941); 44, Noirot-Timothée (1960); 45, Noirot-Timothée (1963); 46, Ogimoto and Imai (1981); 47, Owen and Coleman (1976); 48, Poljansky and Strelkow (1938); 49, Schuberg (1888); 50, Sladecek (1946); 51, Stein (1858); 52, Van Hoven (1975); 53, Wertheim (1933a); 54, Wertheim (1935a); 55, Wertheim (1935b); 56, Wilkinson and Van Hoven (1976a); 57, Zielyk (1961).

believed by Latteur (1968) to be the same as *E. bengalensis* (Banerjee, 1955) and has three small caudal spines and a long curved macronucleus.

4. *Entodinium acutum* was separated from *E. aculeatum* by Kofoid and MacLennan (1930) on the basis of shape and because the spines were never bifurcate. However, Latteur (1968) believed that both species were the same as *E. triacum triacum*. *E. acutum* has not been separately identified by others.

5. *Entodinium alces* was first described from *Alces americanum* (the Alaskan moose) by Dehority (1974). It has only also been found in the giraffe by Kleynhans and Van Hoven (1976) and in the springbok (*Anti-*

dorcas marsupialus) by Wilkinson and Van Hoven (1976a). It is most closely related to *E. dilobum*.

6. *Entodinium alpinum*, which exists in *caudatum* and *loboso-spinosum* forms, has only been described from the ibex (*Capra ibex ibex*) by Christl (1958a).

7. *Entodinium anteronucleatum* was first described in three forms (see Table 2.2) in reindeer by Dogiel (1925d) and has since been found in Dall Mountain sheep (Dehority, 1974) and cattle in Korea (Han, 1984) and in Sri Lanka (Imai, 1986). It is characterized by the anterior placement of the macronucleus. Some authors have described the forms as separate species.

8. *Entodinium attenuatum* is a dubious species described from reindeer in northern European Russia by Dogiel (1935). It was said to resemble *E. elongatum*.

9. *Entodinium augustatum* was first described as a form of *E. furca* that occurred in sheep and cattle in Europe and China by Dogiel (1927). It has rarely been seen since. It is characterized by its long thin body and two short incurved caudal spines.

10. *Entodinium babici* was first described from the ibex by Wertheim (1933a) and was also found in the same host by Christl (1958b). Latteur (1969) believed it to be the same as *E. exiguum*, *E. nanum*, and *E. montanum*.

11. *Entodinium bengalensis*, first described by Banerjee (1955), is the same as *E. acutonucleatum* (Latteur, 1968).

12. *Entodinium bicarinatum* has been found in cattle in the USA (Abou Akkada et al., 1969) and New Zealand (Clarke, 1964), in water buffalo in Brazil (Dehority, 1979), and in Bighorn sheep and wild goats (Latteur, 1969). The protozoon is characterized by the asymmetrical shape of the posterior end.

13. *Entodinium bicaudatum* is found in Bighorn and Dall Mountain sheep (Bush and Kofoid, 1948; Dehority, 1974). It is characterized by its almost spherical shape and two long caudal spines.

14. *Entodinium biconcavum* has been found in water buffalo in Brazil, Indonesia, and Thailand and in cattle, often in high frequency, in Africa, Brazil, Korea, India, Sri Lanka, and Thailand. Although Latteur (1968) prefers the name E. *triacum* forma *biconcavum*, most authors use *E. biconcavum*. It is characterized by its convex dorsal and ventral surfaces.

15. *Entodinium bicornutum* is commonly found in reindeer in the Artic (Dogiel, 1925d; Lubinsky, 1958a; Westerling, 1970) but is rare elsewhere although it has been found by Dogiel (1925a) in African bushbuck (*Tragelaphus scriptus*). It is characterized by its spherical shape and two crossed caudal spines.

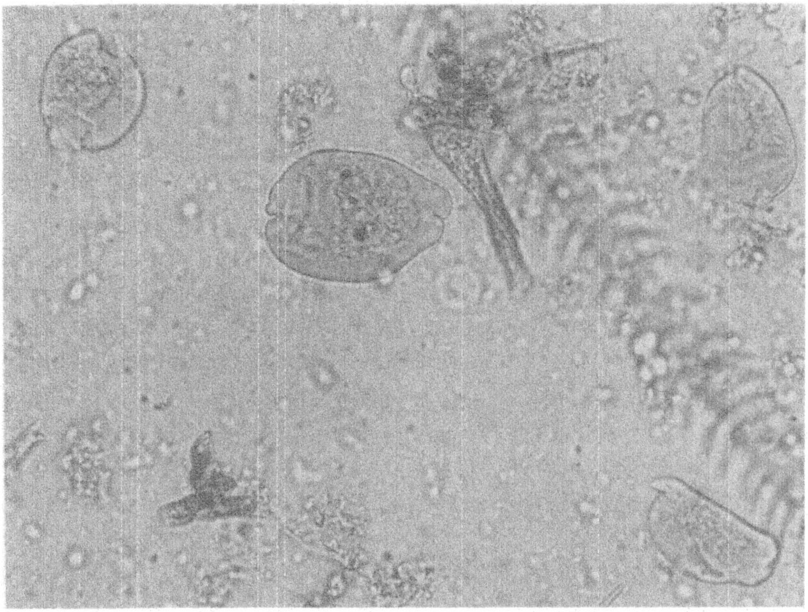

Fig. 2.1 Photomicrograph of *Entodinium bimastus* (center), *E. loboso-spinosum* (one caudal spine), and *E. rostratum* (one small caudal lobe).

16. *Entodinium bidentatum* was mentioned without any other information by Noirot-Timothée (1963).

17. *Entodinium bifidum* has been found fairly commonly in cattle in Korea (Han, 1984), Indonesia (Imai, 1985), India (Kofoid and MacLennan, 1930), Sri Lanka (Imai, 1986), and Thailand (Imai and Ogimoto, 1984), and very rarely in Russia (Dogiel, 1927). This protozoon has had a number of names including *E. rostratum* forma *bifidum* (Dogiel, 1927) and *E. triacum* forma *bifidum* (Latteur, 1968). The ventral side of the body terminates in a wide bifurcated spine that surrounds a small lobe that is not easily seen.

18. *Entodinium bimastus* is widely but not universally distributed in cattle and to a lesser extent in water buffalo. It has been reported in Europe, Russia, Japan, Thailand, Indonesia, the Philippines, and Sri Lanka, but it has not been found in the USA or New Zealand. It is characterized by its posterior end which tapers to a massive roughly triangular caudal lobe (Fig. 2.1).

19. *Entodinium birostratum* is a very occasionally used name for *E. rostratum birostratum* or *E. biconcavum birostratum*. Latteur (1968) believes all are the same as *E. furca* forma *furca*.

20. *Entodinium bovis* was first described by Wertheim (1935a) in Yugoslavian cattle. It is a rare protozoon, although it has been found in Zebu cattle and water buffalo in Thailand (Imai and Ogimoto, 1984), in Sri Lankan cattle (Imai, 1986), and in the springbok (*Antidorcas marsupialis*) by Wilkinson and Van Hoven (1976a). It is spherical in shape and has no caudal spines or lobes. Zielyk (1961) pointed out that it bears a strong resemblance to the late post divisional stages of *E. dubardi dubardi*.

21. *Entodinium brevispinum* was found by Kofoid and MacLennan (1930) in *Bos indicus* from India and Sri Lanka, but it has rarely been reported since. Latteur (1968) believes that it is the same as *E. furca* forma *brevispinum* and quotes a wide distribution for *E. furca* forms. *E. brevispinum* is characterized by its one small, blunt, ventral caudal spine.

22. *Entodinium bubalum* was first described from the water buffalo (*Bubalus bubalis*) by Imai (1981) and has also been found in the same host in Taiwan (Imai et al., 1981b). It has one broad caudal spine which is one-third the length of the body and a second spine close to the first, which is difficult to see.

23. *Entodinium bursa* is the type species of the genus *Entodinium*, but it is probably the largest and least typical member of the genus. It is referred to in the literature equally commonly as *E. bursa* and *E. vorax*. It is widely distributed in cattle and sheep in Britain, Europe, the USA, Korea, Japan, and Sri Lanka, and in goats in Japan. It appears in low numbers in Brazilian cattle (Dehority, 1986b), and it has not been reported in New Zealand cattle. The organism is characterized by its large size in comparison with other entodinia (Fig. 2.2). *E. bursa* is important in vivo because of its ability to induce *E. caudatum* to produce its characteristic caudal spines (Poljansky and Strelkow, 1938; Coleman et al., 1977).

24. *Entodinium caudatum* is widely distributed and is often reported in the absence of *E. bursa*. This suggests that 1) some strains form spines in the absence of *E. bursa*; 2) the process can be stimulated by another protozoal species or a ration component; 3) *E. caudatum* has been wrongly identified as such; or 4) *E. bursa* has wrongly been believed to be absent. It is not known which of the above is true. In the authors' experience, growth of *E. caudatum* in vivo or in vitro in the absence of *E. bursa* always results in the loss of the caudal spines (Figs. 2.3 and 2.4). Some authors have reported the existence of a form *loboso-spinosum*, but it is described here as a separate species. *E. caudatum* has been used extensively for metabolic studies (Coleman, 1980: Hino and Kametaka, 1977).

25. *Entodinium cervi* was originally described from the red deer (*Cervus elaphus*) by Kubikova (1935) but it is believed by Latteur (1968) to be the same as *E. brevispinum* and *E. furca* forma *brevispinum*.

26. *Entodinium chatterjeei* was originally described from the Indian

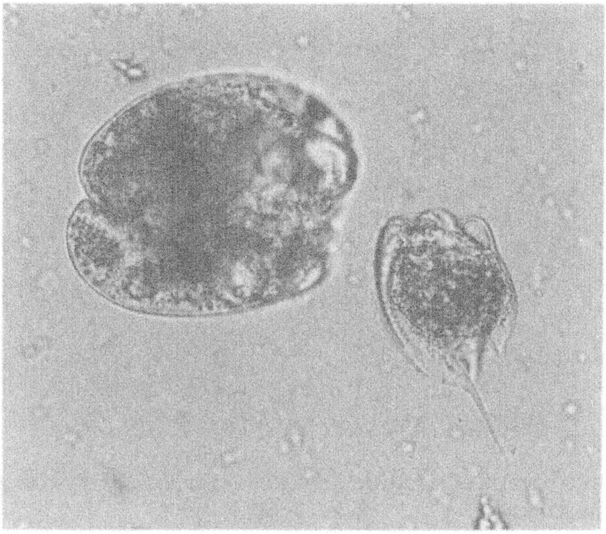

Fig. 2.2 Photomicrograph of *Entodinium bursa* (also known as *E. vorax*) (large cell) and the spined form of *E. caudatum* (small cell).

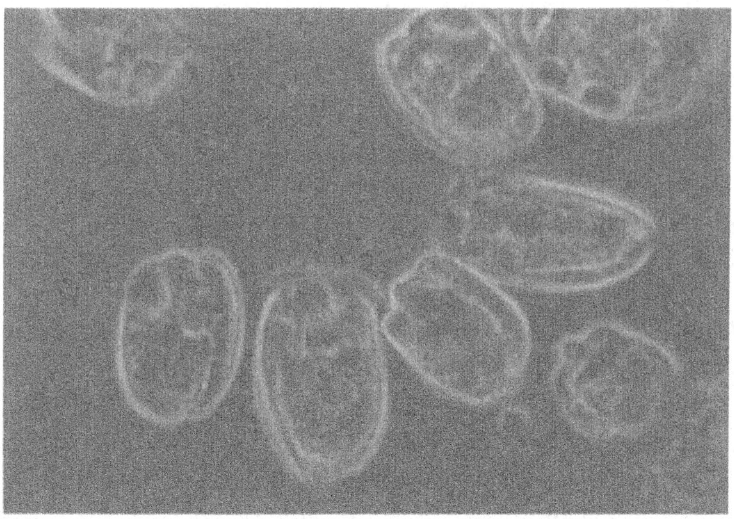

Fig. 2.3 Photomicrograph of the spineless form of *Entodinium caudatum* grown in vitro.

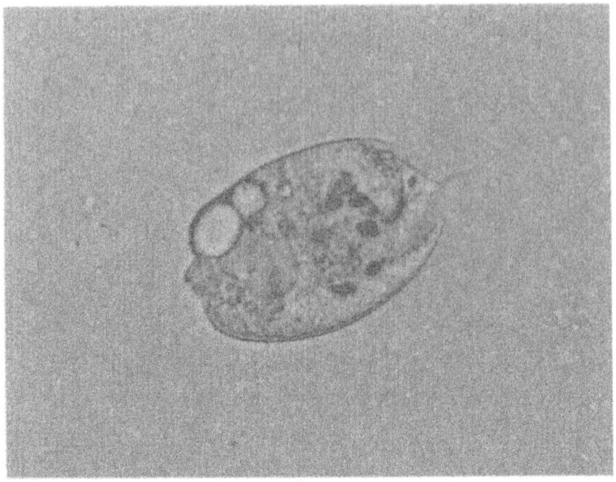

Fig. 2.4 Photomicrograph of a cultured *Entodinium caudatum* cell with an intermediately sized caudal spine.

goat by Das Gupta (1935). Although it is not widely distributed, it has been reported in cattle from Africa, Indonesia, Sri Lanka, and Thailand and from water buffalo in Indonesia and Thailand (Imai, 1985, 1986; Imai and Ogimoto, 1984). This protozoon is believed by Latteur (1968) to be the same as *E. sierrae* and *E. proturberans*. It is very slender and almost rectangular when viewed from the side.

27. *Entodinium ciculum* has only been found in a Brazilian water buffalo by Dehority (1979). It is similar to, but can be distinguished from, *E. rostratum* and *E. pisciculum*. It has two caudal spines.

28. *Entodinium constrictum* has only been described in Dall Mountain sheep (*Ovis dalli*) by Dehority (1974). It resembles, but is different from, *E. exiguum* and *E. orbicularis*.

29. *Entodinium contractum* has only been described from the gaur (*Bos gaurus*) by Kofoid and Christenson (1933).

30. *Entodinium convexum* has only been described from the mule deer (*Odocoileus hemionus*) by MacLennan (1935).

31. *Entodinium costatum* is found in deer (MacLennan, 1935; Ogimoto and Imai, 1981) and occasionally in cattle (Han, 1984; Imai, 1985). Its body is ellipsoidal and it has one small, heavy, caudal spine/lobe.

32. *Entodinium couturieri* has only been described from the ibex (*Capra ibex ibex*) by Christl (1958b).

33. *Entodinium curtum* is called *E. bursa* forma *curtum* by Latteur (1969) who believes it to be a form of *E. bursa* in which the ventral side is extended posteriorly to form a small lobe or blunt spine. It has only been described from the gaur (*Bos gaurus*) by Kofoid and Christenson (1933).

34. *Entodinium dalli* has only been described from Dall Mountain sheep (*Ovis dalli*) by Dehority (1974). The organism appears to be almost spherical in side view and has one long caudal spine and two short spines/lobes like *E. caudatum*.

35. *Entodinium damae* has only been described from Arctic ruminants, such as reindeer (Lubinsky, 1958a; Westerling, 1970) and the musk-ox (Dehority, 1974). There is doubt as to whether it is a separate species, and Lubinsky (1958a) suggested it might be the same as *E. ovalis*.

36. *Entodinium dentatum*, which has six short caudal spines, was described by Stein (1859) and has rarely, if ever, been seen since. It must be a suspect species as, unlike other entodinia, it has the micronucleus on the dorsal surface of the macronucleus.

37. *Entodinium dilobum* was originally described from cattle and sheep by Cunha (1914a) and Dogiel (1927). It has been reported recently in Korean cattle (Han, 1984) and Japanese cattle, sheep and goats (Imai et al., 1978, 1979a), and in Arctic ruminants such as reindeer and musk-oxen (Lubinsky, 1958a; Dehority, 1986c). Latteur (1968) believes that it is the same as *E. furca* forma *furca*, which is the preferred name, and *E. furca dilobum*, *E. furca bilobosum*, *E. rostratum birostratum*, and *E. biconcavum birostratum*. The organism is ellipsoidal with two incurved spines that nearly touch at their tips.

38. *Entodinium dubardi* was originally described by Buisson (1923a) and renamed *E. dubardi* forma *dubardi* by Dogiel (1925a; 1927) to distinguish it from the spined forms. Dogiel (1925a; 1926a) also described *E. simplex* and *E. parvum* and in 1927 decided that they were the same and called the species *E. simplex*. Buisson (1923a) also described an *E. parvum* which was recognised by Dogiel (1927) as a separate species. Since this time there has been much discussion whether or not these were all the same organism. Lubinsky (1957a), Zielyk (1961), and Latteur (1968) believe that *E. dubardi* and *E. simplex* are the same (*E. furca* forma *nanellum*). However, Ogimoto and Imai (1981) separate these two by the shape of the macronucleus, which in *E. simplex* is thicker at the anterior end than the posterior, whereas in *E. dubardi* the two ends are the same thickness with a thinner middle. *E. parvum* can be distinguished by the length of the macronucleus which is over two-thirds of the body length compared with only one-third in the other two species. Many other authors have agreed and *E. dubardi* has been reported to occur widely in different hosts in both Arctic and tropical areas. The two spined forms *crassicaudatum* and *gracilicaudatum* originally described by Buisson (1923a) from buffaloes have rarely been reported since.

39. *Entodinium ekendrae* was originally described from the Indian goat (*Capra hircus*) by Das Gupta (1935) and has also been reported from Japanese cattle, sheep, goats, and serow (Imai et al., 1978, 1979a, 1981a)

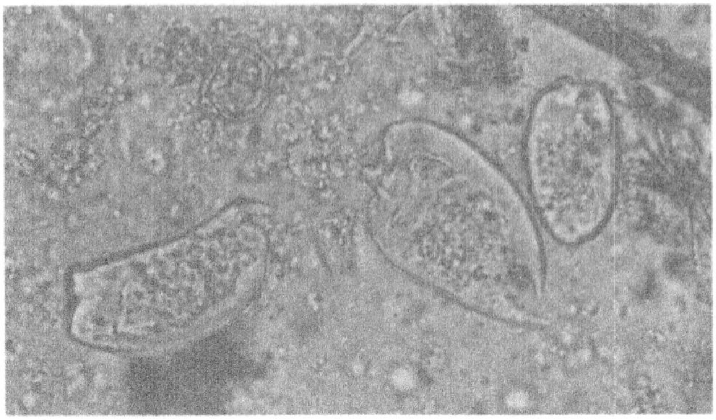

Fig. 2.5 Photomicrograph of *Entodinium furca* (two caudal spines), *E. rostratum* (one small caudal lobe), and *E. simplex* (smooth posterior end).

and in Sri Lankan cattle (Imai, 1986) but rarely elsewhere. This species resembles *E. caudatum* in spination but is smaller and has an ellipsoidal nucleus.

40. *Entodinium ellipsoideum* was separated from *E. bursa* by Kofoid and MacLennan (1930) by the former's smaller oral aperture and its general shape. However it has rarely been seen recently, and it is probably the same as *E. bursa* (Latteur, 1969).

41. *Entodinium elongatum* was originally described from cattle by Dogiel (1927) and is a comparatively large, slender, but irregularly shaped organism without caudal projections. Although it is much larger than *E. dubardi*, Latteur (1968) believes that it is *E. furca* forma *nanellum*. *E. elongatum* has not been reported recently.

42. *Entodinium exiguum* is widely distributed and commonly found in a wide range of ruminants. Latteur (1969) regards it as the same as *E. babici*, *E. nanum*, and *E. montanum*. It is one of the smallest of the entodinia and has no obvious distinguishing feature.

43. *Entodinium fujitai* was first described in water buffalo from Taiwan by Imai (1981) and has also been reported from Brazilian (Dehority, 1986b) and Sri Lankan (Imai, 1986) cattle. The organism is closely related to *E. gibberosum* but lacks the notched macronucleus. It has two caudal spines, and the ventral one is incurved inside the dorsal.

44. *Entodinium furca* was originally described by Cunha (1914a, 1914b) as a fusiform organism with two equivalent caudal spines, the tips of which are incurved towards each other (Fig. 2.5). Subsequently Dogiel (1927) added two forms (*monolobum* and *dilobum*) to the original (forma *furca*) for organisms with one or two lobes. More recently Latteur (1968) has increased the number of forms to six: 1) *nanellum* for species

with no spines including *E. nanellum*, *E. dubardi*, *E. simplex*, and *E. elongatum*; 2) *brevispinum* for species with a small ventral spine or lobe including *E. brevispinum* and *E. laterospinum*; 3) *gracilicaudatum* which includes species with this name that have a single slender ventral spine; 4) *crassicaudatum* which includes *E. dubardi* and *E. parvum* forms with this name and which have a heavy incurved ventral spine; 5) *loboso-spinosum* which includes *E. loboso-spinosum* and *E. dubardi loboso-spinosum* that have one dorsal spine and one ventral lobe; and 6) *furca* which includes *E. furca* (*furca*), *E. furca dilobum*, *E. dilobum*, and *E. rostratum birostratum*, all of which have two caudal spines.

In practice most authors, when describing species in a rumen, have used just "*E. furca* Cunha" to describe the two-spined organism because this is easy to identify. *E. furca* Cunha is not universally distributed but has been found in cattle in the USA (Becker and Talbott, 1927), Indonesia (Imai, 1985), and West Africa (Bonhomme-Florentin et al., 1978), in cattle and water buffalo from Thailand (Imai and Ogimoto, 1984), in reedbuck (Van Hoven, 1983), and in the giraffe (Kleynhans and Van Hoven, 1976).

45. *Entodinium fyferi*, which was first described from the tsessebe (*Damaliscus lunatus lunatus*) by Van Hoven (1975), has also been reported from springbok (*Antidorcas marsupialis*) by Wilkinson and Van Hoven (1976a). The organism is elliptical to nearly circular with two broad flanges terminating in two large flattened lobes. It is more circular than *E. dilobum*.

46. *Entodinium gibberosum* was first described from *Bos indicus* in India and Sri Lanka by Kofoid and MacLennan (1930) and has been reported recently in Brazilian cattle and water buffalo (Dehority, 1979, 1986b) and in Kenyan and Sri Lankan cattle by Imai (1986, 1988). This protozoon has two incurved caudal spines and is characterized by having a notch in the anterior end of the macronucleus.

47. *Entodinium hamatum* was described by Bretschneider (1934) but was regarded as a dubious species by Latteur (1969) as the original description was inadequate.

48. *Entodinium ibicis* has only been reported from the ibex (*Capra ibex ibex*) by Christl (1958b).

49. *Entodinium incurvatum* has only been described from African antelopes by Dogiel (1923a). It has two long, caudal spines, both of which are curved ventrally where they are attached to the body of the protozoon. They then curve again in the opposite direction so that the tips are parallel to the main axis.

50. *Entodinium indicum* is widely but not universally distributed in cattle in Brazil, Indonesia, Kenya, New Zealand, Sri Lanka, Thailand, and the USA (Imai, 1985, 1986, 1988; Imai and Ogimoto, 1984; Abou

Akkada et al., 1969; Clarke, 1964; Dehority, 1986b) but not in Japan (Imai et al., 1978, 1979a). It is also found in water buffalo (Dehority, 1979; Imai, 1985; Imai and Ogimoto, 1984) but has not been reported from sheep. It is characterized by one small ventral spine, one small dorsal spine, and one large spine on the left side at the base of which is the cytoproct.

51. *Entodinium javanicum* has only recently been described in Bali cattle (*Bos javanicus domesticus*) in Indonesia by Imai (1985). It has also been found in *Bos indicus* in Sri Lanka (Imai, 1986). *E. javanicum* is ellipsoidal in shape and is characterized by two heavy, blunt, caudal spines of which the ventral is usually longer than the dorsal.

52. *Entodinium kalaharicus* has been found only in the giraffe and the springbok (Kleynhans and Van Hoven, 1976; Wilkinson and Van Hoven, 1976a). It is ellipsoidal in shape and characterized by a V-shaped cleft in the anterior end of the macronucleus. In this respect only does it resemble *E. gibberosum*.

53. *Entodinium laterale* has only been observed in the past in cattle and goats in India and Sri Lanka (Kofoid and MacLennan, 1930; Das Gupta, 1935). This protozoon has three caudal spines, two of which are short, heavy, and blunt while the other on the dorsal side is longer and thinner. The general morphology is similar to that of the larger *E. rectangulatum*. The caudal spination can resemble that of *E. caudatum* but the two organisms are separated by the position of the contractile vacuole which is in the middle of the left side in *E. laterale* and close to the left side of the anterior end of the macronucleus in *E. caudatum*.

54. *Entodinium laterospinum* is believed by Latteur (1968) to be the same as *E. brevispinum* and that it should be called *E. furca* forma *brevispinum*. However Kofoid and MacLennan (1930), who first described both organisms, believed that they could be separated on their general body shape. Both protozoa have a short stubby ventral lobe/spine. *E. laterospinum* has been reported in *Bos indicus* in India and Sri Lanka and in cattle in the USA (Dehority, 1978).

55. *Entodinium loboso-spinosum,* which also appears in the literature as *E. caudatum* forma *loboso-spinosum,* was renamed *E. furca* forma *loboso-spinosum* by Latteur (1968), but this name has not been generally accepted. *E. loboso-spinosum* appears ascribed to Dogiel (1925a) and Dogiel (1927) because the organism was originally described as a form of *E. dubardi* in 1925 and promoted to a species in 1927. Lubinsky (1957a) examined in detail a very similar organism isolated from sheep and goats in Pakistan which he called *E.* cf. *loboso-spinosum* because of uncertainty over the position of the contractile vacuole in Dogiel's original isolate. There was also considerable size variation between different isolates observed at different times from various sources, but Lubinsky believed that this was due to intraspecific variation. This protozoon is character-

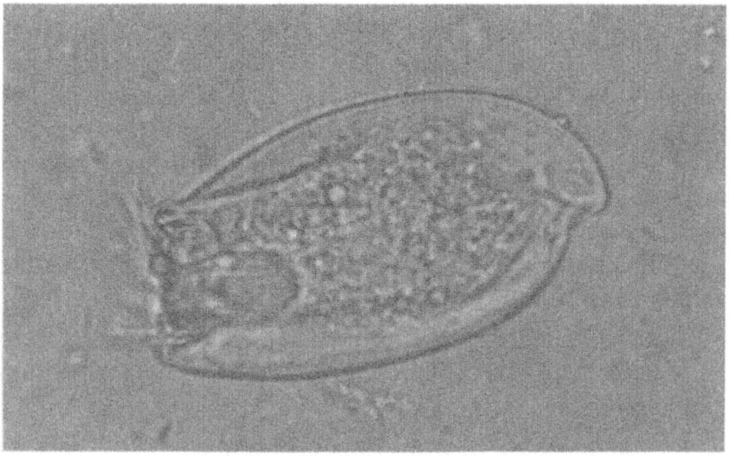

Fig. 2.6 Photomicrograph of *Entodinium longinucleatum* f. *longinucleatum* grown in vitro.

ized by one long, slender, caudal spine and a small ventral lobe. Because of the variability of caudal spination, Lubinsky (1957a) believed that spination should be used to separate forms rather than species. He therefore separated *E. caudatum, E. rectangulatum,* and *E. simulans* into species on the location of the contractile vacuole and upper groove and then described *loboso-spinosum* and *dubardi* (no caudal spine) forms of each.

E. loboso-spinosum is found in Japanese cattle, sheep, goats, and serow (Imai, 1981; Imai et al., 1978, 1979a), the tsessebe (small forms—Van Hoven, 1975), reedbuck (Van Hoven, 1983), and the giraffe (small *dubardi* forms—Kleynhans and Van Hoven, 1976). "*E. caudatum loboso-spinosum*" has been reported in cattle in Korea, Indonesia, Philippines, Sri Lanka, and Thailand (Imai, 1985, 1986; Han, 1984; Imai and Ogimoto, 1984; Shimizu et al., 1983) and in Indonesian water buffalo (Imai, 1985).

56. *Entodinium longicorpum* was first observed in the giraffe by Kleynhans and Van Hoven (1976) but has not been found elsewhere. It is characterized by an elongated form (length to width ratio 3.78).

57. *Entodinium longinucleatum* is characterized by its long rod-like macronucleus that extends almost from the oral region to the cytoproct (Fig. 2.6). It exists in four forms depending on its spination, according to Imai (1984). The form *longinucleatum,* which was described by Dogiel (1925d), is the organism commonly called *E. longinucleatum*. It has a ventral lobe at the posterior end and is widely distributed geographically in many hosts. The form *acutonucleatum* is believed by Kofoid and MacLennan (1930) and Latteur (1968) to be a separate species and has three small

caudal spines. It has only been described in this way in cattle and water buffalo in Southeast Asia (Imai, 1985, 1986; Imai and Ogimoto, 1984). (See also under *E. acutonucleatum*.) The third form, *spinonucleatum*, is believed by Dehority (1979) to be a separate species and has a single posterior left spine and a lobe or short spine on the right side. It is described as such in Sri Lankan cattle (Imai, 1986). The fourth form is *spinolobum*, which was described by Imai (1984), and has a short caudal spine on the right side, a short spine on the upper left side, and a lobe on the lower left side. It has only been observed in Indonesian and Thai cattle (Imai, 1985; Imai and Ogimoto, 1984).

58. *Entodinium longispinosus* was described by Chavarria from cattle in Mexico in 1933 but has not been reported since. It has four long slender caudal spines.

59. *Entodinium lucii* has only been described from the springbok (*Antidorcas marsupialis*) in southern Africa by Wilkinson and Van Hoven (1976a). It has an almost rectangular body with a long thin macronucleus extending almost the length of the body. It is unusual in that the contractile vacuole is situated almost in the middle of the body on the left surface.

60. *Entodinium mamillatum* has only been described from the caecum of the rodent *Cavia aperea* by Cunha (1914a, 1914b). It has a ventral lobe.

61. *Entodinium medium* has only been described from the mule deer (*Odocoileus hemionus*) by MacLennan (1935). It is almost completely oval with no caudal projections. It differs from the similar *E. ovoideum* in the presence of a lateral pellicular line and in other small ways.

62. *Entodinium minimum* is found in Japanese cattle, sheep, goats, and serow (Imai et al., 1978, 1979a, 1981a), in Indonesian Bali cattle and Sri Lankan Zebu cattle (Imai, 1985, 1986), and in Brazilian cattle (Dehority, 1986b). The protozoon has a very characteristic shape as it is roughly triangular with a convex right surface and a flattened or concave left surface.

63. *Entodinium montanum* was described from Sierra Nevada Bighorn sheep (*Ovis canadensis*) in the USA by Bush and Kofoid (1948) but is believed by Latteur (1969) to be the same as *E. exiguum*. *E. montanum* is described as an elongated ellipsoid with, when contracted, a truncated anterior end. The posterior end can be regarded as having a broad ventral lobe.

64. *Entodinium muntiacum* was described from the common muntjak (*Muntiacus muntjak*) in India by Banerjee (1955). This organism is ellipsoidal and has been described as having two caudal lobes of which the ventral is the larger.

65. *Entodinium nanellum* is regarded by Latteur (1968) as being the same as *E. dubardi*, *E. parvum*, *E. simplex*, *E. elongatum*, and *E. ovoideum*,

all of which should be called *E. furca* forma *nanellum*. However, others feel that these are all different species and two or three are often distinguished by those listing protozoal species found in a rumen. Ogimoto and Imai (1981) state that *E. nanellum* and *E. parvum* have relatively elongated bodies whereas those of *E. simplex* and *E. dubardi* are ellipsoidal. The macronucleus of *E. nanellum* is only one third of the body length whereas that of *E. parvum* is over two-thirds of the body length. Further information is given by Wertheim (1935a), Wilkinson and Van Hoven (1976a), and Kleynhans and Van Hoven (1976). *E. nanellum* is widely distributed both geographically and in terms of host species.

66. *Entodinium nanum* was first described from the Bighorn sheep (*Ovis canadensis*) by Bush and Kofoid (1948). It has also been found in Dall Mountain sheep (*Ovis dalli*) by Dehority (1974) and in North Ronsaldsay (Scotland) sheep by Eadie (1957). It is, however, uncommon and is believed by Latteur (1969) to be the same as *E. exiguum*.

67. *Entodinium ogimotoi* was first described in water buffalo from Taiwan by Imai (1981). When viewed from the side, it is rectangular to almost square with one small and one larger, more obvious, triangular lobe at the posterior end.

68. *Entodinium orbicularis* is an almost spherical protozoon described from Bighorn sheep in California, USA, by Bush and Kofoid (1948). It is believed by Latteur (1969) to be the same as *E. muntiacum*, although the illustrations do not suggest identity. It is of interest that Zielyk (1961) pointed out that the post-divisional stages of *E. dubardi* closely resemble *E. orbicularis*.

69. *Entodinium ovale* was the name given by Dogiel (1927) to the protozoon called *E. ovalis* isolated from the mouse deer by Jameson (1925b).

70. *Entodinium ovalis* was first described from the mouse deer (*Tragulus meminna*) by Jameson (1925b). It is the same as *E. ovale*. It is characterized by the shape of the macronucleus, the anterior end of which is bent at right angles to the right. Sladecek (1946) pointed out its similarly to *E. damae*, and the only difference was the side to which the macronucleus was bent. Subsequently Lubinsky (1958a) showed that the nucleus was bent in the same direction in both organisms and that confusion had arisen because of different terminologies used by the different authors. *E. ovalis* may therefore be the same as *E. damae*.

71. *Entodinium ovibos* was first described from musk-oxen (*Ovibos moschatus*) by Dehority (1985) where it accounted for 40–60% of the ciliates in the rumen of seven animals examined. It resembles *E. parvum* in size, shape, and displacement of the cytoproct to the dorsal side, but differs in possessing a club-shaped macronucleus, the anterior end of which is bent ventrally and to the right. In this respect it resembles *E. damae*.

72. *Entodinium ovinum* is widely but not universally distributed, and

is found in Japanese cattle, sheep, and goats (Imai et al., 1978), Thai zebu cattle and water buffalo (Imai and Ogimoto, 1984), Indonesian Bali cattle and water buffalo (Imai, 1985), Brazilian and New Zealand cattle (Dehority 1986b; Clarke, 1964), in musk-ox but rarely in reindeer (Dehority, 1986c), and not apparently in American (USA) cattle or in springbok, reedbuck, or tsessebe. *E. ovinum* is a comparatively large entodinium (53–69 μm long) and, according to Ogimoto and Imai (1981), a "well-regulated oval shape," although Dogiel (1927) shows it with a flattened dorsal surface. The length of the macronucleus is over two-thirds that of the body.

73. *Entodinium ovoideum* is believed by Latteur (1968) to be the same as *E. nanellum, E. dubardi, E. parvum, E. simplex,* and *E. elongatum,* all of which are *E. furca* forma *nanellum*. However, others such as Dehority (1986b) identified *E. nanellum, E. ovoideum,* and *E. parvum* in Brazilian cattle. *E. ovoideum* is similar in shape to *E. ovinum* with the long macronucleus but is smaller in size (30–50 μm long compared with 53–69 μm). It is much rounder than *E. parvum* which is more elongated with flattened dorsal and ventral sides.

74. *Entodinium ovoido-nucleatum* was first described from the Indian goat by Das Gupta (1935) but has not been reported since. It is rectangular in shape with two short and one very long caudal spine. The macronucleus is ovoid. *E. ovoido-nucleatum* is the only protozoon in the dorsopulsatum group of Latteur (1968), and Lubinsky (1957a) believes that it is an exconjugant of *E. caudatum*.

75. *Entodinium ovum-rajae* (called *Amphacanthus ovum-rajae* by Dogiel, 1927) is found in camels and exists in two forms. The form *ovum-rajae* has dorsal and ventral flanges that terminate posteriorly in incurved spines. The form *deficiens* has no flanges and only one small dorsal spine.

76. *Entodinium parvum* as described by Buisson (1923c) (the forma *parvum* of Imai and Ogimoto, 1983) is widely but not universally distributed. It is found in almost all Japanese cattle, sheep, and goats (Imai et al., 1978, 1979a), Thai and Indonesian cattle and water buffalo (Imai and Ogimoto, 1984, Imai, 1985), Sri Lankan and Philippine cattle (Shimizu et al., 1983; Imai, 1986), Brazilian and Yugoslav cattle (Dehority, 1986b; Wertheim, 1935b), the springbok (Wilkinson and Van Hoven, 1976a), and the giraffe (Kleynhans and Van Hoven, 1976). Latteur (1968) believes that *E. nanellum, E. dubardi, E. parvum, E. simplex,* and *E. elongatum* are all the same and should be called *E. furca* forma *nanellum*, but others disagree. Kleynhans and Van Hoven (1976) and Wilkinson and Van Hoven (1976a) considered the description of *E. parvum* in some detail and stated that this protozoon has an average body length greater than 30 μm with a macronucleus lying close to the anterior end of the body and extending to the beginning of the last third of the body. The dorsal and ventral

sides of the protozoon are roughly parallal, and the posterior end is more rounded than that of related protozoa. However, these authors found that in the giraffe *E. parvum* measured only 23.9 μm (range 22.0–28.7 μm).

Imai and Ogimoto (1983) have described a new form *monospinosum* that they found in zebu cattle in Thailand. It has a short, heavy, ventral spine which resembles that of *E. rostratum*. However *E. parvum monospinosum* has a different body shape and a shorter macronucleus. It has also been reported from Brazilian cattle (Dehority, 1986b).

77. *Entodinium pisciculum* was first described and has only been seen in *Bos indicus* from India and Sri Lanka by Kofoid and MacLennan (1930). It has a short, heavy, ventral spine and a dorsal flange that runs from the oral area to a short caudal spine. There is also a very characteristic fusiform projection at the anterior end of the flange. The protozoon has a "fish-like" appearance which inspired its name.

78. *Entodinium pristinum* was described for the first and only time from the mule deer (*Odocoileus hemionus*) by MacLennan (1935). It is ellipsoidal in shape with a distinct cuticular groove on the left side with possibly a small, ventral caudal lobe.

79. *Entodinium protuberans* was described from Bighorn sheep (*Ovis canadensis*) in California, USA, by Bush and Kofoid (1948) but is believed by Latteur (1969) to be the same as *E. chatterjeei*. However, *E. protuberans* was identified as such in Dall Mountain sheep (*Ovis dalli*) by Dehority (1974). The protozoon is elongated, tapering posteriorly with a small ventral caudal lobe.

80. *Entodinium quadricuspis* is commonly found in reindeer (*Rangifer tarandus*) in Russia, Canada, and Finland (Dogiel, 1925d; 1927; Lubinsky, 1958a; Westerling, 1970; Dehority, 1986c) and was at one time believed to be specifically found only in reindeer. However, it has also been found in cattle (Latteur, 1969) and the giraffe (Kleynhans and Van Hoven, 1976), although these protozoa are 30% shorter than those in reindeer. *E. quadricuspis* is almost circular in lateral view and the dorsal and ventral edges terminate posteriorly in short pointed spines. Cuticular folds are also present on the left and right sides of the body and these also terminate posteriorly in spines, although these are shorter than the others. Dogiel (1927) described two forms, one of which, *quadricuspis*, is that described above. The other, *stauracanthum*, which was not seen by Lubinsky (1958a), has longer left and right spines and dorsal and ventral lobes rather than spines. The line drawings of Lubinsky (1958a) for *E. quadricuspis* have features of both forms.

81. *Entodinium rectangulatum* was originally described from *Bos indicus* in India and Sri Lanka by Kofoid and MacLennan (1930) as a heavy, stout, almost rectangular protozoon with three short caudal spines, the dorsal one of which is the end of a body flange. Since then Lubinsky

(1957a) has re-examined this species and found three forms: (a) *caudatum* (which Latteur [1969] believes should be called "*rectangulatum*") has two short and one long caudal spine as in *E. caudatum*; (b) *loboso-spinosum*, which has one ventral lobe and one long dorsal spine; and (c) *dubardi*, in which the spine is missing. These forms all differ from the corresponding forms of *E. caudatum* in the position of the contractile vacuole which is latero- in *E. caudatum* and sinistro- in *E. rectangulatum* (see above #24). In the literature the organism is often referred to as just *E. rectangulatum* (Kofoid and MacLennan) and it is presumed that this refers to the three-spined form. It is not very widely distributed but has been found in Korean native cattle (all three forms—Han, 1984), Thai zebu cattle (Imai and Ogimoto, 1984), USA and Brazilian cattle (Dehority, 1978, 1986b), and in Brazilian water buffalo (Dehority, 1979).

82. *Entodinium rhomboideum* is not very commonly found, although it is widely distributed. It is reported from the mule deer (MacLennan, 1935), and Brazilian, Sri Lankan, Indian, and Thai cattle (Dehority, 1986b; Imai, 1986; Imai and Ogimoto, 1984; Kofoid and MacLennan, 1930). *E. rhomboideum* is similar to *E. bimastus* in that both have a large caudal lobe through which the cytoproct passes. However, the latter is much more spherical in side view and has a much longer macronucleus. Kofoid and MacLennan (1930) pointed out that the posterior end of *E. bimastus* is triangular, whereas that of *E. rhomboideum* is rectangular.

83. *Entodinium rostratum* is widely distributed especially in cattle and water buffalo and is also found in African reedbuck and tsessebe but not springbok. It is much less common in sheep. It is absent from Japanese sheep and goats, although it is found in Japanese cattle (Imai et al., 1978, 1979a). *E. rostratum* has a convex dorsal surface and a concave ventral surface which extends into a short, blunt spine or lobe.

84. *Entodinium rotundum* was described for the first and only time from the mule deer (*Odocoileus hemionus*) by MacLennan (1935). The protozoon is almost circular in lateral view with a small ventral lobe. It has a spherical macronucleus.

85. *Entodinium rupicaprae* was described for the first and only time from the chamois (*Rupicapra rupicapra*) by Christl (1955). It has a long macronucleus that tapers posteriorly.

86. *Entodinium setnai* was described for the first and only time from the Indian goat (*Capra hircus*) by Das Gupta (1935). It is ellipsoidal in shape with a short, incurved, dorsal caudal spine and two short ventral spines. It has a spherical macronucleus.

87. *Entodinium sierrae* was observed for the first and only time from the Bighorn sheep in California, USA, by Bush and Kofoid (1948). Latteur believes that it is the same as *E. chatterjeei*.

88. *Entodinium simplex*. The status of *E. simplex* has been debated at

length for the past sixty years because of the difficulty in deciding whether or not all the small entodinia without very obvious caudal spines or lobes were the same species. Dogiel (1925a, 1927) distinguished *E. dubardi dubardi*, *E. parvum*, *E. simplex*, *E. nanellum* and *E. exiguum*, and Wertheim (1935a) gave criteria for distinguishing between these and *E. babici* and *E. bovis*. However, Latteur (1968) decided that all except the last two, but with the addition of *E. elongatum* and *E. ovoideum*, were the same and should be called *E. furca* forma *nanellum*. This has proved not to be acceptable and the following criteria have been used to separate *E. simplex* from *E. dubardi*, which is the most closely related organism. The distinguishing features for the other species will be found under the appropriate name. *E. simplex* is more elongated than *E. dubardi* (length/width ratio 1.7–1.9 compared with 1.1–1.7). The macronucleus of *E. simplex* is wedge-shaped with the anterior end thicker than the posterior. It is also located in the anterior half of the organism and closely applied to the body wall. In contrast, the macronucleus of *E. dubardi* is more rod-shaped with the middle possibly thinner than the two ends. It is situated in the middle of the organism rather than at the anterior end (Lubinsky, 1958a, Wilkinson and Van Hoven, 1976a; Ogimoto and Imai, 1981). However, it should be noted that Zielyk (1961) believes that *E. simplex* and *E. dubardi dubardi* are the same.

Despite the doubt about its existence as a separate species, *E. simplex* has been widely reported as occurring in many host species with a wide geographical distribution.

89. *Entodinium simulans*, which was first described from sheep in Pakistan by Lubinsky (1957a), superficially resembles *E. caudatum* and *E. rectangulatum* in having two short and one long caudal spine. However, it differs from *E. caudatum* in having a contractile vacuole located sinistro- rather than latero- and its lateral groove extends the length of the body and is slit-like at the anterior end. In *E. rectangulatum* the groove is triangular and terminates anteriorly at the level of the micronucleus. In the literature *E. simulans* is often referred to as such (e.g. in Brazilian water buffalo, Dehority, 1979) and this presumably means the three spined form, although Lubinsky (1957a) called this form *caudatum* to distinguish it from two others. These are *loboso-spinosum*, which has a long dorsal spine but no right ventral lobe/spine, and *dubardi*, which has no long caudal spine or right ventral lobe, although the left ventral lobe is present. Apart from its occurrence in water buffalo, *E. simulans* has only been reported in cattle in the USA (Dehority, 1978).

90. *Entodinium spino-nucleatum* has only been described once and that was from cattle in The Netherlands by Bretschneider (1934). It is rectangular in shape and has a highly characteristic macronucleus which extends from the oral region into the caudal lobe. It has two short caudal spines.

91. *Entodinium spinonucleatum*, which was described from Brazilian water buffalo by Dehority (1979), is different from the previous protozoon in that it is ellipsoidal in shape. The macronucleus extends from the oral region into the base of the dorsal lobe or short spine at the posterior end. It also has one short ventral spine, a finding which distinguishes it from *E. acutonucleatum* which has a ventral lobe carrying two short spines. Imai (1984) believes that this protozoon should be called *E. longinucleatum* forma *spinonucleatum* and it has been reported from Sri Lankan cattle under this name (Imai, 1986). It appears to be a rare organism.

92. *Entodinium triacum* (sensu Buisson, 1923b) has only been reported recently in bushbuck (*Tragelaphus scriptus*) and reedbuck (three *Redunca* spp.—Van Hoven, 1983). If, however, Latteur's (1968) view is correct and some other protozoa are forms of *E. triacum*, then the species is much more common. These forms are (a) *biconcavum*, which is Kofoid and MacLennan's (1930) *E. biconcavum*; (b) *bifidum*, which includes *E. bifidum*, *E. rostratum bifidum*, and *E. biconcavum bifidum*; (c) *triacum* is *E. triacum* (Buisson, 1923a), *E. triacum triacum* (Dogiel, 1927) and *E. acutum* (Kofoid and MacLennan, 1930); (d) *aculeatum*, which is *E. aculeatum* (Kofoid and MacLennan, 1930); and (e) *dextrum*, which is *E. triacum dextrum*. All these except the last one have been described as separate species. *E. triacum dextrum* has three medium-length caudal spines (longer than in forma *triacum*), of which the dorsal and right ventral are incurved while the left ventral is straight.

93. *Entodinium triangulatum* was first described from Brazilian water buffalo by Dehority (1979) and has also been found in Brazilian cattle (Dehority, 1986b). The organism has two heavy caudal spines. The dorsal one is a smooth continuation of the convex curve of the dorsal surface while the ventral one projects slightly outward from the ventral surface before turning at right angles towards the dorsal side. The protozoon obtains its name from the triangular wedge of ectoplasm that contains the macronucleus.

94. *Entodinium tricostatum* was originally described from *Bos indicus* in India and Sri Lanka by Kofoid and MacLennan (1930) and has also been reported from Chinese cattle (Hsiung, 1932) and cattle in Bangladesh (Banerjee, 1955). It has recently been reported from Kenyan cattle by Imai (1988). It is one of the two protozoa in the troncopulsatum group (see earlier, Latteur's seven types) and is characterized by having three ribs running the length of the body. The dorsal and left ventral ribs end in short caudal spines and the thin blade-like right ventral rib ends in a caudal lobe.

95. *Entodinium tricuspis* was described incompletely from Mexican cattle by Chavarria (1933). It has two small caudal spines/lobes and a larger dorsal lobe.

96. *Entodinium tridentatum* was mentioned without any details by Noirot-Timothée (1963) from the giraffe. It was not seen by Kleynhans and Van Hoven (1976).

97. *Entodinium truncatum* has only been described from the Bighorn sheep (*Ovis canadensis*) in California, USA, by Bush and Kofoid (1948). It is a rectangular ellipsoid with a truncated posterior end.

98. *Entodinium tsunodai* was first described from water buffalo in Taiwan by Imai (1981) and has also been reported from zebu cattle in Thailand (Imai and Ogimoto, 1984) and Bali cattle in Indonesia (Imai, 1985). It has four caudal spines. There is one short one and one heavy incurved one on the ventral side, one short spine on the dorsal side, and one in the middle of the right side.

99. *Entodinium vorax* is a widely used alternative name for *E. bursa*, although Dogiel (1927) lists them a separate species and also identifies a form *bispinosum* which has two short, blunt, incurved caudal spines. This form has rarely been seen, although it was reported from Chinese cattle by Hsiung (1932).

100. *Entodinium wertheimi* was described for the first and only time from the ibex (*Capra ibex ibex*) by Christl (1958b).

2.3 The Genus *Eodinium*

This genus, the validity of which has been disputed by Hungate (1978) and Latteur (1966d), contains the small protozoa, 32–60 μm long, which have a dorsal zone on cilia but no skeletal plates. The genus was erected by Kofoid and MacLennan (1932) who further distinguished it from *Diplodinium* spp. by the shape and position of the macronucleus. In protozoa lying flat on a surface, the macronucleus is rod-shaped and at the edge of the organism in eodinia. In diplodinia it is a wedge or rod with the anterior end bent ventrally and tends to be situated near the middle of the cell. The type species is *Eodinium lobatum* (Kofoid and MacLennan, 1932). Eodinia are more common in cattle than sheep. Detailed information about individual species is given below. The reader should beware of the confusion that exists in those papers in the literature that give lists of names of protozoa present in various hosts. Some authors call all the protozoa listed below "*Eodinium*", whereas others call them "*Diplodinium*," and still others call some species by one name and other species by the other. In the list below, alternative names are given in parentheses. Information on *Eodinium* spp. is also given in Table 2.3.

Detailed Descriptions of the Eodinia

1. *Eodinium bilobosum* (*Anoplodinium posteriovesiculatum bilobosum*—Dogiel, 1927; *Diplodinium posteriovesiculatum bilobosum*—Latteur, 1970)

Table 2.3 Characteristics of the *Eodinium* species

Species	Original description/ name[a]	Length (μm)	Width (μm)	Spines	Lobes
1. *E. bilobosum*	2/5	46–60	30–44	0	2
2. *E. bispinosum*	4	23–32	18–24	2	0
3. *E. lobatum*	5	44–60	29–37	0	1
4. *E. monolobosum*	3	42–60	25–33	0	1
5. *E. polygonale*	1/5	32–38	20–24	0	0
6. *E. posterovesiculatum*	2/5	42–62	24–30	0	0
7. *E. rectangulatum*	5	35–70	18–40	0	0

[a]References are as follows: 1, Dogiel (1925a); 2, Dogiel (1927); 3, Hsiung (1932); 4, Kleynhans and Van Hoven (1976); 5, Kofoid and MacLennan (1932); 6, Latteur (1970).

has been observed in cattle in Europe, Russia, New Zealand, and the USA (Dogiel, 1927; Clarke, 1964; Abou Akkada et al., 1969) and in buffalo and antelopes. This protozoon is a slightly compressed ovoid in side view with two fairly prominent caudal lobes.

2. *Eodinium bispinosum* was first described from the giraffe by Kleynhans and Van Hoven (1976) who were uncertain of its taxonomic status, although they placed it in the genus *Eodinium*. However it has only one contractile vacuole instead of the usual two, a broad stumpy macronucleus, and two caudal spines (left and right but both pointed towards the dorsal side) which are atypical of eodinia.

3. *Eodinium lobatum* (*Diplodinium posterovesiculatum lobatum*—Latteur, 1970) has been found in Indian, Brazilian, Japanese (<2%), New Zealand, and Thai (zebu) cattle (Dogiel, 1927; Dehority, 1986; Imai et al., 1978; Clarke, 1964; Imai and Ogimoto, 1984) and in Brazilian water buffalo (Dehority, 1979). It has a small ventral lobe.

4. *Eodinium monolobosum* (*Diplodinium posterovesiculatum monolobosum*—Hsiung, 1932; Latteur, 1970) has been found in the USA, Brazilian, and Chinese cattle (Dehority, 1978, 1986b; Hsiung, 1932) and in Brazilian water buffalo but not in Japanese cattle (Dehority, 1979; Imai et al., 1978, 1979a). It is characterized by a heavy ventral caudal lobe/ spine.

5. *Eodinium polygonale* (*Anoplodinium polygonale*—Dogiel, 1927; *Diplodinium polygonale*—Dogiel, 1925a; Latteur, 1970; Ogimoto and Imai, 1981) has been found in Brazilian water buffalo and cattle (all animals tested—Dehority, 1979, 1986b), Thai zebu cattle and water buffalo (Imai & Ogimoto, 1984), African steinbok (Dogiel, 1927), and in other host species. This protozoon has a flattened dorsal side and a gently convex ventral side that tapers to a blunt point posteriorly. The macronucleus is atypically short and heavy like *E. bispinosum*.

	Contractile vacuoles	Body shape	Macronucleus Size (μm)	Macronucleus Shape	Illustrated in Ogimoto and Imai (1981)	Other reference[a]
1.	2	Ovoid	5 × 28	Rod	–	6
2.	1	Subspherical	5 × 12	Stumpy	–	
3.	2	Ovoid	7 × 30	Rod with depressions	+	6
4.	2	Ovoid	6 × 37	Curved rod	+	6
5.	?	Ovoid	4 × 9	Rod	+	
6.	2	Ovoid	10 × 28	Rod	+	6
7.	2	Rectangular	12 × 30	Rod	+	6

6. *Eodinium posterovesiculatum* (*Anoplodinium posterovesiculatum posterovesiculatum*—Dogiel, 1927; *Diplodinium posterovesiculatum posterovesiculatum*—Latteur, 1970) is found in Japanese cattle (but not sheep or goats), New Zealand, Russian, South African and Brazilian cattle and Thai zebu cattle and water buffalo (Imai et al., 1978, 1979a; Clarke, 1964; Dehority, 1979; Imai and Ogimoto, 1984; Dogiel, 1927; Fantham, 1926). This protozoon, like some other eodinia (those given the specific name *posterovesiculatum* by Latteur, 1970), has one of its two contractile vacuoles at the posterior end of the macronucleus, hence the origin of the name.

7. *Eodinium rectangulatum* (*Diplodinium rectangulatum*—Latteur, 1970) is rare, although Kofoid and MacLennan (1932) found it in *Bos indicus* in India and Sri Lanka, and Ogimoto and Imai (1981) reported it from cattle and sheep.

2.4 The Genus *Diplodinium*

This genus includes the larger species (55–210 μm long; average 100 μm) with a dorsal band of cilia but no skeletal plate, whereas the smaller protozoa are placed in the genus *Eodinium*. The diplodinia also differ from the eodinia in having a rod- or wedge-shaped macronucleus, the anterior third of the dorsal surface of which is bent ventrally at an angle of 30°–90° (Kofoid and MacLennan, 1932). These authors divided the diplodinia into five groups. The *Dentatum* group contains those species with a broad, truncated posterior end of the body and five to six caudal spines. The organisms in the *Anacanthum* group have a more tapering posterior end and from zero to six caudal spines. Although differentiated into species by some, the organisms that differ only in the number of caudal spines may be better described as forms. In clone cultures in vitro, the spines of *Diplodinium pentacanthum* gradually become smaller and tend to disappear, suggesting that the presence and number of spines is determined by environmental

Table 2.4 Characteristics of the *Diplodinium* species

Species (Fig. no)[a]	Original description/ name[b]	Length (μm)	Width (μm)	Spines	Lobes	Contractile vacuoles
1. *D. africanum*	16	150–250	105–155	1	1	2
2. *D. anacanthum*	5/13	70–90	40–60	0	0	2
3. *D. anisacanthum*	5	77–86	53–61	6	0	2
4. *D. archon*	20	39–49	26–33	2	0	1
5. *D. babici*	32	53–84	40–64	0	0	2
6. *D. bubalidis*	9/11	104–195	58–98	1	0	2
7. *D. bursa*	14	100–150	60–90	0	0	2
8. *D. cameli*	10	160–210	92–130	0	0	2
9. *D. ceylonicum*	3			1	0	
10. *D. clevelandi*	2	78–112	44–75	0	0	2
11. *D. consors*	9/18	65–108	35–46	1	0	2
12. *D. costatum*	9/18	80–180	55–110	0	0	2
13. *D. crista-galli*	13/18	77–100	52–70	Fan	0	2
14. *D. dentatum* (2.7)	29/30	65–82	52–62	6	0	2
15. *D. diacanthum*	5/18	70–83	47–60	2	0	2
16. *D. dogieli*	12/18	70–100	48–77	0	0	2
17. *D. elongatum*	13/18	177–205	73–100	0	0	2
18. *D. flabellum*	18					
f. *aspinatum*	6	76–118	53–77	Fan	0	1–2
f. *flabellum* (2.8)	6/18	90–108	58–72	Fan	0	1–2
f. *monospinatum*	6	78–118	55–76	Fan	0	1–2
f. *laterospinatum*	8	78–118	62–90	Fan	0	1–2
19. *D. giraffae*	27					
20. *D. hegneri*	2	See *Ostracodinium obtusum*				
21. *D. helseri*	2	See *Eremoplastron rostratum*				
22. *D. indicum*	1	130–180	85–105	1	0	2
23. *D. italicum*	24	120–147	64–74	0	0	2
24. *D. komareki*	19					
25. *D. laeve*	13/18	77–100	52–70	0	1	2
26. *D. mahidoli*	17					
f. *mahidoli*	17	117–168	77–113	1	1	2
f. *bispinosum*	17	117–168	77–113	2	0	2
27. *D. minor*	9/18	60–90	45–51	0	0	2
28. *D. monacanthum* (2.9)	5/18	60–124	38–72	1	0	2
29. *D. moucheti*	26	74–84	52–59	1	0	2
30. *D. nanum*	16	35–53	28–40	0	1	2
31. *D. pentacanthum*	14/18	67–84	51–60	5	0	2
32. *D. psittaceum*	18					
f. *longi-caudatum*	25	105–155	59–100	1	0	2
f. *psittaceum*	13/18	105–155	59–100	0	1	2
33. *D. quinquecaudatum*	10/18	57–73	47–65	5	0	2
34. *D. rangiferi*	12/18	128–210	100–165	0	0	2
35. *D. sinhalicum*	15	49–59	37–44	0–2	0–2	2
36. *D. tetracanthum*	5/18	72–83	52–61	4	0	2
37. *D. tricanthum*	5/18	70–83	51–64	3	0	2

[a] A number in parentheses indicates that this species is shown in the figure corresponding to that number in this chapter.
[b] References are as follows: 1, Banerjee (1955); 2, Becker and Talbott (1927); 3, Bhatia (1936); 4, Coleman (1979b); 5, Cunha (1914a); 6, Dehority and Potter (1974); 7, Dehority (1985); 8, Dehority (1986b); 9, Dogiel (1925a); 10, Dogiel (1925b); 11, Dogiel (1925c); 12, Dogiel (1925d); 13, Dogiel (1927); 14, Fiorentini (1889); 15,

	Body shape	Macronucleus Size (μm)	Macronucleus Shape	Illustrated in Ogimoto and Imai (1981)	Other reference[b]
1.	Ovoid	21 × 82	Rod, bent anterior end	–	
2.	Ovoid	11 × 35	Rod, bent anterior end	–	
3.	Ovoid	11 × 37	Rod, bent anterior end	+	
4.	Tapering posterior	4–8 × 18	Triangular	–	
5.	Subspherical	6 × 21	Rod	–	
6.	Ovoid	9 × 66	Curved rod	–	18
7.	Ovoid	8 × 74	Rod with hook	–	2
8.	Ovoid	15 × 97	Curved rod	–	
9.				–	
10.	Rectangular	13 × 61	Rod	–	
11.	Ovoid	9 × 29	Rod	–	
12.	Ovoid	9 × 53	Curved rod	–	7, 31
13.	Ovoid	13 × 38	Rod	+	
14.	Short, heavy	14 × 43	Wedge-shaped	+	18, 21, 28
15.	Ovoid		Bent rod	+	
16.	Oval	24–36 × 33–51	Triangular	–	23
17.	Elongated	16 × 107	Bent rod	–	
18.					
	Ovoid	? × 48	Rod	–	
	Ovoid	11 × 44	Wedge-shaped	+	
	Ovoid	? × 49	Rod	–	
	Ovoid	? × 57	Rod	–	
22.	Ovoid	16 × 64	Bent rod	–	
23.	Ovoid	8 × 73	Bent rod	–	
24.	Ovoid		Curved rod	–	
25.	Ovoid	12 × 45	Bent rod	–	22
26.					
	Ovoid	10 × 70	Rod	–	
	Ovoid	10 × 70	Rod	–	
27.	Oval	9–42 × 33	Wedge-shaped	+	
28.	Ovoid	9 × 38	Curved rod	+	
29.	Oval	5 × 11	Ovoid	–	21
30.	Spherical	8 × 22	Rod narrow anteriorly	–	4
31.	Ovoid	11 × 37	Rod bent anteriorly	+	
32.					
	Ovoid	16 × 58	Rod	–	21
	Ovoid	16 × 35	Rod	–	21
33.	Rectangular	8 × 48	Curved rod	–	13,21
34.	Subspherical	18 × 111	Rod bent 90° anteriorly	–	23
35.	Rectangular	6 × 24	Triangular	–	13
36.	Ovoid	11 × 37	Rod bent anteriorly	+	13
37.	Ovoid	11 × 37	Rod bent anteriorly	+	

Imai (1986); 16, Imai (1988); 17, Imai and Ogimoto (1983); 18, Kofoid and MacLennan (1932); 19, Kubikova (1935); 20, Latteur (1966c); 21, Latteur (1970); 22, Lubinsky (1957c); 23, Lubinsky (1958b); 24, Moriggi (1941); 25, Moriggi (1950); 26, Noirot-Timothée (1959); 27, Noirot-Timothée (1963); 28, Ogimoto and Imai (1981); 29, Schuberg (1888); 30, Stein (1858); 31, Van Hoven (1975); 32, Wertheim (1933a).

factors. It is not known whether the number of spines produced in response to a stimulus is genetically determined. The *Bubalidis* group is composed of those species which have a small longitudinal groove extending anteriorly a short distance from the anus. One narrow caudal spine may be present. The *Rangiferi* group all have a distinct, longitudinal line running the length of the dorsal edge of the right lateral surface. The *Crista-galli* group includes those species with a roughly triangular outline, truncated anteriorly, and tapered posteriorly.

In 1970 Latteur revised the classification of the diplodinia and described just 20 species, some as mentioned above but with many forms. He also noted that, unlike all the other species, *D. archon,* which he described for the first time, and *D. flabellum,* had only one contractile vacuole compared with the normal two. Hungate (1978) suggested that this might indicate that they were primitive species related to the entodinia. Detailed information on individual species is given below and in Table 2.4. The type species is *Diplodinium dentatum*.

Detailed Descriptions of the Diplodinia

1. *Diplodinium africanum* was first described by Imai (1988) from Kenyan zebu cattle. It is characterized by its large size and the possession of a large caudal spine.

2. *Diplodinium anacanthum* (*Anoplodinium denticulatum anacanthum*—Dogiel, 1927; *Diplodinium anisacanthum anacanthum*—Latteur, 1970; *Diplodinium anacanthum anacanthum*—Wertheim, 1935b) is the spineless member of the Anacanthum group which is characterized by the tapering posterior half of the body, although this feature is not obvious in the photomicrograph of Ogimoto and Imai (1981). It has been reported in Japanese cattle, sheep, and goats (Imai et al., 1978, 1979a), Thai zebu cattle and water buffalo (Imai and Ogimoto, 1984), Russian cattle (Dogiel, 1927), and Brazilian water buffalo. It is, however, not as common as the forms with spines.

3. *Diplodinium anisacanthum* (*Anoplodinium denticulatum anisacanthum*—Dogiel, 1927; *Diplodinium anisacanthum anisacanthum*—Latteur, 1970) is the six- spined member of the Anacanthum group. It is distinguished from the other six-spined diplodinium (*D. dentatum*) because the spines of *D. anisacanthum* converge with the tapering of the body to give the impression of meeting at a point, whereas those of *D. dentatum* point directly posteriorly. It has been found in Brazilian cattle and water buffalo (Cunha, 1914a; Dehority, 1979, 1986b) and in cattle from New Zealand, Russia, and Thailand (zebu cattle) (Dogiel, 1927; Clarke, 1964; Imai and Ogimoto, 1984) but not in Japanese ruminants.

4. *Diplodinium archon* has only been described from African bushbuck in the Congo by Latteur (1966c). It is not found in the tsessebe, in

the Transvaal, or in reedbuck from Kenya and Tanzania. This protozoon is unusual in having a heavy incurved ventral spine (plus another very small, almost invisible, ventral spine) and only one contractile vacuole instead of the usual two. As it resembles *Entodinium* spp. in the respect, it has been suggested that it might be a primitive form of diplodinium.

5. *Diplodinium babici* has only been described from the Alpine ibex (*Capra ibex*) in Yugoslavia by Wertheim (1933a, 1935b). It is ovoid and has no caudal appendages.

6. *Diplodinium bubalidis* (*Anoplodinium bubalidis bubalidis*—Dogiel, 1927) is found in many African ruminants (Van Hoven, 1975) and particularly in Lichtenstein's hartebeest (*Alcelaphus lichtensteini*), the korrigum (*Damaliscus korrigum*), the beisa (*Oryx beisa*) (Latteur, 1970), the tsessebe (Van Hoven, 1975), and reedbuck (*Redunca* spp.) (Van Hoven, 1983). It has not been reported elsewhere. This protozoon tapers slightly towards the posterior end when seen in side view and has one thin ventral spine incurved towards the dorsal side.

7. *Diplodinium bursa* is a dubious species described by Fiorentini (1889) and mentioned by Kofoid and MacLennan (1932).

8. *Diplodinium cameli* was described originally from the dromedary (*Camelus dromedarius*) by Dogiel (1926b) and has only been found in camelids. This protozoon is ellipsoidal with a very small caudal lobe.

9. *Diplodinium ceylonicum* is a very rare and possibly dubious species that has only been in India by Bhatia (1936).

10. *Diplodinium clevelandii* was described from cattle by Becker and Talbott (1927) and is clearly not a *Diplodinium* as it has a skeletal plate.

11. *Diplodinium consors* (*Diplodinium bubalidis consors*—Dogiel, 1925a; *Anoplodinium bubalidis consors*—Dogiel, 1927) is found in central African antelope (Van Hoven, 1975) and in South African cattle (Fantham, 1926). It has also been reported in the Indian goat (Das Gupta, 1935) and Japanese cattle by Hukui (1940), although it was not found in Japanese cattle, sheep, and goats by Imai et al. (1978, 1979a). It is ovoid in shape with a long, thin, ventral spine curved to the dorsal side. Although Kofoid and MacLennan (1932) believed that *D. consors* should be in the Bubalidis group, Van Hoven (1975) placed it in the Rangiferi group because, like *D. rangiferi*, it has a longitudinal cuticular fold on the right side of the body. The endoplasmic sac also does not extend into the operculum.

12. *Diplodinium costatum* (*Diplodinium costatum major*—Dogiel, 1925a; *Anoplodinium costatum major*—Dogiel, 1927) is also referred to as *Diplodinium major*. It has been found in African antelope (Dogiel, 1925a), reedbuck (Van Hoven, 1983), and the tsessebe where, according to Van Hoven (1975), it possesses a caudal spine. It has also been found in the Indian goat (Das Gupta, 1935), the chamois, the musk-ox, and in sheep

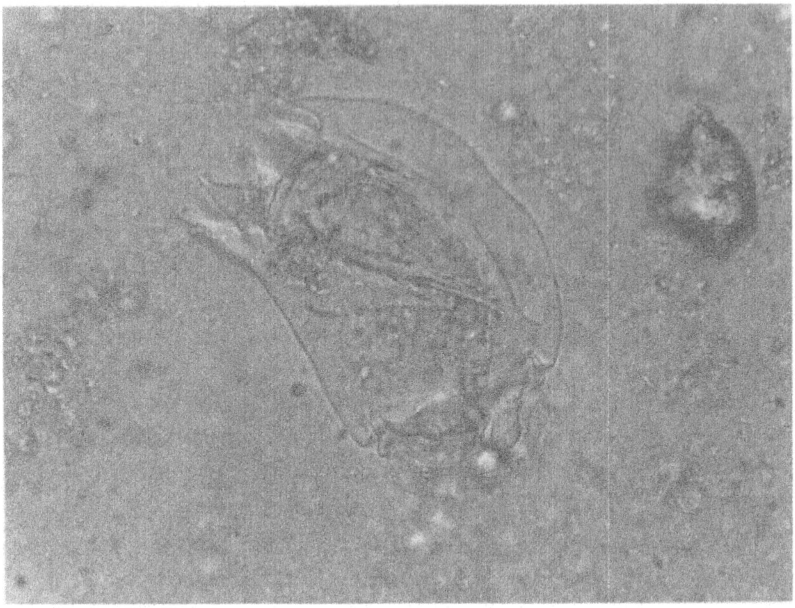

Fig. 2.7 Photomicrograph of *Diplodinium dentatum*.

in China. *D. costatum* has an ovoid body, truncated anteriorly and triangular posteriorly. It has a right cuticular fold and the endoplasmic sac extends into the operculum. Except as indicated above, it has no caudal spines or lobes.

13. *Diplodinium crista-galli* (*Diplodinium crista-galli crista-galli*; *Anoplodinium crista-galli crista-galli*—Dogiel, 1927) is found in goats in Iran, cattle and water buffalo in the Balkans, the wild goat in India and the Balkans (Das Gupta, 1935; Wertheim, 1935b; Latteur, 1970), and in zebu cattle in Thailand (Imai and Ogimoto, 1984). This protozoon tapers posteriorly and has a very characteristic caudal fan with two to seven spines.

14. *Diplodinium dentatum* (*Entodinium dentatum*—Stein, 1858; *Diplodinium denticulatum*—Fiorentini, 1889; *Diplodinium dentatum denticulatum*—Buisson, 1923a; *Diplodinium denticulatum denticulatum*—Dogiel, 1927) is widely distributed although it has not been recorded in New Zealand cattle (Clarke, 1964). The body of this protozoon is short and heavy, almost rectangular in side view (Fig. 2.7). It has six heavy caudal spines which point directly posteriorly unlike those of the other six-spined species, *D. anisacanthum* which are incurved.

15. *Diplodinium diacanthum* (*Anoplodinium denticulatum diacanthum*

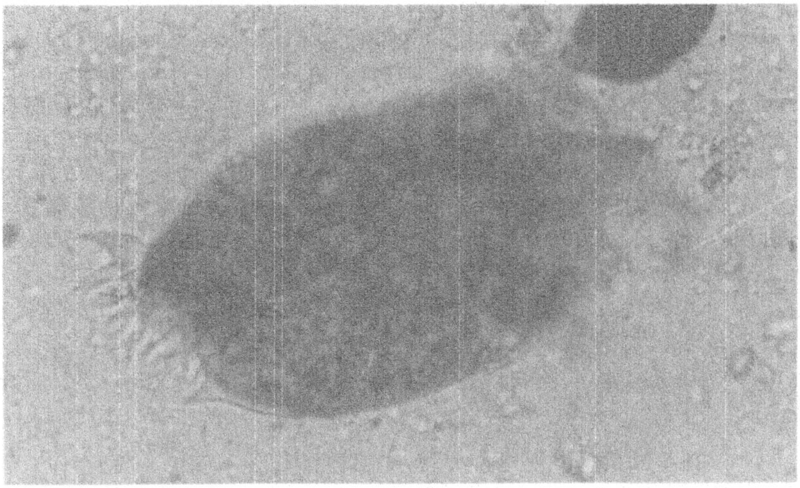

Fig. 2.8 Photomicrograph of *Diplodinium flabellum* f. *flabellum*.

—Dogiel, 1927; *Diplodinium anisacanthum diacanthum*—Latteur, 1970) is widely but not universally distributed and almost invariably occurs with other members of the Anacanthum group. There is doubt whether the members of this group are separate species or just forms induced by different cultural conditions.

16. *Diplodinium dogieli* (*Diplodinium rangiferi minor*—Dogiel, 1925d; *Anoplodinium rangiferi minor*—Dogiel, 1927) is found principally in Artic ruminants such as reindeer (*Rangifer tarandus*) (Lubinsky, 1958b) and musk-ox (Dehority, 1974, 1985), although it has been reported from Japanese cattle (Hukui, 1940). This protozoon is ellipsoidal with a truncated anterior and rounded posterior end. A narrow longitudinal thickening of the pellicle extends along the right dorsal surface from the anterior end to the cytoproct.

17. *Diplodinium elongatum* (*Anoplodinium elongatum*—Dogiel, 1927) has only been found in Russian and Chinese cattle (Dogiel, 1927; Hsiung, 1932). It is one of the largest diplodinia measuring, on average, 195×91 μm. It is elongated with gently convex dorsal and ventral surfaces, has a large operculum into which the endoplasm extends, and has no caudal spines or lobes.

18. *Diplodinium flabellum* is a rare species that has only been found in Indian cattle (Kofoid and MacLennan, 1932) and buffalo (Banerjee, 1955), in Brazilian cattle (Dehority, 1986b), and American (USA) sheep (Dehority and Potter, 1974). This protozoon is ovoid to triangular in side view and has a caudal fan with a variable number of teeth (Fig. 2.8). It

differs from the very similar *D. crista-galli* in two respects. In *D. crista-galli* the fan is an extension of the left body surface, whereas in *D. flabellum* it comes from the right surface. *D. flabellum* also has two small spines on the dorsal side. Dehority and Potter (1974) and Dehority (1986b) have now found and described three new forms which they call subspecies. This means that the classical form described above is subspecies *flabellum*. Subspecies *monospinatum* has only a single, left dorsal spine, subspecies *aspinatum* has no dorsal spine, and subspecies *laterospinatum* has a single dorsal spine on the right side. There is disagreement as to whether *D. flabellum* has one (Kofoid and MacLennan, 1932) or two contractile vacuoles.

19. *Diplodinium giraffae* is just mentioned by name by Noirot-Timothée (1963).

22. *Diplodinium indicum* was described from the Indian cow by Banerjee (1955) but is considered by Latteur (1970) to be *Diplodinium psittaceum longi-caudatum*. It is ovoid in shape with a single, short, ventral spine and a smaller dorsal lobe which is difficult to see.

23. *Diplodinium italicum* was described for the first and only time from cattle in Italy by Moriggi (1941). It is ovoid in shape with no caudal spines or lobes.

24. *Diplodinium komareki* was inadequately described from red deer (*Cervus elaphus*) by Kubikova (1935). Latteur (1970) believes that it may be the same as another species.

25. *Diplodinium laeve* (*Anoplodinium crista-galli laeve*—Dogiel, 1927; *Diplodinium crista-galli laeve*—Latteur, 1970) is a rather rare species only reported from cattle, water buffalo, and the wild goat in the Balkans, in goats in Iran (Dogiel, 1927; Wertheim, 1935b; Moriggi, 1950), and cattle in the USA (Abou Akkada et al., 1969). This protozoon is ovoid with a broad, flattened, ventral lobe.

26. *Diplodinium mahidoli* was first described from zebu cattle (*Bos indicus*) in Bangkok, Thailand by Imai and Ogimoto (1984). It is ovoid in side view and has two caudal processes, a short ventral spine and a dorsal lobe (in forma *mahidoli*) or a spine (in forma *bispinosum*). The size and shape of the caudal processes are quite variable. *D. mahidoli* resembles *D. psittaceum* differing in having a broad ectoplasm and a long slender macronucleus, and in lacking a narrow rib in the posterior quarter of the dorsal midline.

27. *Diplodinium minor* (*Diplodinium costatum minor*—Dogiel, 1925a; *Anoplodinium costatum minor*—Dogiel, 1927) has been found principally in wild African ruminants such as the steinbok (*Rhaphiceros*, sp.), bushbuck (*Tragelaphus scriptus*) (Dogiel, 1925a), the tsessebe (spined form of *D. minor* found—Van Hoven, 1975), the giraffe, and the reedbuck (three *Redunca* spp.). It has also been reported from Indian and American

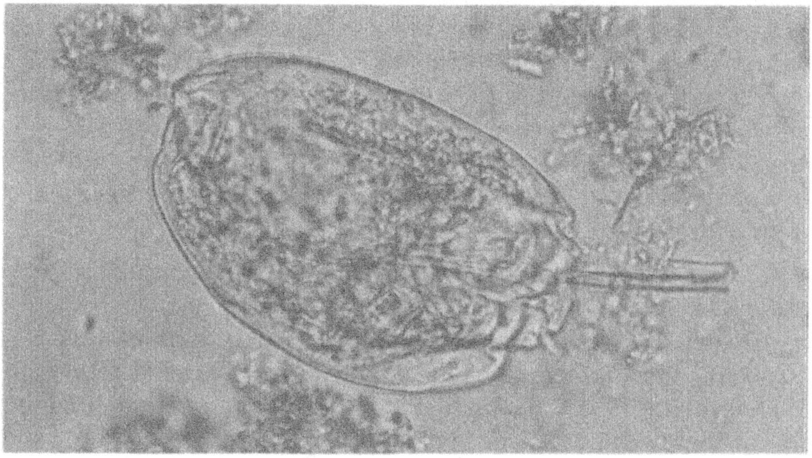

Fig. 2.9 Photomicrograph of *Diplodinium monacanthum* grown in vitro and lacking the single caudal spine. It therefore resembles *D. anacanthum*. Note that it is engulfing a plant fiber.

(USA) cattle (Banerjee, 1955; Abou Akkada et al., 1969). *D. minor* is oval is side view and slightly triangular posteriorly with a narrow longitudinal thickening of the pellicle extending along the right dorsal surface from the anterior end to the cytoproct.

28. *Diplodinium monacanthum* (*Anoplodinium denticulatum monacanthum*—Dogiel, 1927; *Diplodinium anisacanthum monacanthum*—Latteur, 1970) is widely but not universally distributed and usually occurs with the other members of the Anacanthum group, although it occasionally occurs by itself (e.g. in reedbuck—Van Hoven, 1983). The body shape varies with different descriptions from almost rectangular to an oval tapering posteriorly (Fig. 2.9). It has a single ventral spine.

29. *Diplodinium moucheti* was described only once from the pigmy antelope (*Hylarnus batsei*) by Noirot-Timothée (1959). It has an irregularly oval body with a long, thin, ventral spine. The macronucleus is unusual; it is a long curved rod with a swollen anterior end.

30. *Diplodinium nanum* was described from zebu cattle in Kenya by Imai (1988). It is a small organism which could probably be better classied as an *Eodinium* spp. It has a pronounced ventral lobe.

31. *Diplodinium pentacanthum* (*Anoplodinium denticulatum pentacanthum*—Dogiel, 1927; *Diplodinium anisacanthum pentacanthum*—Latteur, 1970) is fairly widely distributed and usually occurs with other members of the Anacanthum group except in Japanese ruminants (Imai et al., 1978, 1979a). The body tapers posteriorly to end in five incurved spines.

32. *Diplodinium psittaceum* (*Anoplodinium psittaceum*—Dogiel, 1927)

is a rare species found in cattle in Russia, Sri Lanka, China, Brazil, and Africa. The exact shape of the body is unclear from the literature. Although Dogiel (1927) and Kofoid and MacLennan (1932) agree that the length to width ratio is about 1.5, the former depicts an elongated organism while the latter shows a stout, round protozoon. It has one ventral caudal lobe as described originally but in 1950 Moriggi found a form with a spine instead of a lobe which he called forma *longi-caudatum*. The first therefore becomes forma *psittaceum*.

33. *Diplodinium quinquecaudatum* (*Diplodinium denticulatum quinquecaudatum*—Dogiel, 1925b; *Anoplodinium denticulatum quinquespinosum*—Dogiel, 1927) is a rare species found only on cattle in Russia and the Balkans and in sheep in Russia and China (Dogiel, 1927; Hsiung, 1931, Wertheim, 1935b). This protozoon has an almost rectangular body with five highly characteristic long caudal spines.

34. *Diplodinium rangiferi* (*Diplodinium rangiferi major*—Dogiel, 1925d; *Anoplodinium rangiferi major*—Dogiel, 1927) is usually found only in reindeer (*Rangifer tarandus*) and caribou (Lubinsky, 1958b; Dehority 1975a), although it has been reported in very low numbers in the giraffe (Kleynhans and Van Hoven, 1976). It has not been reported from other Artic animals such as moose, musk-ox, or Dall Mountain sheep (Dehority, 1974). *D. rangiferi* is one of the largest diplodinia, measuring 128–210 μm long according to Dogiel (1927) and 156–258 μm according to Lubinsky (1958b). In side view the body is a wide ellipse with a truncated anterior end and one inconspicuous caudal lobe. The macronucleus is a curved rod with an expanded anterior end, although the exact shape is variable and can be complex.

35. *Diplodinium sinhalicum* was first described by Imai (1986) from Zebu cattle (*Bos indicus*) in Sri Lanka. It is characterized by the presence of two heavy incurved caudal spines or lobes and in this respect resembles *Eremoplastron magnodentatum*.

36. *Diplodinium tetracanthum* (*Anoplodinium denticulatum tetracanthum*—Dogiel, 1927; *Diplodinium anisacanthum tetracanthum*—Latteur, 1970) is widely but not universally distributed and almost invariably occurs with other members of the Anacanthum group. The body tapers towards the posterior end where there are four small, often incurved, caudal spines.

37. *Diplodinium tricanthum* (*Anoplodinium denticulatum tricanthum*—Dogiel, 1927; *Diplodinium anisacanthum tricanthum*—Latteur, 1970) is widely but not universally distributed and almost invariably occurs with other members of the Anacanthum group. The body tapers towards the posterior end, where there are three small, often incurved, caudal spines.

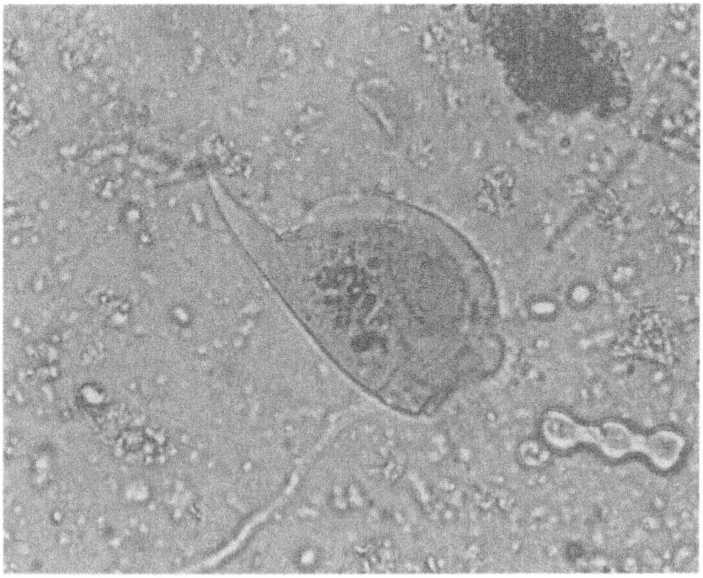

Fig. 2.10 Photomicrograph of *Eremoplastron rostratum*.

2.5 The Genus *Eremoplastron*

This genus was erected by Kofoid and MacLennan (1932) to include those species with a dorsal ciliary zone and a single narrow skeletal plate beneath the right surface. They all have two contractile vacuoles and a triangular or rod-shaped macronucleus, the anterior end of which is often bent ventrally. The micronucleus lies in a small shallow depression on the dorsal surface. The size range of the organisms in this genus is large, ranging from 40 μm long (*E. rostratum*) to 500 μm (*E. giganteum*). The average length is 80–90 μm.

The existence of the genus *Eremoplastron* has been disputed by Hungate (1978) and Latteur (1966d), both of whom place the protozoa in the genus *Eudiplodinium*. This has led to confusion in the literature and the reader should beware. This is compounded because some authors (such as Dehority) believe that *Eudiplodinium*, *Ostracodinium*, or *Eremoplastron*, for example, are subgenera of *Diplodinium* and give names such as *Diplodinium* (*Eudiplodinium*) x.

The *Eremoplastron* type species is *Eremoplastron rostratum* (Fig. 2.10).

Detailed information on individual species is given below and in Table 2.5.

Table 2.5 Characteristics of the *Eremoplastron* species

Species (Fig. no.)[a]	Original description/ name[b]	Length (μm)	Width (μm)	Spines	Lobes	Contractile vacuoles
1. *E. asiaticus*	1	390–456	199–273	0	1	
2. *E. bovis* (2.11)	5/8	52–100	36–57	0	0	2
3. *E. brevispinum*	8	72–92	42–53	0	2	2
4. *E. bubalus*	2	35–56	24–37	0	1	2
5. *E. dilobum*	5/8	68–102	40–70	0	2	
6. *E. giganteum*	3/8	256–500	150–240	0	0	2
7. *E. impalae*	3/8	74–105	50–72	0	0	2
8. *E. insigne*	4/8	124–222	65–125	0	0	2
9. *E. kenyensis*	7	195–290	125–175	0	1 variable	2
10. *E. magnodentatum*	8	58–82	30–50	0	2	2
11. *E. monolobum*	5/8	58–83	41–70	0	1	2
12. *E. neglectum*	3/8	81–124	40–68	0	1	2
13. *E. rostratum* (2.10)	6/8	46–63	29–47	1	0	2
14. *E. rotundum*	8	70–95	51–63	0	0	2
15. *E. rugosum*	4/8	69–90	44–63	0	1	2
16. *E. spectabile*	4/8	115–150	83–101	0	1	2
17. *E. tarandi*	8	74–105	54–77	0	1	2

[a] A number in parentheses indicates that this species is shown in the figure corresponding to that number in this chapter.
[b] References are as follows: 1, Banerjee (1955); 2, Dehority (1979); 3, Dogiel (1925a); 4, Dogiel

Detailed Descriptions of *Eremoplastron* Species

1. *Eremoplastron asiaticus* was described for the first and only time from cattle in India by Banerjee (1955). It is a very large organism (408 × 236 μm), ellipsoidal in shape with a characteristic protruding, almost spherical ventral caudal lobe.

2. *Eremoplastron bovis* (*Eudiplodinium neglectum bovis*—Dogiel, 1927 is widely but not universally found in cattle and rare in sheep (Dogiel, 1927; Clarke, 1964; Banerjee, 1955; Imai et al., 1978, 1979a; Imai and Ogimoto, 1984) although the authors have grown it as a single species in sheep for long periods. In side view the body is an elongated oval with an inconspicuous ventral lobe (Fig. 2.11). It has a smaller operculum than the very similar *E. neglectum* with which some think that it is synonymous.

3. *Eremoplastron brevispinum* is a rare organism that has only been found in Sri Lankan cattle (Kofoid and MacLennan, 1932) and in New Zealand cattle (Clarke, 1964). The protozoon is ellipsoidal in shape and is very similar to *E. bovis* but has two small inconspicuous caudal lobes or blunt spines.

4. *Eremoplastron bubalus* (*Diplodinium* [*Eudiplodinium*] *bubalus*—Dehority, 1979) has been found in Brazilian cattle and water buffalo

	Body shape	Macronucleus Size (μm)	Macronucleus Shape	Illustrated in Ogimoto and Imai (1981)	Other reference[b]
1.	Ovoid	26 × 152	Rod	−	
2.	Ovoid	10 × 41	Rod	+	8
3.	Ovoid	9 × 37	Rod	−	
4.	Ovoid	6 × 28	Rod	+	
5.	Ovoid	12 × 41	Rod	+	
6.	Elongated	10 × 190	Bent rod	−	11
7.	Subspherical	10 × 38	Wedge/rod	−	5, 10
8.	Elongated	11 × 87	Curved rod	−	5
9.	Semi-ovoid	16 × 148	Curved rod	−	
10.	Rectangular	10 × 37	Curved rod	−	
11.	Subspherical	10 × 37	Rod	+	5
12.	Ovoid	10 × 46	Rod	−	5, 9
13.	Ovoid	6 × 24	Rod	+	5
14.	Broad oval	10 × 43	Rod	−	
15.	Semi-ovoid	5 × 45	Rod, anterior notch	−	5, 11
16.	Subspherical	12 × 59	Notched rod	−	5, 10
17.	Subspherical			−	10

(1925d); 5, Dogiel (1927); 6, Fiorentini (1889); 7, Imai (1988); 8, Kofoid and MacLennan (1932); 9, Kraschenninikow (1955); 10, Lubinsky (1958b); 11, Van Hoven (1975).

(*Bubalus bubalis*) (Dehority, 1979, 1986b), and in Thai zebu cattle and water buffalo by Imai and Ogimoto (1984). *E. bubalus* is ellipsoidal in shape and closely resembles *E. bovis* except that it is appreciably smaller (35 × 56 compared with 52 × 100 μm) and has a macronucleus which is parallel with the axis of the organism rather than the skeletal plate.

5. *Eremoplastron dilobum* (*Eudiplodinium neglectum dilobum*—Dogiel, 1927) is found infrequently in cattle in Brazil, Japan, and Thailand (Dehority, 1986; Imai et al., 1978, 1979a; Imai and Ogimoto, 1984) and is also found in sheep (Dogiel, 1927; Ogimoto and Imai, 1981). The organism is ellipsoidal in shape, tapering toward the posterior end where there are two fairly prominent caudal lobes.

6. *Eremoplastron giganteum* (*Diplodinium neglectum giganteum*—Dogiel, 1925a; *Eudiplodinium neglectum giganteum*—Dogiel, 1927) was first described from Coke's hartebeast (*Bubalus cokei*) and the dik dik (*Madoqua* sp.) in East Africa. It has more recently been seen in the tsessebe (*Damaliscus lunatus lunatus*) and the blesbok (*D. dorcas phillipsi*) by Van Hoven (1975, 1978) and in cattle in Kenya (Imai, 1988). This is one of the largest entodiniomorphid protozoon, measuring 330 × 174 μm (Dogiel, 1927) and 330 × 138 μm (Van Hoven, 1975). The body is elongated with no caudal appendages but the dorsal zone of cilia is displaced posteriorly a one-quarter of the body length.

7. *Eremoplastron impalae* (*Diplodinium neglectum impalae*—Dogiel,

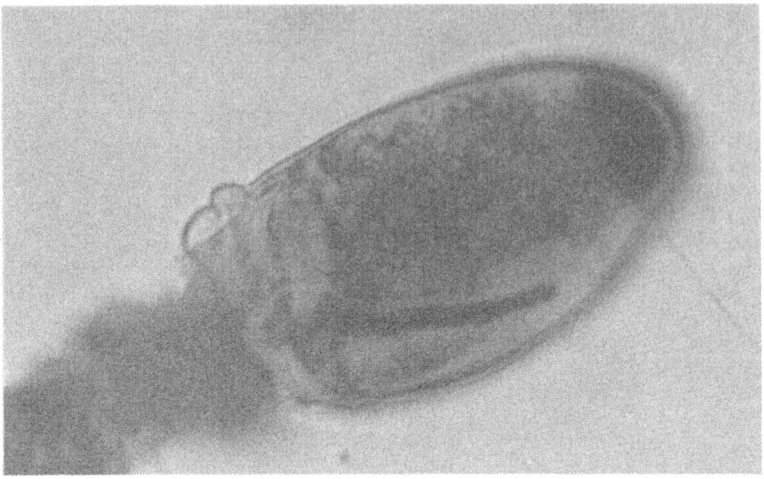

Fig. 2.11 Photomicrograph of *Eremoplastron bovis* attached to plant material.

1925a; *Eudiplodinium neglectum impalae*—Dogiel, 1927) was originally described by Dogiel from the Impala antelope (*Aepyceros melampus*). In 1927 Dogiel described an organism which he believed to be the same. Subsequently Kofoid and MacLennan (1932) produced evidence that the organism from the reindeer was a new species (*E. tarandi*), although this has been disputed by Lubinsky (1958b). More recently *E. impalae* (sensu lato) has been found in the musk-ox (*Ovibos moschatus*) (Dehority, 1974, 1985) and three reedbuck species (*Redunca* spp.). *E. impalae* is ellipsoidal and has an inconspicuous caudal lobe. It is very similar to *E. spectabile* but is smaller (78 × 56 compared with 135 × 88 μm). For details of other differences see Lubinsky (1958b).

8. *Eremoplastron insigne* (*Diplodinium neglectum insigne*—Dogiel, 1925a; *Eudiplodinium neglectum insigne*—Dogiel, 1927) was described from Coke's hartebeast (*Bubalis cokei*) and the dik dik (*Madoqua* sp.) by Dogiel (1925a) but has rarely been seen since. It is similar in shape and proportions to *E. giganteum* and *E. kenyensis* and was believed to be a connecting link in evolution between the smaller eremoplastrons and these large species (Kofoid and MacLennan, 1932). However it is possibly a small or division form of *E. giganteum*.

9. *Eremoplastron kenyensis* was first described from zebu cattle in Kenya by Imai (1988) who called it *Eudiplodinium kenyensis*. However, it clearly does not belong to this genus as it lacks the characteristically shaped macronucleus and is, by all criteria, an eremoplastron. It has a short ventral lobe which is variable in shape and may be notched.

10. *Eremoplastron magnodentatum* has only been found in the cow

and buffalo in India and Sri Lanka (Kofoid and MacLennan, 1932; Banerjee, 1955). It has a highly characteristic shape with a rectangular body and two massive, blunt caudal spines.

Eremoplastron microlateralis is referred to as such by Kleynhans and Van Hoven (1976) but is *Eudiplodinium microlateralis* (Van Hoven, 1983).

11. *Eremoplastron monolobum* (*Eudiplodinium neglectum monolobum*—Dogiel, 1927) is found principally in cattle and has been found in a small percentage of animals in Brazil, Japan, and Thailand (Dehority, 1986; Imai et al., 1978, 1979a; Imai and Ogimoto, 1984). It has also been found in 3–9% of sheep in Japan (Imai et al., 1978, 1979a). This protozoon resembles *E. bovis*, with its elliposidal body, but is distinguished by the presence of a prominant ventral lobe.

12. *Eremoplastron neglectum* (*Diplodinium neglectum neglectum*—Dogiel, 1925a; *Eudiplodinium neglectum neglectum*—Dogiel, 1927) was first described by Dogiel (1925a) from Coke's hartebeast (*Bubalis cokei*) and the dik dik (*Madoqua sp.*) Fantham (1926) also found it in South African cattle and Dogiel (1927) stated that it was one of the most commonly occuring members of the genus. Van Hoven (1975) found that it was the most common eremoplastron in the tsessebe, although it was not found in African reedbuck (Van Hoven, 1983). Hungate (1942) used *E. neglectum* in experiments done in the USA, but otherwise there have been few reports of its presence. It is very similar to the widely reported *E. bovis*; it has a larger operculum and is generally larger overall (100×58 compared with 88×44 μm). Therefore there must be doubt as to whether they are separate species. Van Hoven (1975), for example, reported that his *E. neglectum* was 10 μm shorter than Dogiel's (1927) organisms.

13. *Eremoplastron rostratum* (*Diplodinium rostratum*—Fiorentini, 1889; *Eudiplodinium rostratum*—Dogiel, 1927; *Diplodinium helseri*—Becker and Talbott, 1927) is widely found in cattle throughout the world and less commonly found in water buffalo. It is not found in Japanese sheep, although it is common in Japanese cattle (Imai et al., 1978, 1979a). Although *E. rostratum* is the type species, it is in many ways atypical because of its small size (only 54×34 μm) and dorsal flange. It has a characteristic heavy, curved ventral spine (Fig. 2.10).

14. *Eremoplastron rotundum* is a rare species originally found in Indian and Sri Lankan cattle by Kofoid and MacLennan (1932) and seen recently in Brazilian cattle by Dehority (1986b). This protozoon, although oval in shape, is more nearly spherical than similar organisms such as *E. bovis*. It has no caudal spines or lobes.

15. *Eremoplastron rugosum* (*Eudiplodinium neglectum rugosum*—Dogiel, 1927) was originally described from cattle in Leningrad by

Dogiel (1927). It is rare, although Van Hoven (1975) found a protozoon in the tsessebe which he claimed was *E. rugosum*, but it was a different shape from Dogiel's organism. *E. rugosum* is characterized by a complex ventral lobe with its dorsal border cut by 8–10 shallow identations and a pellicular fold running the length of the dorsal side of the organism.

16. *Eremoplastron spectabile* (*Diplodinium neglectum spectabile*—Dogiel, 1925d; *Eudiplodinium neglectum spectabile*—Dogiel, 1927) is found in reindeer (*Rangifer tarandus*) (Dogiel, 1925d; Lubinsky, 1958b) but not in other Arctic ruminants (Dehority, 1974, 1985). It is ellipsoidal in shape with a truncated anterior end and an inconspicuous ventral lobe. It resembles *E. impalae* but is larger (135 × 88 compared with 78 × 56 μm in Lubinsky's reindeer) and has a longer and wider skeletal plate.

17. *Eremoplastron tarandi* is a disputed species which Kofoid and MacLennan (1932) regard as separate from *E. impalae* but which Lubinsky (1958b) regards as the same.

2.6 The Genus *Eudiplodinium*

If it is accepted that the genus *Eremoplastron* exists, then the genus *Eudiplodinium* erected by Dogiel (1927) and amended by Kofoid and MacLennan (1932) contains only two species. Eudiplodinia are characterized by having a dorsal ciliary zone, a thin narrow skeletal plate beneath the right surface, and a hook-shaped macronucleus. The micronucleus is in the eye of the hook. There are two contractile vacuoles which differ from those in *Eremoplastron* species. They have thick membranes and open via pores that form a conspicuous pit on the surface. Kofoid and MacLennan state that the pellicle of *E. maggii* is "heavy and resistant" and the authors have found this protozoon very resistant to disruption with ultrasound. Information about the two true eudiplodinia is given below. Information about all species is in Table 2.6.

8. *Eudiplodinium maggii* (*Diplodinium maggii*—Fiorentini, 1889; *Diplodinium bursa*—Becker and Talbott, 1927) is widely distributed both geographically and in host range and is found in the hindgut of the capybara (*Hydrochoerus hydrochaeris*). It is a member of the B-type protozoal population. In side view *E. maggii* has a triangular appearance, an inconspicuous operculum, and an almost flat anterior end (Figs. 2.12 and 2.13).

10. *Eudiplodinium microlateralis* was first described from the bohor and common reedbuck (*Redunca redunca wardi; R. arundinum arundinum*) by Van Hoven (1983). This organism has a prominent pellicular fold that runs the entire length of the dorsal side of the body and terminates posteriorly in a small lobe. There is also another dorsal lobe and a more prominent ventral one.

Table 2.6 Characteristics of the *Eudiplodinium* species

Species	Original description/name	Length (μm)	Width (μm)	Spines	Lobes	Contractile vacuoles
1. *E. affine*	See *Diploplastron affine*					
2. *E. banksi*	See *Metadinium banksi*					
3. *E. bovis*	See *Eremoplastron bovis*					
4. *E. bubalus*	See *Eremoplastron bubalus*					
5. *E. impalae*	See *Eremoplastron impalae*					
6. *E. insigne*	See *Eremoplastron insigne*					
7. *E. kenyensis*	See *Eremoplastron kenyensis*					
8. *E. maggii* (2.12, 2.13)	1/2	104–198	63–125	0	0	2
9. *E. medium*	See *Metadinium medium*					
10. *E. microlateralis*	3	51–64	30–39	0	1	2
11. *E. neglectum*	See *Eremoplastron neglectum*					
12. *E. rostratum*	See *Eremoplastron rostratum*					
13. *E. sablei*	See *Metadinium sablei*					
14. *E. spectabile*	See *Eremoplastron spectabile*					

References are as follows: 1, Fiorentini (1889); 2, Kofoid and MacLennan (1932); 3, Van Hoven (1983)

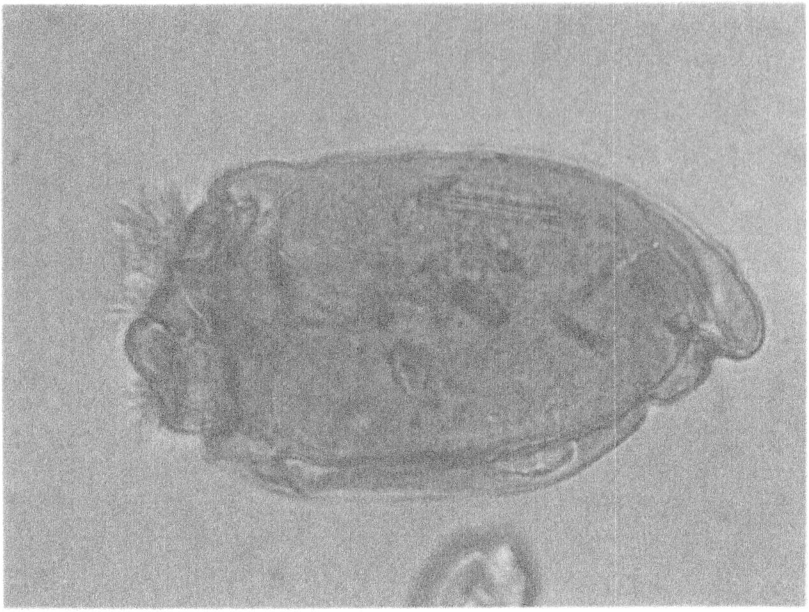

Fig. 2.12 Photomicrograph of *Eudiplodinium maggii*.

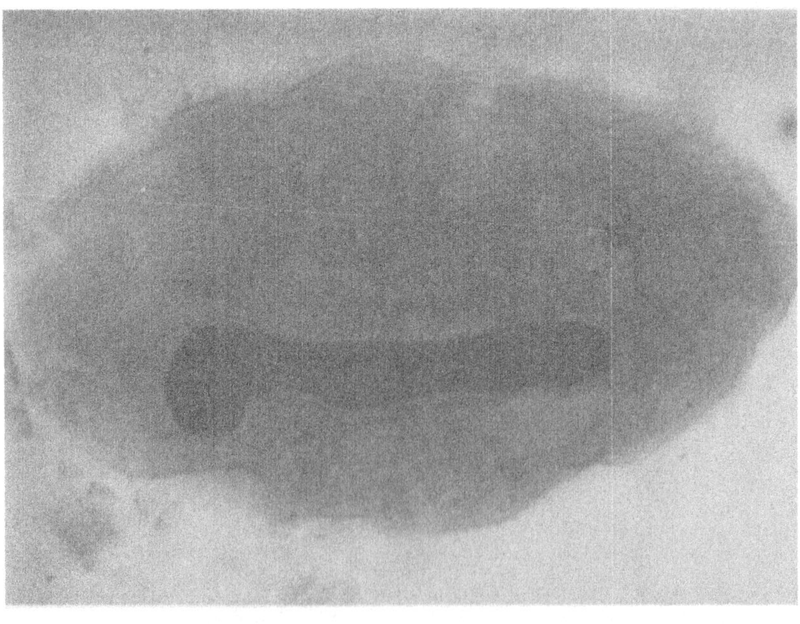

Fig. 2.13 Photomicrograph of *Eudiplodinium maggii* stained to show the characteristic hook-shaped macronucleus.

2.7 Genus *Ostracodinium*

This genus is characterized by the presence of a very distinctive broad skeletal plate under the right surface. The plate runs into the body of the protozoon by the macronucleus and extends round the right side to the middle of the ventral side. This genus was first erected by Dogiel (1927) to distinguish these protozoa from the remainder of the subfamily Diplodiniinae with which they had previously been placed. The ostracodinia have 2–6 contractile vacuoles arranged in a longitudinal row rear the dorsal midline. In the opinion of Kofoid and MacLennan (1932), the number is constant in any one species.

Detailed information on individual species is given below and in table 2.7.

1. *Ostracodinium bohor* was first described by Van Hoven (1983) from three reedbuck species (*Redunca* spp.) in southern Africa. The body is triangular and the macronucleus is bent anteriorly toward the ventral side; the micronucleus is in the angle of the bend. There is a deep pellicular fold running longitudinally the length of the body on the dorsal side. There are two contractile vacuoles.

2. *Ostracodinium brazili* was first described by Dehority (1979) from Brazilian water buffalo. The skeletal plate is comparatively narrow and short. The nucleus has two flattened lobes on the dorsal side with the micronucleus between the lobes. *O. brazili* has three contractile vacuoles. The body has a small ventral lobe.

4. *Ostracodinium clipeolum* is widely distributed in cattle, although only a few hosts are infected in each area. It is not found in Japanese sheep or goats although it is present in cattle (Imai et al., 1978, 1979a) but Dehority (1978) successfully introduced it from cattle into ciliate-free sheep. It is found in Brazilian but not Thai water buffalo (Dehority, 1979; Imai and Ogimoto, 1984). *O. clipeolum* is ellipsoidal with a characteristic small shield-shaped lobe on the posterio-ventral end of the body.

5. *Ostracodinium crassum* (*Diplodinium crassum*—Dogiel, 1925a) is a rare protozoon first described in steinbok (*Rhaphiceros* sp.) in East Africa by Dogiel (1925a) and subsequently seen in South African cattle by Fantham (1926) and in the giraffe by Kleynhans and Van Hoven (1976). *O. crassum* has a short, broad body with a comparatively narrow skeletal plate and no caudal spines or lobes. The macronucleus is wider posteriorly than anteriorly.

6. *Ostracodinium crustaceum* (*Diplodinium crustaceum*—Buisson, 1924) is a rare organism first described from the duiker (*Cephalopus grimmia*) and the bushbuck (*Tragelaphus scriptus*) by Buisson (1924) but not seen in reedbuck by Van Hoven (1983). *O. crustaceum* is a long, slender organism without lobes or spines.

Table 2.7 Characteristics of the *Ostracodinium* species

Species	Original description/ name[b]	Length (μm)	Width (μm)	Spines	Lobes	Contractile vacuoles
1. *O. bohor*	13	73–88	51–64	0	0	2
2. *O. brazili*	4	65–104	44–74	0	1	3
3. *O. caudatum*	See *Metadinium caudatum*					
4. *O. clipeolum*	12	92–128	50–65	0	1	3
5. *O. crassum*	5/6	120–142	80–100	0	0	2
6. *O. crustaceum*	2/6	136	50	0	0	2
7. *O. dentatum*	9					
8. *O. dilobum*	8/12	88–140	54–78	0	2	5
9. *O. dogieli*	6/12	92–130	48–63	0	1	2
10. *O. dyurum*	2/12	60–80	50–60	2	0	
11. *O. esalqum*	4	65–96	47–66	0	0	2
12. *O. gauri*	11	44–70	33–50	0	2	Variable
13. *O. gladiator*	5/12	78–112	40–55	1	0	2
14. *O. gracile*	5/12	92–125	42–60	0	0	2
15. *O. iwawoi*	10	60–90	40–60	0	1	4
16. *O. mammosum*	9/12/14	53–97	33–58	0	2	3
17. *O. minorum*	3	95–131	65–89	0	0	2
18. *O. monolobum*	6/12	105–150	55–77	0	1	5
19. *O. mysorei*	11	42–53	32–40	0	1	Variable
20. *O. nanum*	5/12	47–70	30–41	1	0	2
21. *O. nucleolobum*	4	45–79	35–55	0	0	2
22. *O. obtusum* (2.14)[a]	7/12	118–148	55–80	0	0	6
23. *O. quadrivesiculatum*	12	92–112	43–56	0	0	4
24. *O. rugoloricatum*	12	84–125	37–58	0	1	3
25. *O. stokyi*	2/6	70–105	50–76	2	0	2?
26. *O. tenue*	5/12	59–76	28–38	0	0	2
27. *O. tiete*	4	51–80	33–57	0	1	3
28. *O. trivesiculatum*	12	78–100	42–60	0	0	3
29. *O. ventricosum*	1/12	60–80	50–60	1	0	
30. *O. venustum*	12	76–115	41–60	0	1	2

[a]The number in parentheses indicates that this species is shown in Fig. 2.14.
[b]References are as follows: 1, Buisson (1923a); 2, Buisson (1924); 3, Dehority (1975); 4, Dehority (1979); 5, Dogiel (1925a); 6, Dogiel (1927); 7, Dogiel and Fedorowa (1925b); 8, Fantham (1926); 9, Fiorentini

7. *Ostracodinium dentatum* is a name over which there is confusion. Fiorentini (1889) first produced an illustration of a protozoon that he called *Diplodinium dentatum*. In 1890 Railliet (1890) accurately described an organism that he called *Diplodinium mammosum*. Subsequently, Dogiel and Fedorowa (1925b) described *Diplodinium dentatum obtusum* and pointed out that Fiorentini's protozoon could be the same. Hungate (1978) accepted this argument, pointing out that the name *dentatum* had priority over *obtusum*, and that with *Ostracodinium* accepted as the generic name, *Ostracodinium obtusum* should be called *O. dentatum* and was the type species. In contrast, others, including Kofoid and MacLennan (1932) and Ogimoto and Imai (1981), believe, firstly, that Railliet (1890) produced the first accurate description, and secondly, that *O. mammo-*

	Body shape	Macronucleus Size (μm)	Macronucleus Shape	Illustrated in Ogimoto and Imai (1981)	Other reference[b]
1.	Triangular	2–6 × 57	Triangular, bent anteriorly	–	
2.	Ovoid	9 × 50	Lobed rod	–	
3.					
4.	Ovoid	7 × 70	Rod	+	
5.	Broad oval	6–12 × 80	Rod	–	12
6.	Elongated	12 × 90	Rod	–	12
7.					
8.	Elongated	6 × 58	Rod	–	6
9.	Elongated	8 × 70	Rod	–	
10.	Subspherical	short		–	6
11.	Semiovoid	6–9 × 51	Lobed rod	–	
12.	Broad oval	5 × 33	Rod	–	
13.	Elongated	4–10 × 50	Lobed rod	–	
14.	Triangular	6–13 × 64	Lobed rod	+	11
15.	Ovoid	8 × 42	Rod	–	
16.	Rectangular	6 × 56	Rod	+	
17.	Ovoid	5–11 × 67	Lobed rod	–	
18.	Elongated	8 × 70	Rod	–	
19.	Broad oval	7 × 30	Rod	–	
20.	Ovoid	6 × 24	Rod	–	6
21.	Ovoid	3–9 × 36	Lobed rod	–	
22.	Elongated	10 × 65	Rod	+	6, 13
23.	Triangular	8 × 66	Rod	–	
24.	Rectangular	7 × 60	Rod	–	
25.	Semiovoid	6–9 × 32	Rod	–	10, 12
26.	Elongated	2–6 × 34	Lobed rod	–	
27.	Ovoid	6–10 × 39	Rod	–	
28.	Triangular	9 × 61	Rod	+	11
29.	Subspherical	5 × 27	Rod	–	
30.	Triangular	4–9 × 59	Lobed rod	–	

(1889); 10, Imai (1988); 11, Kofoid and Christenson (1933); 12, Kofoid and MacLennan (1932); 13, Lubinsky (1958b); 14, Railliet (1890); 15, Van Hoven (1983).

sum and *O. obtusum* are different organisms. If this argument is followed, then *O. mammosum* is the type species.

8. *Ostracodinium dilobum* (*Ostracodinium obtusum dilobum*—Dogiel, 1927) is a rare species but has been found in New Zealand cattle (Clarke, 1964), Russian cattle (Dogiel, 1927), and British cattle by the authors. *O. dilobum* is similar to *O. mammosum* but is larger (117 × 66 compared with 82 × 47 μm), has five contractile vacuoles rather than three, and does not have a scoop-shaped ventral caudal lobe.

9. *Ostracodinium dogieli* (*Ostracodinium gracile monolobum*—Dogiel, 1927) is a rare species that has been found in Russian cattle (Dogiel, 1927) and Brazilian water buffalo (Dehority, 1979). It is similar to *O. clipeolum* with an ellipsoidal body and a shield-shaped ventral lobe, but

it has two contractile vacuoles rather than three and a flat rather than a curved ventral surface. In view of the similarity of the two species, they could easily be confused by those scanning rumen contents for different species.

10. *Ostracodinium dyurum* (*Diplodinium ventricosum dyurum*—Buisson, 1924; *Ostracodinium ventricosum dyurum*—Dogiel, 1927) is a very rare species that has only been found in Lichtenstein's hartebeest (*Bubalus lichtensteini*) in central Africa. It has an irregularly circular body with a straight dorsal spine and a dorsally curved ventral spine.

11. *Ostracodinium esalqum* was first described from Brazilian water buffalo by Dehority (1979). It has a Y-shaped skeletal plate, typical of some species of *Metadinium* rather than *Ostracodinium*, and will be considered under this genus.

12. *Ostracodinium gauri* was described from the gaur (*Bos gaurus*) by Kofoid and Christenson (1933) but has not been reported since. It is the two-lobed form of *O. mysorei*. It has an oval body, is smaller than the similar *O. dilobum* (54 × 39 compared with 117 × 66 μm) and has two contractile vacuoles rather than five.

13. *Ostracodinium gladiator* (*Diplodinium gracile gladiator*—Dogiel, 1925a; *Ostracodinium gracile gladiator*—Dogiel, 1927) is found in African antelopes such as Coke's hartebeast (*Bubalus cokei*), the dik dik (*Madoqua* sp.) (Dogiel, 1925a), the tsessebe (*Damaliscus lunatus lunatus*) (Van Hoven, 1975), and in two reedbuck species (*Redunca* spp.) (Van Hoven, 1983). *O. gladiator* has a slender body and a macronucleus with a dorsal lobe at the anterior end and a characteristic long, very narrow, ventral spine.

14. *Ostracodinium gracile* (*Diplodinium gracile gracile*—Dogiel, 1925a; *Ostracodinium gracile gracile*—Dogiel, 1927) is widely but not universally distributed in cattle, sheep, water buffalo, and the musk-ox. It is present in Japanese cattle, but not in Japanese sheep or goats (Imai et al., 1978, 1979a). *O. gracile* is roughly triangular in shape with a flat ventral surface but no caudal spines or lobes. It is separated from the similar *O. tenue* by its larger size (101 × 54 compared with 67 × 31 μm) and from *O. trivesiculatum* and *O. quadrivesiculatum* by its two rather than three or four contractile vacuoles.

15. *Ostracodinium iwawoi* was first described from Kenyan zebu cattle by Imai (1988) where it was present in the majority of the animals examined. It resembles *O. tiete* in general shape but has an extra contractile vacuole and a much more prominant ventral lobe.

16. *Ostracodinium mammosum* (*Diplodinium dentatum*—Fiorentini, 1889; *Diplodinium mammosum*—Railliet, 1890; *Metadinium dentatum*—Crawley, 1923; *Ostracodinium dentatum*—Dogiel, 1927). For history of these names see *O. dentatum*. *O. mammosum* has been fairly commonly

found in the USA, Russia, India, Sri Lanka, and New Zealand but not in Japan or the Far East. *O. mammosum* has a rather rectangular body and is similar to *O. dilobum*, but has larger caudal lobes, the ventral one of which is hollow on the dorsal side. *O. mammosum* is also slightly smaller (82 × 47 compared with 117 × 66 μm) and has only three contractile vacuoles compared with five.

17. *Ostracodinium minorum* was first described from sheep in the USA by Dehority (1975). As this organism has a Y-shaped skeletal plate, it would be classified as belonging to the genus *Metadinium* under the classification used in this book.

18. *Ostracodinium monolobum* (*Ostracodinium obtusum monolobum*—Dogiel, 1927) is a rare species described from Russian cattle by Dogiel (1927). It is similar to *O. obtusum* and *O. dilobum* in size, shape, and number of contractile vacuoles but differs in having just one large ventral lobe.

19. *Ostracodinium mysorei* is a rare species originally described from the gaur (*Bos gaurus*) by Kofoid and Christenson (1933) and since then found only in Brazilian water buffalo by Dehority (1979). It has a short stout body with one ventral lobe, but differs from *O. monolobum*. It is smaller (49 × 35 compared with 129 × 64 μm), has two contractile vacuoles rather than five, and has a posterior diverticulum to the endoplasm.

20. *Ostracodinium nanum* (*Diplodinium gracile nanum*—Dogiel, 1925a; *Ostracodinium gracile nanum*—Dogiel, 1927) is found principally in African antelopes such as Coke's hartebeest (*Bubalus cokei*), the dik dik (*Madoqua* sp.) (Dogiel, 1927), the tsessebe (*Damaliscus lunatus lunatus*—Van Hoven, 1975), and reedbuck (*Redunca* spp.—Van Hoven, 1983). It is relatively small (59 × 36 μm), ovoid in shape with a short macronucleus, two contractile vacuoles, and a short, slender, curved, ventral, caudal spine.

21. *Ostracodinium nucleolobum* was first described from Brazilian water buffalo by Dehority (1979) and has since been reported in Indonesian water buffalo and in Brazilian, Indonesian, and Thai cattle (Dehority, 1986; Imai, 1985, 1988; Imai and Ogimoto, 1984). It is a small (59 × 42 μm) protozoon with an ellipsoidal to triangular body and a macronucleus with two dorsal lobes. Some organisms have a short heavy, ventral caudal spine.

22. *Ostracodinium obtusum* (*Diplodinium dentatum obtusum*—Dogiel and Fedorowa, 1925b; *Diplodinium obtusum*—Dogiel, 1925d; *Ostracodinium obtusum obtusum*—Dogiel, 1927) (Fig. 2.14). This protozoon is found in cattle in the USA, Japan, and Thailand (Becker and Talbott, 1927; Imai et al., 1978, 1979a; Imai and Ogimoto, 1984), in reindeer (Dogiel, 1927; Lubinsky, 1958b), in mountain reedbuck (*Redunca fulvor-*

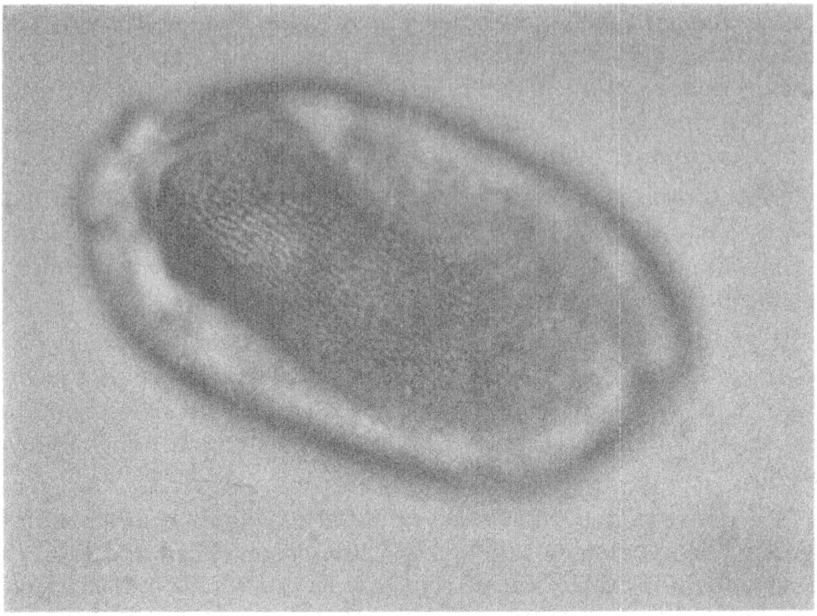

Fig. 2.14 Photomicrograph of *Ostracodinium obtusum* stained to show the broad characteristic skeletal plate.

ufula fulvorufula), and occasionally in Japanese sheep (Imai *et al.*, 1978, 1979a). *O. obtusum* has an elongated oval body with no caudal lobes or spines and is regarded as the simplest of the group also containing *O. monolobum* and *O. dilobum*, although *O. obtusum* has six contractile vacuoles rather then five.

23. *Ostracodinium quadrivesiculatum* is a rare species originally described from *Bos indicus* from India and Sri Lanka (Kofoid and MacLennan, 1932). It has a triangular body in the side view with no caudal lobes or spines but with four contractile vacuoles. It is closely related to *O. gracile* with two vacuoles and *O. trivesiculatum* with three.

24. *Ostracodinium rugoloricatum* has been reported from cattle in the USA (Dehority, 1978), New Zealand (Clarke, 1964), and India (Kofoid and MacLennan, 1932; Banerjee, 1955) and in water buffalo (Banerjee, 1955). Its body is rectangular in side view with an exceptionally large inturned skeletal plate and a small ventral lobe.

25. *Ostracodinium stokyi* (*Diplodinium stokyi*—Buisson, 1924; *Ostracodinium stokyi*—Dogiel, 1927) is a very rare species described by Buisson from the antelope *Hippotragus equinus*. It has a short, broad body with two ventral spines.

26. *Ostracodinium tenue* (*Diplodinium gracile tenue*—Dogiel, 1925a;

Ostracodinium gracile tenue—Dogiel, 1927) is found principally in African antelopes such as Coke's hartebeest (*Bubalus cokei*—Dogiel, 1927), the tsessebe (*Damaliscus lunatus lunatus*—Van Hoven, 1975) and three species of reedbuck (*Redunca* spp.—Van Hoven, 1983). It was also found in South African cattle by Fantham (192b). *O. tenue* has an elongated body with no caudal spines or lobes and is related to the larger (76 × 31 compared with 101 × 51 μm) *O. gracile*.

27. *Ostracodinium tiete* was first described from Brazilian water buffalo (*Bubalus bubalis*) by Dehority (1979) and has since been reported from water buffalo in Indonesia and the Philippines (Imai, 1985; Shimizu et al., 1983). It has also been found in cattle from Kenya and Thailand (Imai and Ogimoto, 1984; Imai, 1988). This organism has an ovoid body with an exceptionally large, inturned skeletal plate resembling that of *O. rugoloricatum*. However, it differs from *O. rugoloricatum* is being smaller (65 × 42 compared with 100 × 48 μm) and in being much stouter.

28. *Ostracodinium trivesiculatum* is found in cattle and water buffalo in Brazil (Dehority, 1979; 1986b), in cattle and gaur in India and Sri Lanka (Kofoid and MacLennan, 1932; Kofoid and Christenson, 1933) and in the musk-ox (*Ovibos moschatus*) in the Arctic (Dehority, 1985). *O. trivesiculatum* has a triangular body in side view with no caudal spines or lobes. It is very similar to *O. gracile* and *O. quadrivesiculatum* differing in having three rather than two or four contractile vacuoles.

29. *Ostracodinium ventricosum* (*Diplodinium ventricosum*—Buisson, 1923a; *Ostracodinium ventricosum ventricosum*—Dogiel, 1927) is a very rare species described from Lichtenstein's hartebeest (*Bubalus lichtensteini*) by Buisson (1923a, 1923b). It has a short, roundish body with a single stout ventral spine.

30. *Ostracodinium venustum* is a very rare species found in cattle in India and Sri Lanka by Kofoid and MacLennan (1932). It is triangular in side view and differs from *O. gracile* in having a small ventral lobe.

2.8 The Genus *Polyplastron*

This genus which was erected by Dogiel (1927) is characterized by the presence beneath the right surface of two long, almost parallel, skeletal plates, which can in rare species be completely or partially fused together, and by the presence of two or three short plates under the left surface. The reader should be warned that, although these left skeletal plates are of great taxonomic importance, they are very difficult to see and are absent from the genera *Metadinium* and *Diploplastron*. The five skeletal plates have all been given names and these are shown in Fig. 2.16. There is disagreement over the importance of the "anticarina" in determing whether or not a species belongs to the genus *Polyplastron*. Kofoid and MacLennan (1932) believed

Table 2.8 Characteristics of the *Polyplastron* species

Species (Fig. no.)	Original description/ name[b]	Length (μm)	Width (μm)	Spines	Lobes	Contractile vacuoles
1. *P. alaskum*	3	110–147	78–113	0	1	5
2. *P. articum*	7	147–214	123–175	0	0	8–9
3. *P. californiense*	1	130–200	110–155	0	1	5
4. *P. fenestratum*	4/6			0	0	8
5. *P. monoscutum*	4/6			0	0	8
6. *P. multivesiculatum* (2.15)[a]	4/5	120–190	78–140	0	0	9

[a] This species is shown in fig. 2.15.
[b] References are as follows: 1, Bush and Kofoid (1948); 2, Coleman et al. (1972); 3, Dehority (1974); 4, Dogiel (1927); 5, Dogiel and Fedorowa (1925b); 6, Kofoid and MacLennan (1932); 7, Lubinsky (1958b).

that only species with the anticarina could be included in *Polyplastron* and that Dogiel's (1928) *Polyplastron bubali* without the anticarina should be placed in a new genus *Elytroplastron*. This has been universally accepted. However, the anticarina is also absent in *Polyplastron articum* first described by Lubinsky (1958b). He suggested that the presence or absence of the anticarina is unimportant and that the genus *Elytroplastron* should fall to be replaced by *Polyplastron*. This suggestion has not been followed.

Detailed information on individual species is given below and in Table 2.8.

Detailed Descriptions of *Polyplastron* Species

1. *Polyplastron alaskum* was first described from Dall sheep (*Ovis dalli*) by Dehority (1974). Its body is ellipsoidal in side view and it is smaller than the other species (127 × 85 compared with 157 × 104 to 183 × 141 μm). It differs from *P. articum* in the presence of the anticarina and a small ventral caudal lobe. It has a long, narrow tergum compared with the short, very narrow one of *P. multivesiculatum*. It differs from *P. californiense* in that it has a smaller caudal lobe and does not have a lobe like half an arrowhead on the "carina" skeletal plate.

2. *Polyplastron articum* was described from the reindeer (*Rangifer tarandus* by Lubinsky (1958b). In side view the body is almost circular with a prominant operculum and no caudal lobes. This species, unlike the others, lacks the middle skeletal plate (anticarina) under the left side. The tergum is a long thin plate extending almost the entire length of the organism. For differences from other species see under *P. alaskum*.

3. *Polyplastron californiense* was described by Bush and Kofoid (1948) from the Sierra Nevada Bighorn sheep (*Ovis canadensis sierrae*) in California, USA. In side view the body is stout with a heavy ventral caudal lobe. The carina skeletal plate has a characteristic "half arrowhead"

	Body shape	Macronucleus Size (μm)	Macronucleus Shape	Illustrated in Ogimoto and Imai (1981)	Other reference[a]
1.	Broad oval	9 × 60	Curved rod	−	
2.	Broad oval	20 × 115	Club-shaped	−	3
3.	Ovoid			−	
4.	Ovoid			−	
5.	Oval	5–13 × 77	Lobed rod	+	2, 3, 6, 7

structure at the posterior end and the tergum is long and narrow extending almost the length of the body. For differences from other species see under *P. alaskum*.

4. *Polyplastron fenestratum* is a very rare species described from cattle in Russia by Dogiel (1927). It is characterized by partial fusion of the two right skeletal plates leaving a window of cytoplasm between the two.

5. *Polyplastron monoscutum* is a very rare species described from cattle in Russia by Dogiel (1927). It is characterized by fusion of the two right skeletal plates into one broad plate.

6. *Polyplastron multivesiculatum* (*Diplodinium multivesiculatum*—Dogiel and Fedorowa, 1925b) is widely distributed in some areas and found in Japanese cattle, sheep, and goats (Imai et al., 1978, 1979a), American (USA) cattle and sheep (Bush and Kofoid, 1948; Abou Akkada et al., 1969; Dehority, 1978), and British cattle and sheep (Eadie, 1962b, 1967; Coleman et al., 1972). It is not found in New Zealand (Clarke, 1964), Brazil (Dehority, 1986b), or India (Kofoid and MacLennan, 1932; Banerjee, 1955). *P. multivesiculatum* is the key member of the A-type protozoal population (Eadie, 1962b, 1967) where it usually occurs with *Diploplastron affine*. Although *D. affine* is usually smaller (105 × 59 μm compared with 161 × 95 μm), there can be confusion between the two which are otherwise structurally similar. However, whereas the two skeletal plates in *P. multivesiculatum* run parallel to each other, those of *D. affine* converge, without joining, toward the posterior end.

P. multivesiculatum (Figs. 2.15 and 2.16) has an oval to triangular body with a small ventral caudal lobe. The two skeletal plates under the right side are easily seen in semi-starved protozoa, but the three small plates under the left side are very difficult to see. *P. multivesiculatum* has been described as *Metadinium medium* (Sugden, 1953) because of this difficulty. The size of *P. multivesiculatum* also depends markedly on the growth conditions measuring 123 μm long when grown *in vitro* in the absence of other protozoa, 173 μm long when grown in the rumen as

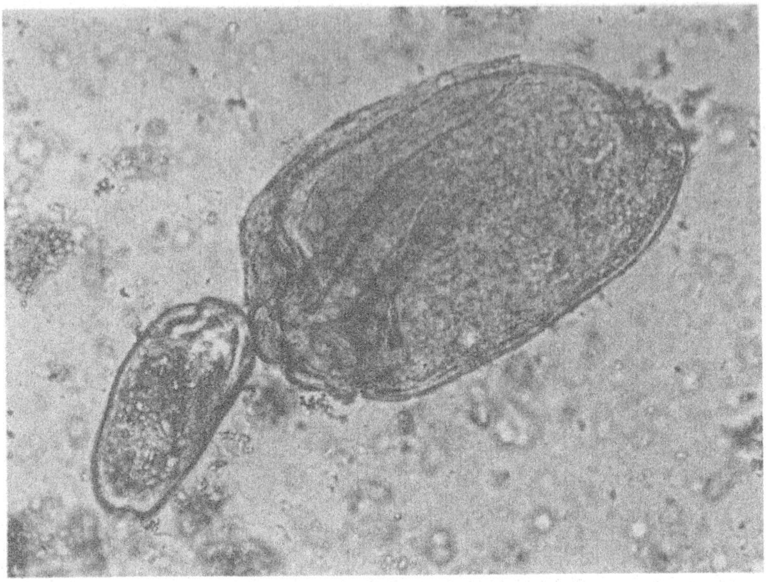

Fig. 2.15 Photomicrograph of *Polyplastron multivesiculatum* (larger organism), showing the two characteristic skeletal plates, and of *Epidinium caudatum* (small organism).

a member of an A-type population, and 205 μm long when grown in presence of epidinia *in vivo* or *in vitro*. It has a large operculum.

2.9 The Genus *Metadinium*

This genus was first erected by Awerinzew and Mutafowa (1914) to include those species of diplodinia in which the dorsal and oral ciliary zones were not connected. It was shown subsequently that these zones were not connected in the diplodinia either and "metadinia" were included in the eudiplodinia (Dogiel, 1927) and diplodinia (Buisson, 1923a; Becker and Talbott, 1927). Kofoid and MacLennan (1932) re-erected the genus to include those large and heavy (180–272 × 136–170 μm) protozoa which have two narrow skeletal plates that can be fused together posteriorly. The oral area is comparatively broad and the operculum not particularly prominent. According to Kofoid and MacLennan (1932), but not Latteur (1966d) or Hungate (1978), the shape of the macronucleus, which has two or three lobes on the dorsal surface like an inverted E or F, is a characteristic feature. Metadinia also have thicker pellicles and ectoplasm than other genera. The type species is *Metadinium medium*. Information on the various species is given below and in Table 2.9.

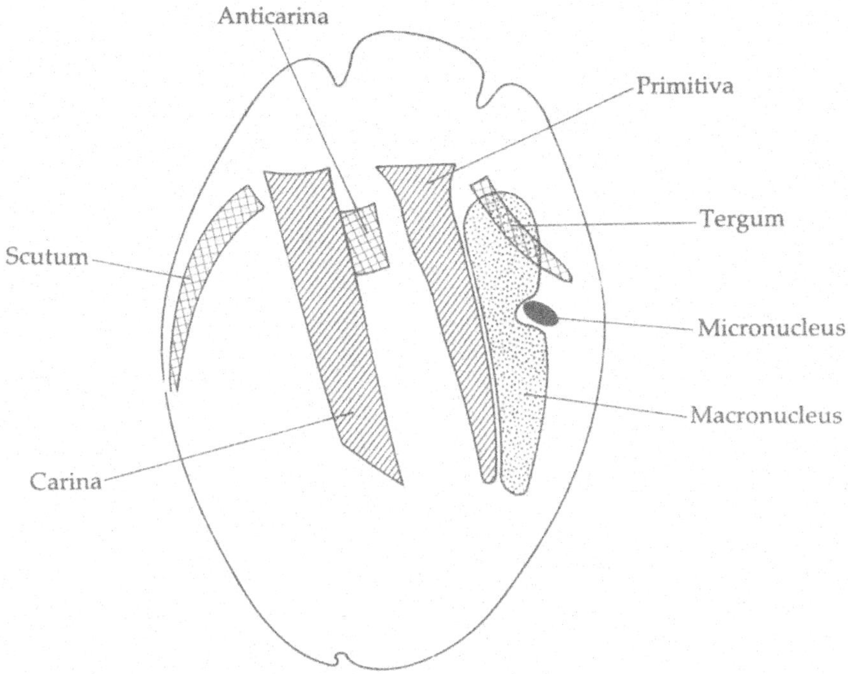

Fig. 2.16 Bchematic representation of *Polyplastron multivesiculatum* showing the names given to the skeletal plates.

Detailed Descriptions of *Metadinium* Species

1. *Metadinium affine*. See *Diploplastron affinc*

2. *Metadinium banksi* (*Diplodinium* [*Eudiplodinium*] *banksi*—Dehority, 1985) was first described from musk-oxen (*Ovibos moschatus*) from Banks Island, Northwest Territory, Canada, by Dehority (1985). It was not found in musk-oxen from other places (Dehority, 1986c). This protozoon is ellipsoidal with a smoothly rounded posterior end and has two converging skeletal plates which are usually, but not invariably, fused at the posterior end. It differs from the very similar *M. ypsilon*; it is longer and wider and has a larger cytoproct.

3. *Metadinium caudatum* (*Ostracodinium caudatum*—Dehority, 1985) was first described from a musk-ox in northern Canada by Lubinsky (1963) and has also been found in the same species at another site by Dehority (1985). In side view the body is an almost perfect ellipse with a rounded posterior end a short scoop-shaped caudal lobe. It is one of the two caudate species of the genus; the other is *M. sablei*. The skeletal plate is Y-shaped with a cleft anterior third but only a narrow fissure between the two arms.

Table 2.9 Characteristics of the *Metadinium* species

Species (Fig. no.)	Original description/ name[a]	Length (μm)	Width (μm)	Spines	Lobes	Contractile vacuoles
1. *M. affine*	See *Diploplastron affine*					
2. *M. banksi*	4	118–162	77–118	0	0	2
3. *M. caudatum*	11	175–250	115–180	0	1	2
4. *M. esalqum*	3	65–96	47–66	0	0	2
5. *M. magnum*	5/10	156–231	92–162	0	0	2
6. *M. medium* (2.17)	1	180–272	92–170	0	0	2
7. *M. minorum*	2	95–131	65–89	0	0	2
8. *M. rotundatum*	9	52–73	35–51	0	0	2
9. *M. sablei*	12	76–102	52–69	0	1	2
10. *M. tauricum*	8/10	185–288	70–160	0	0	2
11. *M. ypsilon* (2.18)	6/10	110–152	60–72	0	0	2

[a] References are as follows: 1, Awerinzew and Mutafowa (1914); 2, Dehority (1975b); 3, Dehority (1979); 4, Dehority (1985); 5, Dogiel (1925d); 6, Dogiel (1925e); 7, Dogiel (1927); 8, Dogiel and Fedorowa (1925b); 9, Kofoid and Christenson (1933); 10, Kofoid and MacLennan (1932); 11, Lubinsky (1963).

4. *Metadinium esalqum* was first described from Brazilian water buffalo by Dehority (1979) who called it *Ostracodinium esalqum*. However, in the classification used in this book, the possession of a Y-shaped skeletal plate places it in the genus *Metadinium*. This protozoon is comparatively small and is a flattened ellipsoid shape. It has no caudal spines or lobes.

5. *Metadinium magnum* (*Diplodinium ypsilon magnum*—Dogiel, 1926d; *Eudiplodinium ypsilon magnum*—Dogiel, 1927; *Ostracodinium magnum*—Dehority, 1974, 1986c) is found only in Arctic ruminants such as reindeer (Lubinsky, 1958b; Westerling, 1970) and musk-oxen (Dehority, 1974; Lubinsky, 1963; Dehority, 1985, 1986c). There is considerable confusion over the genus in which this organism should be placed. Those authors who believe that there are two plates in the (evolutionary) process of fusing place it in *Metadinium*, whereas those who believe that there is one plate dividing in two place it in *Ostracodinium*. Lubinsky (1958b) still places this protozoon in *Metadinium* while believing that it is one plate splitting into two. *M. magnum* has a Y-shaped skeletal plate and no caudal lobe. It is very similar to *M. ypsilon* except that it is larger (183 × 111 compared with 123 × 70 μm). It is almost identical with *M. caudatum* except that it does not have the caudal lobe.

6. *Metadinium medium* (*Diplodinium medium*—Buisson, 1923a; *Eudiplodinium medium medium*—Dogiel, 1927) is fairly widely distributed and found in cattle in Japan (Imai et al., 1979a), Brazil (Dehority, 1986a), Thailand (Imai and Ogimoto, 1984), the USA (Abou Akkada et al., 1969), and New Zealand (Clarke, 1964). It is also found in Brazilian water buffalo (Dehority, 1979), African reedbuck (Van Hoven, 1983), and musk-oxen (Dehority, 1985). This protozoon has an ellipsoidal to triangular

	Body shape	Macronucleus Size (μm)	Macronucleus Shape	Illustrated in Ogimoto and Imai (1981)	Other reference[a]
1.					
2.	Ovoid	9–17 × 86	Lobed rod	–	
3.	Oval	4–16 × 140	Lobed rod	–	
4.	Semi-ovoid	6–9 × 51	Lobed rod	–	
5.	Ovoid	5–14 × 106	Lobed rod	–	7, 11
6.	Ovoid	7–14 × 128	Lobed rod	+	4, 7, 10
7.	Ovoid	5–11 × 67	Lobed rod	–	
8.	Ovoid	5–7 × 38	Rod	–	
9.	Ovoid	8 × 58	Rod	–	
10.	Rectangular	7–23 × 146	Lobed rod	–	7
11.	Ovoid	4–10 × 68	Lobed rod	+	7

body with a small operculum and an apparently truncated anterior end. It has two separate parallel skeletal plates (Fig. 2.17).

7. *Metadinium minorum* (*Diplodinium* [*Ostracodinium*] *minorum*—Dehority, 1975) was first described from domestic sheep in Ohio, USA by Dehority (1975b, 1978). In side view the organism is ellipsoidal to rectangular with a rounded posterior end which may have a small flattened

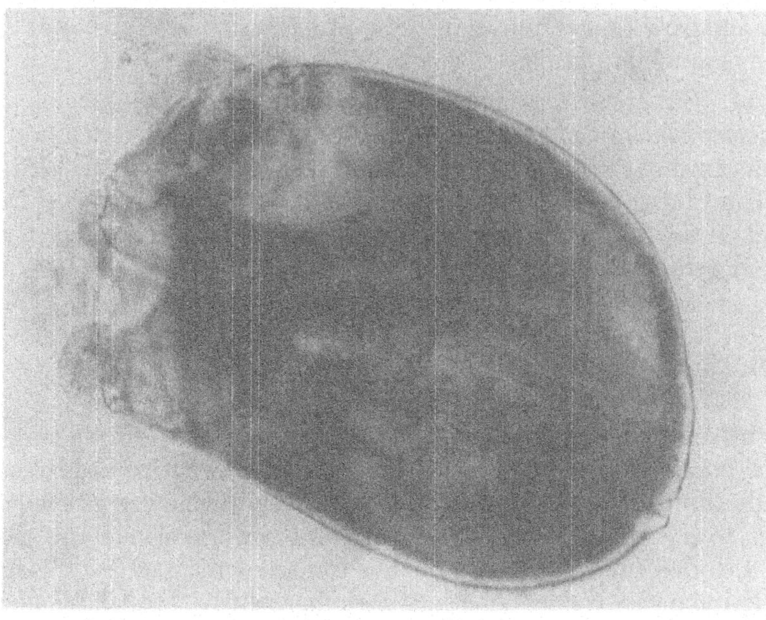

Fig. 2.17 Photomicrograph of *Metadinium medium*.

lobe. *M. minorum* has a Y-shaped skeletal plate with a narrow fissure between the two arms. It is shorter and wider than the similar *M. ypsilon*, has a large rectum and anus, and has no lobe at the anterior end of the macronucleus. It differs also in the general shape of the skeletal plate, which has a wide fissure between the arms in *M. ypsilon*.

8. *Metadinium rotundatum* was first described from the gaur (*Bos gaurus*) by Kofoid and Christenson (1933) and has also been seen in domestic cattle in India by Banerjee (1955). *M. rotundatum* is similar to *M. ypsilon* but appreciably smaller (65×37 compared with 123×70 μm) and has a Y-shaped skeletal plate which is concave on the ventral side.

9. *Metadinium sablei* was first described from the sable antelope (*Hippotragus niger*) by Van Hoven et al. (1979) who called it *Diplodinium* (*Eudiplodinium*) *sablei*. The two skeletal plates are fused posteriorly and the dorsal arm is curved to be almost U-shaped to the right of the macronucleus. The cytoproct is surrounded by a "cover" along which run striations as found in *Eremoplastron rugosum*. Otherwise, *M. sablei* generally resembles *M. ypsilon*.

10. *Metadinium tauricum* (*Diplodinium medium tauricum*—Dogiel and Fedorowa, 1925b; *Eudiplodinium medium tauricum*—Dogiel, 1927; *Ostracodinium tauricum*—Dehority, 1974) is found principally in sheep (in Russia—Dogiel, 1927; in Dall Mountain sheep—Dehority, 1974; in Sierra Bighorn sheep—Bush and Kofoid, 1948) but rarely in goats or cattle (Dogiel, 1927; Clarke, 1964). *M. tauricum* is similar to *M. medium* except that the skeletal plates are fused over two-thirds of their length with only a narrow fissure between the non-fused arms. The posterior lobe of the macronucleus is also smaller in *M. tauricum*.

11. *Metadinium ypsilon* (*Diplodinium ypsilon ypsilon*—Dogiel, 1925e; *Eudiplodinium ypsilon ypsilon*—Dogiel, 1927) is a rare species found principally in cattle from Russia (Dogiel, 1927), Indonesia (Imai, 1985), and Thailand (Imai and Ogimoto, 1984) and also in Indonesia water buffalo (Imai, 1985). *M. ypsilon* is similar to *M. tauricum* but is smaller, narrow, and more delicate (Fig. 2.18).

2.10 The Genus *Epidinium*

The members of this genus are characterized by their relatively long cylindrical bodies (e.g. 109×46 μm) with a dorsal ciliary zone displaced posteriorly from the oral zone such that in actively motile organisms the impression is given that the anterior 20% of the body is ciliated. The epidinia, like the other rumen ciliates, swim in a corkscrew manner, but this form of motion is more obvious than in some other genera. Most *Epidinium* spp. have three skeletal plates described as dorsal, median, and ventral.

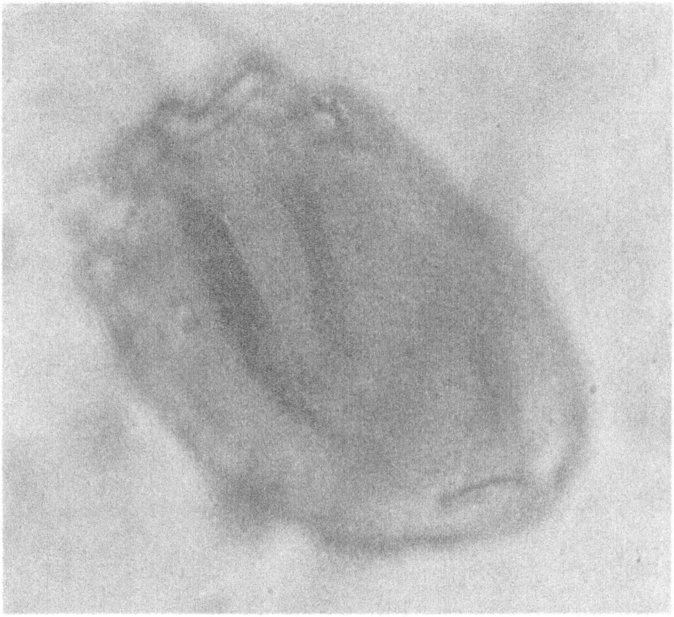

Fig. 2.18 Photomicrograph of *Metadinium ypsilon* stained to show the Y-shaped sketetal plate.

Although these plates are characteristic of the genus, they are not always easy to see and the epidinia are more easily identified by their shape. *Epidinium eberleini* is unusual as it has a fourth skeletal plate in the caudal spine.

Although epidinia were first described by Fiorentini (1889), they were classified as *Diplodinium* spp. and *Ophryoscolex* spp. before Crawley (1923) erected the genus *Epidinium* which has been used by almost all authors since. There has been much discussion since then over whether similar epidinia differing only in the number of caudal spines are different species, as advocated by Kofoid and MacLennan (1933), or are different forms of one species, as suggested by Dogiel (1927). Both systems are to be found in the current literature. In culture, clones tend to lose their caudal spines over a period of many months, but it not known why this occurs or what conditions are necessary for the redevelopment of the spines. Although the authors feel that organisms with different numbers of spines are probably "forms" rather than "species," this is unproven and in the discussion of individual species given below the less cumbersome use of separate species will be followed.

All epidinia have two contractile vacuoles. The type species is *Epidinium ecaudatum* (Fiorentini, 1889). Information about the various species is given in Table 2.10.

Table 2.10 Characteristics of the *Epidinium* species

Species (Fig. no.)	Original description/ name[a]	Length (μm)	Width (μm)	Spines	Lobes	Number of contractile vacuoles
1. *E. bicaudatum*	16	82–144	39–60	2	0	2
2. *E. bovis*	2	150–200	70–120	3	0	2
3. *E. bulbiferum*	6	98–150	40–55	0	1	2
4. *E. camerounense*	13					
5. *E. capricornisi*	9	90–130	50–70	5	0	2
6. *E. cattanei* (2.19)	8/4	78–120	42–65	5	0	2
7. *E. caudatum*	8/4	85–140	37–54	1	0	2
(Ovine) (2.15, 2.20)	3	98–120	41–52	1	0	2
(Bovine) (2.21)	3	126	46	1	0	2
8. *E. dactylodonta*	12	68–122	32–58	1–5	0	2
9. *E. eberleini*	7/11	85–118	42–65	1	0	2
10. *E. ecaudatum*	8/4	122–132	40–45	0	0	2
11. *E. fasciculus*	1/6	85–110	45–60	5	0	2
12. *E. gigas*	5/11	135–230	53–100	0	0	2
13. *E. hamatum*	15/11	104–150	43–61	1	0	2
14. *E. lunatus*	17	53–69	23–31	1	0	2
15. *E. parvicaudatum*	1/11	82–160	38–54	5	0	2
16. *E. quadricaudatum*	16/4	110–119	48–52	4	0	2
17. *E. rusa*	2	80–105	40–50	5	0	2
18. *E. tricaudatum* (2.22)	16/4	98–140	20–77	3	0	2

[a]References are as follows: 1, Awerinzew and Mutafowa (1914); 2, Banerjee (1955); 3, Coleman et al. (1972); 4, Crawley (1923); 5, Dogiel (1925d); 6, Dogiel (1927); 7, Eberlein (1895); 8, Fiorentini (1889); 9, Imai et al. (1981a); 10, Kleynhans and Van Hoven (1976); 11, Kofoid and MacLennan (1933); 12, Latteur (1966a); 13, Noirot-Timothée (1963); 14, Ogimoto and Imai (1981); 15, Schulze (1924); 16, Sharp (1914); 17, Van Hoven (1975).

Detailed Descriptions of *Epidinium* Species

1. *Epidinium bicaudatum* (*Diplodinium ecaudatum bicaudatum*—Sharp, 1914; *Epidinium ecaudatum bicaudatum*—Dogiel, 1927) which has two caudal spines, is probably the rarest of the various spined forms. Where present it is usually found in only a few animals and at low population density. It has been reported from cattle in Russia (Dogiel, 1927), the USA (Sharp, 1914; Becker and Talbott, 1927), Thailand (Imai and Ogimoto, 1984), and New Zealand (Clarke, 1964) and in the giraffe (Noirot-Timothée, 1963) and musk-ox (*Ovibos moschatus*) (Dehority, 1985). This protozoon is very similar to *E. ecaudatum* except in the possession of a large incurved ventral spine and a small dorsal one.

2. *Epidinium bovis* has only been described from the rumen of the Indian cow by Banerjee (1955). Its body is similar to that of *E. ecaudatum* and it has three caudal spines like *E. tricaudatum*. However, *E. bovis* is much larger than *E. tricaudatum* (174 × 87 compared with 118 × 39 μm) and has three spines all of the same length.

3. *Epidinium bulbiferum* (*Epidinium ecaudatum bulbiferum*—Dogiel,

	Extra skeletal plate	Body shape	Macronucleus Size (μm)	Macronucleus Shape	Illustrated in Ogimoto and Imai (1981)	Other reference[a]
1.	0	Elongated	8 × 67	Club-shaped	+	4, 6, 11
2.	0	Elongated	11 × 90	Curved club	−	
3.	0	Elongated	7 × 66	Rod	+	11
4.	0				−	
5.	0	Elongated	9 × 54	Rod	+	14
6.	0	Comparatively short	11 × 55	Club-shaped	+	6, 11
7.	0	Elongated	12 × 53	Rod	+	6, 10, 11
	0	Elongated			−	
	0	Elongated			−	
8.	0	Elongated	6 × 44	Rod	−	
9.	1	Elongated	5–9 × 56	Club-shaped	+	6
10.	0	Elongated	8 × 67	Rod	+	6, 10, 11
11.	0	Comparatively short	5–17 × 66	Lenticular	−	
12.	0	Elongated	12 × 90	Rod	−	18
13.	0	Elongated	11 × 63	Rod	+	6
14.	0	Elongated	4–9 × 36	Club-shaped	−	
15.	0	Elongated	10 × 65	Rod	+	
16.	0	Elongated	9 × 69	Rod	+	6, 10
17.	0	Elongated	5–10 × 65	Lobed rod	−	6, 10, 11
18.	0	Elongated	10 × 60	Rod	−	6, 10, 11

1927) has only been found in one domestic cow in Russia by Dogiel (1927). It is similar to *E. hamatum*, but it has a rounded ventral lobe instead of a spine.

4. *Epidinium camerounense* was mentioned but not described by Noirot-Timothée (1963).

5. *Epidinium capricornisi* (*Epidinium ecaudatum capricornisi*—Imai et al., 1981a) was first described from the Japanese serow (*Capricornis crispus*) by Imai et al. (1981a). Like *E. parvicaudatum* and *E. rusa*, it has five smallish caudal spines but it differs in that they are all approximately the same length and two of them bifurcate and one trifurcates as do those of *Ophryoscolex* spp.

6. *Epidinium cattanei* (*Diplodinium cattanei*—Fiorentini, 1889; *Diplodinium ecaudatum cattanei*—Becker and Talbott, 1927; Buisson, 1923a; *Ophryoscolex cattane(o)i*—Railliet, 1890; Cunha, 1914b; *Ophryoscolex fasciculus*—Awerinzew and Mutafowa, 1914; *Ophryoscolex ecaudatus cattaneoi*—Dogiel, 1926; *Epidinium (e) caudatum fasciculus*—Dogiel, 1927). The organism described by Dogiel (1927) as *Epidinium ecaudatum cattanei* is *Epidinium parvicaudatum* rather than *Epidinium cattanei*. This species is widely but not universally distributed occurring in a small proportion of animals examined at each site. It occurs in Japanese sheep (Imai et al., 1978), Thai zebu cattle (Imai and Ogimoto, 1984), the Indian goat (*Capra hircus*) (Das Gupta, 1935), musk-oxen in Canada (Tener, 1965) and cattle

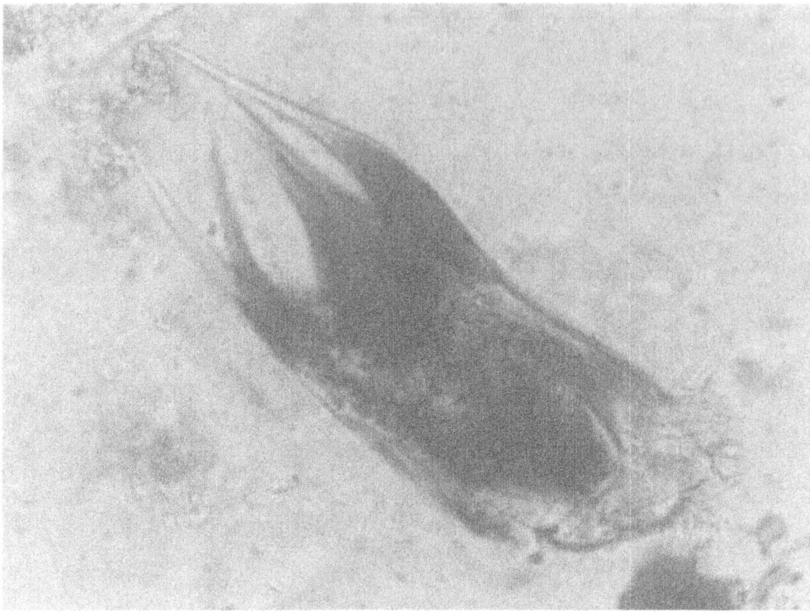

Fig. 2.19 Photomicrograph of *Epidinium cattanei*.

from India (Kofoid and MacLennan, 1933; Banerjee, 1955), Russia (Dogiel, 1927), USA (Becker and Talbott; 1927) and Brazil (Cunha, 1914a, 1914b) *E. cattanei* is characterized by its relatively short body and five long straight caudal spines (Fig. 2.19). These features separate it from all other epidinia.

7. *Epidinium caudatum (Diplodinium caudatum*—Fiorentini, 1889; *Diplodinium ecaudatum caudatum*—Sharp, 1914; Becker and Talbott, 1927; *Ophryoscolex intermixtus*—Awerinzew and Mutafowa, 1914; *Ophryoscolex inermis caudatus*—Cunha, 1914b; *Epidinium ecaudatum caudatum*—Dogiel, 1927) is one of the most widely distributed protozoal species and was found in all the B-type populations of Japanese cattle, sheep, and goats by Imai et al. (1979a). It is not found in A-type populations and may therefore be found in a variable proportion of all animals investigated. This protozoon has the same general structure as the others members of the "ecaudatum" group with an elongated body tapering posteriorly. In the authors' experience there are two forms of this species: an "ovine" form from sheep with a thin slender incurved ventral spine (Fig. 2.20) and a "bovine" form from cattle with a heavier, longer, straight ventral spine (Fig. 2.21).

8. *Epidinium eberleini* (*Diplodinium caudatum*—Eberlein, 1895; *Diplodinium eberleini*—Cunha, 1914b; Sharp, 1914: *Diplodinium longispinum*—

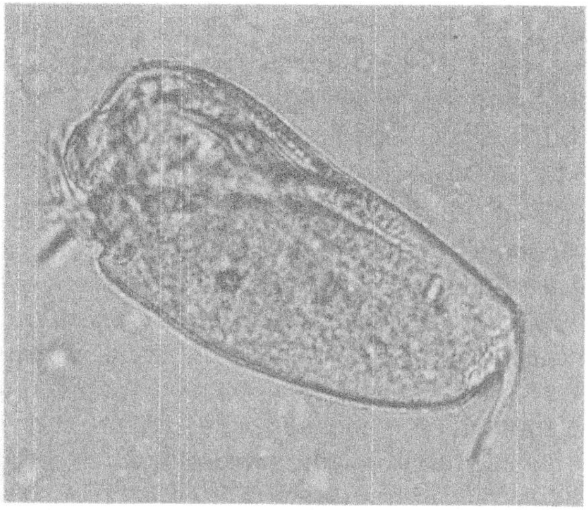

Fig. 2.20 Photomicrograph of *Epidinium caudatum* (ovine form with the small curved caudal spine) grown in vitro.

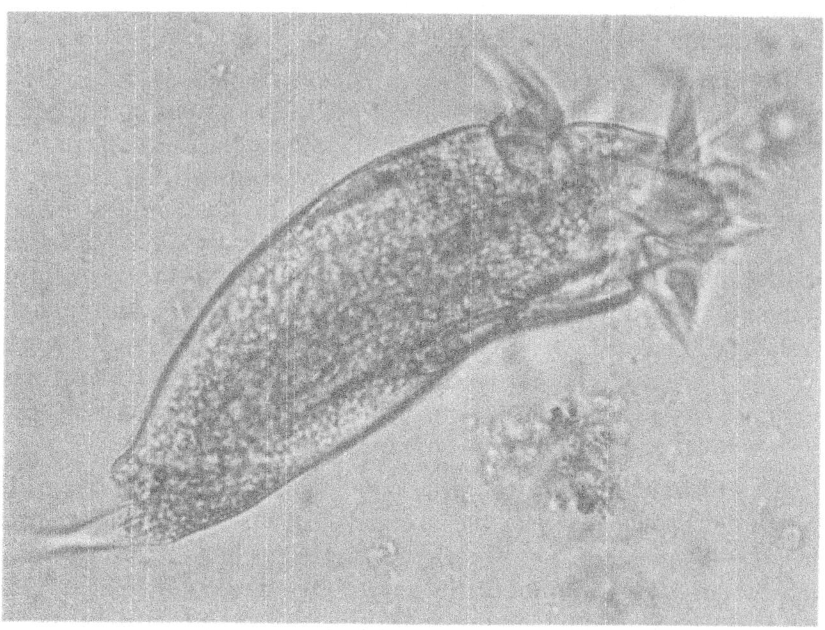

Fig. 2.21 Photomicrograph of *Epidinium caudatum* (bovine form with the long straight caudal spine) grown in vitro.

Schulze, 1924; *Epidinium lobatum*—Dogiel, 1928) has been found in the past in German cattle (Eberlein, 1895; Schulze,1924) and Californian sheep (Bush and Kofoid, 1948), but more recently it has only been found in cattle in India (Banerjee, 1955), Thailand (Imai and Ogimoto, 1984), and Sri Lanka (Imai, 1986). This species is characterized by a flattened caudal lobe on the right side, a blunt caudal lobe on the left side, and a long, heavy, ventral spine containing an additional skeletal plate.

9. *Epidinium ecaudatum* (*Diplodinium caudatum*—Fiorentini, 1889; *Diplodinium ecaudatum ecaudatum*—Sharp, 1914; *Ophryoscolex inermis*—Cunha, 1914a; *Ophryoscolex labiatus*—Awerinzew and Mutafowa, 1914; *Ophryoscolex ecaudatus ecaudatus*—Dogiel, 1925d; *Epidinium ecaudatum ecaudatum*—Dogiel, 1927) is a widely distributed species and was found in all the B-type populations of Japanese cattle, sheep, and goats by Imai et al. (1979a). It usually occurs with other epidinia, especially *E. caudatum*, although it was absent from New Zealand cattle (Clarke, 1964) and Thai zebu cattle (Imai and Ogimoto, 1984). It has been found in bushbuck (*Tragelaphus scriptus*), giraffe (Kleynhans and Van Hoven, 1976), sambar (*Rusa unicolor*—Banerjee, 1955), and reindeer and musk-oxen (Dehority, 1986c). It is the first member of the "ecaudatum" group of species characterized by elongated bodies tapering posteriorly. *E. ecaudatum* has no caudal spines.

10. *Epidinium fasciculus* (*Ophryoscolex fasciculus*—Awerinzew and Mutafowa, 1914; *Epidinium ecaudatum fasciculus*—Dogiel, 1927) is a non valid name for *Epidinium cattanei*.

11. *Epidinium gigas* (*Ophryoscolex ecaudatus gigas*—Dogiel, 1925d; *Epidinium ecaudatum gigas*—Dogiel, 1927) is found almost solely in reindeer (*Rangifer tarandus*) (Westerling, 1970), although it has been reported in one musk-ox (Tener, 1965). It was not found in sheep and cattle in the reindeer breeding area (Westerling, 1969). *E. gigas* has a similar shape to *E. ecaudatum* but is larger (172 × 73 compared with 122 × 51 μm), although Westerling (1970) found that *E. gigas* was only larger in Finnish reindeer in the winter and the same size as *E. ecaudatum* in summer. The left skeletal plate of *E. gigas* is larger than in *E. ecaudatum* and the posterior part is expanded dorsally. The anterior part of the macronucleus is also bent ventrally. *E. gigas* ingests lichen and was reported by Kopperi (1929) to form cysts, but this has not been confirmed.

12. *Epidinium hamatum* (*Diplodinium hamatum*—Schulze, 1924; *Ophryoscolex ecaudatus hamatus*—Dogiel, 1926b; *Epidinium ecaudatum hamatum*—Dogiel, 1927) is a rare species that has been found in various domestic animals throughout the world (Schulze, 1924; Dogiel, 1927; Bush and Kofoid, 1948; Ogimoto and Imai, 1981). It has also been found in the camel (*Camelus dromedarius*) (Dogiel, 1927). *E. hamatum* differs from *E. caudatum* in that the posterior end of the body is thick and heavy

rather than tapering. The dorsal lobe extends beyond the level of the cytoproct. It has one thin ventral spine.

13. *Epidinium lunatus* was first described from the tsessebe (*Damaliscus lunatus lunatus*) by Van Hoven (1975) and has also been seen in two reedbuck species (*Redunca arundinum arundinum* and *R. fulvorufula fulvorufula* (Van Hoven, 1983). There must be some doubt about ascribing this organism to the genus *Epidinium* as it is smaller than the other species and has only one skeletal plate. It does, however, have the same general body shape as the epidinia and has a long tapering caudal spine.

14. *Epidinium parvicaudatum* (*Diplodinium ecaudatum cattanei*—Sharp, 1914; *Ophryoscolex fasciculus parvicaudata*—Awerinzew and Mutafowa, 1914; *Epidinium ecaudatum cattaneoi*—Dogiel, 1927). There was some confusion in the early years, up to and including Dogiel's (1927) monograph on the naming of the five-spined species of *Epidinium*. The problem was resolved by Kofoid and MacLennan (1933) who stated that the long-spined species was *E. cattanei* and the short-spined species *E. parvicaudatum*. This latter organism is widely but not universally distributed being found in cattle in the USA (Sharp, 1914), Russia (Awerinzew and Mutafowa, 1914; Dogiel, 1927, 1932), New Zealand (Clarke, 1964), and Indonesia (Imai, 1985), in sheep from India (Banerjee, 1955) and the USA (Bush and Kofoid, 1948), in the gaur (*Bos gaurus*) (Kofoid and Christenson, 1933), in musk-oxen (Dehority, 1985), and in the giraffe (Kleynhans and Van Hoven, 1976). The organisms from the giraffe were thought to have an extra skeletal plate in the caudal region (Kleynhans and Van Hoven, 1976) but this is probably just a loose mass of polysaccharide granules. The general shape of *E. parvicaudatum* is the same as that of *E. ecaudatum*, but it has five short, heavy, caudal spines of which the ventral is the largest, and all are incurved.

15. *Epidinium quadricaudatum* (*Diplodinium ecaudatum quadricaudatum*—Sharp, 1914; *Ophryoscolex ecaudatus quadricaudatus*—Dogiel, 1926b; *Epidinium ecaudatum quadricaudatum*—Dogiel, 1927) is widely but not universally distributed. It is, for example, common in cattle in California (Sharp, 1914) but absent from those in Iowa (Becker and Talbott, 1927). It is found in cattle from Sri Lanka (Kofoid and MacLennan, 1933; Imai, 1986), Thailand (Imai and Ogimoto, 1984), Indonesia (Imai, 1985), and New Zealand (Clarke, 1964), in sheep from the USA (Bush and Kofoid, 1948), in the musk-ox (Dehority, 1985), in the gaur (Kofoid and Christenson, 1933), and the giraffe (Kleynhans and van Hoven, 1976). *E. quadricaudatum* resembles *E. ecaudatum* except that it has four incurved caudal spines of which the ventral spine is the longest.

16. *Epidinium rusa* was described for the first and only time from the sambar (*Rusa unicolor*) by Banerjee (1955). It is very similar to *E. parvicaudatum* except that all five caudal spines are the same length.

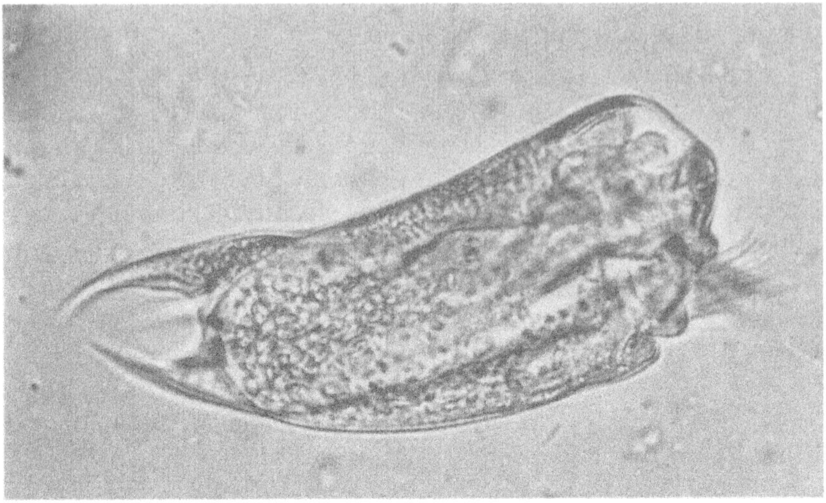

Fig. 2.22 Photomicrograph of *Epidinium tricaudatum*.

17. *Epidinium tricaudatum* (*Diplodinium ecaudatum tricaudatum*—Sharp, 1914; *Ophryoscolex ecaudatus tricaudatus*—Dogiel, 1926b; *Epidinium ecaudatum tricaudatum*—Dogiel, 1927) is one of the rarer of the epidinia related to *E. ecaudatum* (Fig. 2.22). It occurs in cattle in the USA (Sharp, 1914; Becker and Talbott, 1927; Abou Akkada et al., 1969), New Zealand (Clarke, 1964), Russia (Dogiel, 1927), and Sri Lanka (found by Kofoid and MacLennan, 1933, but not by Imai, 1986). It has not been found in Japanese cattle. *E. tricaudatum* is also found in the giraffe (Kleynhans and Van Hoven, 1976), musk-ox (Dehority, 1985), reindeer (Dogiel, 1925d; 1935), and the Kaffir buffalo (*Syncerus caffer*) (Dogiel, 1932). *E. tricaudatum* resembles *E. ecaudatum* except that it has three incurved caudal spines of which the ventral is the longest.

Table 2.11 Characteristics of the *Enoploplastron* species

Species (Fig. no.)	Original description/ name[a]	Length (μm)	Width (μm)	Spines	Lobes	Contractile vacuoles
1. *E. confluens*	3/7	120–157	72–103	0	0	2
2. *E. garstangi*	9	76–95	43–52	1	0	2
3. *E. stokyi*	1/4					
f. *stokyi*	5	55–75	35–47	0	2	2
f. *buissoni*	5	55–80	40–50	0	1	2
4. *E. triloricatum* (2.23)	2/7	60–112	37–70	0	0	2

[a] References are as follows: 1, Buisson (1924); 2, Dogiel (1925a); 3, Dogiel (1925d); 4, Dogiel (1927); 5, Imai (1988); 6, Kleynhans and Van Hovan (1976); 7, Kofoid and MacLennan (1932); 8, Lubinsky (1958b); 9, Van Hoven (1975).

2.11 The Genus *Enoploplastron*

This genus is characterized by the presence of a dorsal zone of cilia near the anterior end of the body and three narrow skeletal plates. The dorsal and median plates lie beneath the right side of the body between the macronucleus and the ventral surface. The ventral plate lies adjacent to the median plate and extends beneath the ventral surface. The plates touch and can be fused together; however, the plates or the arms of the single plate are separate in the middle of the anterior half of the body leaving two windows of clear cytoplasm. The fibers around the cytopharynx are heavy. All species have two contractile vacuoles.

When *E. triloricatum* was first described by Dogiel (1925a) he put it in the genus *Diplodinium* and then in 1927 transferred it to the genus *Ostracodinium*. Subsequently Kofoid and MacLennan (1932) erected a new genus *Enoploplastron*, which has been universally accepted. The type species is *Enoploplastron triloricatum* (Dogiel, 1925a) Information on the various species is given below and in Table 2.11.

Detailed Descriptions of *Enoploplastron* Species

1. *Enoploplastron confluens* (*Diplodinium triloricatum confluens*—Dogiel, 1925d; *Ostracodinium triloricatum confluens*—Dogiel, 1927; *Ostracodinium confluens*—Dehority, 1986c) is only found in Arctic ruminants such as reindeer (*Rangifer tarandus*) (Lubinsky, 1958b; Westerling, 1970; Dehority, 1986c; Dogiel, 1925d) and the musk-ox (*Ovibos moschatus*) (Tener, 1965; Dehority, 1985). *E. confluens* has one very long and wide skeletal plate extending from the base of the operculum nearly to the posterior end and from the macronucleus under the right surface to the ventral side and under part of the left surface. There are two pear-shaped windows in the plate at the anterior end, although it is not always easy to see both if the orientation of the organism is wrong.

		Macronucleus		Illustrated in	
	Body shape	Size (μm)	Shape	Ogimoto and Imai (1981)	Other reference[a]
1.	Broad ovoid	10 × 74	Lenticular	–	4, 8
2.	Broad ovoid	3–9 × 41	Elongated wedge	–	
3.					
	Ovoid	4–7 × 25	Rod	–	
	Ovoid	4–7 × 30	Rod	–	
4.	Ovoid	6 × 50	Rod	+	4, 6

2. *Enoploplastron garstangi* was first described from the tsessebe (*Damaliscus lunatus lunatus*) by Van Hoven (1975). It is ellipsoidal with the three skeletal plates fused together posteriorly but with the anterior arms free and not fused together to leave windows. The dorsal zone of cilia is much smaller than the oral zone and is situated slightly posteriorly to the latter. *E. garstangi* is much smaller than *E. confluens* (85 × 46 μm compared with 138 × 86 μm) and has a slender ventral spine.

3. *Enoploplastron stokyi* (*Diplodinium stokyi*—Buisson, 1924; *Ostracodinium stokyi*—Dogiel, 1927; Kofoid and MacLennan, 1932) is apparently found only in African ruminants such as the antelope (*Hippotragus equinus*) (Buisson, 1924) and Kenyan zebu cattle (*Bos indicus*) (Imai, 1988) but is not universally present. It is absent from the tsessebe and reedbuck (Van Hoven, 1975, 1983) and from West African zebu cattle (Bonhomme-Florentin et al., 1978). *E. stokyi* has two slender and one larger skeletal plate. The dorsal plate is bent ventrally at the anterior end to touch the ventral plate and form with the median plate a pear-shaped to almost round window. Imai (1988) divided this species into two forms. Forma *stokyi* has two elongated dorsal posterior lobes, the right one of which is sometimes bifurcated, and forma *buissoni* which has a single blunt dorsal spine.

4. *Enoploplastron triloricatum* (*Diplodinium triloricatum*—Dogiel, 1925a; *Diplodinium triloricatum triloricatum*—Dogiel, 1925d; *Ostracodinium triloricatum triloricatum*—Dogiel, 1927) is widely but not universally distributed. It is found in Bighorn sheep (Bush and Kofoid, 1948), the musk-ox (Tener, 1965), Japanese sheep but not cattle or goats (Imai et al., 1979), Thai and Indonesian zebu cattle (Imai, 1985; Imai and Ogimoto, 1984), African zebu cattle (Bonhomme-Florentin et al., 1978; Imai, 1988), and the giraffe (Kleynhans and Van Hoven, 1976). It is not found in cattle in India (Banerjee, 1955), the USA (Becker and Talbott, 1927; Dehority, 1978), or New Zealand (Clarke, 1964). *E. triloricatum* has an ellipsoidal body with three separate skeletal plates (Fig. 2.23) which touch at the posterior end but are separated at the anterior end to leave a large lenticular window on the dorsal side and a much smaller, sometimes nonexistent, window on the ventral side. The macronucleus has a marked depression on the dorsal side in which the micronucleus is situated.

2.12 The Genus *Ophryoscolex*

Ophryoscolex spp. were first described by Stein (1858) as members of the type genus that he erected at the same time. These organisms are the most complex and distinctive of the entodiniomorphid protozoa found in the rumen and are usually characterized by a formidable array of caudal spines.

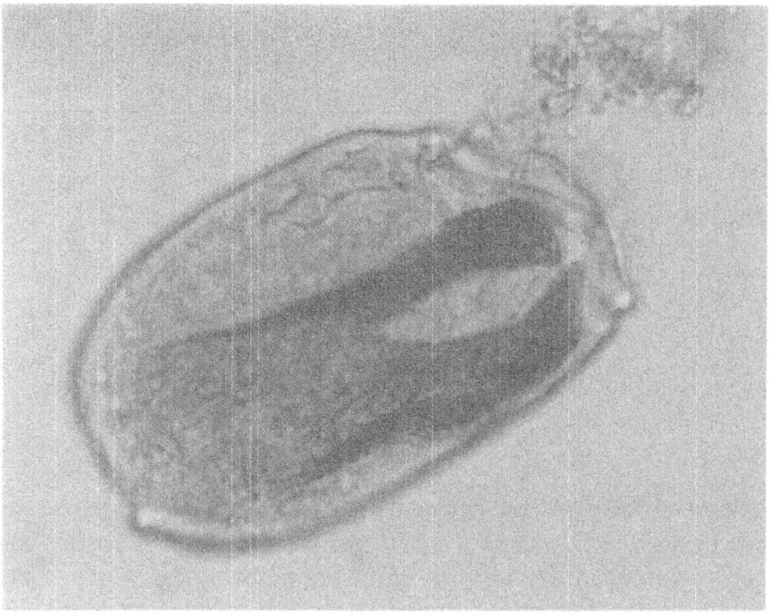

Fig. 2.23 Photomicrograph of *Enopoloplastron triloricatum* stained to show the three characteristic skeletal plates.

The dorsal zone of cilia is displaced posteriorly about a third of the length of the body and forms a girdle round about three-quarters of the circumference of the body, which is roughly circular in cross section. There are three long skeletal plates lying beneath the right surface of the body, but they are often difficult to see because of the cylindrical shape of the body. Some species also have accessory plates in some of the secondary spines.

As with the epidinia, the genus *Ophryoscolex* has been separated either into many species (Kofoid and MacLennan, 1933) or a few species and many forms (Dogiel, 1927) on the basis of the number and distribution of the caudal spines which can be arranged in up to four circles often round a main long spine. A spineless form, *O. inermis*, has been described inadequately by Stein (1858) and reported by Becker and Talbott (1927), but it is of doubtful validity. Unfortunately Buisson (1924) selected it as the type species, although *O. purkynjei* (Stein, 1858) also has a claim. However, the argument may be academic because Mah (1964, and 1962, quoted by Hungate, 1978) and Coleman and Reynolds (1982a) have shown that during culture in vitro the number and size of the caudal spines decreases and that, for example, *O. purkynjei* can come to resemble *O. inermis*. With *O. caudatus*, at least, the main caudal spine returns to its original size when cultured organisms are reinoculated into a defaunated rumen, but the reason for this is not known. Latteur (1966b) states, without elaboration,

Table 2.12 Characteristics of the *Ophryoscolex* species

Species (Fig. no.)	Original description/ name[a]	Length (μm)	Width (μm)	Length main spine	Circles of spines	Spines in anterior circle
1. *O. buissoni*	3/5	125–165	60–92	Long	1	5
2. *O. bicinctus*	3/5	130–170	68–87	Long	2	5
3. *O. bicoronatus*	3/5	120–170	81–90	Long	2	6
4. *O. caudatus* (2.24)	4	137–162	80–98	Long	3	6
5. *O. inermis*	7	170–190	65–100	None	0	0
6. *O. purkynjei*	7	155–215	80–110	Short	3	6
7. *O. quadricoronatus*	3/5	128–180	86–100	Long	4	6
8. *O. spinosus*	5	122–160	63–82	Short	2	6

[a] References are as follows: 1, Coleman (1979b); 2, Coleman and Reynolds (1982a); 3, Dogiel (1927); 4, Eberlein (1895); 5, Kofoid and MacLennan (1933); 6, Ogimoto and Imai (1981); 7, Stein (1858).

that there are two variable and two stable species of *Ophryoscolex*. The author believes that *O. purkynjei* and *O. caudatus* are distinct species and that the other described species and forms are probably just variants of these two, but the whole matter is open for reinvestigation. In the descriptions that follow, however, all the organisms described in the literature will be assumed to be separate species. See also Table 2.12.

Ophryoscolex spp. have from 9 to 15 contractile vacuoles arranged round the body in two transverse rows.

Detailed Descriptions of *Ophryoscolex* Species

1. *Ophryoscolex buissoni* (*Ophryoscolex buissoni unicinctus*—Dogiel, 1927; *Ophryoscolex buissoni unicinctus simplicispinosus*—Dogiel, 1927) is a rare species and has only been reported from sheep and cattle from Russia (Dogiel, 1927), from sheep in the USA (Bush and Kofoid, 1948), and in Dall Mountain sheep (*Ovis dalli*—Dehority, 1974). *O. buissoni* has a relatively stout body with a long, slender, main caudal spine and a single circle of five secondary spines. It is therefore one of the simplest of the genus. It has fifteen contractile vacuoles of which ten are in the anterior row.

2. *Ophryoscolex bicinctus* (*Ophryoscolex buissoni bicinctus*—Dogiel, 1927) is another rare species which has only been reported from sheep in Turkestan (Dogiel, 1927), from sheep in the USA (Bush and Kofoid, 1948), and in Dall Mountain sheep (*Ovis dalli*—Dehority, 1974). *O. bicinctus* closely resembles *O. buissoni* but has two circles of secondary spines surrounding the long main spine, and the anterior circle has five spines.

3. *Ophryoscolex bicoronatus* (*Ophryoscolex caudatus bicoronatus*—Dogiel, 1927; *Ophryoscolex caudatus bicoronatus atavus*—Dogiel, 1927) is another rare species which has only been reported in sheep from Russia

	Contractile vacuoles	Body shape	Macronucleus Size (μm)	Macronucleus Shape	Illustrated in Ogimoto and Imai (1981)	Other reference[a]
1.	15	Stout	12 × 60	Rod	–	
2.	15	Stout	14 × 77	Rod	–	
3.	9	Stout	12 × 58	Rod	–	
4.	9	Stout	15 × 67	Rod	+	1, 2, 3, 5, 6
5.		Slender	16 × 110	Rod	–	3, 5
6.	9	Slender	17 × 74	Rod	+	3, 5
7.	9	Stout	26 × 65	Ovoid	–	
8.	10	Slender	10 × 73	Rod	–	

(Dogiel, 1927), China (Hsiung, 1931), and the USA (Bush and Kofoid, 1948), and in Dall Mountain sheep (*Ovis dalli*—Dehority, 1974). *O. bicoronatus* has a relatively stout body with a long, slender, main caudal spine surrounded by two circles of secondary spines, the anterior of which has six spines compared with five in *O. bicinctus*.

4. *Ophryoscolex caudatus* (*Ophryoscolex caudatus tricoronatus*—Dogiel, 1927) is the most widely distributed species of the genus, although it is not universally found. It has been reported in sheep from Russia (Dogiel, 1927), China (Hsiung, 1931), the USA (Bush and Kofoid, 1948), and Japan (Imai et al., 1979a), in goats in India (Das Gupta, 1935) and Japan (Imai et al., 1979a), and in cattle in the USA (Becker and Talbott, 1927) and Africa (Bonhomme-Florentin et al., 1978). It is also found in Dall Mountain sheep (Dehority, 1974). *O. caudatus* is the middle member of three species having two (*O. bicoronatus*), three (*O. caudatus*) or four (*O. quadricoronatus*) circles of secondary caudal spines with the anterior circle having six spines (Fig. 2.24). All species have a long main spine and nine contractile vacuoles.

5. *Ophryoscolex inermis* is a dubious species with a smooth posterior end. Kofoid and MacLennan (1933) believed that reports of its presence were misidentified *Epidinium* spp. or *Polyplastron multivesiculatum*. *O. inermis* has, however, been reported in German (Eberlein, 1895), Brazilian (Cunha, 1914a), and American (USA) (Becker and Talbott, 1927) cattle and American sheep (Bush and Kofoid, 1948). There is no good description of the species.

6. *Ophryoscolex purkynjei*, also spelled *O. purkynei* (*Diplodinium vortex*—Fiorentini, 1889), is one of the more common species although it is not widely distributed. It has been reported in cattle from Germany (Eberlein, 1895), Russia (Dogiel, 1927), Brazil (Cunha, 1914b), South Africa (Fantham, 1926), the USA (Mah, 1964), Japan (Imai et al., 1978), and Africa (Bonhomme-Florentin et al., 1978). It is also found in American (USA) sheep (Bush and Kofoid, 1948). *O. purkynjei* has a short,

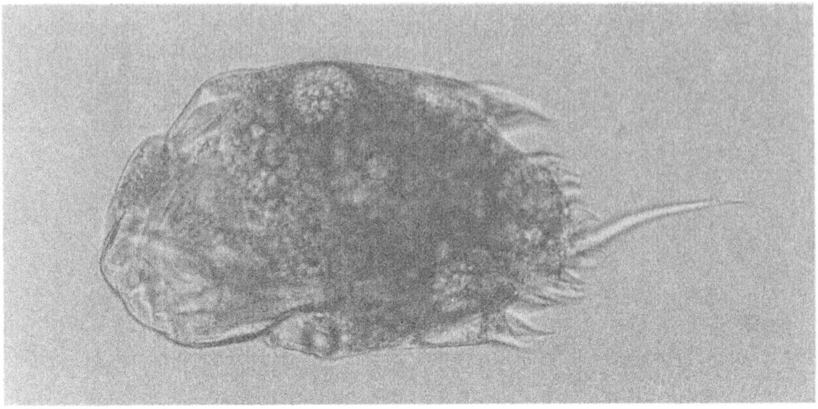

Fig. 2.24 Photomicrograph of *Ophryoscolex caudatus* isolated from the rumen.

stout, main caudal spine with three accessory spines projecting from it. It has three circles of secondary spines, compared with only two in *O. spinosus*, and in the anterior circle the spines are usually trifurcated, whereas in *O. spinosus* they are usually simple.

7. *Ophryoscolex quadricoronatus* (*Ophryoscolex caudatus quadricoronatus*—Dogiel, 1927) is a very rare species that has only been reported from sheep in Turkestan and northern Iran by Dogiel (1927) and from sheep in the USA by Bush and Kofoid (1948). It is very similar to *O. caudatus* with a long main caudal spine. *O. quadricoronatus* has four circles of secondary spines with six spines in the anterior circle. It is, however, far from easy to decide how many circles of spines are present on individual protozoa, and it may be more common than appears from the literature. Equally well we do not know if the number of circles of spines is a constant feature between parent and daughter organisms as it is not a constant feature in clone cultures.

8. *Ophryoscolex spinosus* is a very rare species only described in small numbers from *Bos indicus* in India by Kofoid and MacLennan (1933). It is very similar to *O. purkynjei* and has a short main caudal spine but only two circles of secondary spines instead of three and has simple rather than trifurcated spines in the anterior row.

2.13 The Genus *Diploplastron*

This genus has only one species. It was erected by Kofoid and MacLennan (1932) for those protozoa with two non-parallel skeletal plates beneath their right surfaces. These plates come close together at the posterior end but do not fuse. The pellicle and ectoplasm form only a thin layer and the cyto-

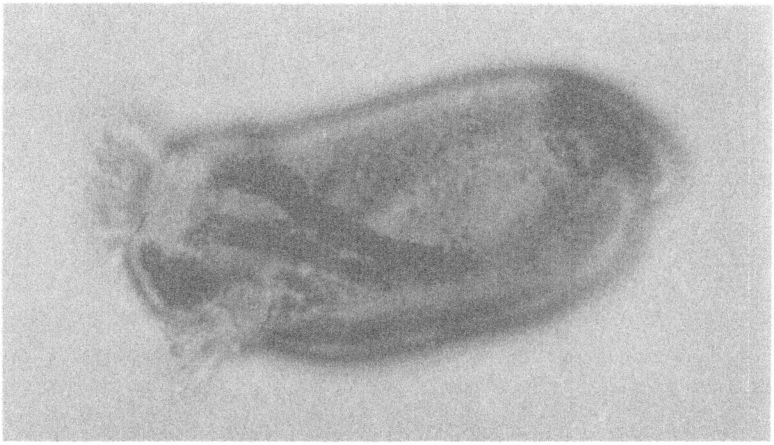

Fig. 2.25 Photomicrograph of *Diploplastron affine* stained to show the two skeletal plates that converge posteriorly.

proct is small with thin walls in contrast to the more complex structures in *Metadinium* spp.

This is another genus which is disputed by some, and the reader will find *Diploplastron affine* referred to as *Metadinium affine* (Hungate, 1978) and *Eudiplodinium affine* (Dehority, 1978, 1986b) *Diploplastron affine* (*Diplodinium affine*—Dogiel and Fedorowa, 1925b; *Eudiplodinium affine*—Dogiel, 1927) is widely distributed in some areas of the world but not others. It is a typical member of the A-type protozoal population. In sheep in Britain it is almost always found wherever *Polyplastron multivesiculatum* is present. The same is true in sheep and goats in Japan, but in Japanese cattle *D. affine* is present in only 40% of animals containing *P. multivesiculatum* (Imai et al., 1979a). Eleven per cent of zebu cattle in Thailand contain *D. affine*, although *P. multivesiculatum* in never present. Dogiel (1927) also found *D. affine* in Russia. Although *D. affine* is normally much smaller than *P. multivesiculatum* (105 × 59 compared with 161 × 95 μm), there is a size overlap. Since the species usually occur together, it can be difficult to distinguish one from the other; they are morphologically similar, both having a large operculum. They can only be separated with certainty for semistarved organisms where the parallel skeletal plates of *P. multivesiculatum* can be distinguished from the converging plates of *D. affine* (Figs. 2.15 and 2.25).

2.14 The Genus *Elytroplastron*

There is a single species, *Elytroplastron bubali* (Kofoid and MacLennan, 1932), which was originally described as *Diplodinium* (*Polyplastron*) *bubali* by

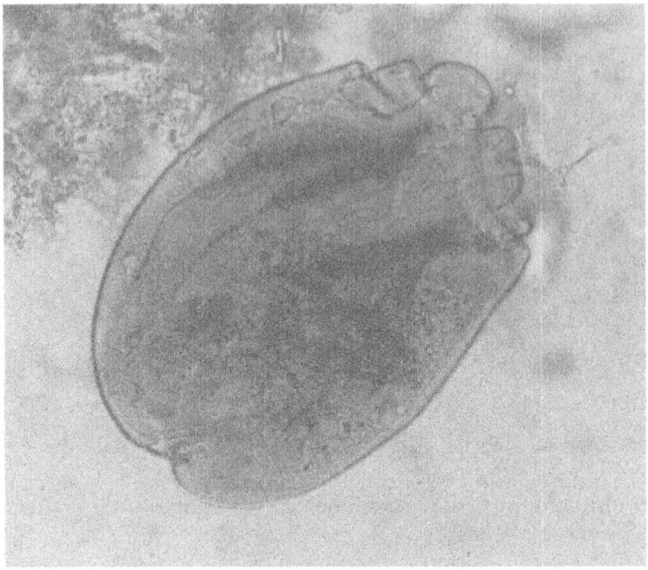

Fig. 2.26 Photomicrograph of *Elytoplastron bubali* stained to show the skeletal plates.

Dogiel (1928). It is characterized by a dorsal zone of cilia at the anterior end of the body, two narrow skeletal plates (Fig. 2.26) beneath the right surface (similar to those in *Polyplastron multivesiculatum* and *Metadinium medium*), and, unusually, a long narrow plate beneath the left surface. There is also a small, inconspicuous, ventral plate at the anterior end. *E. bubali* is smaller than *P. multivesiculatum* and *M. medium* (110–160 μm long compared with 120–190 μm and 180–224 μm respectively) but is most easily separated from these two by the arrangement of the three main skeletal plates. *E. bubali* has a smoothly rounded posterior end and four contractile vacuoles along the dorsal midline. It is found in Japanese cattle but not in Japanese sheep or goats (Imai et al., 1979a), and in cattle but not buffalo from Indonesia (Imai, 1985), Thailand (Imai and Ogimoto, 1984), and Philippines (Shimizu et al., 1983). However, it is found in Indian cows, sheep, goats, and buffalo (Kofoid and MacLennan, 1932; Das Gupta, 1935; Banerjee, 1955), and in Brazilian cattle and buffalo (Dehority, 1979, 1986b). It is also found in the Japanese serow (*Capricornis crispus*—Imai et al., 1981) and in the hindgut of the capybara (*Hydrochoerus hydrochaeris*—Dehority, 1987).

2.15 The Genus *Epiplastron*

This genus was erected by Kofoid and MacLennan (1933) and contains only two species which have only been found in some African antelopes. *Epi-*

plastron spp. have relatively long bodies with length/width ratios of 2.3–2.6. They have a superficial resemblance to *Epidinium* spp. but have five skeletal plates rather than three. Four of these can be fused together posteriorly but there are windows between them anteriorly. The dorsal band of cilia lies slightly behind the anterior end of the body. There are two contractile vacuoles beneath the dorsal surface.

1. *Epiplastron africanum* (*Ophryoscolex ecaudatus africanus*—Dogiel, 1925a; *Epidinium africanum*—Dogiel, 1927) was first found in antelope (*Rhaphiceros* sp.) in East Africa by Dogiel (1925a) but was not found in the tsessebe from the Transvaal (Van Hoven, 1975) or reedbuck species from Kenya or Tanzania (Van Hoven, 1983). *E. africanum* has a smoothly rounded posterior end.

2. *Epiplastron spinosum* (*Epidinium africanum spinosum*—Dogiel, 1932) has only been reported from the impala (*Aepyceros melampus*) from Africa (Dogiel, 1932). It is similar to *E. africanum* differing only in the possession of a long, narrow, caudal spine.

2.16 The Genus *Opisthotricum*

The one species of this genus, *Opisthotricum janus*, has only been found in African antelopes such as *Bubalis cokei*, *Bubalis lichtensteini*, and *Madoqua* sp., and was first described by Buisson (1923a, 1923b). *O. janus* has a single skeletal plate which completely encircles the body at the anterior end but which tapers posteriorly and is present only under the right surface. The dorsal zone of cilia is displaced posteriorly to almost the middle of the body. *O. janus* has an elongated body (length to width ratio 2.2) and a long narrow caudal spine (Dogiel, 1927).

The only study on the numbers of *O. janus* is that of Van Hoven (1978) who found that in the blesbok (*Damaliscus dorcas phillipsi*) the population density of *O. janus* varied from 1.5×10^4/ml in winter to $11.4 \times {}^4$/ml in summer. The pattern followed that of *Entodinium* spp. so it was suggested that *O. janus* might also require starch for growth.

2.17 The Genus *Caloscolex*

This genus was first described by Dogiel (1926b) and is found only in old world camelids. These protozoa have three skeletal plates and a dorsal zone of cilia which is displaced posteriorly about one-quarter of the length of the body. These cilia completely encircle the body in a slight spiral so that the two ends of the ciliary band are not joined together. The various species have from one to seven contractile vacuoles arranged transversely round the middle of the body.

1. *Caloscolex camelinus leavis* is found in the dromedary (*Camelus dromedarius*) and the camel (*Camelus camelinus laevis*) in Turkestan and Taschkent (Dogiel, 1926b; 1927). It has a smooth posterior end and an indeterminate number of contractile vacuoles.

2. *Caloscolex camelinus cuspidatus* is found in *Camelus camelinus laevis*. It has a single, blunt, caudal spine and probably one contractile vacuole.

3. *Caloscolex camelinus bisuspis* is found in *Camelus camelinus cuspidatus*. It has two blunt caudal spines (Dogiel, 1926b; 1927).

4. *Caloscolex camelinus tricuspis* is found in *Camelus camelinus laevis*. It has four short, incurred, blunt caudal spines and seven contractile vacuoles (Dogiel, 1926b; 1927).

5. *Caloscolex camelinus quadricuspis* is found in *Camelus camelinus laevis*. It has four short, blunt caudal spines and seven contractile vacuoles (Dogiel, 1926b; 1927).

6. *Caloscolex camelinus quinquecuspis* is found in *Camelus camelinus laevis*. It has five short, blunt caudal spines and two(?) contractile vacuoles.

2.18 The Genus *Rhinozeta*

Van Hoven et al. (1988) have described seven *Rhinozeta* spp. that they found in the caecum of the white and black rhinoceros. These protozoa belong to the order entodiniomorphida and have three, four, or five ciliary bands on the lateral body surfaces in addition to adoral zone ciliature carried on a rectractable cone. They will not be described in detail as they have never been found in the rumen.

2.19 The Genus *Parentodinium*

Until recently all the ciliates of the order Entodiniomorphia found in the rumen have belonged to the family Ophryoscolecidae. Those of the family Cycloposthiidae most commonly occur in the hindgut of non-ruminant herbivores although Thurston and Noirot-Timothée (1973) described two new genera and three new species from the stomach of the hippopotamus (*Hippopotamus amphibius*). One of these new species was *Parentodinium africanum* which has now been found in the rumen of Brazilian cattle by Dehority (1986b).

P. africanum is oval when seen in lateral view and the only cilia are on an extrudable cone which can be protruded much further than the peristome of *Entodinium* spp. It has no skeletal plates and one contractile vacuole.

2.20 Evolution of entodiniomorphid protozoa

There is good evidence on which to build an evolutionary tree for the development of structurally more complex ruminants from their simpler ancestors. There is, of course, no such evidence available for the rumen ciliates. We only know the structure of organisms living today. Dogiel (1947) argued that, as the entodinia are the simplest entodiniomorphs, they must have arisen first, and as they are found in camelids, they must have been present in the primitive animals living before their differentiation, say, 40 million years ago. Similarly Dogiel argued that, as epidinia and eudiplodinia are found in deer, cattle, and sheep, these species must have arisen 25 million years ago. As ostracodinia are found in cattle but not deer, they must have appeared 10–15 million years ago. As Opisthotricum is found in antelopes but not in the sheep or cattle and, as the reverse is true for Ophryoscolex and Polyplastron, these protozoal species must have developed about 5 million years ago. Finally, as *Ophryoscolex caudatus* occurs only in sheep and goats, whereas *O. purkynjei* is found only in cattle, this suggests that these protozoal species separated 2 million years ago at the end of the Pliocene period. It is of interest that the hippopotamus (*Hippopotamus amphibius*), which has fermentative digestion in its stomach but does not ruminate, contains protozoa of the Entodiniomorphid genus *Parentodinium* but not *Entodinium* spp. As the Ruminantia and Tylopoda (camelids) separated from the other Artiodactyla about 50 million years ago, the entodinia must have arisen after this.

Dogiel (1947) suggested that the primitive hosts first became infected with protozoa by taking up free-living protozoa when drinking and that these would survive stomach acid due to inadequate mixing. Such protozoa could survive by predation upon other protozoa and by engulfing plant material which was comparatively cellulose-free. However, with the change in climate toward cooler and drier conditions during the Oligocene period, and with the appearance of grasses in the early Miocene period, the animals would have to adapt toward foods containing more and more fibrous material (Janis, 1976). Hume and Warner (1980) have suggested that under these conditions the hindgut developed first in response to a need not only to contain the larger quantity of material being eaten but also to enable the host to continue to absorb water, minerals, and nutrients when the rate of passage of material was gradually increasing. They believe that foregut fermentation then developed possibly initially in spontaneously forming diverticula of the stomach. These diverticula gradually developed as separate organelles that did not produce acid and eventually came completely separated from the stomach by the reticulo-omasal orifice. This effectively separated the reticulo-rumen from the lower tract and prevented the passage of large particles into the omasum.

It would be during the slow development of these diverticula and the decrease in acid production that the protozoa taken up by chance would be able to establish themselves. Dogiel (1947) postulated that these protozoa would have resembled the present entodinia and diplodinia and would have lived on starch grains, spores, bacteria, and other protozoa. He believed that, as the protozoa evolved, most lost their ability to engulf other protozoa and developed the ability to engulf and ingest large plant fragments. He further argued that the skeletal plates developed in the larger entodiniomorphs to protect the larger protozoa from internal damage by engulfed spines and cellulose fibers and that the plates were on the right side because the esophagus points in that direction. In the most highly developed species the plate almost completely encircles the esophagus.

Lubinsky (1957a, 1957b) developed this theme for the evolution of the higher ophryoscolecids from the entodinia. Unfortunately this involves a confusing change in the orientation when passing from the entodinia to the more complex protozoa. Traditionally the side nearest the macronucleus has been called dorsal (Sharp, 1914; Dogiel, 1927) in all ciliates with left and right following as in vertebrate anatomy, although it must be recognized that dorsal has nothing to do with orientation with respect to gravity. When swimming the ciliates move with a corkscrew motion. In an examination of the structure of entodinia and diplodinia, Lubinsky (1957b, 1957c) noted the presence in certain species of a lateral groove running from near the cilia to the caudal spine. This is particularly pronounced in *Entodinium simulans* (Lubinsky, 1957a), *E. rectangulatum* (Kofoid and MacLennan, 1930), and in diplodinia and eudiplodinia. Lubinsky noted that in entodinia and diplodinia the contractile vacuole, longitudinal groove, micronucleus, and macronucleus always occurred in the same order from left to right and that the left body side of *Entodinium simulans* resembled the right body side of *Diplodinium dentatum*. It was therefore suggested that the contractile vacuole, the lateral groove, and the nuclei were moved ventrally during the evolution to diplodinia from the entodinia. The anterior end of the groove separated the "dorsal" zone of cilia from the oral zone and finally formed the operculum. The movement of the vacuoles, etc., is progressive from *Entodinium caudatum*, *E. simulans*, *Diplodinium dentatum*, and *Eudiplodinium maggii* until in *Metadinium* spp. the groove disappears and the micronucleus is compressed between the vacuoles and the macronucleus. The skeletal plates form in the "tract" of the macronucleus. The reasons for these torsional changes is not clear, although Lubinsky (1957b) suggested that it was related to the clockwise swimming motion of the protozoa. As the displacements are anticlockwise they can be described as lagging behind the rest of the body.

In descriptions of the morphology of ciliates, Lubinsky (1958a, 1958b) proposed that the side closest to the macronucleus should be called right in entodinia and left in the more complex ciliates, using the terms upper and lower instead of dorsal and ventral. Readers of the literature should beware as Hungate (1978) followed the old system and Ogimoto and Imai (1981) followed the Lubinsky system.

The present authors find it difficult to accept that while in the rumen the simple entodinia developed a second ciliary zone when the need to move must be less than in fresh water. There is no evidence that the skeletal plate has a protective function, although it may have a structural one (see Section 7.1). It is not clear what advantage there is in the storage of starch in a plate compared with random groups of granules. While the appearance of the larger cellulolytic ciliates may have coincided with the increased uptake of more fibrous material by the host, there is no evidence that the larger ciliates evolved from primitive entodinia by the acquirement of cellulase genes from, for example, bacteria. If the larger entodiniomorphs have evolved in this way from entodinia, then it is surprising that they became less well adapted to their environment in the process. In the event of any physiological trauma, the larger protozoa always disappear before the entodinia and reappear later when rumen conditions return to normal. The larger species can also disappear for no apparent reason from an animal kept on a constant ration under constant conditions; this does not occur with entodinia. Dogiel (1947) believed that the presence of a maximum array of caudal spines was the final stage in the evolutionary process, but there is no evidence except with *Entodinium caudatum* (Coleman et al., 1977) that the presence of caudal spines is any advantage. *E. caudatum* only develops nutritionally expensive spines to protect itself from predation by *E. bursa*. In many years of growing pure and mixed cultures of entodiniomorphs in vitro the authors have obtained no evidence for any other advantages of caudal spination. It is therefore possible that instead of primitive entodinia acquiring a second ciliary zone, skeletal plates, larger size, and cellulase genes which apparently make them less well adapted to their environment, these protozoa may have acquired these characteristics while still free-living and have colonized the rumen later than the entodinia. Such protozoa are therefore still adapting to their new warm anaerobic environment which is very rich in readily accessible food.

3

The Holotrich Ciliates

Holotrich ciliate protozoa have been observed frequently in samples obtained from the principal fermentative regions in the digestive tracts of herbivorous mammals. The descriptive term "holotrich" initially arose from the terminology used in the original major systems of protozoal classification. Prior to the taxonomic revisions of Levine et al. (1980), the families of holotrich ciliates were classified in two orders of the subclass Holotrichia; the entodiniomorphid (oligotrich) ciliates were classified in the order Entodiniomorphida (Corliss, 1979). In subsequent revisions the order Entodiniomorphida has been retained while the subclass Holotrichia has been superceded (see Section 3.1), although the term holotrich is still widely used for collective description of fifteen genera of ciliates that are classified within two separate taxonomic orders. The classification of the rumen holotrich ciliates is summarized in the following section; subsequent sections discuss their occurrence and morphological characteristics. The term holotrich has been retained here because of its current widespread usage in the literature; the limitations of its use with this diverse group of ciliates should, however, be recognized.

3.1 Classification of the Rumen Holotrich Ciliates

The rumen ciliates that are currently described as holotrichs occur within one of the four families Isotrichidae, Blepharocorythidae, Buetschliidae, or Paraisotrichidae. The classification schemes published in earlier reviews on the rumen ciliate protozoa reflect the taxonomic status of the period.

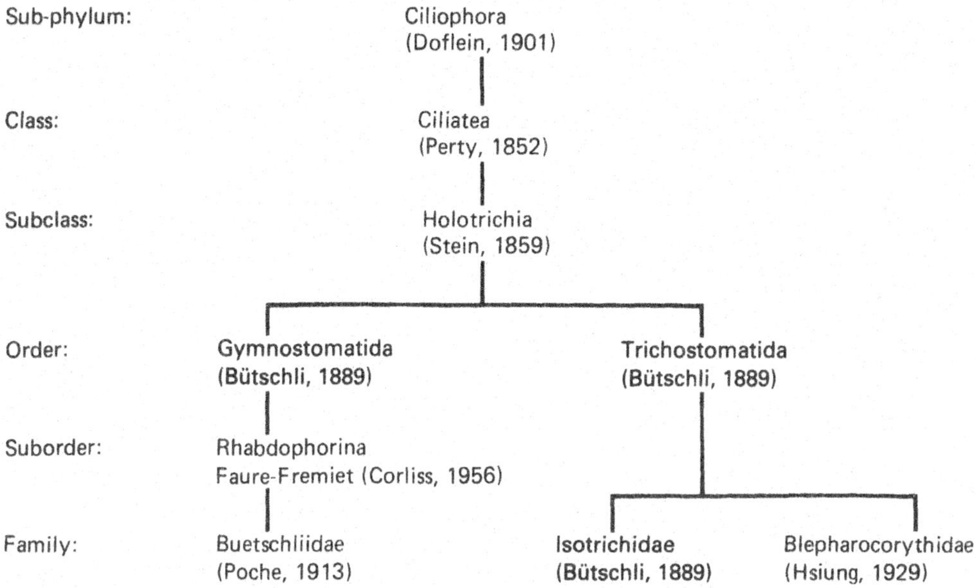

Fig. 3.1 Protozoal classification based on Honigberg et al. (1964).

However, taxonomic revisions have occurred in the intervening periods and consequently the classification of the holotrichs in the literature appears inconsistent as different schemes have been published. The taxonomic classification of Honigberg et al. (1964) was adopted by reviewers in the pre-1980 period (Clarke, 1977; Hungate, 1966, 1978). The holotrich genera were, therefore, placed in the two orders Gymnostomatida and Trichostomatida of the subclass Holotrichia (Fig. 3.1). The classification of the protozoa was subsequently revised by Levine et al. (1980) and this format was used by Ogimoto and Imai (1981). In this revision the rumen holotrichs were placed in two subclasses, the Vestibuliferia and Gymnostomatia; the family Buetschliidae was located in the order Gymnostomatida, while the Isotrichidae and Blepharocorythidae remained within the Trichostomatida (Fig. 3.2). In a more recent reappraisal (Lee et al., 1985) the Buetschliidae and Blepharocorythidae have been placed within the order Entodiniomorphida. The Isotrichidae belong to the Vestibuliferida; both orders are in the subclass Trichostomatia (Fig. 3.3). These various classifications are summarized in Table 3.1.

3.2 Distribution of Holotrich Ciliates

Holotrich ciliates occur widely in the reticulorumen of both domesticated and wild ruminants. Protozoa from the holotrich families are also found at the active fermentation sites in the foregut and hindgut of other nonrumi-

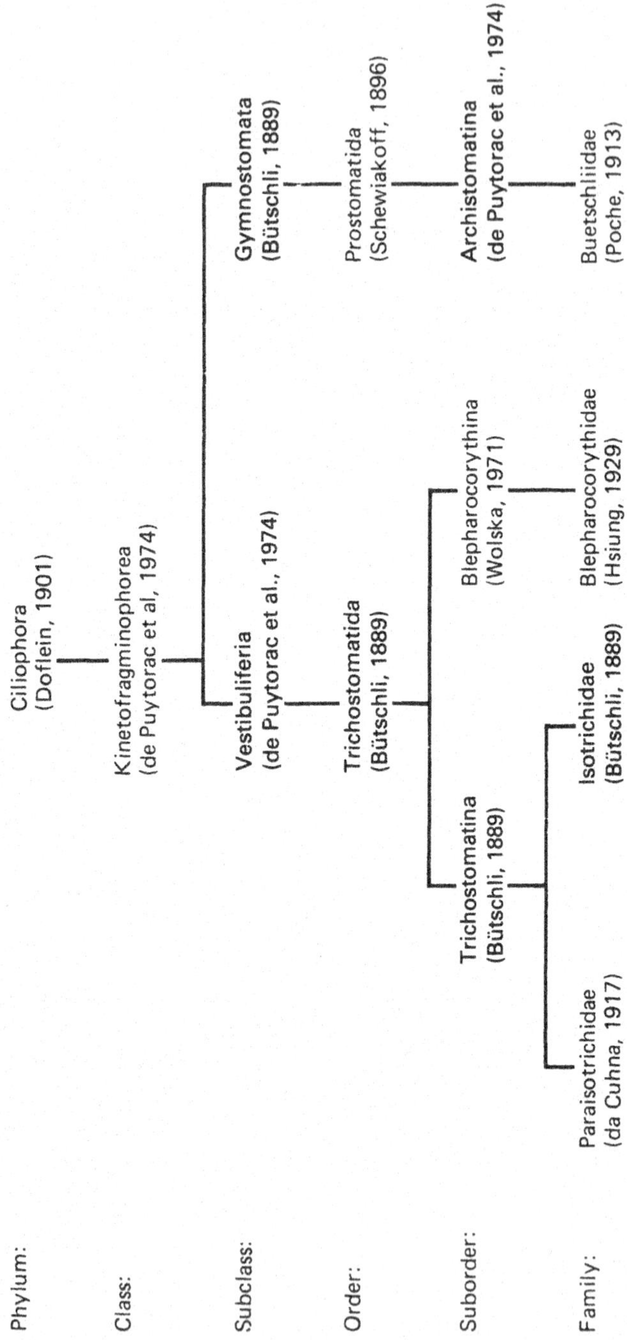

Fig. 3.2 Protozoal classification based on Levine et al. (1980).

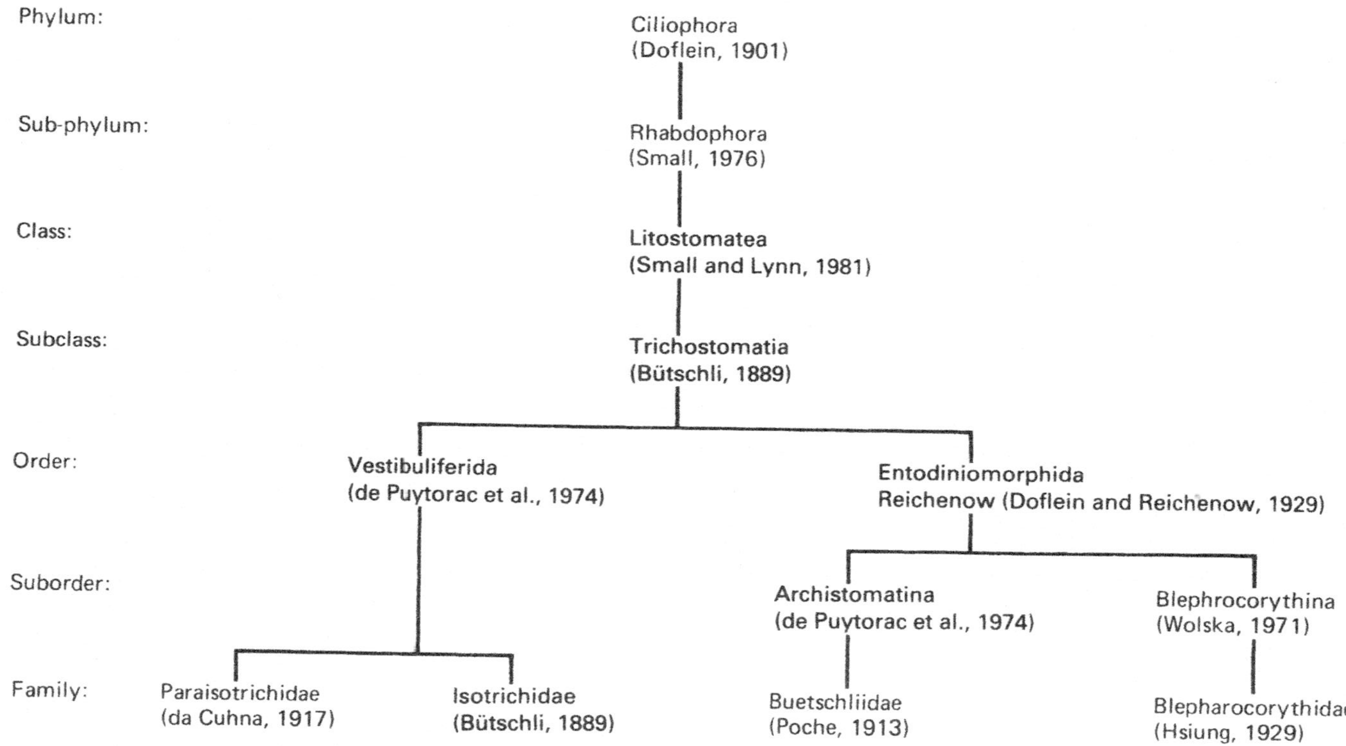

Fig. 3.3 Protozoal classification based on Lee et al. (1985).

Table 3.1 Classification of the rumen holotrich ciliates

Scheme	Phylum	Subphylum	Class	Subclass	Order	Suborder	Family
1[a]	Ciliophora		Ciliatea	Holotrichia	Trichostomatida		Isotrichidae Paraisotrichidae
2[b]	Ciliophora		Kinetofragminophorea	Vestibuliferia	Trichostomatida	Trichostomatina	
3[c]	Ciliophora	Rhabdophora	Litostomatea	Trichostomatia	Vestibuliferida		
1	Ciliophora		Ciliatea	Holotrichia	Trichostomatida		Blepharocorythidae
2	Ciliophora		Kinetofragminophorea	Vestibuliferia	Trichostomatida	Blepharocorythina	
3	Ciliophora	Rhabdophora	Litostomatea	Trichostomatia	Entodiniomorphida	Blepharocorythina	
1	Ciliophora		Ciliatea	Holotrichia	Gymnostomatida	Rhabdophorina	Buetschliidae
2	Ciliophora		Kinetofragminophorea	Gymnostomatia	Prostomatida	Archistomatina	
3	Ciliophora	Rhabdophora	Litostomatea	Trichostomatia	Entodiniomorphida	Archistomatina	

[a] From Honiberg et al., 1964.
[b] From Levine et al., 1980.
[c] From Lee et al., 1985.

nant mammalian herbivores (Dehority, 1986a). Hindgut fermenters of the Equidae and Tapiridae (order Perissodactyla), the Elephantidae (order Proboscida), and the Caviidae and Hydrochocridae (order Rodentia) possess a holotrich-containing protozoal population (Buisson, 1923c; Cunha and Muniz, 1928; Dehority, 1986a; Eloff and Van Hoven, 1980; Hsiung, 1930a, 1930b; Kopperi, 1935; Latteur, 1967; Wolska, 1967b, 1968). Some nonruminant members of the Camelidae and Hippopotimidae (order Artiodactyla) have an active foregut fermentation; ciliate protozoa including certain members of the holotrich families are present in these specifically adapted pre-gastric regions of the digestive tract (Buisson, 1923c; Dogiel, 1926b, 1928; Thurston and Grain, 1971).

It is not unusual for an individual protozoal genus to be host-specific, and many of the genera and individual species of rumen holotrichs only occur in ruminants. There are of course exceptions; *Dasytricha ruminantium, Isotricha prostoma, I. intestinalis,* and *Buetschlia parva* have been reported in some camelids (Buisson, 1923c), although Dehority (1986a) did not detect members of the Isotrichidae in the stomach contents of the New World camelids examined in his studies. Protozoa similar in appearance to *Dasytricha* and *Isotricha* spp. have been observed in the forestomach of several species of macropodid marsupials (Dellow et al., 1988; Obendorf, 1984). In addition, *Charonina equi* has been observed in the bovine rumen (Clarke, 1964) and horse large intestine (Hsiung, 1930b), while *Blepharoconus krugerensis* has been found in the rumen of Brazilian cattle (Dehority, 1986b) and the large intestines of the elephant (Eloff and Van Hoven, 1980). However, attempts to transfer protozoa from the horse colon to the caprine rumen were not successful (Becker and Hsiung, 1929; Dogiel and Winogradowa-Fedorowa, 1930).

Holotrich ciliates are known to occur in hosts from the five families in the suborder Ruminantia (order Artiodactyla), including the Giraffidae (Kleynhans and van Hoven, 1976; Noirot-Timothée, 1963) and the most primitive of the families, the Tragulidae (Jameson, 1925b), which differ from other ruminants in that their stomach comprises three, not four, compartments. The species of holotrich ciliates and the various rumen ecosystems they inhabit are given in Table 3.2a, 3.2b, and 3.2c. It is apparent that there are pronounced variations in the occurrence of the individual holotrich species in the various host ruminants examined. The families Buetschliidae (Table 3.2b), Blepharocorythidae (Table 3.2c), and Paraisotrichidae occur infrequently while the Isotrichidae (Table 3.2a) are more widespread and can represent up to 40% of the total ciliate population (Clarke, 1964). Dehority (1986a) collated the data from 17 surveys of rumen ciliate populations of both domesticated and feral animals and concluded that the Isotrichidae usually constituted less than 5% of the total ciliates. There are reports of the occurrence of a single genus or the absence of all holotrichs in a particular host species; some of these studies are included in Table 3.3. However, many of these surveys used a limited number of animals and further detailed surveys of these hosts may reveal a wider range of protozoal types.

Table 3.2a References to the occurrence of Isotrichidae in various host animals[a]

Host animal	Reference number			
	Isotricha prostoma	*Isotricha intestinalis*	*Dasytricha ruminantium*	*Oligoisotricha bubali*
Domesticated sheep (*Ovis aries*)	4, 25, 27, 38, 62	4, 25, 38, 62	1, 4, 20, 25, 27, 38, 62	
North Ronaldshay sheep (*Ovis aries*)	22		22	
Nevada big horn (*Ovis canadensis*)	5		5	
Domesticated goat (*Capra hircus*)	2, 4, 7, 38	4, 38	2, 7, 38	
Domesticated cattle (*Bos taurus*)[b]	4, 6, 19, 20, 25, 26, 38, 42	4, 6, 20, 25, 26, 38, 42	4, 6, 19, 20, 25, 26, 38, 42, 56, 57	14, 42, 52
Bali cattle (*Bos javanicus domesticus*)	33	33	33	33
Zebu cattle (*Bos indicus*)	2, 28, 34, 35, 40, 48	28, 34, 35, 40	2, 28, 34, 35, 40, 48	34, 35, 40
Water buffalo (*Bubalus bubalis*)	11, 18, 37, 50	11, 18, 37, 50	11, 18, 24, 33, 37, 50	18, 31, 33, 37, 40, 41
Buffalo (*Bison bison bison*)	25, 53	25	25, 53, 57, 58	
Musk-ox (*Ovibos moschatus*)			9, 13	
Gaur ox (*Bos gaurus*)	2, 47	2, 47	2, 47	
Yak (*Poephagus poephagus*)	20			

[a]References are as follows: 1, Asada et al. (1980); 2, Bhatia (1936); 3, Brüggeman et al. (1967); 4, Buisson (1923c); 5, Bush and Kofoid (1948); 6, Clarke (1964); 7, Das Gupta (1935); 8, Dehority (1970); 9, Dehority (1974); 10, Dehority (1975a); 11, Dehority (1979); 12, Dehority (1986b); 13, Dehority (1986c); 14, Dehority et al. (1983); 15, Dehority and Mattos (1978); 16, Dogiel (1926a); 17, Dogiel (1926b); 18, Dogiel (1928); 19, Dogiel (1932); 20, Dogiel (1934); 21, Dogiel (1935); 22, Eadie (1957); 23, Fiorentini (1890); 24, Fujita et al. (1979); 25, Giesecke (1970); 26, Han (1984); 27, Hsiung (1931); 28, Hsiung (1932); 29, Hukui and Nisida (1954); 30, Hungate (1943); 31, Imai (1981); 32, Imai (1984); 33, Imai (1985); 34, Imai (1986); 35, Imai (1988); 36, Imai et al. (1981a); 37, Imai et al. (1981b); 38, Imai et al. (1978);

3.3 Population Determinants

The generic composition and ultimate size of the holotrich population is influenced by several interacting factors, the more important of which are the type of host, its geographical location, the nature of the diet consumed, and the frequency of feeding. Seasonal variations in the availability and the quality of the feed have more impact on the wild ruminants because the feed available to domesticated animals is more rigidly controlled. Surveys

Table 3.2a *Continued*

Host animal	Reference number			
	Isotricha prostoma	*Isotricha intestinalis*	*Dasytricha ruminantium*	*Oligoisotricha bubali*
Mouse deer (*Tragulus meminna*)		2, 44		
Japanese serow (*Capricornus crispus*)			36	
Reindeer (*Rangifer tarandus*)	4, 10	4, 10	4, 10, 13, 21, 49, 61	
Caribou (*Rangifer tarandus*)			13	
Red deer (*Cervus elaphus*)	3, 25	3, 25	3, 25	
Mountain reedbuck (*Redunca fulvorufulla fulvorufulla*)	59		59	
Sable antelope (*Hippotragus niger*)	60		60	
African antelope spp.			19	
Giraffe (*Giraffa camelopardalis*)	46, 51	46, 51	46, 51	
Camel (*Camelus bactrianus*)	4	4	4	
Dromedary (*Camelus dromedarius*)	17			
Llama (*Auchenia lama*)	4	4	4	

39, Imai and Ogimoto (1983); 40, Imai and Ogimoto (1984); 41, Imai et al. (1981c); 42, Imai et al. (1982); 43, Jameson (1925a); 44, Jameson (1925b); 45, Jirovec (1933); 46, Kleynhans and Van Hoven (1976); 47, Kofoid and Christenson (1933); 48, Kofoid and MacLennan (1933); 49, Kopperi (1926); 50, Michalowski (1975a); 51, Noirot-Timothée (1963); 52, Orpin and Mathiesen (1986); 53, Pearson (1967); 54, Schuberg (1888); 55, Syrjala et al. (1976); 56, Towne and Nagaraja (1989); 57, Towne et al. (1988a); 58, Towne et al. (1988b); 59, Van Hoven (1983); 60, Van Hoven et al. (1979); 61, Westerling (1970); 62, Williams, A.G., unpublished observat; 63, Wolska (1967a).

[b] *D. hukuokaensis* has also been reported in cattle: ref. no. 29.

on the occurrence of holotrichs in various hosts indicate that the genera *Dasytricha* and *Isotricha* occur most frequently, and when both genera are present, *D. ruminantium* is frequently more numerous than *Isotricha* spp. Holotrichs have been observed more regularly in domesticated animals than in wild ruminants, and their numbers in the rumen are increased when the diet contains a readily available source of soluble carbohydrates such as fresh temperate grasses (Clarke, 1965a) or sugar cane (Valdez et al., 1977). The number of holotrichs typically present in the rumen of domesticated animals ranges up to 10^5/ml of rumen fluid and on forage diets the

Table 3.2b References to the occurrence of Buetschliidae in various host animals[a]

Host animal	Reference number											
	B. parva	*B. neglecta*	*B. lanceolata*	*B. omnivora*	*B. nana*	*Hsiungia* (*syn. Buetschlia triciliata*)	*Parabundleia ruminantium*	*Polymorphella bovis*	*Blepharoconus krugerensis*	*Blepharoprosthium parvum*	*Pingius minutus*	*Holophryozoon bovis*
Domesticated sheep (*Ovis aries*)	4, 8, 30											
Domesticated goat (*Capra hircus*)	4											
Domesticated cattle (*Bos taurus*)	4, 6, 19, 54	54	23									45
Zebu cattle (*Bos indicus*)	48					28, 35	34, 39, 40	32, 34, 35, 40	12	28	28	
Water buffalo (*Bubalus bubalis*)	11											
Musk-ox (*Ovibos moschatus*)	9, 13											
Reindeer (*Rangifer tarandus*)	4, 13, 21											
African antelope spp.	19											
Camel (*Camelus bactrianus*)	4			18	18							
Dromedary (*Camelus dromedarius*)	17, 18											
Llama (*Auchenia lama*)	4											

[a] The references are listed in footnote [a] of Table 3.2a.

Table 3.2c References to the occurrence of the Blepharocorythidae, Paraisotrichidae, and *Microcetus lappus* in various host animals[a]

Host animal	Reference number				
	Charonina ventriculi	*C. equi*	*C. nuda*	*Paraisotricha sp.*	*Microcetus lappus*
Domesticated sheep (*Ovis aries*)	27, 38, 43, 55, 62				
Domesticated cattle (*Bos taurus*)	15, 16, 19, 20, 38, 42, 43, 52, 56, 63	6, 38			52, 56
Bali cattle (*Bos javanicus domesticus*)	33				
Zebu cattle (*Bos indicus*)	2, 28, 34, 35, 40, 48		28	12	
Water buffalo (*Bubalus bubalis*)	11, 18, 24, 33, 37				
Buffalo (*Bison bison bison*)	58				56, 58
Giraffe (*Giraffa camelopardalis*)				46	

[a] The references are listed in footnote [a] of Table 3.2a.

Table 3.3 References to protozoal population surveys of various ruminant hosts containing no holotrich ciliates

Host	Reference
Dall Mountain sheep	Dehority, 1974
Ibex	Giesecke, 1970
Water buffalo	Ogimoto et al., 1983
Musk-ox	Dehority, 1985, 1986c
Caribou	Dehority, 1975a, 1986c
Moose	Dehority, 1974; Giesecke, 1970
Reindeer	Giesecke, 1970
Red deer	Clarke, 1968
Fallow deer	Giesecke, 1970; Prins and Geelen, 1971
Roe deer	Giesecke, 1970; Prins and Geelen, 1971
White-tailed deer	Giesecke, 1970; Pearson, 1965; Wilkinson and Van Hoven, 1976a, 1976b
Barking deer	Banerjee, 1955
Mule deer	Giesecke, 1970; Maclennan, 1935
Chamois	Giesecke, 1970; Wertheim, 1934
Antelope	Dogiel, 1932
Bohor reedbuck	Van Hoven, 1983
Southern reedbuck	Van Hoven, 1983
Springbok	Wilkinson and Van Hoven, 1976a, 1976b
Blesbok	Van Hoven, 1978
Sambar	Banerjee, 1955
Tsessebe	Van Hoven, 1975
Guanaco	Lubinsky, 1964

holotrichs represent some 20% (12 to 40%) of the total ciliate population (Clarke, 1964).

The holotrich population is thus influenced by the diet and feeding practices of the animal. In addition, changes in the composition or physical characteristics of the diet and the level and frequency of feeding produce changes in the rumen ciliate population. Different genera are encouraged by different feeds. These effects will be discussed in Chapter 10 and have been reviewed in detail elsewhere (Clarke, 1977; Hungate, 1966; Warner, 1965). Although the number of holotrich protozoa is generally increased in animals receiving diets high in available carbohydrates, there are reports of the holotrich population being increased by dietary constraint (Dehority and Purser, 1970) and in some seaweed-eating sheep the proportion of *D. ruminantium* in the ciliate population was 36.5% (Orpin et al., 1985). Fluctuation in holotrich numbers did not apparently correlate with rumen fluid osmolality (Dehority and Males, 1974), and the pattern of variation of holotrich numbers in the rumen does not necessarily parallel the changes in the entodiniomorphid population; the proportion of holotrichs can increase under certain conditions in a declining protozoal population (Eadie et al., 1970).

Host Specificity In addition to these dietary effects, studies on the holotrich populations of various hosts have indicated that variations in the geographical distribution can also occur. Such effects are exemplified by surveys conducted on musk-oxen in Alaska and Canada (Dehority, 1974; 1986c) and water buffalo (*Bubalus bubalis*) from diverse locations worldwide including Brazil, Taiwan, Japan, and Korea (Dehority, 1986b; Han, 1984; Imai et al., 1981b, 1981c, 1982). Holotrichs, however, were not present in the animals examined in west Asia (Ogimoto et al., 1983). It is likely that these variations are related to the diet and the extent of contact with other ruminants. Animal isolation and dietary effects are important determinants for arctic ruminants (Dehority, 1986c). In Europe the variety of ciliates is larger in grazing than browsing animals and the holotrich species predominate in grazers (Giesecke, 1970). However, in Africa *Dasytricha* and *Isotricha* spp. occur more frequently in browsers (Van Hoven, 1983). This effect is not apparent in arctic ruminants (Dehority, 1974).

It is apparent from Table 3.2 that many of the rumen holotrichs have only been observed in a limited range of ruminant hosts and the question of host or protozoal specificity has been considered to explain the disparity of protozoal distribution between ruminant hosts (Dehority, 1974, 1978, 1986a, 1986c). In the wider context the specificity of protozoal genera or species for a particular host and site in the intestinal tract remains unresolved. Although early attempts to transfer ciliates from the horse colon to goat rumen were unsuccessful (Becker and Hsiung, 1929; Dogiel and Winogradowa-Fedorowa, 1930), ciliate species of domesticated ruminants appear to be less specific (Becker, 1932; Dehority, 1978; Naga et al. 1969) and may be transferred between hosts. *Dastytricha* and *Isotricha* spp. did not exhibit host specificity in either Alaskan ruminants (Dehority, 1975a) or domesticated hosts (Dehority, 1978). Conversely however, *D. ruminantium* did not survive cross-inoculation between cow and buffalo calves (Naga et al., 1969) and has failed to establish following transfaunation between sheep within the same flock (Dehority and Purser, 1970). It was confirmed that the establishment of these Isotrichidae was diet-related; following transfaunation the environment was modified in a manner which facilitated their retention. Further population and cross-inoculation studies are necessary before any definitive conclusions may be made regarding the host specificity of the rumen holotrich ciliates.

Diurnal Cycle The number of holotrichs in the rumen is not constant throughout the day and pronounced changes in the population size occur in the pre-feed and postprandial periods. The diurnal variation in the number of holotrich and entodiniomorphid ciliates in the rumen is different. The entodiniomorphid population decreases for some 16 hours after feeding before increasing to the pre-feed level (Warner, 1962b, 1966a). The numbers of *D. ruminantium* and the *Isotricha* spp., however, begin to increase 0.5–4 hours before feeding and subsequently decrease when feeding has ceased, although the point at which the population increase and decline begin is not always the same in the surveys undertaken. In some studies using animals fed once daily, the holotrich numbers increased prior to feeding and either began to fall as soon as the animals had been fed (Michalowski, 1975a, 1977; Michalowski and Muszynski, 1978; Warner, 1962a, 1962b, 1966a) or began to decline soon thereafter without increasing further (Purser, 1961). The numbers increased markedly, however, in the two hours immediately prior to feeding, and dividing forms were evident. However, in other studies this pre-feed increase continued during the feeding period so that the population was highest some 1 or 2 hours after the feed had been consumed (Dehority, 1970; Dehority and Mattos, 1978; Dehority and Tirabasso, 1989; Dennis et al., 1983). With cattle fed red clover (Clarke, 1964, 1965a) or sugar cane (Valdez et al., 1977) the number of *Dasytricha* and *Isotricha* spp. increased approximately threefold in the 2 hour period after feeding. The numbers of *D. ruminantium* in bison increased after feeding and were highest in the 4 hour post-feed sample (Towne and Nagaraja, 1989; Towne et al., 1988a). The pronounced increase in numbers prior to feeding was not apparent in the populations of some animals monitored (Abe et al., 1981; Clarke, 1965a; Murphy et al., 1985). In these studies the numbers of holotrichs increased approximately fourfold within 1 hour of feed consumption and then declined rapidly. The timing of the decline in the *Dasytricha* population differed however in cattle and bison receiving the same ration (Towne et al., 1988a). These three distinctive variations in the number of holotrichs in the rumen fluid are summarized in Fig. 3.4. It would appear that *Buetschlia parva*, *D. ruminantium*, and *Isotricha* spp. have a

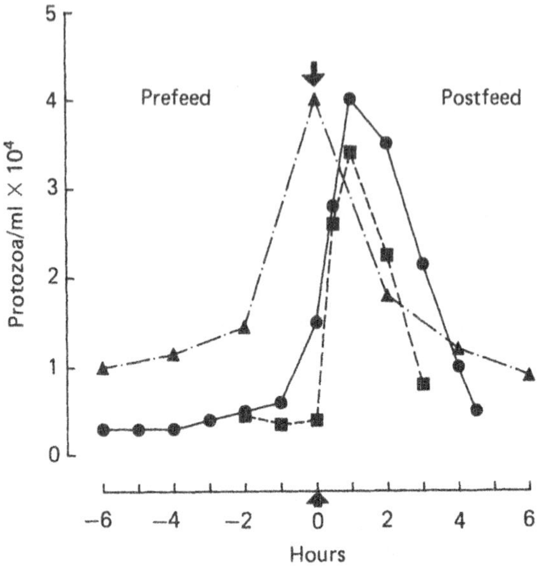

Fig. 3.4 Variation in the numbers of holotrich ciliates present in rumen contents in the period prior to feeding and during the postfeed period. The arrows indicate the time at which the animals were fed. Data from Abe et al. (1983) (■), Michalowski and Muszynski (1978) (▲), and Murphy et al. (1985) (●).

similar diurnal distribution pattern (Dehority, 1970), whereas the variation in the number of *Charonina ventriculi* and *Microcetus lappus* in the rumen during the diurnal cycle is different and resembles that of the entodiniomorphid ciliates (Dehority and Mattos, 1978; Towne and Nagaraja, 1989). In animals fed more than once daily the protozoal population increases (Abe et al., 1981; Warner, 1966b) and the diurnal variation typical of animals fed once daily is repeated in a shorter time period at each feeding (Dehority and Tirabasso, 1989; Michalowski and Muszynski, 1978; Warner, 1966b). The diurnal variation in grazing animals is similar to that of animals fed a restricted diet once daily (Warner, 1966c).

The marked decrease in holotrich numbers after feeding has been attributed to postfeed increases in rumen outflow rates (Michalowski and Muszynski, 1978; Warner, 1966a), protozoal settlement in the rumen (Dehority and Tirabasso, 1989; Minor et al., 1977; Valdez et al., 1977), disintegration as a consequence of an over-accumulation of storage polysaccharide (Clarke, 1965a, 1965b), and sequestration on food particles (Orpin, 1985; Orpin and Letcher, 1978) or the reticulum wall (Abe et al., 1981). The subsequent migration of the holotrichs into the rumen is in response to chemical and physical stimuli. The contractions of the reticulum wall in anticipation of and during feeding and the presence of glucose in the rumen immediately after feeding will stimulate the migration of the holotrich protozoa into the rumen (Abe et al., 1983; Murphy et al., 1985). Dehority and Tirabasso

(1989) proposed that the Isotrichidae sequestered by settling in the ventral rumen and that the migration was in response to storage polysaccharide depletion from the cell. In animals fed once daily, chemotactic migration occurred 12 h after feeding whereas after 22–24 h more extensive polymer utilization invoked a depletion response and migration into the rumen. Abe and Iriki (1989) determined holotrich numbers in the cranial, ventral, and caudal regions of the reticulorumen in the post-feed period. They detected differences in dilution rates, volatile fatty acids (VFA), and reducing sugar concentrations at these sites and concluded that migration between the sites could be correlated with reducing sugar concentrations.

Although the mechanism of sequestration is unresolved, it is apparent that the holotrich ciliates are extensively retained within the rumen ecosystem and do not pass out of the rumen with the digesta (Ankrah et al., 1990; Leng et al., 1986). The holotrich genera present and the population size are thus influenced by many varied factors. Fluctuations in the holotrich population are normally monitored through changes in protozoal concentration. It should be borne in mind that when the isotrichid holotrichs represent only 5% of the ciliate community they would still account for some 35% of the total protozoal volume or 40% of the protozoal nitrogen (Abe et al., 1981; Holler and Harmeyer, 1964), although the contribution made to the total protozoal biomass by the smaller holotrich genera (such as *Microcetus lappus*) will be less significant. Any assessment of their role in the rumen ecosystem should consider the contribution in terms of both cell size and numbers present (Clarke et al., 1982; Dehority et al., 1983). However, determinations of the holotrich population and biomass will be complicated because the cell size is not constant throughout the diurnal cycle (e.g. Michalowski, 1977) and the distribution of the cells in the rumen is uneven due to settlement (Dehority and Tirabasso, 1989) and sequestration to both plant material (Orpin, 1985) and the rumen wall (Abe et al., 1981). The sequestration is almost complete 2–3 hours after feeding and the numbers of *Dasytricha* and *Isotricha* are maximal at, or shortly after, feeding time. When the ration is fed once daily, the total number of Isotrichidae is best determined in unfed animals approximately 3 hours after the usual feeding time (Dehority and Tirabasso, 1989).

3.4 Species Description

The fifteen genera of rumen ciliates generally regarded as being holotrichs are members of the Isotrichidae, Buetschliidae, Blepharocorythidae, or Paraisotrichidae. The taxonomic position of one genus (*Microcetus*) has yet to be established although it does have some features typical of other holotrich genera. The purpose of this section is to outline the principal morphological and ultrastructural characteristics of these genera. Many of the features that are of value for identification purposes are summarized in Table 3.4. Some

Table 3.4 Characteristics of the species of the rumen holotrich ciliates

Species	Morphology	Size range, average (μm)	Length : width range; average	Somatic ciliature
Isotricha prostoma	Elongated ovoid to elipsoidal	80–200 × 50–120; 135 × 70	1.69–2.55; 2.03	Complete; rows parallel to body axis
Isotricha intestinalis	Elongated ovoid to ellipsoidal	90–200 × 45–150; 110 × 60	1.65–1.93; 1.76	Complete; rows parallel to body axis
Dasytricha ruminantium	Ovoid	35–75 × 20–40; 57 × 27	1.70–2.70; 2.11	Complete; rows oblique to body axis
Dasytricha hukuokaensis	Ovoid	120–180 × 68–122; 151 × 95	1.47–1.76; 1.59	Complete
Oligoisotricha bubali	Ovoid	12–20 × 8–15; 16 × 12	1.07–1.60; 1.30	80% of surface; parallel to body axis
Buetschlia parva	Ovoid	30–67 × 20–48; 55 × 35	1.58–2.38; 1.91	Complete; spiral around body
Buetschlia neglecta	Ovoid	40–60 × 20–30	2.0	As *B. parva* with zones at anterior and posterior
Buetschlia lanceolata	Spear-shaped (lanceolate)	48 × 20	2.4	Complex; zone on anterior constriction
Buetschlia omnivora	Oval/spherical	Variable; 35–110 × 27–97		Complete
Buetschlia nana	Ovoid	17–21 × 12–17; 19 × 15		Well developed at anterior
Buetschlia (syn. *Hsiungia*) *triciliata*	Ovoid/ellipsoidal	40–70 × 33–52; 50 × 45.5	1.4–1.6; 1.48	Incomplete; restricted to five specific zones
Parabundleia ruminantium	Ovoid	37.5–50 × 27.5–32.5; 42.5 × 30.5	1.25–1.54 1.39	Two anterior and one posterior zones
Polymorphella bovis	Ovoid to bottle-like	26–37.5 × 20–26; 34 × 22	1.30–1.80 1.56	Anterior zone, posterior tuft
Pingius minutus	Elliptical	22–28 × 13–18; 24.5 × 15		Complete, except at cytoproct
Blepharoprosthium parvum	Pyriform	26–32 × 16–20; 29 × 18		Zones at anterior, middle, and cytoproct
Blepharoconus krugerensis	Ovoid with anterior knoblike protuberence	30–65 × 21–60; 46 × 35	1.10–1.80; 1.34	Incomplete; at anterior and other zones

Macronucleus	Contractile vacuole	Concretion vacuole	Comments	Figure number
Elongated	6–12	–	Posterior vestibulum	3.6 3.9
Ovoid, 30 × 20	6–12	–	Lateral vestibulum	3.7 3.9
Elongated/ ellipsoidal, 16–18 × 8–9	1	–	Posterior vestibulum; no karyophore	3.8 3.9
Ellipsoidal, 24–38 × 16–20 31 × 18	1–11	–	Lateral vestibulum; no karyophore	3.9
Spherical–elliptical	1	–	No karyophore	3.9 3.10
Spherical	1	+	Anterior end truncated	3.14 3.15
Spherical	Several	+	As *B. parva* with posterior indentations	3.15
Large	+	+ (posterior)	Anterior constriction in form of collar around cell	3.15
Elongated	12	+	Host species camel	3.15
Spherical	1	+	Host species camel	3.15
Disc-shaped	1	+		3.14 3.15
Elliptical, 16 μm long	1	+		3.14 3.16
Subspherical, 2.5 μm long	1	+	Anterior of cell tapered	3.14 3.16
Ovoid	1	Not reported	Anterior and posterior end truncated	3.16
Spherical	1	Not visible	Anterior of body constricted	3.16
Disc-shaped 7–15 × 4–8; 11 × 5.5 μm	2–3	+		3.14 3.16

Table 3.4 *Continued.*

Species	Morphology	Size range, average (μm)	Length : width range; average	Somatic ciliature
Holophryozoon bovis	Ovoid	13–18 × 10–12		Complete, but sparse
Buetschliella bovis	Pear-shaped, tapered anterior	30 × 15–20		Restricted to anterior
Charonina ventriculi	Elongate, one side more convex	24–41 × 12–19; 35 × 16.5	1.92–2.42; 2.16	Anterior zone, two posterior tufts
Charonina equi	Lancet-like (with a pronounced convex side)	30–48 × 10–18; 39 × 12	2.41–2.74; 2.59	Anterior zone, two posterior tufts
Charonina nuda	Lancet-like	35–45 × 15–19; 40.5 × 16		Anterior zone only
Paraisotricha sp.	Ellipsoidal	14–23 × 9–14; 19 × 12	1.45–1.80; 1.61	Complete in 8 spiral rows
Microcetus lappus	Ovoid/elongate	18–29 × 7.5–18; 23.6 × 13		90% coverage (thinly); absent from cytoproct area

references to original studies are included for readers requiring more detailed information on cellular morphology and structures.

Isotrichidae The holotrich genera occurring most frequently in ruminant animals are *Isotricha* and *Dasytricha* (Table 3.2a). The genera are usually found together, although there are some reports of the presence of a single genus or the absence of both from an individual host or group of animals (Table 3.3). They are easily seen during microscopic examination of rumen contents and when present are readily discernable in view of their size, motility, and uniform somatic ciliature (Fig. 3.5). A third genus, *Oligoisotricha,* is observed less frequently in rumen samples. The features of the three genera will be considered separately.

The Genus Isotricha The genus *Isotricha* was created by Stein who described two individual species *I. prostoma* and *I. intestinalis* (Stein, 1858, 1859). Other species have been described in various hosts; for example *I. ferrum equinum* was reported in the dromedary (Schedrina, 1926), *I. magna* in the antelope (Buisson, 1924), and *I. bubali* in the buffalo (Dogiel, 1928). Subsequent studies and comparisons with these original descriptions have resulted in the confirmation of only two species of *Isotricha: I. prostoma* and

Macronucleus	Contractile vacuole	Concretion vacuole	Comments	Figure number
Spherical, 2–3 μm diameter	1	Not reported		3.16
Spherical 7–8 μm diameter	Not specified	Not observed	Not a member of the Buetschliidae	3.16
Spherical/ elliptical	1	–		3.17 3.18
Elongated ellipsoidal	1	–		3.17 3.18
Spherical	1	–		3.18
Ovoid	1	Not observed	Not assigned to species	3.9
Spherical	1	Not reported	Not assigned to taxon; two distinctive cytopharyngeal rods	3.10

I. intestinalis (Table 3.5; Grain, 1966a; Imai, 1981). In addition to these studies there have been two unsubstantiated observations of this genus. Dogiel (1926b) observed the occurrence of a new species of *Isotricha* in samples from two camels. However, a description was not given in the original article or published subsequently (Dehority, 1986a). In addition, there has been a single observation of *I. caulleryi* in the cockroach digestive tract. (Weill, 1929). Morphological descriptions will, therefore, be limited to the recognized rumen species *I. prostoma* and *I. intestinalis*. Both species usually occur together and reports of the natural occurrence of a single species are unusual (e.g., Prins and Van Hoven, 1977).

Isotricha prostoma Stein and *Isotricha intestinalis* Stein are elongated ovoid to ellipsoidal organisms; the posterior end is rounded and the anterior end pointed (acuminate). The body is flattened dorso-ventrally. The somatic ciliature is complete; the cilia are of uniform length (10–12 μm) and are arranged in at least eighty rows which are parallel to the body axis (Campbell, 1929). The protozoa are similar in size although *I. prostoma* is the larger ranging from 80–200 μm in length and from 50–120 μm in width (average dimensions 135 μm × 70 μm). The equivalent values for *I. intestinalis* are 90–200 μm by 45–150 μm (average 110 × 60 μm). The location of the vestibulum is different in the two species and is the most obvious distinguishing

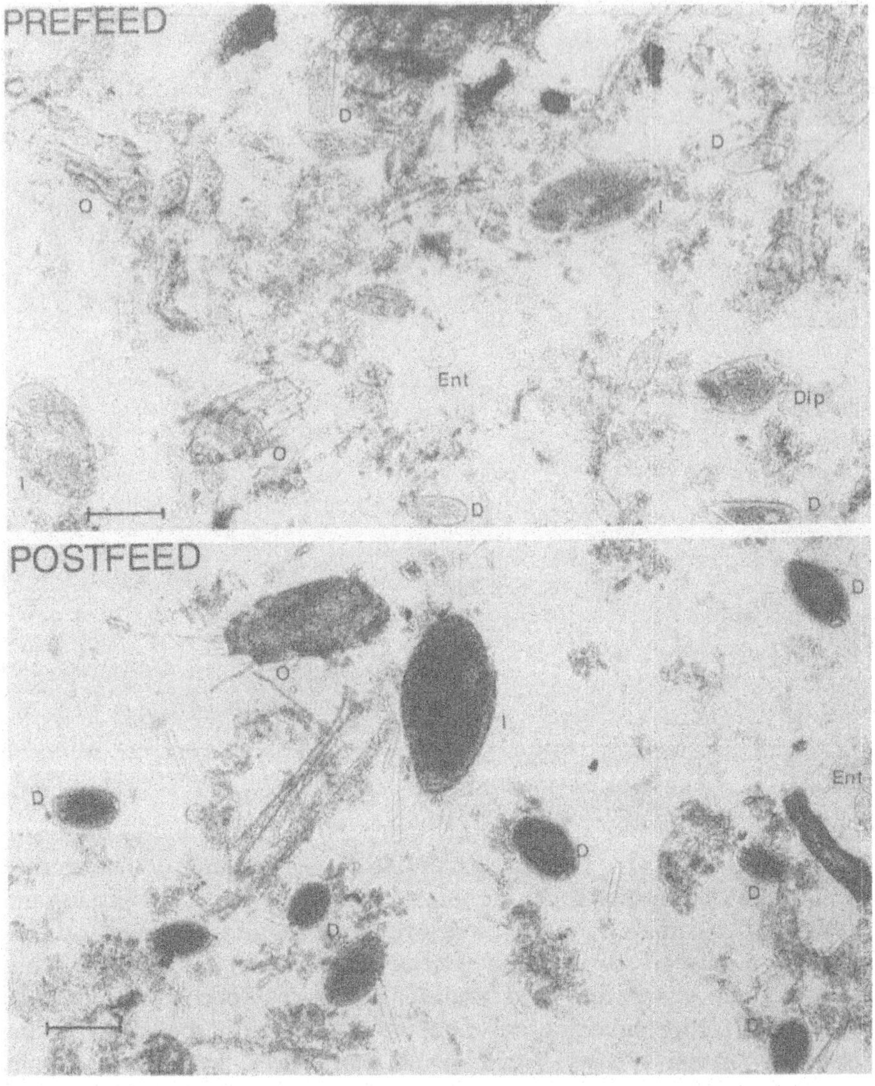

Fig. 3.5 Holotrich ciliates in rumen contents withdrawn immediately before and 2 hours after the sugar beet pulp-hay ration had been consumed by the host sheep. Protozoa present include *Dasytricha ruminantium* (D), *Isotricha* spp. (I), *Diplodinium* spp. (Dip), *Entodinium* spp. (Ent), and *Ophryoscolex* spp. (O). Bar = 100 μm.

criterion. The vestibulum of *I. prostoma* is at the posterior end of the body (Fig. 3.6) whereas the vestibular opening of *I. intestinalis* is on the ventral surface approximately one-third of the body length away from the posterior end of the cell (Fig. 3.7). The vestibulum in both species has the shape of an elongated cone 30 μm in length; at the apex the cytostome and cytopharynx are close to the adjacent nuclear apparatus. The cytoproct (3 μm diameter) is close to the opening of the vestibulum. Although the cytoproct is more evident in *I. intestinalis* it is not always identifiable in *I. prostoma*. There is, however, confusion in the literature concerning mouth location. Many of the original publications referred to the oral end as anterior, and commented that the oral apparatus was to the rear of swimming cells (e.g., Bhatia, 1936; Campbell, 1929; Schuberg, 1888). However, Grain (1966a) concluded from his study of vestibular morphogenesis that the end opposite to the vestibulum should be designated anterior. The orientation of Grain has been adopted for this summary description of the morphological features of the genus. Readers requiring a more detailed morphological characterization are referred to the extensive monograph of Grain (1966a, 1966b).

The cytoplasm is comprised of an ectoplasm and endoplasm which are separated by a stratified fibrillar layer. The ectoplasm, which extends inward for approximately 3 μm, is homogeneous and contains numerous kinetosomes. The endoplasm is more heterogeneous and has a large number of inclusions. There are 6–12 (ave. 10) contractile vacuoles which are arranged in three or four longitudual rows in the posterior two-thirds of the cell. The number, size, and distribution of the numerous food vacuoles is more variable. Ingested starch grains (4–15 μm in diameter) and granules of reserve polymer (0.5–2 μm in length) are also abundant in the endoplasm. Both species have extensive fibrillar systems that comprise the ecto-endoplasmic boundary and connect to the kinetosome network. A Golgi apparatus has not been detected (Grain, 1966a).

The size, shape, and location of the macronucleus are constant in non-dividing cells of both species. Its orientation is parallel to the long axis of the cell with its posterior end close to the vestibulum. The macronucleus of *I. intestinalis* is ovoid (approximately 30 μm wide by 20 μm long), whereas that of *I. prostoma* is more elongated with dimensions approximately one-third of the cell length and one-seventh of the cell diameter (Campbell, 1929; Grain, 1966a). The finely granular homogenous structure includes obvious nucleoli and vacuoles. Rickettsia-like microorganisms have been observed in the nucleus of *I. intestinalis* (Bretschneider and Van Vorstenbosch, 1964). The micronucleus is oval (3–6 μm long $\times$ 3–4 μm wide) and located at a variable position within a depression on the ventral side of the macronucleus. The micronucleus of *I. intestinalis* may be more rounded or triangular (Bhatia, 1936; Ogimoto and Imai, 1981). Both nuclei are enclosed and supported by a fibrillar nucleo-suspensory apparatus, the karyophore (Campbell, 1929; Grain, 1966a; Ten Kate, 1928). This interconnects with the ecto-endoplasmic boundary. During the division process, which is completed within 2 h, both nuclei elongate and become fibrillar. These division processes have been described in detail by Campbell (1930) and Grain (1966a). Multiple macronuclei (2 or 4) have been observed in what were believed to be ex-conjugant cells (Grain, 1966a).

Table 3.5 List of the described species of rumen holotrich ciliate protozoa

Species	Synonym	References
Isotrichidae		
Genus *Isotricha*		
I. prostoma		Buisson, 1924; Campbell, 1929; Dogiel, 1928; Grain, 1966a; Schedrina, 1926; Schuberg, 1888; Stein, 1859
I. intestinalis		
I. bubali	*Oligoisotricha bubali*	
I. magna	*Isotricha intestinalis*	
I. ferrum-equinum	*Dasytricha ruminantium*	
I. ruminantium	*Dasytricha ruminantium*	
Genus *Dasytricha*		
D. ruminantium		Grain, 1966a; Hukui, 1940; Hukui and Nisida, 1954; Jirovec, 1933; Schuberg, 1888
D. hukuokaensis		
D. ozakii	*Isotricha intestinalis*	
D. rectum	*Isotricha prostoma*	
D. nipponicum	*Dasytricha ruminantium*	
D. bovis	*Dasytricha ruminantium*	
D. elongatum	*Dasytricha ruminantium*	
Genus *Oligoisotricha*		
O. bubali	*Isotricha bubali*	Dehority et al., 1983; Dogiel, 1928; Imai, 1981
Buetschliidae		
Genus *Buetschlia*		
B. parva		Becker and Talbott, 1927; Dehority, 1970; Dogiel, 1928; Fiorentini, 1890; Hsiung, 1932; Imai, 1988; Schuberg, 1888
B. neglecta		
B. triciliata	*Hsiungia triciliata*	
B. lanceolata		
B. nana		
B. omnivora		
Genus *Parabundleia*		
P. ruminantium		Imai and Ogimoto, 1983

[a]The genus *Microcetus* has not yet been classified and its taxonomic position is uncertain.

The Genus Dasytricha This genus was initially described by Schuberg (1888). Two species *Dasytricha ruminantium* Schuberg and *Dasytricha hukuokaensis* Hukui and Nisida are recognized; *D. ruminantium* occurs more regularly. Other proposed species *D. ozakii, D. rectum, D. nipponicum* (Hukui, 1940), *D. bovis*, and *D. elongatum* (Jirovec, 1933) were not considered to be morphologically distinctive and Grain (1966a) reassigned these "species" as *I. intestinalis, I. prostoma*, or *D. ruminantium*.

Dasytricha ruminatium (Fig. 3.8) resembles *Isotricha* spp. in shape but it is smaller; its size range is usually quoted at 35–75 μm × 20–40 μm (Clarke, 1977; Grain, 1966a) although Ogimoto and Imai (1981) give figures of 68–100 μm × 38–53 μm. The protozoon is ovoid in shape. The posterior end is more rounded and slightly flattened dorso-ventrally. The somatic ciliature is complete with rows of cilia (approximately 40) arranged obliquely to the

Table 3.5 *Continued*

Species	Synonym	References
Genus *Polymorphella*		
P. bovis		Imai, 1984
Genus *Pingius*		
P. minutus		Hsiung, 1932
Genus *Hsiungia*		
H. triciliata	*Buetschlia triciliata*	Imai, 1988
Genus *Blepharoprosthium*		
B. parvum		Hsiung, 1932
Genus *Blepharoconus*		
B. krugerensis		Dehority, 1986b; Eloff and Van Hoven, 1980
Genus *Holophryozoon*		Jirovec, 1933
H. bovis		
Paraisotrichidae		
Genus *Paraisotricha*		
Paraisotricha sp.		Dehority, 1986b
Blepharocorythidae		
Genus *Charonina*		
C. ventriculi	*Blepharocorys bovis*/ *B. ventriculi*	Dehority and Mattos, 1978; Dogiel, 1926a; Hsiung, 1930a, 1930b, 1932; Jameson, 1925a; Schumacher, 1915
C. equi		
C. nuda		
Buetschliellidae		
Genus *Buetschliella*		
B. bovis		Jirovec, 1933
Genus *Microcetus*[a]		
M. lappus		Orpin and Mathiesen, 1986

body axis. The vestibular opening is located at the posterior end of the cell (Grain, 1966a), although many authors consider that, by definition, the vestibulum must be at anterior of the cells and hence to the rear while the protozoa are swimming. The vestibulum is about 8 μm in diameter and extends 10 μm into the cell; it is ciliated on the ventral side only. The cytoproct (2 μm long, 1.5 μm diameter) also opens at the posterior of the cell. The cytoproct is adjacent to the vestibulum and dorsal to it.

The cytoplasm comprises ectoplasm and endoplasm. These are separated by a fibrillar boundary which does not extend internally as in the *Isotricha* spp. The absence of transverse fibres and the karyophore are important distinguishing features. The ectoplasm is a narrow zone 1.5–2 μm wide that is homogenous and contains numerous kinetosomes. There are numerous inclusions in the endoplasm, the most obvious of which in post-

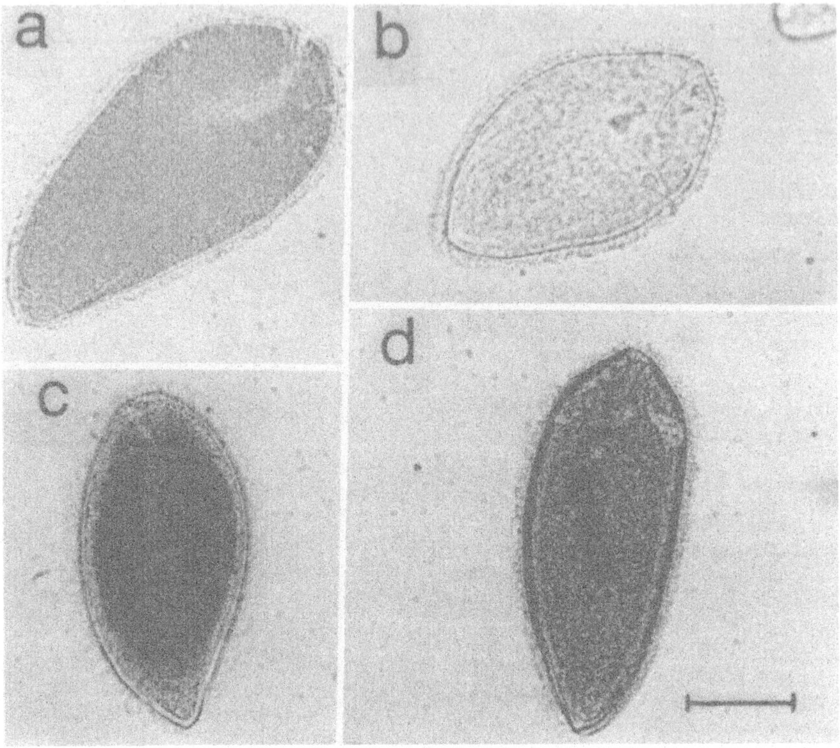

Fig. 3.6 Photomicrographs of prefeed (b) and postfeed (a, c, d) cells of *Isotricha prostoma*. The position of the vestibulum is evident. The nucleus of the cell in micrograph (d) was stained with methyl green. Bar = 50 μm.

feed cells are the amylopectin granules. These polysaccharide storage granules are depleted by starvation for 48 h, and this depletion process is exemplified in a series of electron micrographs published by Ogimoto and Imai (1981). Also present within the endoplasm are many food vacuoles. These arise from the cytopharynx which is a short inward extension of the vestibulum. There is, however, only one contractile vacuole which is situated close to the vestibulum in the posterior part of the cell.

The macronucleus is elongated and elliptical in shape and rounded at its extremities. In nondividing cells it measures 16–18 μm in length and 8–10 μm in width. It is generally homogeneous in appearance although vacuoles and nucleoli are detectable. The micronucleus (2–4 μm in length) is located in a depression on the surface of the macronucleus, the position of which is variable. Although the nuclear layer is enveloped by a layer of ergastoplasm (i.e., a discernable layer) there is no karyophore. Multiple nuclei have been observed in some non-dividing cells (Grain, 1966a). During cell division both the nuclei elongate and 4 chromosomes have been

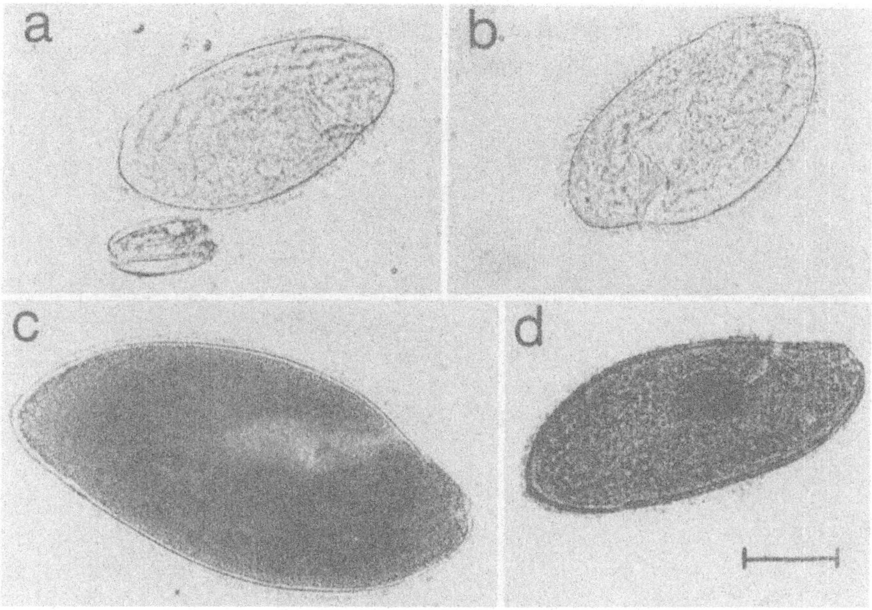

Fig. 3.7 Photomicrographs of prefeed (a, b) and postfeed cells of *Isotricha intestinalis*. (c, d). The nucleus of the cell in micrograph (d) was stained with methyl green. The lateral position of the vestibulum is a characteristic of this species. The raised attachment area on the anterior dorsilateral surface is evident in the cells shown in micrographs (a) and (b). Bar = 50 μm.

distinguished in the micronucleus. It is not unusual for conjugant cells to contain smaller, spherical (8 μm diameter) macronuclei (Dogiel and Fedorowa, 1925a; Grain 1966a). Reproduction in both *Dasytricha ruminantium* and *Isotricha* spp. and the subsequent morphogenesis and development of the ciliature has been examined in some detail (e.g., Campbell, 1930; Dogiel and Fedorowa, 1925a; Schuberg, 1891). The processes are well-illustrated and described for both genera in the review article by Grain (1966a).

Dasytricha hukuokaensis was first described by Hukui and Nisida (1954) and although it is recognized as a separate species there have been no further reports on its occurrence. It is larger than *D. ruminantium* with a body length of 120–180 μm and width of 68–122 μm (average 151 × 95 μm). The somatic ciliature is complete and the oral apparatus conspicuous. The vestibular opening is on the ventral surface approximately one-third of the body-length away form the posterior end of the cell (Fig. 3.9). The spherical micronucleus (2 μm in diameter) is located in a central depression on the surface of the ellipsoidal macronucleus which is some 24–38 μm long and 16–20 μm wide (average 31 × 18 μm). The number of contractile vacuoles present varies between 1 and 11. Although some supporting fibers

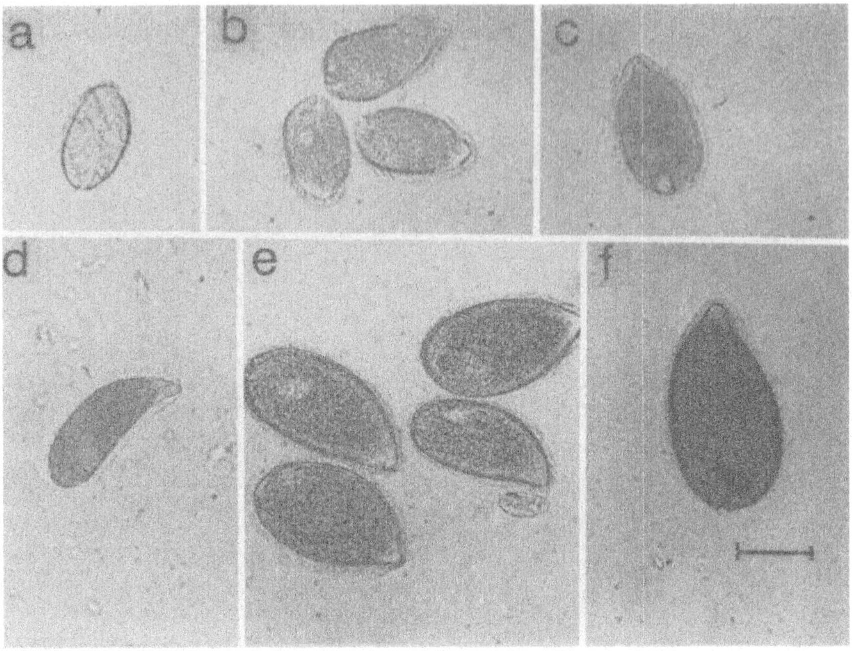

Fig. 3.8 Photomicrographs of prefeed (a, b) and postfeed cells (c–f) of *Dasytricha ruminantium*. The nucleus of the cell in micrograph (f) was stained with methyl green. Distinctive features of this genus (e.g., location of mouth and contractile vacuole; rows of cilia arranged obliquely to the body axis) are visible. Bar = 50 μm.

were observed, no karyophore was described. In many aspects the protozoon is morphologically similar to *I. intestinalis*.

The Genus Oligoisotricha The ciliate *Isotricha bubali* was originally observed in rumen contents from water buffalo and described by Dogiel (1928) who compared its characteristics with those of *Dasytricha*. The ciliate was not reported again for over 50 years until it was detected in water buffalo in Taiwan (Imai, 1981). However, the extent and arrangement of the somatic ciliature and the absence of a karyophore were not compatible with its inclusion in the genus *Isotricha* or *Dasytricha*, and Imai proposed that a new genus, *Oligoisotricha*, should be invoked.

Oligoistricha bubali Dogiel has now been observed in host species other than the water buffalo (Table 3.2). Surveys on the protozoal populations of water buffalo (Imai et al., 1981b), zebu cattle (Imai and Ogimoto, 1984), and Bali cattle (Imai, 1985) indicated that *O. bubali* was usually present in 20–40% of the animals examined. In one survey of Japanese water buffalo *O. bubali* was the only holotrich representative observed (Imai et al., 1981c). In

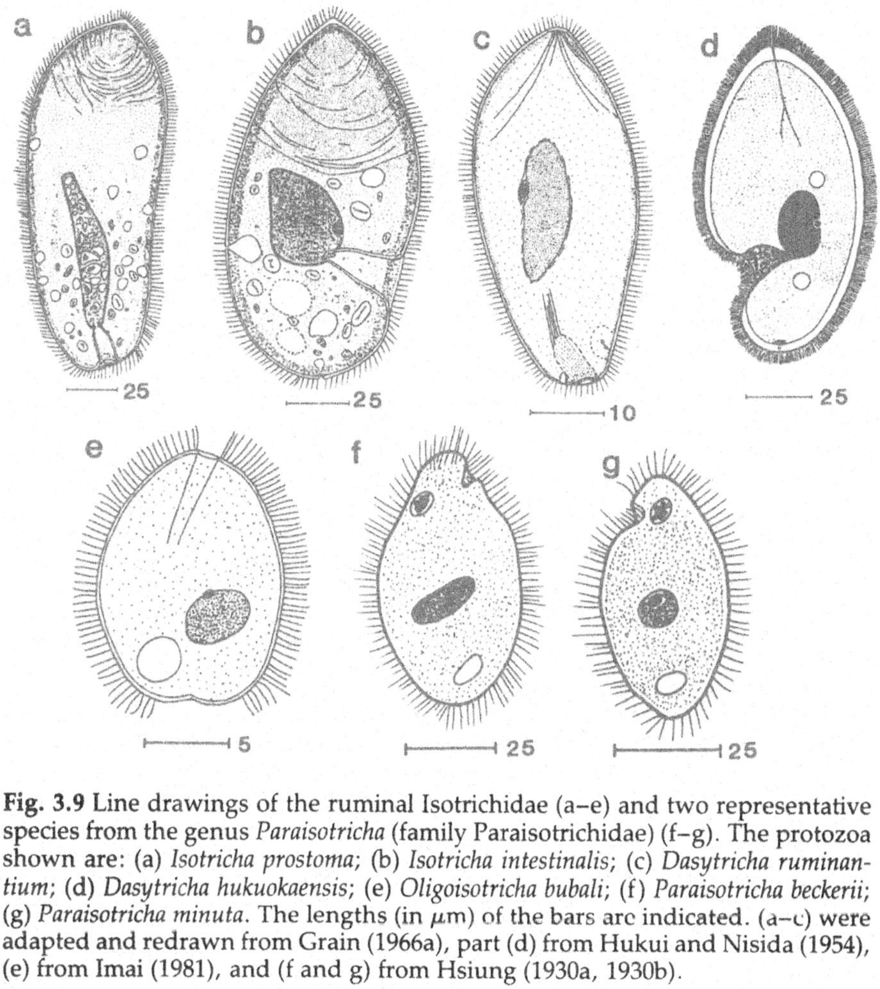

Fig. 3.9 Line drawings of the ruminal Isotrichidae (a–e) and two representative species from the genus *Paraisotricha* (family Paraisotrichidae) (f–g). The protozoa shown are: (a) *Isotricha prostoma;* (b) *Isotricha intestinalis;* (c) *Dasytricha ruminantium;* (d) *Dasytricha hukuokaensis;* (e) *Oligoisotricha bubali;* (f) *Paraisotricha beckerii;* (g) *Paraisotricha minuta.* The lengths (in μm) of the bars are indicated. (a–c) were adapted and redrawn from Grain (1966a), part (d) from Hukui and Nisida (1954), (e) from Imai (1981), and (f and g) from Hsiung (1930a, 1930b).

other water buffalo studied, *O. bubali* comprised up to 20% of the total ciliate population (Imai, 1985), whereas in domestic cattle in Tennessee, USA., the concentrations ranged from as high as 35% of the protozoal population in unweaned calves to a maximum of 72% in mature feedlot animals (Dehority et al., 1983). Unlike the other holotrichs present, the proportion of *O. bubali* increased when the cattle were fed a corn silage ration.

Oligoisotricha bubali is ovoid in shape, flattened bilaterally, and ranges in length from 12–20 μm and in width from 8–15 μm (average dimensions 16×12 μm; see Fig. 3.10). The vestibular opening is subterminal on the ventral anterior surface. The posterior end is not uniformly convex but appears slightly concave as a shallow depression is present. The somatic

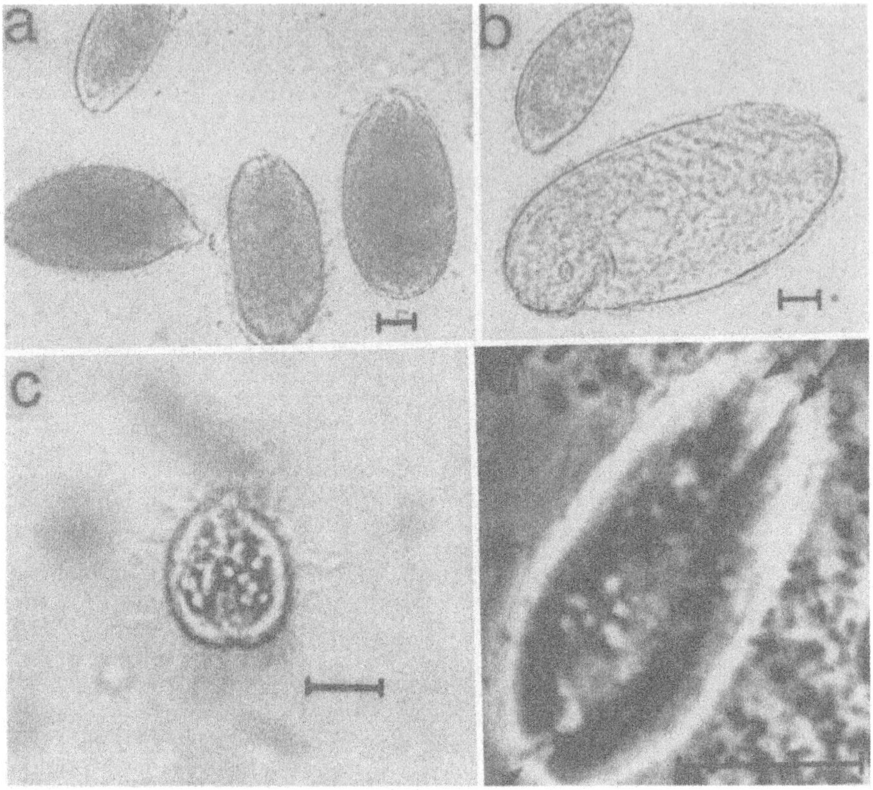

Fig. 3.10 Photomicrographs of the three genera of the rumen Isotrichidae and of a protozoon which has not been assigned to any taxon. (a) *Dasytricha*. (b) *Isotricha*. (c) *Oligoisotricha*. (d) *Microcetus lappus*. (a–b) Bar = 20 μm. (d–d) Bar = 10 μm. (c) Courtesy of S. Imai. (d) From Orpin and Mathiesen (1986).

ciliature is arranged in rows parallel to the body axis over approximately 80% of the body surface, but it is absent from the vestibular opening and the posterior one-sixth of the body surface, including the depression.

The ectoplasm is well-developed but clearly separated from the homogeneous granular endoplasm. The vestibulum is clearly visible and may penetrate up to one-third of the body length into the endoplasm. A single contractile vacuole is usually found at the posterior end of the cell although this location may vary. In fact, Dogiel (1928) noted that the vacuole was situated close to the cytopharynx. The macronucleus is spherical to elliptical in shape and is inconsistently located within the endoplasm, although in many cells it is within the middle section but close to one side. A spherical micronucleus is positioned close to the anterior margin of the macronucleus. The nuclei are not supported by a karyophore. During transverse division the micronucleus divides initially and the portions separate to the opposite poles of the dividing macronucleus. Ex-conjugants with two spherical macronuclei and a single micronucleus have also been observed (Dogiel, 1928).

Orpin and Mathiesen (1986) observed a small ciliate in the ruminal contents of Norwegian red cattle that resembled *Oligoisotricha* spp. This ciliate and *O. bubali* have a striated vestibulum but lack cytopharyngeal rods. However, unlike *O. bubali* the posterior end was smoothly rounded and not concave (Dogiel, 1928; Imai, 1981). These differences suggest that the ciliate observed by Orpin and Mathiesen may represent a previously undescribed species. The protozoon described as *O. bubali* by Dehority et al. (1983) likewise had a smoothly rounded posterior end.

Ultrastructure of the Isotrichidae Electron microscopic studies have revealed details of the complex ultrastructure and surface structures of *Dasytricha ruminantium* and the two species of *Isotricha* (Buckelew and Kontir, 1977; Gerassimova, 1981; Grain, 1966b; Ogimoto and Imai, 1981; Paul et al., 1987, 1989; Stern et al., 1977). An extensive study of the ultrastructure of the endocommensal ciliates of the digestive tract of ruminants and equids was undertaken by Grain (1966b) and detailed descriptions of the general organization and ultrastructure of *D. ruminantium* have been published more recently by Paul et al. (1989).

Many features are evident in sectioned cells (Figs. 3.11, 3.12). In addition to the principal cytoplasmic components such as nuclei and polysaccharide storage granules, numerous vacuoles containing bacteria and ingested food particles are distributed throughout the cytoplasm. Many electron-dense bodies are also present, but as yet there is no clearly defined role for all of the observed subcellular structures. Both the ectoplasm and endoplasm are rich in ribosomes, although the macronucleus, micronucleus (Fig. 3.11f), contractile vacuole, and cytoproct are all endoplasmic.

The ectoplasm and endoplasm are clearly differentiated. The outer layer of the ectoplasm contains the kinetosomes and their accessory fibers. The kinetosomes, found at the base of each cilium, are about 450 nm long and formed by the prolongation of the peripheral fibers of the cilium. The cilium itself has the typical 9:2 arrangement of peripheral and axial fibers that are 17–25 nm in diameter. The outer ectoplasmic layer of *Dasytricha* and *Isotricha* are similar, although the pores occupying the interkinetal ridges between the kinetosomes in *Dasytricha* are smaller and more numerous.

The ecto-endoplasmic boundary of *Isotricha* spp. was originally described by Noirot-Timothée (1958). It is a continuous double-layered fibrillar system composed of 4 nm-diameter filaments (Vigues et al., 1984). The two microfibrillar layers are parallel to the cell wall but orientated perpendicularly to each other. In *Isotricha* the average thickness of each layer is 380 nm (Grain, 1966b). Platelet-like structures of unknown function are associated with this layer in both genera (Grain, 1966b; Noirot-Timothée, 1958). In

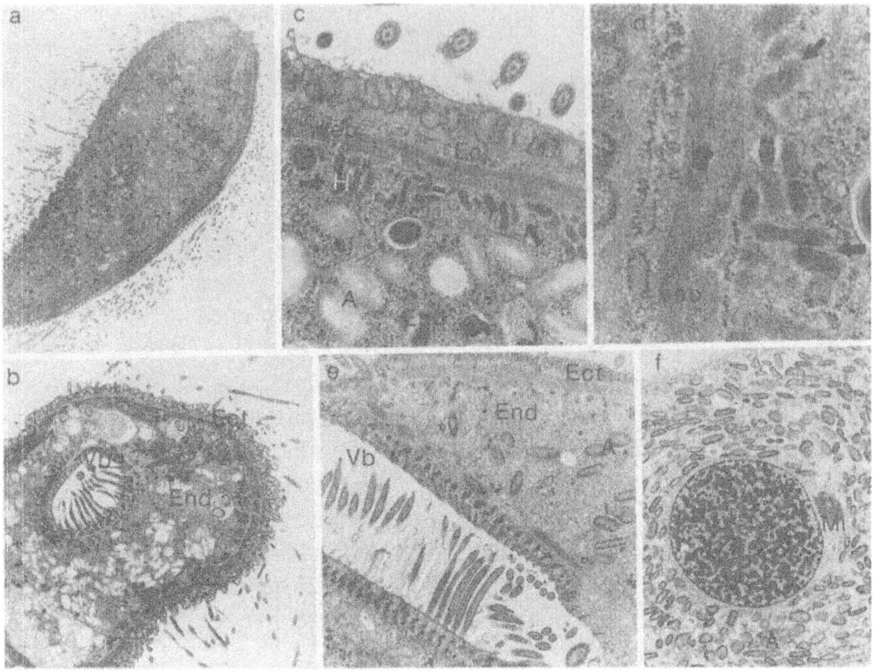

Fig. 3.11 Transmission electron micrographs of longitudinal and transverse sections of *Dasytricha ruminantium*. Subcellular features indicated are: End, endoplasm; Ect, ectoplasm; A, amylopectin granule; H, hydrogenosome; Ma, macronucleus; Mi, micronucleus; Vb, vestibulum; eeb, ecto-endoplasmic boundary layer.

Isotricha only the inner layer is extended internally into the endoplasm to form the karyophore. This structure does not exist in *Dasytricha*. Biochemical and immunological characterization of the cortical cytoskeleton of *I. prostoma* has confirmed that the kinetosome interconnecting lattice and the ecto-endoplasmic boundary are separate systems with differing filament sizes (6 and 4 nm respectively) and protein compositions (Vigues et al., 1985). These principal microfibrillar structures of the cortical cytoskeleton are non-actin systems; two proteins with molecular weights of 58 and 63 kDa have been prepared from the kinetosome-associated filamentous reticulum, while cortical proteins with molecular weights < 23 kDa were shown to originate from the ecto-endoplasmic boundary (Vigues et al., 1984, 1985). One such low-molecular-weight calcium-binding protein which differed from calmodulin was shown to be present in both *I prostoma* and the entodiniomorphid ciliate *Polyplastron multivesiculatum* (Vigues and Grolière, 1985).

Numerous inclusions are present in the endoplasm. Digestive

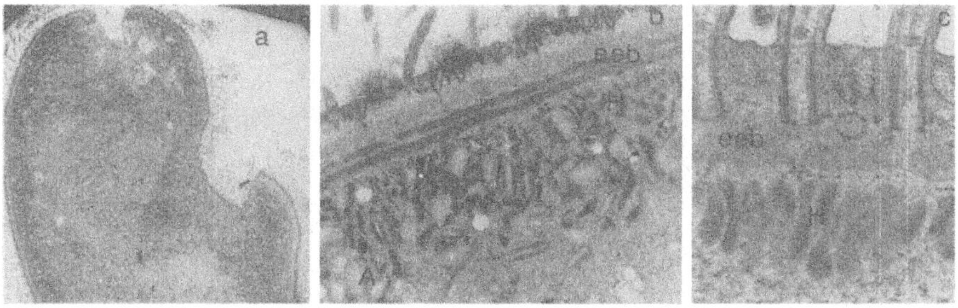

Fig. 3.12 Transmission electron micrographs of sections of *Isotricha intestinalis*. Structural features shown include: H, Hydrogenosome; eeb, ecto-endoplasmic boundary layer; A, amylopectin deposits.

vacuoles, formed at the cytopharynx, are limited by a single or double membrane in *Isotricha* and are distributed within the endoplasm (Fig. 3.11e). Many of the hydrolytic enzymes in rumen ciliates are located in pleomorphic membrane-bound vesicles 0.2–1.0 μm in diameter (Williams et al., 1986; see also Chapter 8). Preliminary studies to characterize this organelle further have only been undertaken with *D. ruminantium* (Yarlett et al., 1985) and *Polyplastron multivesiculatum* (Williams and Ellis, 1985). The organelle of *D. ruminantium* after isopycnic centrifugation on a sucrose gradient had a mean equilibrum density of 1.13 g/ml. However, the location of these vesicles within the cell ultrastructure has yet to be established.

Many electron-dense bodies are also present in the endoplasm although, with the notable exception of the hydrogenosome, their functions have not been elucidated. Both species of *Isotricha* and *D. ruminantium* contain this granular microbody-like organelle, which is generally located in the endoplasm close to the ecto-endoplasmic boundary (Figs. 3.11c, 3.11d, 3.12c, 3.12d; Constantinescu and Dragos, 1984; Paul et al., 1989; Yarlett et al., 1981, 1983a). The organelle is approximately 500 nm in diameter (Figs. 3.11, 3.12) but on isolation the hydrogenosomes of *D. ruminantium* and *Isotricha* spp. have differing equilibrium densities of 1.18 and 1.23 g/ml respectively (Yarlett et al., 1981, 1983a). Preliminary reports on the occurrence of mitochondria in these ciliates were not correct. It is evident on the basis of subcellular location and morphology that the organelles described as mitochondria by Grain (1966a, 1966b) and Gaumont and Grain (1967) were in fact hydrogenosomes. The organelle is also present in certain entodiniomorphid ciliates (Paul et al., 1990; Yarlett et al., 1984). It is also probably present in other genera of ciliates from the digestive tract of nonruminants as similar structures (i.e., "the chondrioma") were present in many of the species examined by Grain (1966a, 1966b).

A structure that has not been observed during either light or electron microscopic examination of the rumen Isotrichidae is the Golgi apparatus

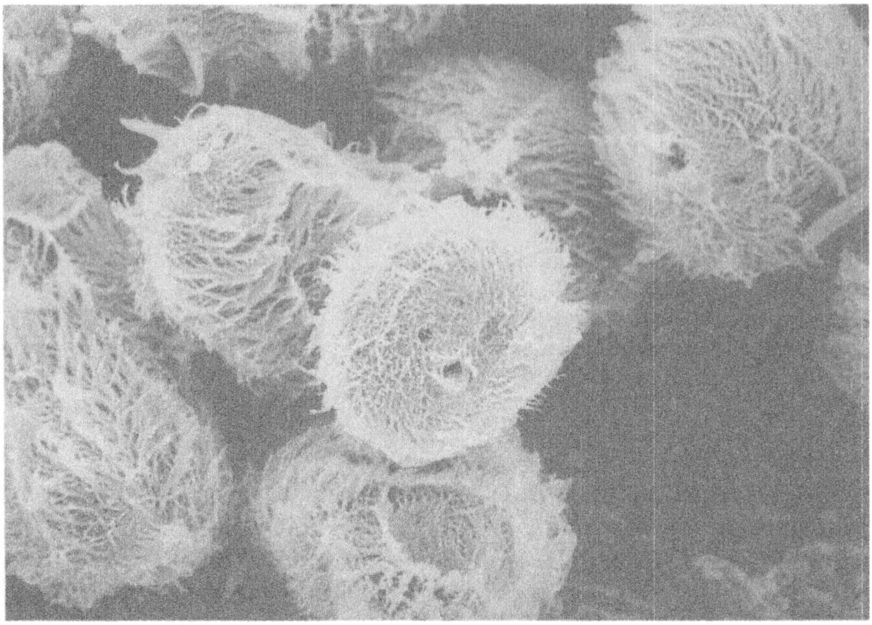

Fig. 3.13 Scanning electron micrograph of *Dasytricha ruminantium*. The cytostome, cytoproct, and surface pores are clearly visible.

(Grain, 1966a, 1966b), although there are reports of its occurrence in some rumen entodiniomorphid ciliates from the family Ophryoscolecidae (Delfosse, 1977; Noirot-Timothée, 1960). The internal organization of the holotrichs is complex, however, and further efforts are required to define the subcellular structures present and identify their biochemical functions.

Scanning and transmission electron microscopic examination of the surface of *Isotricha intestinalis* showed that it was composed of predominantly longitudinal cytoplasmic ridges between the rows of cilia (Imai and Tsunoda, 1972; Orpin and Hall, 1983). On the anterior dorsilateral surface these ridges are modified and extend up to 9 μm from the cell surface. There is an extensive sheet of microtubules beneath these processes. The projection is an attachment organelle and appears as a longitudinal ridge approximately 24–37 μm long and 5 μm wide which terminates at or close to the anterior extremity of the cell (Orpin and Hall, 1977, 1983). This region of the cell surface is used for the attachment of the cell to plant material and inert surfaces. A similar attachment zone has been observed on the anterior of *I. prostoma* but not *D. ruminantium* (Orpin and Letcher, 1978) although all three ciliates are attracted to, and will attach themselves to, plant particles (Orpin, 1985; Orpin and Letcher, 1978). Scanning electron microscopy has also been used to study structural features of *Dasytricha ruminantium* (Paul et al., 1989; see also Fig. 3.13).

Buetschliidae Eight genera from the family Buetschliidae have been observed in the ruminal contents of various hosts (Table 3.2). Reports of their occurrence are infrequent, however, and their importance in the rumen ecosystem unknown.

The Genus Buetschlia The genus *Buetschlia* (syn. *Bütschlia*) was established by Schuberg (1888). In addition to the type species *Buetschlia parva* Schuberg three other species (i.e., *B. neglecta, B. triciliata,* and *B. lanceolata*) have been reported in bovine rumen contents (Fiorentini, 1890; Hsiung, 1932; Schuberg, 1888). *Buetschlia nana* and *B. omnivora* were described by Dogiel (1928) in the stomach contents of the camel (Table 3.5). However, only *B. parva* and *B. triciliata* (syn. *Hsiungia triciliata*) have been reported on more than one occasion and Corliss (1979) regards the genus as consisting of a single species. Therefore, the taxonomic position of the other species is doubtful. With the exception of *B. parva,* all lack the characteristic complete somatic ciliature (Schuberg, 1888).

Although *B parva* was initially observed only in the rumen contents of cattle (Clarke, 1964; Hungate, 1943; Kofoid and MacLennan, 1933; Schuberg, 1888; Sharp, 1914), the protozoon was subsequently shown to be present in a wider range of hosts (Table 3.2). In sheep (Dehority, 1970), musk-oxen (Dehority, 1974), and cattle (Clarke, 1964), the population size of *B. parva* was only 10^3–10^4 cells/ml representing 0.2–0.6% of the total ciliate population, although values of 2–3% were reported for some cattle in New Zealand (Clarke, 1964). The numbers of *B. parva* in ovine rumen contents were highest in the immediate post-feed period and exhibited a diurnal pattern of variation which was similar to that of the other holotrich species present (i.e., *D. ruminantium* and *Isotricha* spp.).

Buetschlia parva is ovoid in shape; the anterior is truncate and the posterior end more rounded (Figs. 3.14, 3.15). The size of the organism is variable and it is reported to range in length from 30–67 μm and in width from 20–48 μm (Clarke, 1964; Dehority, 1970; Ogimoto and Imai, 1981; Schuberg, 1888). The average dimensions given by Clarke (1964) and Dehority (1970) were 47×25 μm and 55×35 μm respectively. The ectoplasm is particularly dense at the anterior end and distinct from the homogeneous endoplasm. The mouth is situated at the anterior end of the cell and is surrounded by cilia which are longer than those which uniformly cover the remainder of the body surface. The somatic ciliature is arranged in rows which spiral around the long axis of the cell. The macronucleus is spherical in shape and located in the mid-region of the cell but towards the periphery. Schuberg (1888) was able to observe cells undergoing transverse fission; conjugation was not observed. A contractile vacuole is not clearly evident. There are reports of the presence of a single vacuole (Becker and Talbott, 1927; Ogimoto and Imai, 1981), although some authors were unable to see a contractile vacuole (Schuberg, 1888). A concretion vacuole is present in the anterior part of the cell.

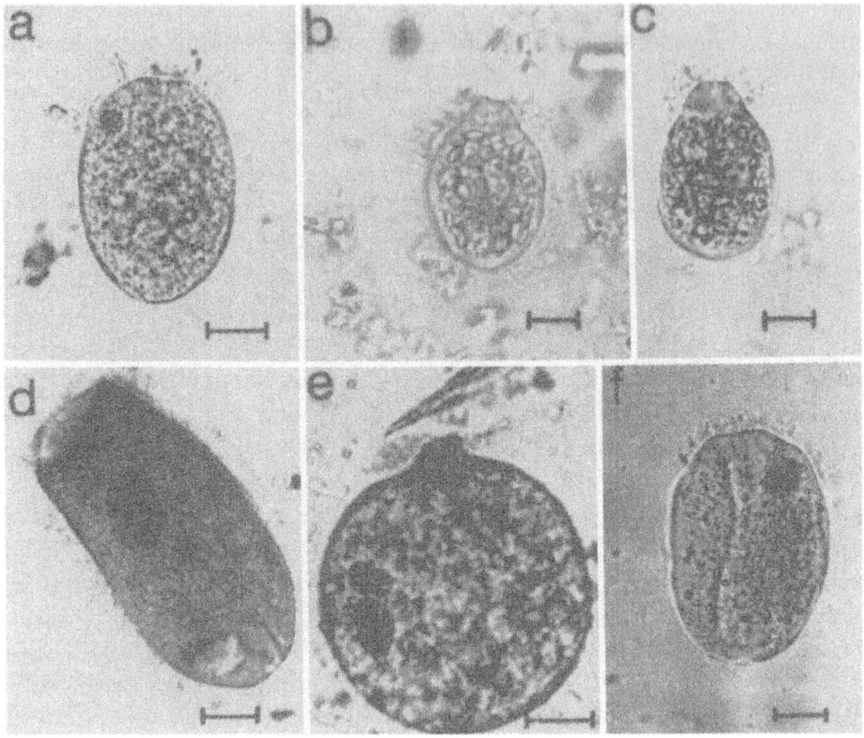

Fig. 3.14 Photomicrographs of some of the rumen Buetschliidae. (a) *Parabundleia ruminantium;* (b and c) *Polymorphella bovis;* (d) *Buetschlia parva;* (e) *Blepharoconus krugerensis;* (f) *Hsiungia triciliata* (syn. *Buetschlia triciliata*). Bar = 10 μm. (a–c) Courtesy of S. Imai. (d) From Dehority (1970). (e) From Dehority (1986b). (f) From Imai (1988).

Buetschlia neglecta Schuberg was present in low numbers in the rumen contents of cattle examined by Schuberg (1888). The organism like *B. parva* is ovoid in shape although the posterior end is more pointed and has four indentations, such that in cross section the posterior region would present the form of a cross with rounded inner angles. The protozoon is 40–60 μm in length and 20–30 μm in width. The anterior thickening of the endoplasm was not apparent and in this species contractile vacuoles were evident in the middle and posterior part of the cell; a concretion vacuole was again present in the anterior portion. The anterior location of the mouth and conical "pharynx" (that is, the cytopharyngeal apparatus) were the same as for *B. parva,* and as with this organism no anus or micronucleus were evident; the spherical nucleus was larger in *B. neglecta.* The somatic ciliation

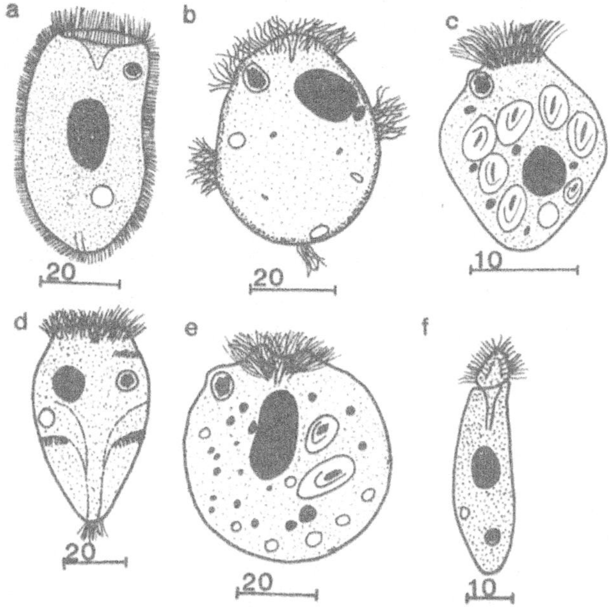

Fig. 3.15 Drawings of the species of *Buetschlia* that have been described in rumen contents (a, b, d, and f), and in the stomach contents of camel (c and e). The protozoa shown are: (a) *Buetschlia parva;* (b) *B. triciliata;* (c) *B. nana;* (d) *B. neglecta;* (e) *B. omnivora;* (f) *B. lanceolata.* The figures were adapted and redrawn from Schuberg (1888) (*B. parva, B. neglecta*); Fiorentini (1890) (*B. lanceolata*); Dogiel (1928) (*B. omnivora, B. nana*); and Hsiung (1932) (*B. triciliata*—syn. *Hsiungia triciliata*). The lengths (in μm) of the bars are indicated.

was apparently more complex in that, in addition to the anterior zone, specific ciliation was evident at the posterior end, close to the concretion vacuole and in the area of the posterior depressions (Fig. 3.15). The ciliary fringes were not apparent to Fiorentini (1890).

Buetschlia lanceolata Fiorentini has been reported on one occasion in the bovine rumen (Fiorentini, 1890). The organism is lanceolate (i.e., spear-shaped) in form measuring 48×20 μm. The oral aperture is at the anterior of the cell; the pharynx extends for approximately one-third of the length of the body. A large nucleus is present and in this species the concretion vacuole is located in the posterior part of the cell. A distinctive feature on the upper one-fifth of the body is a pronounced constriction in the form of a collar which is surmounted by very fine cilia.

Buetschlia omnivora and *Buetschlia nana* were observed by Dogiel (1928) in the stomach of the camel, and brief descriptions are given for comparison with the rumen species. *Buetschlia nana* is a small organism measuring only

17–21 μm long and 12–17 μm wide (average dimensions 19 × 15 μm). The body is ovoid although the anterior end is more pointed and slightly truncated. The concretion vacuole is located close to the oral assembly in the anterior of the cell. The cytoplasm of the anterior region was devoid of granular inclusions, although a fibrillar system was apparent. The remainder of the protoplasm was finely granular and contained a single posterior contractile vacuole and a spherical macronucleus and its accompanying micronucleus. A characteristic of this species is the presence of approximately 20 oval starch grains. The somatic ciliature is not well-developed with the exception of the anterior area. The somatic cilia are not particularly evident and there are no definite girdles or posterior ciliation zones (Fig. 3.15).

Buetschlia omnivora is an oval to spherical organism with an anterior mouth. The edges of the mouth are extended and enveloped with cilia. A cytoproct is located at the posterior of the cell. The concretion vacuole is in the anterior part of the cell and approximately twelve randomly occurring contractile vacuoles are present. The micronucleus is situated in a shallow indentation on a lateral surface of the elongated macronucleus. The endoplasm contains starch grains and this species is apparently able to ingest smaller protozoa. The dimensions of the protozoon are extremely variable as exemplified by the measurements quoted for three separate populations by Dogiel (1928). The average dimensions for each population were 90 × 80 μm, 64 × 54 μm and 41 × 31 μm. The overall range in length was 35–110 μm and in width 27–97 μm.

Buetschlia triciliata Hsiung (syn. *Hsiungia triciliata*) was present in the rumen of cattle in Tientsin in China and was described by Hsiung (1932). The ciliate was subsequently detected frequently in Kenyan zebu cattle by Imai (1988). However, the absence of complete somatic ciliation, a characteristic of the genus *Buetschlia,* and other morphological features resulted in Imai (1988) placing this protozoon in the newly created genus *Hsiungia*. It should be noted that within the rumen species of Buetschlia only *B. parva* has complete somatic ciliation (Schuberg, 1888). The ciliates do, however, possess other characteristics typical of the family Buetschliidae.

Hsiungia triciliata is ovoid or ellipsoidal in shape and both the anterior and posterior ends are rounded (Figs. 3.14, 3.15). The organism is 42–70 μm long and 33–52 μm wide, with average dimensions of 58.6 ± 5.7 × 40.1 ± 5.5 (Imai, 1988). Hsiung (1932) gave an average size of 50 × 45.5 μm. A funnel-shaped cytostome is located at the anterior of the cell and a cytoproct at the posterior. The concretion vacuole is situated in the anterior of the cell close to the surface and adjacent to the posterior end of the cytopharynx. The ectoplasm is generally thin but is thickened in the anterior region and over the concretion vacuole. The endoplasm is granular and contains numerous small food particles. A contractile vacuole is present in the endoplasm close to the cytoproct; other vacuoles may occur within the cytoplasm. The position of the large biconvex ellipsoidal macronucleus is variable, and cells containing two macronuclei have been observed. The micronucleus is ellipsoidal and located at the margin of the macronucleus.

The somatic ciliation is incomplete in this species and restricted to specific areas around the cytostome and over the concretion vacuole. Three other distinctive groups of cilia occur; one is located around the cytoproct and the other two occur close to the mid-line of the cell but on opposite sides of the body.

The Genus Parabundleia Imai and Ogimoto (1983) reported the occurrence of a ciliate in the rumen contents of zebu cattle in the Bangkok region of Thailand that resembled certain *Buetschlia, Bundleia,* and *Hsiungia* spp. However, its morphological characteristics differed sufficiently for these authors to place the ciliate into a separate genus. *Parabundleia ruminantium* is the only species of the genus described to date, and occurred in 40% of the zebu cattle examined by Imai and Ogimoto (1983).

Parabundleia ruminantium has an ovoid body, the outer surface of which is elevated at the anterior extremity and adjacent to the concretion vacuole. The cell is 37.5–50 μm in length and 27.5–32.5 μm in width (average dimensions 42.5×30.5 μm) with two anterior zones of cilia around the cytostome and concretion vacuole and one limited posterior zone beside the cytoproct (Figs. 3.14, 3.16). A spherical micronucleus is positioned close to the anterior margin of the elliptical macronucleus, which has an average length of approximately 16 μm. The position of the macronucleus, although usually toward the center of the cell, can be variable within the endoplasm. The cell contains a single posterior contractile vacuole and a spherical concretion vacuole close to the anterior extremity of the body.

The Genus Polymorphella The only species of this genus, *Polymorphella* (syn. *Polymorpha*) *ampulla* was observed in samples from the large intestine of the horse (Hsiung, 1930a). A morphologically similar, albeit distinctive, ciliate was subsequently detected in the rumen contents of zebu cattle (*Bos indicus*) from Bangkok, Thailand (Imai, 1984). This ciliate named *Polymorphella bovis* was present in almost 40% of the cattle surveyed.

Polymorphella bovis has a body that is generally ovoid although the anterior one-third is tapered giving an overall impression of a flask or bottle-like shape (Figs. 3.14, 3.16). The cell length is in the range 26–37.5 μm (average 34 μm) and the width between 20–26 μm (average 22 μm). Uniform cilia are present in two zones, the larger of which is at the tapered anterior while a smaller tuft is located at the posterior extremity close to the cytoproct. The ectoplasm, although generally thin, is thicker in the region of the concretion vacuole. The endoplasm is thickened in the posterior two-thirds of the body and contains numerous spherical inclusions that were believed to be food granules. The position of the subspherical macronucleus is usually toward in the center of the cell although its position can

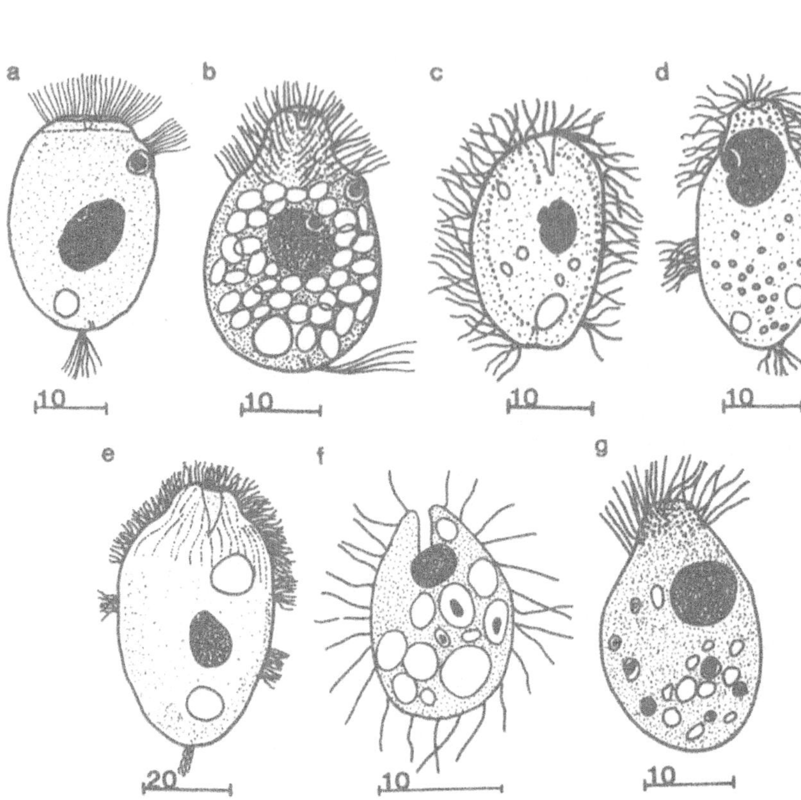

Fig. 3.16 Drawings of the genera (other than *Buetschlia*) from the family Buetschliidae that have been reported to occur in the rumen. The protozoa represented are: (a) *Parabundleia ruminantium* (Imai and Ogimoto, 1983); (b) *Polymorphella bovis* (Imai, 1984); (c) *Pingius minutus* (Hsiung, 1932); (d) *Blepharoprosthium parvum* (Hsiung, 1932); (e) *Blepharoconus krugerensis* (Eloff and Van Hoven, 1980); (f) *Holophryozoon bovis* (Jirovec, 1933); (g) *Buetschliella bovis* (Jirovec, 1933). The figures were adapted and redrawn from the references given in parentheses. The lengths (in μm) of the bars are indicated.

vary. A spherical micronucleus is located close to the margin of the macronucleus which has an average length of 2.5 μm. Two types of vacuole are again present; the contractile vacuole is in the posterior part of the cell and the concretion vacuole is close to the body surface in the anterior region of the cell just below the tapered zone.

The Genus Pingius This genus is represented by a single species and was described by Hsiung (1932). The ciliate *Pingius minutus* was present in the rumen of cattle in the Tientsin region of China.

Pingius minutus is elliptical in shape measuring 22–28 μm in length and 13–18 μm in width (average dimensions 24.5 × 15 μm). The body is flattened dorso-ventrally and the two lateral sides are almost parallel to each other; both anterior and posterior ends are truncated. The cytostome is at the anterior of the cell and the cytoproct at the posterior end. The spherical micronucleus is located at the periphery of the ovoid macronucleus. A contractile vacuole is found in the posterior part of the cell close to the cytoproct. The cytoplasm contains other large granules. Unlike many other genera of the Buetschliidae, the surface ciliature is complete with the exception of a small area near to the cytoproct (Fig. 3.16).

The Genus Blepharoprosthium Although the genera of the Buetschliidae tend to be host-specific, this genus, like *Polymorphella,* has been observed in both equines and ruminants. *Blepharoprosthium pireum* (syn. *musculus*) has been reported in the large intestine of horses (Hsiung, 1930a) and *B. parvum* was observed on a limited number of occasions in the rumen of cattle (Hsuing, 1932).

Blepharoprosthium parvum is pyriform in shape as the anterior one-third of the body is somewhat constricted. Both the anterior and posterior ends are rounded and the organism ranges in length from 26–32 μm and in width from 16–20 μm, with average dimensions of 29 × 18 μm. The cytostome is at the anterior end of the cell and the cytoproct at the posterior extremity. The cytoplasm appears granular with a spherical nucleus located in the anterior half of the body. A small spherical micronucleus occurs at the periphery of the macronucleus. A contractile vacuole is located near to the cytoproct and other vacuoles are also present in the cytoplasm. Although there is a surface distension on the anterior surface of the cell, a concretion vacuole was not reported. A concretion vacuole is present in the anterior portion of *B. pireum* (Hsiung, 1930a). The somatic ciliature of *B. parvum* is limited to three areas. The anterior one-third of the cell is ciliated and the remaining ciliature is restricted to an area at the middle of the body below the raised surface and in the region of cytoproct (Fig. 3.16).

The Genus Blepharoconus *Blepharoconus krugerensis* was observed in the ruminal contents of one animal during a study of the rumen ciliates of Brazilian cattle by Dehority (1986b). The protozoon had been reported previously in the intestinal contents of the African elephant (*Loxodonta africana*) by Eloff and Van Hoven (1980). Other species of this genus are present in the equine large intestine (Hsiung, 1930a).

Blepharoconus krugerensis (Figs. 3.14 and 3.16) has a regular-shaped ovoid body that is flattened dorso-ventrally. Its posterior end is rounded while the anterior end is constricted to produce a knob-like protuberance. This is encircled by a slight depression which extends around the whole body. A conical-shaped cytostome is located at the anterior extremity of the cell. The rumen isolates were more rounded than those present in the elephant intestines having body measurements of 46(30–65) × 35(21–60) μm. The ectoplasm is homogeneous while the endoplasm is granular and contains food particles. The disc-shaped macronucleus measures 11(7–

15) × 5.5(4–8) μm and has a variable position within the cell. An elliptical micronucleus is located in a depression on the surface of the macronucleus. A concretion vacuole is situated close to the anterior constriction while the position of the two or three contractile vacuoles in the endoplasm is more variable. The somatic ciliature is incomplete. The anterior of the body is covered with regularly spaced longitudinal rows of thin cilia; the remaining ciliature comprises two short bands on opposite sides of the mid-region of the body and a bundle around the cytoproct at the posterior extremity (Figs. 3.14, 3.16).

The Genus Holophryozoon *Holophryozoon bovis* is a small ovoid protozoon that was observed in the rumen contents of cattle by Jirovec (1933). It measures 13–18 μm in length and 10–12 μm in width and has a complete but sparse somatic ciliature (Fig. 3.16). The oral aperture is pronounced at the anterior end of the cell. A large contractile vacuole was located at the posterior end of the body and the cytoplasm often contained other vacuoles and starch granules. A spherical macronucleus, 2–3 μm in diameter and the micronucleus occurred close together at a variable position within the anterior of the cell. Some cells contained two macronuclei and were considered to be ex-conjugants by Jirovec (1933).

The Genus Buetschliella A ciliate present in low numbers in bovine rumen contents was provisionally named *Buetschliella bovis* by Jirovec (1933) who compared the organism with previously described species of *Buetschlia*. Jirovec considered that, although *Buetschliella bovis* had many similarities with the *Buetschlia* spp., it was morphologically distinctive. The genus *Buetschliella*, however, is not in the family Buetschliidae, but is included in this section because of its many similarities with the rumen ciliates of this family and its morphological relatedness to *Polymorphella bovis* (Imai, 1984). The actual classification of the genus *Buetschliella* is given by Corliss (1979) as:

Class	Oligohymenophora	de Puytorac et al.
Subclass	Hymenostomata	Delage and Herouard
Order	Astomatida	Schewiakoff
Family	Buetschliellidae	de Puytorac
Genus	*Buetschliella*	Awerinzew

Buetschliella bovis is pear-shaped with a tapered anterior and more rounded posterior end (Fig. 3.16). The ciliate has an average length of 30 μm and is 15–20 μm wide. The anterior endoplasm is homogeneous although the remainder is vacuolated and contains numerous inclusions. The somatic ciliature is restricted to the anterior end of the cell. The spherical macronucleus, 7–8 μm in diameter, was situated in the middle of the cell. A micronucleus, concretion vacuole, mouth, and anus were not observed.

Blepharocorythidae Only one genus of this family is represented in the rumen. Jameson (1925a) observed a small ciliate in ruminal contents from sheep and cattle and assigned the organism to a new genus, *Charon*, commenting that it was closely related to the genus *Blepharocorys*, although more simply organized. The following year Dogiel (1926a) described the new species *Blepharocorys bovis* from bovine rumen contents but on recognizing the similarity with *C. ventriculi* he subsequently proposed (Dogiel, 1934) that the form should be called *Blepharocorys ventriculi* Jameson. The generic name was also deemed to be unsuitable in that it was a *nomen praeoccupatum* and already in use as an arachnid genus. The genus was therefore renamed *Charonina* (Strand, 1928); an alternative suggestion of *Charonella* (Bhatia, 1936) was not adopted. It has subsequently been established by Wolska (1967a) after detailed studies of the infraciliature and its morphogenesis that the ciliate should be included in the more primitive genus, *Charonina*. In addition to this type species two other members of the genus have been observed in rumen contents (Table 3.5), i.e., *C. equi* (Clarke, 1964) and *C. nuda* (Hsiung, 1932).

Charonina ventriculi has been observed on a limited number of occasions in the ruminal contents from sheep, cattle, and water buffalo (Table 3.2). When present the protozoon can attain a population size of 10^4–10^5 cells per ml. Surveys of different host species indicated that its frequency of occurrence in the various hosts was approximately 20–25%. The ciliate was present, for example, in 22% of the bison monitored by Towne et al. (1988b) and in 25% of the Bali cattle examined by Imai (1985). The ciliate was also detected in the protozoal populations of 17%, 23%, and 25% of the water buffalo examined in other studies (Dehority, 1979; Imai, 1985; Imai et al., 1981b). In Bali cattle and bison *C. ventriculi* represented 0–3% of the total ciliate population, whereas the corresponding values for Indonesian water buffalo were 0–9% (Imai, 1985; Towne et al., 1988b). However, in a Brazilian water buffalo studied by Dehority (1979) the proportion ranged from 1.9–14.6% and was dependent upon the diet the animal received. In the bison examined by Towne et al. (1988b) numbers were highest in animals fed an all-forage ration. The population size is evidently affected by the diet and *Charonina* has been reported to represent over 50% of the rumen ciliate population of sheep fed grass silage (Syrjala et al., 1976) and over 30% of that in a cow at pasture or receiving a grass hay ration (Dehority and Mattos, 1978). However, the population was reduced to approximately < 1% in this latter animal and bison (Towne et al., 1988b) when concentrates were fed. The number of *C. ventriculi* in the rumen also exhibit a postprandial variation. However, the numbers of this species in the rumen follow a diurnal cycle which is more closely related to that of the entodiniomorphid protozoa than the other holotrich ciliates present (Dehority and Mattos, 1978).

Charonina ventriculi has a very characteristic shape in that one side is

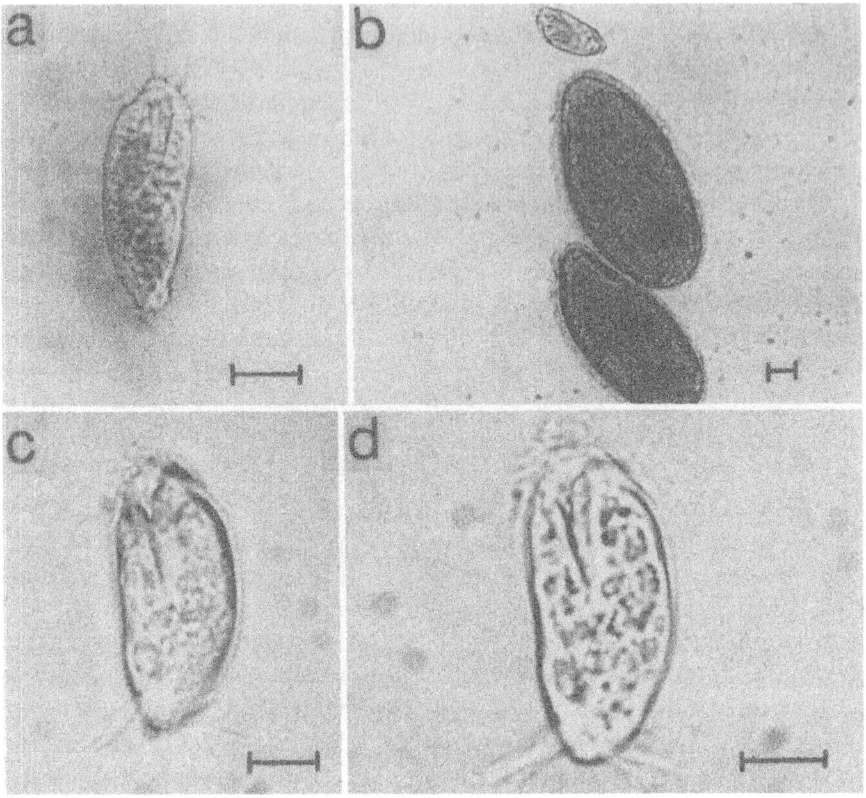

Fig. 3.17 Photomicrographs of some ruminal members of the family Blepharocorythidae. (a) *Charonina equi*. (b) *C. ventriculi* from the ovine rumen and *Dasytricha ruminantium*. (c and d) *C. ventriculi*. (a) From Imai et al. (1978). Bar = 10 μm.

almost straight and the other convex (Figs. 3.17, 3.18); the anterior end is bluntly pointed and the posterior end more finely tapered. The length of body varies between 24–41 μm and the width between 12–19 μm; the average dimensions are 35 × 16.5 μm. The oral aperture is round to ovoid in shape and located on the ventral surface of the body close to the base of the anterior tip. The cytopharyngeal apparatus is prominent and extends for half of the length of the cell. The position of the spherical to elliptical coarsely granular macronucleus in the cell is variable (e.g., Dehority and Mattos, 1978); the spherical micronucleus is usually located in a depression in the macronucleus. The cytoproct and a single large contractile vacuole are situated at the posterior of the cell. The somatic ciliature is restricted to two posterior tufts and an anterior zone. Electron microscopic examination of *C. ventriculi* isolated from the ovine rumen has revealed that the protozoan has features found in the Isotrichidae and Ophryoscolecidae (Fig. 3.18; R.G. Paul, R.D. Butler, and A.G. Williams, unpublished results). The

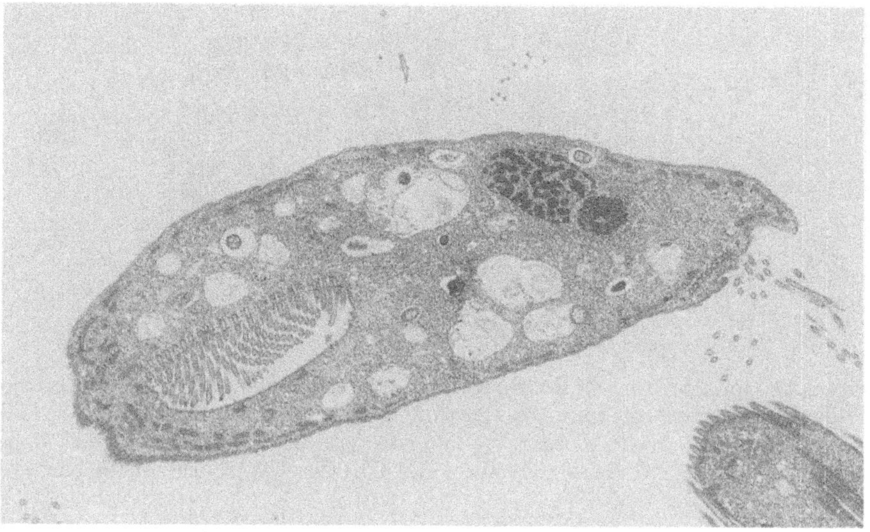

Fig. 3.18 Transmission electron micrograph of a thin section through a cell of ovine *Charonina ventriculi*.

cytoplasm is not separated into ecto- and endoplasm and contains food vacuoles, two types of endobionts, and granular hydrogenosome-like organelles. A cilial band lines one side of the vestibulum which is supported by microtubular ribbons and dense bands derived from the oral ciliature. The oral monokinetid has a kinetodesmal fiber, a transverse ribbon, and post-ciliary microtubules. In a few instances these give rise to small nematodesmata. The somatic ciliature is similar in structure, although the anterior tuft contains some apparent dikinetids. The cortex is lined by bundles of microtubules and contains some paired barren kinetosomes.

Charonina equi was orignally described in the feces and material from the colon of the horse (Hsiung, 1930b). *Blepharocorys equi* had been described previously by Schumacher (1915). The ciliate has been observed subsequently on two occasions in bovine rumen contents (Clarke, 1964; Imai et al., 1978). In the Japanese survey the population of *C. equi* was only 200 cells/ml representing less than 0.1% of the total ciliate population. However, in cattle fed grass hay in New Zealand the distribution of *C. equi* in the rumen ciliate population ranged from 0.2–12.8% (Clarke, 1964). Although morphologically similar to *C. ventriculi* there are certain characteristic features by which *C. equi* may be distinguished: body size, anal ciliature, shape and location of nuclei, and length of cytopharynx.

Charonina equi again is lancet-like in shape as one side appears to be more convex than the other (Figs. 3.17, 3.19). The anterior end is bluntly pointed and projecting while the posterior end is more rounded. The dimensions of *C. equi* given by Hsiung (1930b) are 30–48 μm in length and

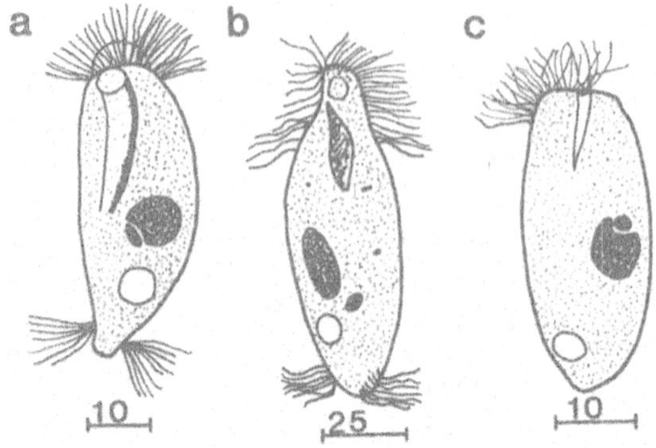

Fig. 3.19 Line drawings of the three species of *Charonina* (family Blepharocorythidae) that have been found in the rumen. The protozoa represented are (a) *Charonina ventriculi*; (b) *C. equi*; (c) *C. nuda*. The drawings are adapted from Jamieson (1925a) and Hsiung (1930b, 1932). The lengths (in μm) of the bars are indicated.

10–14 μm in width with average values of 39.5 × 12 μm. Clarke (1964) gives average figures of 38 × 15 μm. As with *C. ventriculi* both dorsal and ventral surfaces are slightly convex and the organism is compressed dorsoventrally. Jameson (1925a) estimated that the thickness of *C. ventriculi* was only some 3 μm. The ectoplasm of *C. equi* is homogeneous and the finely granular endoplasm contains food particles and among other inclusions are the nuclei and a single contractile vacuole in the posterior part of the body. The position of the large, elongated, ellipsoidal macronucleus can vary, although it is usually located in the posterior half of the cell. The ovoid micronucleus is located some distance away from macronucleus. The triangular opening of the mouth is on the ventral surface of the anterior protuberance or "knob". The cytopharyngeal apparatus is prominent and extends for one-third of the length of the cell. The mouth opening is surrounded by three zones of cilia. Apart from the anterior cilia, the somatic ciliature is restricted to two bundles on either side of the cytoproct at the posterior end of the body.

Charonina nuda resembles *C. ventriculi* but differs principally in that its posterior end is not ciliated (Fig. 3.19). The protozoon was originally described by Hsiung (1932) in rumen samples from Chinese cattle. Strelkov (1939) proposed that the organism should be transferred to the genus *Charonnautes*. *Charonina nuda* is 35–45 μm long and 15–19 μm wide (average dimensions 40.5 × 16 μm) with a lancet-shaped body, which is again flattened dorso-ventrally. The buccal aperture is on the ventral surface at the base of the anterior knob and a long narrow cytopharynx extends into the cell. A distinct cytoproct is located at the posterior of the cell. The somatic ciliature is restricted to the anterior knob. The ectoplasm is thicker along the margins of the cytopharynx and the endoplasm contains many large granules. A single contractile vacuole is situated at the posterior end of the cell close to the cytoproct, whereas the location of the spherical macronucleus in the cytoplasm is variable. A small spherical micronucleus is located at the side of the macronucleus.

Paraisotrichidae There has only been one report of the occurrence of this family within the rumen. A small ciliate with many characteristics of the family Paraisotrichidae was observed in rumen contents of cattle from Brazil (Dehority, 1986b). The ciliate was present in only 1 of the 4 animals surveyed and comprised 8.5% of the total ciliate population. The body was ellipsoidal averaging 19(14–23) × 12(9–14) μm and was covered by long cilia that were arranged in eight spiral rows. The anterior end of the body was slightly tapered while the posterior end was smoothly rounded. The mouth was located at the anterior end of the cell with the anus at the posterior end. An ovoid macronucleus was close to the centre of the body and one contractile vacuole was present in the posterior portion of the cell. Although an anterior concretion vacuole is characteristic of the family, none was observed in the granular endoplasm. Superficially similar *Paraisotricha* spp. have been observed in the giraffe and horse (Hsiung, 1930a; Kleynhans and Van Hoven, 1976) (Fig. 3.9). There are, however, morphological differences between the rumen isolate and the previously described equine species and it is likely that the rumen ciliate represents a previously undescribed species. The cytology and ultrastructure of the equine *Paraisotricha* spp. were examined in detail by Grain (1966a, 1966b).

Ciliates Not Assigned to any Specific Family A small ciliate was present in some of the Norwegian cattle examined by Orpin and Mathiesen (1986). The protozoon had certain distinctive characteristics and the name *Microcetus lappus* was proposed. It was detected subsequently both in cattle and in the rumen ciliate population of the bison (Towne and Nagaraja, 1989; Towne et al., 1988b).

The Genus *Microcetus* contains a single species which has been described in bovine ruminal contents (Orpin and Mathiesen, 1986). *Microcetus lappus* was not present in all of the cattle examined but in two animals represented 12 and 15% of the total ciliate population. It has been suggested by Orpin and Mathiesen (1986) that *M. lappus* was confused with *O. bubali* in earlier studies of protozoa in North American cattle (Dehority et al., 1983). In this latter study the concentrations of the ciliate ranged up to 35% of the protozoal population in unweaned calves and as high as 72% in mature feedlot animals. *M. lappus* was absent from feedlot bison but represented an average 2% (0.9–8.4%) of the total ciliate population of animals given forage or forage/grain rations (Towne et al., 1988b). In heifers the *M. lappus* concentration was variable and numbers exceeded 50% of the total protozoal population (Towne and Nagaraja, 1989). The protozoa were not evenly distributed in the rumen ecosystem; they were most numerous in the reticulum and occurred in low numbers in the mid-dorsal sac. Dietary protein and energy levels did not affect the population size and, although described as a holotrich, the diurnal variation in numbers was similar to that of the entodiniomorphs (Towne and Nagaraja, 1989).

The cells are ovoid to elongate ranging from 18–29 μm long and 7.5–18 μm wide (average dimensions 23.6×12.7 μm). Dimensions of protozoa from bison and cattle are not significantly different. The buccal cavity is apical and the cytoproct is subterminal on the ventral surface or terminal. The somatic ciliature is absent from the cytoproct region and covers approximately 90% of the cell surface. The rows of cilia are inclined at an angle to the body axis (approx. 20°). The buccal cilia occur on the dorsal side of the cavity. The macronucleus is spherical to ovoid and located close to the centre of the cell. Some cells containing two adjacent spherical macronuclei were observed. A single contractile vacuole occurs close to the cytoproct. Two characteristic cytopharyngeal rods are present. Many cells contain ingested starch grains and bacteria (Fig. 3.10d).

The taxonomic position of the genus *Microcetus* is uncertain. Orpin and Mathiesen (1986) proposed that the genus should be included in the class Kinetofragminophorea (Levine et al., 1980). However, the cytopharyngeal apparatus of *Microcetus* does not correspond with the structures described for either of the two subclasses which possess cytopharyngeal strengthening assemblies (viz. the Vestibuliferia and the Hypostomatia) and the ciliate may therefore represent a previously undescribed taxon. The protozoon has yet to be assigned to a taxon (see Table 3.4).

3.5 Evolution of the Holotrich Ciliates

The evolution of the ciliate groups has been considered in detail by other reviewers (e.g., Corliss, 1979; Dogiel, 1947; Grain, 1966b). It is assumed that the ancestors of the rumen ciliates were originally taken in with the drinking water (Dogiel, 1947). The phylogenetic affinities as proposed by Corliss (1979) for the Trichostomatida are indicated below:

Other ciliate lines
↗
Primociliatida → Prostomatida → Trichostomatida { Trichostomatina / Blepharocorythina }
↓
Entodiniomorphida

Grain (1966b) discussed at length the phylogeny of the endocommensal holotrichs of the digestive tract of both ruminants and equids. He considered that both the gymnostomes and trichostomes were polyphyletic and that the two orders had a common ancestor. He proposed the interrelationships that are shown in Fig. 3.20. During the evolution of the gymnostomes, diversification primarily affected the extent of the somatic ciliature

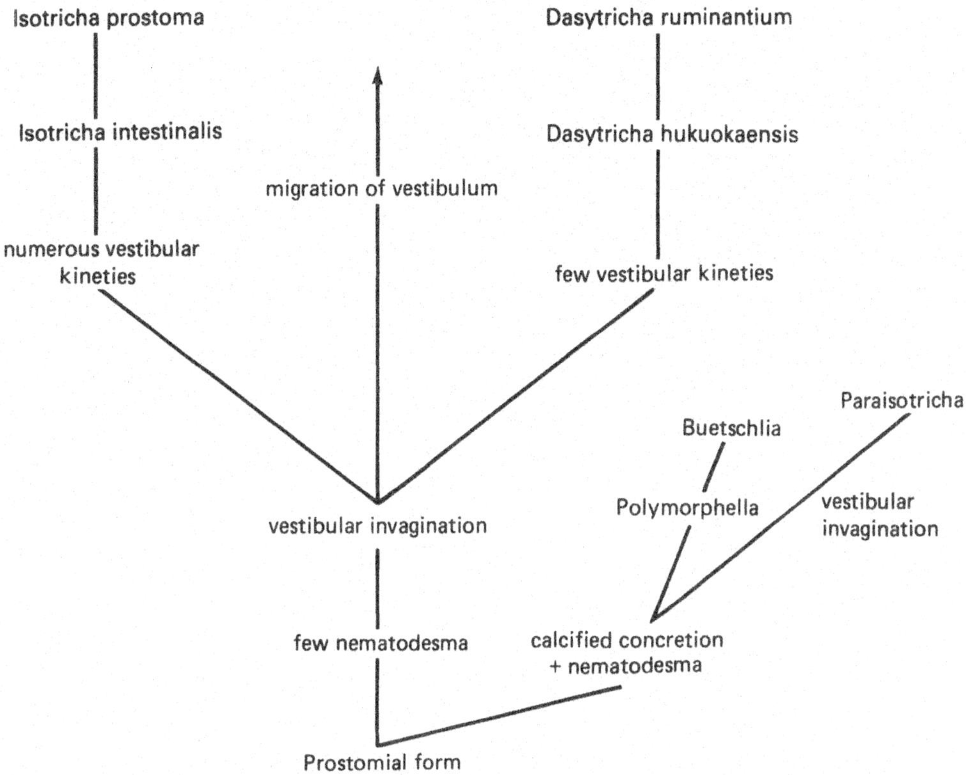

Fig. 3.20 An evolutionary tree of the rumen holotrich ciliates. Adapted from Grain (1966b).

and the nature and arrangement of the elements of the pharyngeal architecture. In the trichostomes there have been several directions of evolution in which the ends of all or part of the somatic kineties were invaginated into a vestibulum, which may have been repositioned by subsequent migration. Vestibular migration has attained its maximum effect in the Isotrichidae.

Within the Isotrichidae the two species of *Isotricha* show analogies in terms of general organization, course of morphogenesis, and ultrastructure. It is probable, starting from a primitive prostomal form, that evolution in the genus occurred as a result of the progressive retreat of the vestibulum at the same time as morphogenetic processes were being simplified. Grain (1966b) thus concludes that *I. prostoma* is more highly evolved than *I. intestinalis*.

Although there are marked similarities between *Dasytricha* and *Isotricha*, there are also some pronounced differences in cellular organization and vestibular ciliature. The two species of *Dasytricha* therefore constitute a different evolutionary line to that of *Isotricha*, and have evolved as a result of the vestibular migration. *Dasytricha hukuokaensis* has a degree of evolution similar to that of I. intestinalis, and *D. ruminantium* has a degree of evolution equivalent to that of *I. prostoma* (Fig. 3.20).

The Isotrichidae and Paraisotrichidae belong to two different evolutionary lines in which the migration of the vestibulum toward the back has not proceeded at the same speed, although the vestibulum has attained the same degree of organization in the two lines. A critical divergence was the appearance of calcareous concretions. Grain (1966b) proposed that the Paraisotrichidae are derived from the primitive Buetschliidae, or at least from the forms that have given rise to them. In the family Buetschliidae, evolution consists of a regression of the somatic ciliature (Strelkov, 1939). *Buetschlia* is thus more highly evolved than *Polymorphella* (Fig. 3.20) (Grain, 1966b). The family Blepharocorythidae is distanced from other endocommensal trichostomes by a number of characteristics. However, further studies on morphogenesis are necessary before the exact position of the Blepharocorythidae can be determined with any certainty.

4

Methods Used for the Separation and the Cultivation of Rumen Protozoa

4.1 Techniques for the Isolation of Ciliates from Rumen Contents

In the following section the recovery of protozoa from conventional animals or those containing specifically manipulated protozoal populations will be considered.

Isolation of a Mixed Ciliate Population from Rumen Contents The procedures that have been described for the isolation of rumen ciliates from rumen contents are all based on the use of simple filtration and centrifugation techniques. The rumen liquor sample is first strained through one or more layers of cheesecloth or similar material (i.e., muslin, surgical gauze), to remove the large particulate plant debris. The particulate material retained by filtration is thoroughly washed with an anaerobic salts solution and the washings combined with the strained rumen liquor (Oxford, 1951); the overall dilution of the rumen liquor should be approximately twofold. Pretreatment of rumen contents in a mechanical homogenizer or blender to release protozoa from the particulate material should be avoided; the ciliates are damaged by these procedures (Dehority, 1984). The diluted liquor is then transferred to a separating funnel where it is allowed to stand for approximately 30–60 minutes at 39°C. All buffer solutions and glass utensils used in the separation are gassed with an appropriate O_2-free gas (e.g., N_2 and CO_2) and pre-warmed to 39°C before use. The composition of the anaerobic salts solutions used by different workers does vary but the most common components are NaCl, $NaHCO_3$, CH_3COONa, KH_2PO_4, K_2HPO_4,

$MgSO_4 7H_2O$, and $CaCl_2 2H_2O$ producing a pH in the range 6.7–7.2 (Coleman, 1978a; Heald et al., 1952; Oxford, 1951; Williams and Harfoot, 1976). The reducing agents used most regularly are cysteine hydrochloride, sodium sulphide, and dithiothreitol.

During sedimentation the protozoa settle to the bottom of the separating funnel and may be collected in fresh buffer (Figs. 4.1 and 4.3). The microbial fermentative activity in the liquor is sufficient to cause flocculant debris to rise to the surface as a scum layer which can be removed by suction. The addition of glucose to stimulate this activity has been advocated (Heald and Oxford, 1953; Heald et al., 1952; Oxford, 1951), although in the experience of the authors this addition is unnecessary and adequate separation has been obtained using rumen samples from animals fed a range of grass, hay, concentrate, or silage-based diets. It is sometimes advantageous, however, to increase the sedimentation period to 60–90 minutes with rumen liquor samples obtained more than 6h after the host animal has been fed.

The liquor fraction remaining after sedimentation does contain protozoa that have not settled, principally some of the smaller *Entodinium* spp. and the more actively motile types (e.g., *Dasytricha* and *Isotricha* spp.) which may be recovered by centrifugation (1000 g for 10 min) or by filtration and collection on a 10 μm porosity filter (Williams and Strachan, 1984). The retentate is combined with the sedimented protozoal fraction. The protozoal preparation thus obtained will be contaminated with bacteria, flagellate protozoa/fungal zoospores, and some small particulate debris, especially if centrifugation has been used in the initial recovery procedures. Any residual plant debris is readily removed by floatation. The protozoal suspension is held at 39°C in fresh buffer for a further 10–15 min. During this period the contaminating material forms a scum layer which may be removed by suction. Bacterial contamination is minimized by thorough washing with fresh, and preferably sterile, anaerobic buffer. Flagellate protozoa and fungal zoospore contamination, if present, is also removed at this stage. The washing procedure consists of successive resuspension and sedimentation in buffer. Sedimentation can be under gravity or by centrifugation (200 g, 30 sec) (Coleman and Sandford, 1979a). The supernatant is removed by suction and discarded. Contaminating plant residues are also effectively removed during this washing procedure. The bacterial contamination can also be minimized by thorough washing of the protozoal preparation with sterile buffer on a 10 μm porosity filter (Williams and Strachan, 1984). Contaminating microorganisms, unlike the protozoa, are not retained by the filter (Fig. 4.1).

The ciliate preparations obtained in this manner have negligible free bacterial contamination and are suitable for use in biochemical, morphological, or other studies. Older methods that advocate protozoal and bacterial recovery from rumen contents by differential centrifugation are, however, unreliable. The protozoal and bacterial fractions are recovered sequentially

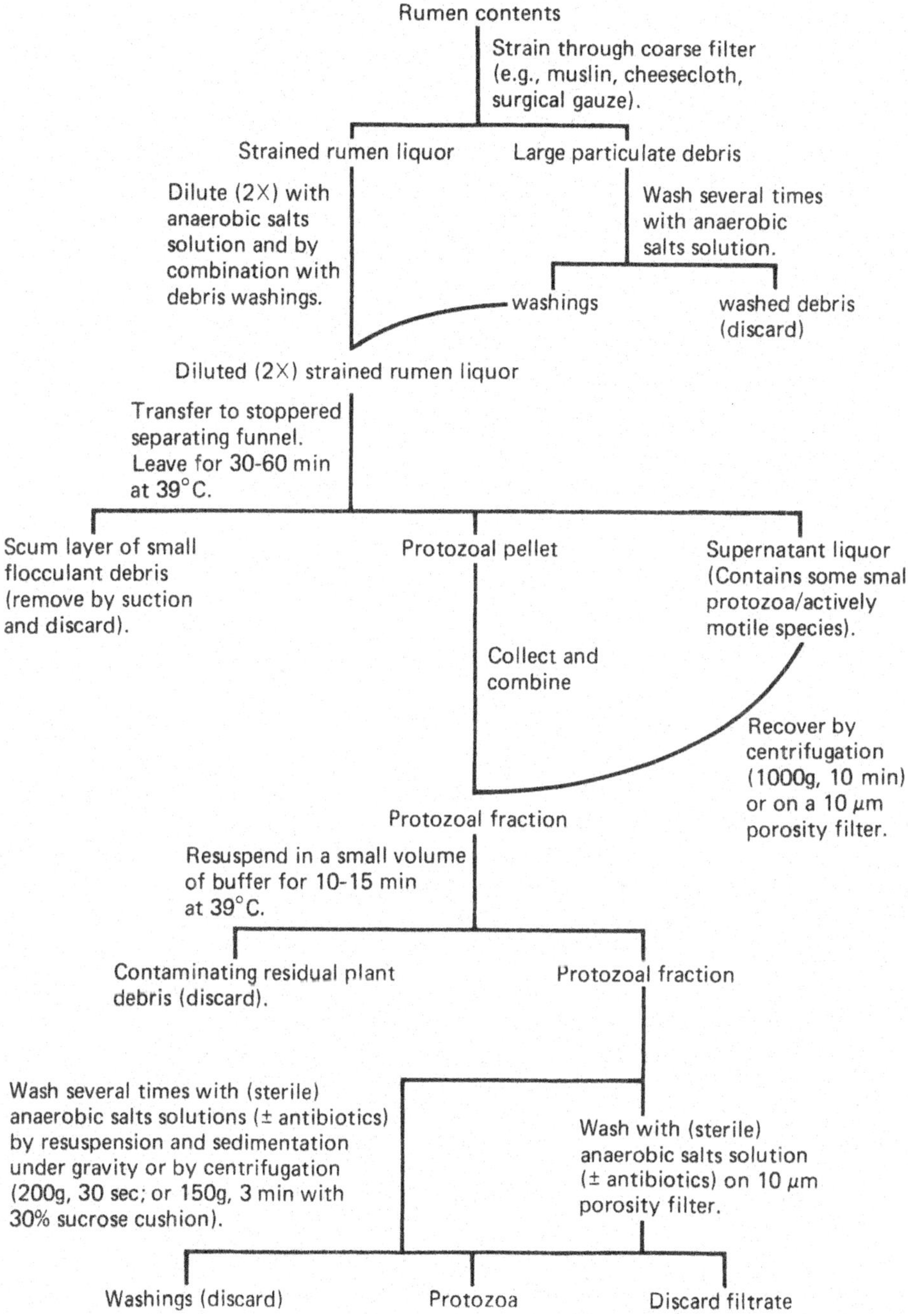

Fig. 4.1 Flow diagram of a procedure for the separation of protozoa from a sample of rumen contents.

by centrifugation at 1000 g for 10 min and 20,000 g for 30 min respectively. From the foregoing discusssion it is obvious that the "protozoal" suspensions obtained by this method would be heavily contaminated and unsuitable for any meaningful studies. Other methods that have been suggested for removing bacteria from protozoal preparations involve the use of sucrose-density gradient centrifugation (Mangan and West, 1977) or the use of sucrose-enriched centrifugation buffers (Ogimoto and Imai, 1981). In the latter method the "protozoal fraction" is overlayed on a buffer containing sucrose (30% w/v) and centrifuged at 150 g for 3 min. The ciliates are sedimented while the bacteria do not enter the sucrose cushion. Ciliates separated by this method are claimed to be particularly suitable for taxonomic and morphological examination.

A method of separation which has yet to be evaluated and developed involves the use of electromigration (Masson et al., 1952; Wagener et al., 1986). Using this technique Wagener and his coworkers were able to extract anaerobic ciliates from sludge samples. They also noted that some rumen ciliates (e.g., *Dasytricha, Isotricha, Diplodinium*, and *Epidinium*) were stimulated by the applied electric current to swim to the cathode. This method may, therefore, be of value in the isolation of ciliates from rumen contents or more specifically in the recovery of specific organisms from mixed populations. Masson et al. (1952) were able to separate holotrich, large, and small entodiniomorphid ciliates at the cathode by varying both the current (5–25 mA, 10 V) and application time.

Although the free bacterial contamination may be effectively minimized, adherent and ingested bacteria will still be present in the protozoal suspension. This contamination may be further reduced by treatment with antibiotics. In earlier studies free bacterial contamination was reduced by post-isolation treatment of the protozoa with antibiotics for 36–48 h (Heald and Oxford, 1953; Heald et al., 1952). This incubation period was also of value in depleting the amylopectin reserves of cells isolated in the presence of added glucose. More recent attempts to isolate axenic preparations of the ciliates have included antibiotics in the washing buffers and in extensive post-isolation treatments (e.g., Bonhomme et al., 1982a, 1982b; Hino and Kametaka, 1977; Jarvis, 1974; Onodera and Henderson, 1980; Onodera et al., 1977c; Quin et al., 1962). Antibiotics regularly used include penicillin and streptomycin derivatives, chloramphenicol, kanamycin, spiramycin, and cephaloridine. Heald and Oxford (1953) suggested that streptomycin was toxic to certain of the entodiniomorphid protozoa. More information on the direct and indirect effects of antibiotics on the rumen protozoa is undoubtedly required.

Isolation of Specific Protozoa from a Mixed Ciliate Population It is possible to isolate individual genera or specific protozoal groups from rumen contents or a mixed ciliate preparation. The procedures used are generally based on a size fractionation by differential centrifugation, sedimentation, or filtration. It is relatively easy to separate large or small organisms from a mixed population; it is not practicable, however, to separate organisms whose size ranges overlap, although mixed preparations occurring within a specific size range may be separated by differential centrifugation (Coleman and Sandford, 1979a), density gradient centrifugation (Mangan and West, 1977), sedimentation through buffered gradients (Mould and Thomas, 1958), or more specifically by filtration (Newbold et al., 1987). Preparations of entodiniomorphid protozoa may be freed from contamination by holotrichs by treatment with mannose. Mannose and certain hexosamines are toxic to the holotrich ciliates causing cell lysis (Sugden and Oxford, 1952). Overnight treatment of a large protozoal preparation with mannose enabled Oxford (1958) to obtain a holotrich-free suspension of *Epidinium*.

Other entodiniomorphid ciliates have been isolated directly from rumen contents for biochemical or other studies. The larger species *Metadinium medium* (Sugden, 1953) and *Epidinium ecaudatum* (Oxford, 1958) were separated by sedimentation and washing while *Ophryoscolex and Diplodinium* spp. were recovered from a mixed large protozoal pellet by differential sedimentation in a column containing buffer (Williams et al., 1961). *Entodinium* spp. are smaller and settle much more slowly and may, therefore, be recovered from the supernatant after sedimentation and centrifugation procedures as described by Coleman (1969a). Preparations of *Entodinium* spp. are readily obtained by filtration using sintered glass (Bonhomme-Florentin, 1967) or defined porosity textile filters (Williams et al., 1984). *Polyplastron multivesiculatum* has also been isolated from a mixed protozoal suspension by differential filtration using a 100 μm porosity polyester textile filter (Williams and Ellis, 1985).

The holotrich ciliates synthesize a storage polysaccharide and the increased cell mass imparted by this material has been used to aid cell separation. Protozoa were separated from rumen liquor by sedimenting strained rumen contents (Heald et al., 1952; Masson and Oxford, 1951) or by repeatedly washing the residue remaining after rumen contents had been filtered through surgical gauze (Oxford, 1951). Glucose was included to encourage polysaccharide formation by the holotrichs and facilitate a more rapid sedimentation. *Isotricha* and *Dasytricha* were subsequently recovered by differential sedimentation of the mixed holotrich preparation in buffer (Heald et al., 1952) or on gradients containing rumen liquor (Gutierrez, 1955) or glycerol (Mould and Thomas, 1958). Species separation has also been achieved by adhesion to glass surfaces (Gutierrez, 1955) and by sucrose gradient centrifugation (Mangan and West, 1977). Holotrich preparations obtained by these methods were treated with antibiotics to reduce bacterial contamination and starved for 48 h to deplete the storage material in the cells (e.g., Heald and Oxford, 1953).

These older methods have many drawbacks, and separation of the holotrich genera by differential sedimentation is frequently incomplete. The preparations are contaminated by bacteria and entodiniomorphid protozoa. Methods relying on cell adhesion are also unreliable and produce poor yields, and have been replaced by filtrations procedures using sintered glass filters (Williams and Harfoot, 1976) or, more effectively, defined aperture nylon and polyester textiles (Williams and Yarlett, 1982). Filtration techniques are particularly useful in the preparation of protozoal suspensions for metabolic studies because of the sensitivity of these anaerobic organisms to excessive manipulation, and because the metabolic status of the cell is not affected. Preliminary fermentation or starvation procedures are unnecessary and thorough washing of the cell preparation with sterile anaerobic buffer ensures negligible contamination by free bacteria. However, it is not possible to separate *Isotricha prostoma* and *I. intestinalis* by filtration and in consequence mixed *Isotricha* spp. preparations are obtained. This problem can be overcome by the reintroduction of a single species to a host animal that has been treated chemically or physically to remove its natural protozoal population (i.e., defaunated; see Section 4.2). Partial defaunation and reintroduction of single holotrich species has been used as a means of obtaining individual holotrichs for biochemical studies (Eadie and Oxford, 1957; Howard, 1957).

The range and uniformity of the precision textiles available and their ease of use offer considerable potential for precise separations where the fractionation of complex populations into species or groups on a size basis is feasible. Sintered glass and steel filters give inferior separation and lower cell yields (Williams and Yarlett, 1982). The textile filter material is supported in a Hartley filter from which the perforated back plate has been removed and replaced by two sealing rings which hold and tension the material (Fig. 4.2). The filtration process is completed more quickly with the textiles and the use of reduced pressure to achieve filtration is unnecessary. A consequence is that air is not drawn into the suspension remaining on the filter. The surface area of the filter is also gassed with an oxygen-free gas mixture during the procedures and the usual precautions are observed to maintain anaerobiosis (e.g., pre-warmed deoxygenated buffers are used). The maintenance of anaerobic conditions during the separation procedure is aided in that the filtration proceeds rapidly without recourse to reduced pressure, and as a result the protozoa isolated are morphologically normal and metabolically active. The technique is particularly useful in the preparation of protozoal suspensions for metabolic studies because of the sensitivity of the organisms to excessive manipulation and because the metabolic status of the cells is not affected by preliminary fermentation or post-isolation starvation procedures. If desired the initial sedimentation and filtration stages can be performed in an anaerobic cabinet.

Some applications of differential filtration for the isolation of rumen ciliates are shown in Fig. 4.3. It is probable that the careful selection of filter combinations will permit the separation of other protozoal fractions from the rumen contents of conventional or partially defaunated animals (e.g., Lockwood et al., 1988).

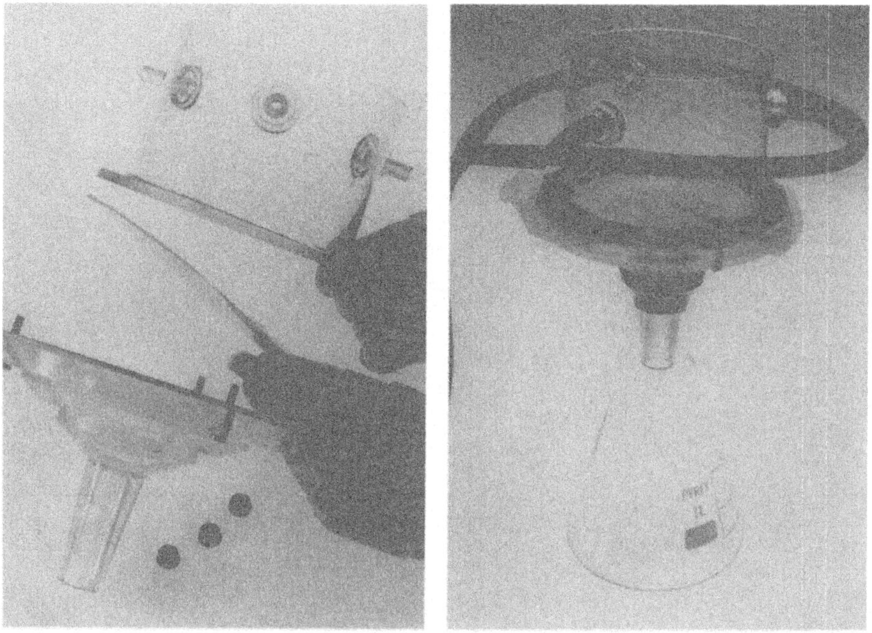

Fig. 4.2 Filter assembly used for the separation and isolation of rumen protozoa.

4.2 In situ Culture of Selected Protozoa

Protozoa isolated directly from the rumen have not been exposed to the imposed nutritional constraints and environmental effects of in vitro cultivation. In addition high cell densities can be established and maintained in vivo. The rumen contents of animals maintaining a monoculture or predetermined limited protozoal population thus represent an excellent source of protozoa whose biochemical and structural characteristics have not been affected by growth in the natural environment. The establishment of specific protozoal populations is a two-stage process. It is first necessary to completely eliminate the original protozoal population from the host animal (defaunation). When this has been achieved the desired protozoa may be reintroduced into the rumen. The cells for the refaunation can be isolated directly from the rumen contents of a donor animal or alternatively a laboratory culture may be used as the inoculum.

Defaunation Procedures Chemical or physical treatments to remove the protozoa should have no adverse effects upon the host animal or the other rumen microorganisms. Although a variety of chemical agents and treatments have been used to achieve defaunation, with varying degrees of success, as yet no completely satisfactory method has been described. Many are either ineffective or injurious to the health of the animal (Hobson

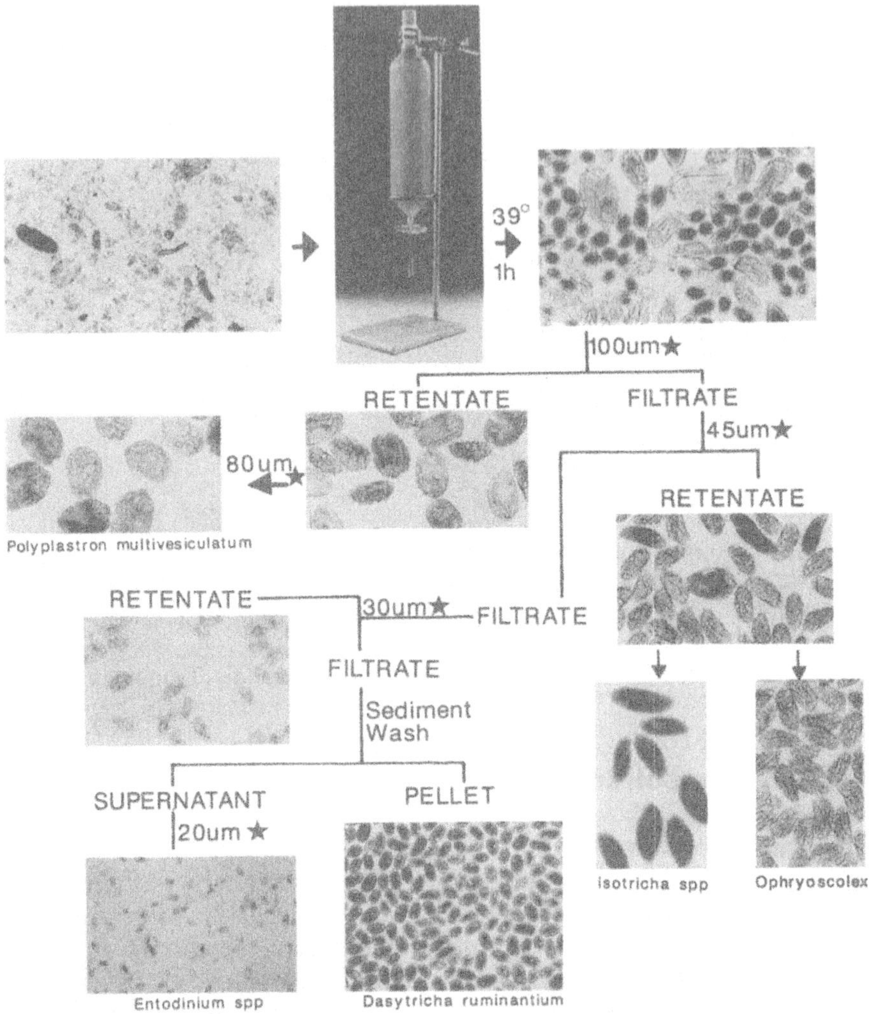

Fig. 4.3 Outline of a procedure used to isolate specific protozoal groups from a mixed protozoal population by differential filtration and sedimentation. The filter porosity size is indicated at each filtration stage by a star.

and Jouany, 1989; Jouany et al., 1988; Lovelock et al., 1982). In addition the omasal protozoa can be more difficult to eliminate, particularly in cattle, and may reinoculate the treated rumen (Towne and Nagaraja, 1990). Some of the procedures advocated for defaunation are listed in Table 4.1.

Defaunation has been used widely in studies to evaluate the role of the protozoa in the nutrition and productivity of the host ruminant, and methods of protozoal removal were evaluated initially for this purpose. Early attempts to eliminate the protozoa included the addition of hydrochloric acid, acetic acid, or copper sulphate to the rumen (Becker, 1929;

Table 4.1 Various procedures for defaunation—the removal of ciliate protozoa from the rumen

Method	Reference
Physical treatment	
Rumen emptying + heat treatment of contents	Eadie and Oxford (1957)
Rumen emptying + sequential washing	Jouany and Senaud (1979a)
Isolation of newborn animal from birth	Abou Akkada and El Shazly (1964); Bryant and Small (1960); Eadie (1962a)
Dietary treatment:	
Starvation	Mangold and Schmidt-Kramer (1927); Mangold and Usuelli (1930); Warner (1962b)
Feeding milk	Dogiel and Winogradowa-Fedorowa (1930); Kreuzer and Kirchgessner (1987a)
Feeding oils	Ikwuegbu and Sutton (1982); Newbold and Chamberlain (1988); Sutton et al. (1983); Van Nevel and Demeyer (1981)
Feeding of high starch diet to induce mild acidosis	Whitelaw et al. (1984)
Feeding of high energy diets to young animals	Kreuzer and Kirchgessner (1987b)
Feeding a diet with a high zinc content (1 g/kg DM)	Durand and Kawashima (1980)
Chemical treatment	
Copper sulfate	Becker (1929); Becker et al. (1929)
Hydrochloric acid	Gunther (1899)
Acetic acid	Becker (1929); Becker et al. (1929)
Calcium peroxide	Demeyer (1982)
Anionic surfactants	
Dioctyl sodium sulfosuccinate (manoxol OT/aerosol OT)	Abou Akkada et al. (1968a); Orpin (1977c); Wright and Curtis (1976)
Alkanates[a]	Burggraaf and Leng (1980)
Nonionic surfactants	
Terics[b]	Bruggraaf and Leng (1980); Lovelock et al. (1982); Wright and Curtis (1976)
Pluronics	Wright and Curtis (1976)
Poloxalene	Bartley (1965); Davis and Essig (1972)

[a]Some of the surfactant derivatives tested include: calcium alkyl benzene sulfonate, alkyl phenoxy polyoxyethylene sulfate, and sodium lauryl diethoxy sulfate.
[b]Some of the terics include: ethoxylates of nonylphenol, cetyl oleyl alcohol, polypropylene glycol, and straight-chain synthetic alcohols (teric GN9, Rexol 25J, Synperonic NP9).

Becker et al., 1929; Gunther, 1899). Subsequently it was shown that various derivatives and certain organic infusoricides were toxic to the rumen ciliates in vitro (Bailey and Howard, 1962; Eadie and Oxford, 1954; Eadie et al., 1956). The anionic surfactant teepol (sodium salt of a secondary alkyl sulphate) was also used to disrupt the rumen ciliates (Oxford, 1951) and currently the most widely used chemical defaunating agents are surfactant detergents. Anionic (e.g., dioctyl sodium sulfosuccinate, alkanates) and non-ionic (terics, pluronics, poloxalene) surfactants have been evaluated both in laboratory (Campbell et al., 1982; Wright and Curtis, 1976) and animal trials (Burggraaf and Leng, 1980; Davis and Essig, 1972; Orpin, 1977c).

Bis-(2-ethylhexyl)-sodium sulfosuccinate (i.e., dioctyl sodium sulfosuccinate, DSS) is an anionic detergent that was first used successfully as a defaunating agent by Abou Akkada et al. (1968a). The lethal concentration for the holotrich and entodiniomorphid ciliates in vitro is between 20 and 30 μg/ml (Orpin, 1977c). However, the quantity of DSS required to remove these protozoa from the rumen of sheep was approximately 30 times higher than this lethal concentration as the surfactant interacts with the particulate material in the rumen. The efficiency of defaunation is improved if food is withheld from the animal for 24 h before treatment with the defaunating agent. Although Abou Akkada et al. (1968a) believed that DSS treatment was not injurious to the health of the host animal, it is now realized that treatment with DSS can adversely affect the host (Lovelock et al., 1982). An effective regime for defaunating sheep that has been used successfully on several occasions by the authors is outlined below.

Two doses of DSS are administered; the first treatment (10 g) is added in solution to the rumen through the cannula some 24 h after the animal was last fed. A second treatment (5 g) is administered after a further 24 h and food is then offered to the animal. The most effective surfactants, however, automatically reduce the voluntary feed intake of sheep. The animals are separated from other untreated (conventionally faunated) animals during, and subsequent to, the defaunation procedures.

Willard and Kodras (1967) screened over 170 compounds for antiprotozoal activity and concluded that certain anionic surface active agents were potentially useful, the more effective derivatives being long-chain sulfates and phosphates. The antiprotozoal properties of several nonionic and anionic detergents were also evaluated by Burggraaf and Leng (1980) who likewise concluded that anionic detergents were more effective. Although the anionic alkanates appeared to have the most potential, several nonionic surfactants were shown in the latter survey to exhibit considerable antiprotozoal properties. The Teric group of nonionic surfactants are long chain alcohols condensed with various amounts of ethylene oxide per mole of alcohol (i.e., alcohol ethoxylates). Two separate studies (Burggraaf and Leng, 1980; Wright and Curtis, 1976) demonstrated that the most effective Teric compound for defaunation was Teric GN9 (nonylphenol with 9 moles of ethylene oxide/mole alcohol). Nonylphenol ethoxylate (Teric GN9, Synperonic NP9) has been used to defaunate both cattle and sheep (Bird and Leng, 1978; Bird et al., 1979; Eadie and Shand, 1981), and although it proved to be an effective defaunating agent, the dose required for complete defaunation had adverse effects on some animals (Burggraaf and Leng, 1980; Eadie and Shand, 1981; Lovelock et al., 1982). The bacterial population is also affected by both anionic and nonionic surfactants (Orpin, 1977c).

Other types of neutral nonionic surfactants that are toxic to rumen ciliates are the Pluronic detergents (Wright and Curtis, 1976). These surfacants are condensation products of polypropylene glycol and polyoxyethylene. The Pluronic surfactant L62 was shown to be most active against

Epidinium spp. but was less harmful to the larger holotrichs. When used as a bloat-preventing agent L62 reduced, but did not completely eliminate, the protozoal population of a bovine rumen (Oxford, 1959). Treatment with another nonionic bloat-preventing surfactant poloxalene (Bartley, 1965; Bartley et al., 1965) also resulted in a partial defaunation of grazing cattle (Davis and Essig, 1972). The surfactant poloxalene is a polyoxypropylene-polyoxyethylene block polymer. Other nonsurfactant feed additives that may have some potential as defaunating agents include calcium peroxide (Demeyer, 1982) and free oils (Broudiscou et al., 1988; Newbold and Chamberlain, 1988; Sutton et al., 1983; Van Nevel and Demeyer, 1981).

It is also possible to defaunate the rumen without recourse to potentially toxic chemical additives. Jouany and Senaud (1979a) described a rumen washing procedure for fistulated sheep. The rumen was emptied and sequentially washed with water, a dilute formaldehyde solution, and water again. The procedure was repeated on 3 successive days and the animal fasted for 48 hours before the first treatment. Animals defaunated by this method have been maintained ciliate-free for periods in excess of six months duration by the author. Other workers have not been able to obtain effective long-term defaunation by this method and there are reports of a low incidence of animal losses (Jouany and Senaud, 1979a; Jouany et al., 1988; Lovelock et al., 1982). The most effective means of obtaining ciliate-free animals is by rearing the ruminant in isolation from soon after birth so that contact with, and inoculation from, faunated animals is prevented (Abou Akkada and El Shazly, 1964; Bryant and Small, 1960; Eadie, 1962a). Isolating newborn animals from adult ruminants provides an effective and safe means of obtaining ciliate-free ruminants.

Partial Defaunation of the Rumen Although there are obvious advantages in achieving a complete defaunation prior to reinoculation with the selected protozoa or protozoal genera, it is not always necessary to eliminate all the ciliates. Partial or selected defaunation is equally effective provided that the inoculated species is readily separated from the residual protozoal population. Thus the elimination of the larger protozoa or a specific group (e.g., the holotrichs) would permit the reestablishment of a single large species (e.g., *Polyplastron, Eudiplodinium*) or an individual holotrich representative. Partial defauntation can be achieved by either physical or chemical means.

Eadie and Oxford (1957) described a method for the removal of the holotrich ciliates from the rumen of fistulated sheep. The method required that the rumen should be emptied and thouroughly washed. The removed ruminal contents were strained through surgical gauze and the filtrate heated (50°C, 15 min.) to kill the protozoa. The treated contents were returned to the animal. Although some small entodiniomorphid ciliates were not removed by the treatment, it was possible to reestablish and maintain an active population of *Dasytricha ruminantium* in the animal.

The holotrich ciliates have also been removed by treatment with 1,2-dimethyl-5-nitroimidazole (Clarke et al., 1969).

The larger protozoa and holotrichs are, in general, more susceptible to the effects of the surfactant detergents (Burggraaf and Leng, 1982; Orpin, 1977c; Wright and Curtis, 1976) and it is therefore feasible to eliminate these species with the detergents using lower dosing rates than prescribed for complete defaunation. Some antibiotics also have potential for selective defaunation, and many compounds have been screened for activity against the protozoal population (e.g., Gyulai and Hudouskova, 1987; Kubo et al., 1981; Newbold et al., 1988; O'Connor et al., 1970, 1971; Olumeyan et al., 1986; Panjarathinam et al., 1973; Purser et al., 1965). The effects of the coccidiostatic ionophore antibiotics monensin and lasalocid on the protozoal population are influenced by the diet (Dennis et al., 1986; Dinius et al., 1976; Gyulai and Barani, 1988; Katz et al., 1986; Poos et al., 1979; Richardson et al., 1978). The holotrich ciliates are less sensitive to the effects of monensin than the entodiniomorphid ciliates (Dennis et al., 1986; Hino, 1981), and the differing sensitivity of the protozoal types to various antimicrobial agents may offer some potential for the selective manipulation of the generic composition of the protozoal population.

Management of Animals with a Manipulated Protozoal Population
The maintenance of defaunated animals and those possessing defined reintroduced protozoal populations requires some specific management practices. Careful housing and handling of the animals is essential to prevent physical contact between the animals at all times in order to eliminate any possibility of ciliate transfer in saliva or rumen liquor during normal handling and feeding procedures. Inadvertent transfer by people working with the animals must also be avoided.

It is advisable, although not essential, to accommodate the animals in areas not used to house normally faunated animals. The individual refaunated animals are kept in well-separated high-sided pens which can be extended at the front if the animals are not tethered (Fig. 4.4). Restrained animals are usually arranged in a nose-to-tail manner (Eadie, 1962a). All feed utensils and instruments necessary for the care and sampling of the animals should be designated exclusively for individual animals. Other precautions advocated for animal handlers include the wearing of specific protective clothing, footwear, and sterile disposeable rubber gloves and the use of footbaths prior to entering the pens (Lovelock et al., 1982). The enforcement of these simple precautions will enable both defaunated or partially refaunated animals to be maintained for long periods.

The defaunation-reinoculation technique is thus a particularly effective means of maintaining stable monocultures or cocultures of rumen ciliates, or for evaluating the effects of individual protozoa in the rumen ecosystem. The protozoal inoculum introduced into the rumen of a defaunated animal establishes a stable population of up to 10^5 cells/ml within 2–4 weeks. Indi-

Fig. 4.4 Pens modified to accommodate either defaunates or sheep that have been defaunated and subsequently refaunated with a single protozoal species or defined protozoal population. The sides of the pens have been heightened and extended to prevent inter-animal contact and undesired transfaunation.

vidual species of protozoa can be recovered easily by differential filtration or sedimentation techniques (Gyulai, 1983; Lockwood et al., 1988; Williams et al., 1984; Fig. 4.5). Several species of protozoa have been cultivated in this manner for studies on their biochemistry (Coleman, 1980, 1986c; Williams, 1989; Williams and Coleman, 1989) or role (Coleman, 1986a, 1986b; Jouany and Senaud, 1982, 1983; Jouany et al., 1981a, 1981b).

4.3 Cultivation of Entodiniomorphid Protozoa In vitro

Entodiniomorphid protozoa have now been cultivated in vitro with a mean generation time of 6–48 h for periods of from one to greater than 27 years in the presence of bacteria in a reduced buffered salts medium, but it took many years of work before the present culture media were derived. Coste (1864) and Edwards (1864) used hay infusions and Becker and Talbott (1927) used media based on serum and egg fractions, all without success. Knoth (1928) used a hay infusion under CO_2/CH_4 and managed to keep the ciliates alive for 107 h. Margolin (1930) was more successful and kept the ciliates alive for 24 days, even without incubation anaerobically, by the use of a hay

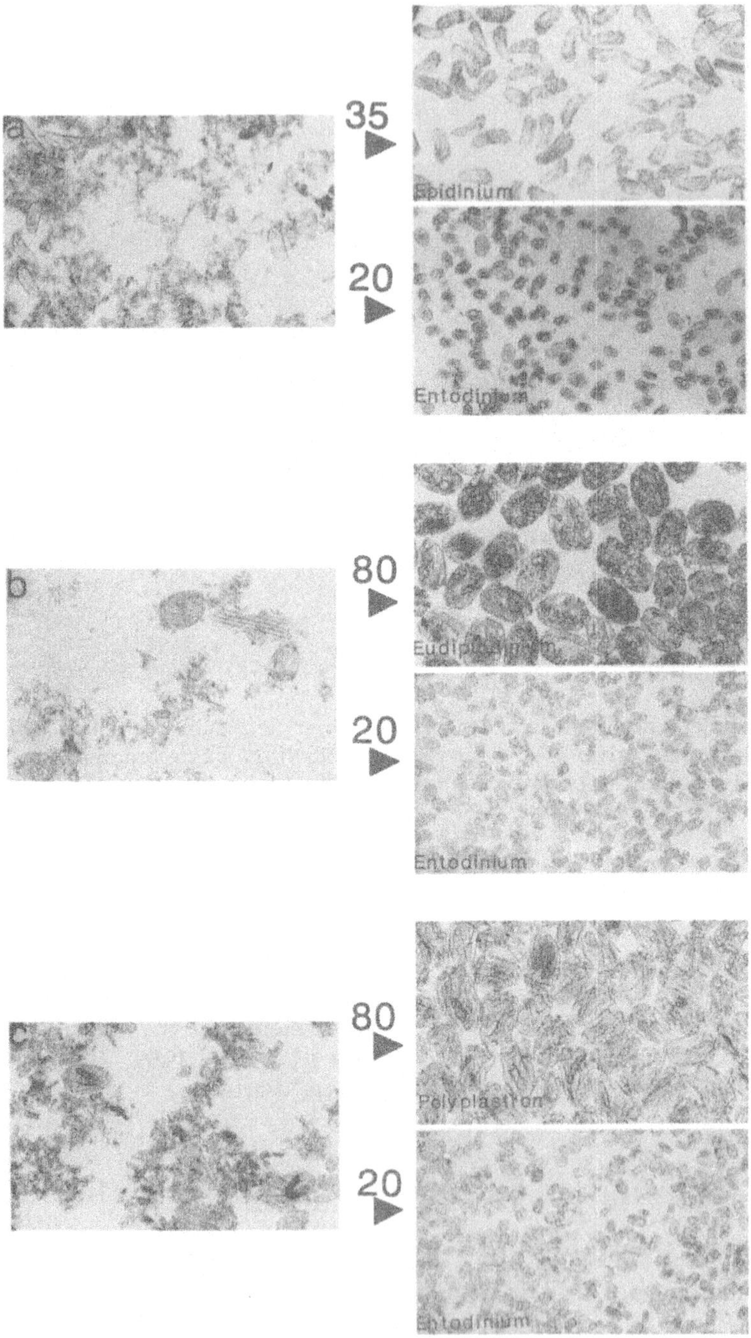

Fig. 4.5 Examples of the application of differential filtration with defined-aperture textile filters to recover specific protozoal types from the rumen contents of selectively refaunated donor animals. The cocultures introduced and established in the otherwise ciliate-free hosts were: (a) *Epidinium caudatum* and *Entodinium* spp.; (b) *Eudiplodinium maggii* and *Entodinium* spp.; (c) *Polyplastron multivesiculatum* and *Entodinium* spp. The filter sizes in (μm) required to separate these protozoa from washed protozoal preparations are indicated.

infusion with filter paper and rice starch previously incubated with bacteria. Westphal (1934) made further progress by showing the advantage of removing dissolved oxygen by bubbling inert gas through the medium and by showing that protozoa-free rumen fluid was stimulatory and substances such as peptone and beef extract were detrimental to protozoal growth.

The first really successful attempt to culture entodiniomorphid protozoa was made by Hungate (1942) who used 0.6% NaCl adjusted to pH 6.1–7.6 from which oxygen was removed by the passage of $N_2 + CO_2$ (95:5). The only sources of organic matter necessary for the growth of the cellulolytic protozoon *Eudiplodinium (Eremoplastron) neglectum* were powdered dried grass and cellulose. Hungate found that it was necessary to maintain the temperature at 38°C and to transfer the ciliates to fresh medium every other day. Average generation times were 1–2 days and the protozoa were kept alive at population densities of 100–1000/ml, depending on the species, for up to 22 months. Hungate (1943) extended these studies to other species and found that the addition of ground wheat was essential for the cultivation of *Entodinium caudatum* which reached a population density of 5000/ml.

Unfortunately after this, a number of workers (Kandatsu and Takahashi, 1955a, 1955b, 1956; Oxford, 1958; Sugden, 1953) could not repeat these successful experiments and a belief grew up that it was not possible to culture these ciliates. However in 1958 Coleman (1958, 1960a) showed that it was possible to culture *Entodinium caudatum* apparently indefinitely at population densities of about 20000/ml with twice-weekly dilution of the culture with fresh medium. Although it is not possible to determine exactly why so many had failed to grow entodinomorphid protozoa, the following are the most important points.

Anaerobiosis All media must be rendered oxygen-free and maintained under oxygen-free gas. The removal of oxygen can be done by autoclaving in the conventional way but it is quicker and easier to bubble any oxygen-free gas through the medium for 2–3 min. For preference the cultures should be maintained in almost completely filled long thin tubes, such as centrifuge tubes, which are sealed with a rubber bung after all air has been flushed out of the gas space with CO_2. Carbon dioxide is preferable to N_2 as the former tends to dissolve in the medium and create a slight vacuum which pulls the bung in. With N_2 the bungs tend to be displaced on placing in an incubator. The shape of the culture vessel is important because it is desirable to have as much reduced medium as possible between any air that might reach the surface during feeding, for example, and the protozoa at the bottom of the vessel.

Low Redox Potential The ciliates will not survive, even in oxygen-free media, unless the redox potential is lowered by the addition of 0.02% (w/v) L-cysteine hydrochloride (neutralized) and/or 10% (v/v) fresh rumen fluid from which the protozoa have been removed.

Carbon Dioxide Some CO_2 dissolved in the medium is essential for the growth of all entodiniomorphid protozoa. With some species 5% (v/v) CO_2 in the gas used to remove dissolved oxygen is adequate and this does not alter the pH of the phosphate-rich medium used. With other species it is necessary to use 100% CO_2 in the gas phase, in which case it is necessary to replace some phosphate in the medium by bicarbonate.

pH This should be maintained between 6.5 and 7.0 and certainly not below 6.0 or above 7.5. For this reason the authors have found that phosphate-rich (Coleman, 1960a) rather than chloride-rich (Hungate, 1942) media are preferable.

Rumen fluid Fresh rumen fluid from which the protozoa have been removed (by centrifugation at 500 g for 5 min and standing for 24 h at 4°C) should be added to all media at a concentration not exceeding 10% (v/v) (Coleman, 1987a). In the authors' experience higher concentrations tend to be toxic, although Michalowski (1975b) used undiluted rumen fluid to culture entodinia and diplodinia from the buffalo for 3 months. Many of the larger and cellulolytic protozoa, once established in culture, grow well on media in the absence of rumen fluid (Coleman et al, 1976) whereas with the entodinia the addition of rumen fluid is either essential or stimulatory. *Entodinium longinucleatum* dies in 17 days in the absence of rumen fluid (Owen and Coleman, 1976) whereas the population density of *E. simplex* increased from 8000 to 37000/ml on addition of 10% rumen fluid (Coleman, 1969a). There is evidence (Coleman, 1969a) that it is the particulate matter in the rumen fluid that is important although Tompkin et al. (1966) found the reverse. Tompkin et al. (1966) also found that the source of rumen fluid used in the culture was important and that, for entodinia, fluid taken from animals containing a large population of entodinia gave better protozoal growth than fluid from an animal with a low population density of entodinia. For established cultures of *Entodinium* spp. the rumen fluid can be sterilized by autoclaving before use without detriment to the cultures (Coleman, 1969a; Owen and Coleman, 1976; Tompkin et al., 1966) but cannot be used as a replacement for fresh rumen fluid with other genera. Fresh rumen fluid can be stored for 2 weeks at 4°C (Coleman, 1987a).

Green Plant Material Some form of green plant material has been used in the culture medium by almost all who have attempted to grow rumen ciliates (e.g., Coleman, 1960a; Margolin, 1930; Michalowski, 1975; Onodera and Henderson, 1980; Oxford, 1958). The most popular and widely used material has been ground dried grass, although Michalowski (1975) used dried carrot. Dried grass when used as sole added source of food will support the growth of the larger cellulolytic protozoa such as *Eudiplodinium maggii, Diplodinium pentacanthum, Enoploplastron triloricatum,* and *Ostracodinium dilobum* (Coleman, 1987a). If the powdered dried grass is sterilized by dry heat (160°C for 1 h), some charring occurs and the resulting material will not support protozoal growth. Plant material has, however, been successfully sterilized, without detriment to the culture by autoclaving anaerobically (Hino and Kametaka, 1977). As will be shown below, the grass acts not only as a source of cellulose and hence energy, for the protozoa, but also as a source of growth factors such as sterols. It is essential to add fresh grass (where this is the only added source of food) to the cultures each day even though there is a mass already present.

Starch In the authors' experience, all the entodiniomorphid protozoa tested, including the small *Entodinium simplex* (Coleman, 1969a), engulf and utilize starch grains as a source of energy. Some species such as *Entodinium caudatum* and *Eudiplodinium maggii* will utilize rice starch, which has small grains, for growth, but most will only use an unpurified starch such as ground wheat (Coleman, 1987a). Clarke (1963), however, obtained good growth of mixed ciliates using rice starch as the sole source of starch. Fresh starch must be added to the cultures each day and the quantity added is critical to the success of the experiment. If too little is added, there is insufficient starch for the protozoa to grow on. If too much is added, the non-engulfed starch remains in the bottom of the culture vessel and is attacked by the bacteria with the production of toxic and acidic products which are detrimental to the protozoa. In a range of 0.1 to 0.5 mg starch/ml, the protozoal population density is proportional to the amount of starch added (Coleman, 1969a; Rahman et al., 1964). However, the optimal range does depend on other factors such as the presence or absence of fresh or autoclaved rumen fluid, with much less starch being needed in the absence of rumen fluid or in the presence of autoclaved fluid than is needed in the presence of fresh rumen fluid (Coleman, 1960a, 1987a). Intact starch grains cannot be replaced by soluble starch, amylose, or amylopectin because these materials are soluble and readily attacked by bacteria with the production of toxic products. As with the dried grass, the starches cannot be sterilized by dry heat (160°C for 1 h) without producing a material which will not support protozoal growth. This was probably the cause of the failure of Sugden (1953) to culture these protozoa. In addition to being a source of starch, the rice starch used for *Entodinium caudatum* cultures also provides choline which is essential for protozoal growth (Broad and Dawson, 1976).

Soluble Complex Organic Materials In the authors' experience, the addition to the basal medium of soluble complex organic materials such as yeast extract, peptone, casein, glucose, or serum invariably results in the death of the culture within a few days or weeks, depending on how much is added. Any stimulation of protozoal growth is nullified by the massive growth of bacteria. However, some authors found the reverse. For example, Clarke (1963) routinely added 40 μg casein/ml/day and Hino et al. (1973a,b) used whole egg protein, peptone, and yeast extract in their media for entodinia although they found that the last two could be omitted. Einszporn (1961) also claimed that gluten stimulated the growth of *Entodinium longinucleatum* but the results tend to suggest that the stimulation was only of short duration. Onodera and Henderson (1980) found that, in antibiotic-treated culture of *Entodinium caudatum,* heat-treated yeast and cell-free extract of rumen bacteria bound onto charcoal were essential for limited protozoal survival. Hino and Kametaka (1977) used peptone and yeast extract in axenic cultures of the same protozoon. Gredzinski (1975) found that urea might stimulate growth of entodinia. From all the evidence available, it appears likely that the protozoa have a very limited biosynthetic ability and require a large number of organic compounds for growth. These are normally supplied by engulfed bacteria and plant particles but, as suggested by Onodera and Henderson (1980), the problem in axenic culture may be the delivery of these compounds in the "right" state and concentration to the protozoa.

Growth Factors Broad and Dawson (1976), Hino et al. (1973c), and Onodera and Henderson (1980) have shown that the green plant material commonly added to cultures of entodinia can be replaced by certain sterols and particularly by β-sitosterol. As these sterols are not soluble in water, they are fed coated on starch grains (Coleman, 1987a; Coleman and Reynolds, 1982a; Onodera and Henderson, 1980). Coleman and Reynolds (1982a) showed using this technique that the dried grass could be replaced by β-sitosterol, stigmasterol, ergosterol, or α-spinasterol but not by cholesterol, dihydrocholesterol, or lanosterol. With some species, for example *Epidinium caudatum,* the population density increased from 155 to 730/ml when the starch was coated with β-sitosterol and the dried grass omitted. *Choline,* which is normally present in commercial rice starch, is essential for *Entodinium caudatum* (Broad and Dawson, 1973). It cannot be replaced by ethanolamine or by its mono- or dimethyl derivatives. *Haemin* has a stimulatory effect on the growth of *Epidinium caudatum* and *Ophryoscolex caudatus* (Coleman and Reynolds, 1982a). Vitamin B_{12} was first observed to stimulate the growth of ciliates from the goat rumen by Kandatsu and Takehashi (1955a, 1955b; 1956) although Coleman (1960a) could not confirm this finding. More recently Bonhomme et al. (1982a) found that inorganic cobalt (0.1–1.0 μg/ml) or vitamin B_{12} (0.12–0.24 μg/ml) increased the population density of *Entodinium* spp. from 21000 to 55000/ml. The growth of ciliates treated with antibiotics was stimulated by the addition of bacteria and this effect was more pronounced when the bacteria had been preincubated with cobalt salts.

Temperature Rumen ciliates are very temperature sensitive. The optimum is 39°C and they will not grow above 42°C or below 35°C. Growth of *Entodinium caudatum* at 41°C produces chains of individuals that do not separate properly. A few individuals of this species will also survive for 12 h at room temperature but this treatment kills all other species.

Other Rumen Ciliates When grown under certain conditions in vitro some isolates of *Polyplastron multivesiculatum* have a growth requirement for *Epidinium* spp. which are engulfed (Coleman, 1987a; Coleman et al., 1972). *Entodinium bursa* will not grow in vitro in the absence of the spineless form of *Entodinium caudatum* which, when present, is engulfed avidly (Coleman et al., 1977).

Practical Details for the Cultivation of Entodiniomorphid Protozoa The following method can be used to cultivate any species of entodiniomorphid protozoa. For greater detail, see Coleman (1987a).

It is first necessary to make up Coleman-type salt solution. This can be prepared in bulk and will keep indefinitely. It is important during the preparation not to add the calcium salt until the phosphates have been dissolved in the bulk of the water.

Coleman-type Salt Solution (g/liter)

K_2HPO_4	12.7
KH_2PO_4	10.0
NaCl	1.3
$MgSO_4 \cdot 7HO$	0.18
$CaCl_2$ (dried)	0.09

This is then used to prepare the following two solutions which should be placed in suitable tubes (see below), gassed as indicated, and the tubes sealed with a rubber bung. The cysteine hydrochloride solutions should be neutralized with NaOH solution before use.

Caudatum-type Salt Solution

Coleman-type salt solution	50.0 ml
15% (w/v) $CH_3 \cdot COONa$	0.5 ml
2% (w/v) L-cysteine hydrochloride	1.0 ml
H_2O	50.0 ml

This solution should be bubbled vigorously for 2–3 min with 95% (v/v) $N_2 + 5\%$ (v/v) CO_2.

Simplex-type Salt Solution

Coleman-type salt solution	40.0 ml
5% (w/v) $NaHCO_3$	15.0 ml
2% (w/v) L-cysteine hydrochloride	1.1 ml
H_2O	60.0 ml

This solution should be bubbled vigorously with CO_2 for 2–3 min.

Cultivation of Mixed Ciliates Anybody attempting to grow rumen ciliates for the first time should inoculate a few drops of crude rumen contents into 35 ml caudatum or simplex salt solution (contained in a 50 ml centrifuge tube) containing 10% (v/v) prepared fresh rumen fluid (from which the protozoa have been removed) and feed initially and each day thereafter with powdered dried grass (a few mg) with and without 0.1 ml aqueous suspension of wholemeal flour (1.5% w/v). The tubes should be gassed with CO_2, sealed with a rubber bung, and incubated at 39°C. After 3–7 days, depending on how well the protozoa are growing, the tube should be mixed, half poured into another tube, and both made up to the original volume with fresh medium. This dilution of the cultures 1–2 times a week, combined with daily feeding with grass and starch, should be continued until it is apparent which protozoa are growing well under each of the four conditions. These can then be picked out individually and grown as clone cultures under the optimum conditions.

Isolation of Individual Species Individual ciliates can be picked out of diluted rumen contents or a mixed culture with a micropipette attached to a micromanipulator and inoculated into 3 ml of the appropriate salt solution containing fresh prepared rumen fluid as above and contained in a 125 × 12 mm tube gassed with CO_2 and sealed with a rubber bung. Powdered dried grass (a few mg) and if required, wholemeal flour (minimal amount, for example 10–100 μg), should be added initially and daily thereafter. As the numbers of protozoa increase, the amount of flour added each day should also be increased. Once or twice a week the volume of medium should be doubled by the addition of fresh medium, the exact rate depending on how well the protozoa are growing. If the growth is poor, the medium above the protozoa should be removed and replaced by fresh without increasing the volume. The size of the tube used should be increased as the volume of medium increases until the cultures are contained in 50 ml centrifuge tubes. Table 4.2 gives details of the media and food materials on which, in the authors' experience, the different protozoal species grow best. More information on the details of the cultivation of individual species is given in Coleman (1987a).

Life of Protozoal Cultures Although *Entodinium caudatum* apparently grows indefinitely in culture (> 27 years—Coleman, 1987a) and E. *simplex* has been cultured for 8 years, most of the other species have a life of only about two years in vitro. The reason for this is unknown, but most species appear to suffer a loss of vigour after 12–18 months in culture and then gradually decline until they die. This loss of vigour can be reversed by inoculation of such a culture into the rumen of a defaunated sheep. Although not all protozoal species grow when reintroduced in this way, those that do can be reinoculated into media in vitro and will then grow for a further period of two years. The reason for this re-invigoration is not known but it could be due to conjugation which only occurs in the rumen (Coleman, 1980).

The Preparation and Survival of Axenic and Monoxenic Cultures of Entodiniomorphid Protozoa Although it is comparatively easy to separate entodiniomorphid protozoa from free bacteria by differential centrifugation, it is much more difficult to rid the protozoa of their intracellular bacteria and this can only be achieved by treatment with antibiotics. In order to prepare axenic cultures of a rumen ciliate, it is necessary to consider four questions: 1) which antibiotic or combination of antibiotics and the concentration is required to kill the intracellular bacteria in a reasonable period of time without harming the protozoa; 2) what growth media and other techniques should be used to detect the presence of residual intracellular bacteria; 3) how should the solid food materials such as starch be sterilized when autoclaving and dry heat cannot be used; and 4) what is the range and concentration of supplementary nutrients that are required by protozoa growing in the absence of the bacteria which usually supply the complex organic molecule requirements of the protozoa?

Treatment of Protozoa with Antibiotics Penicillin is not toxic to *Entodinium caudatum* even at high concentrations (1 mg/ml). In cultures penicillin apparently kills starch-digesting bacteria, as protozoal suspensions which would normally have to be fed each day will, in the presence of the antibiotic, survive for 14 days without further addition of starch (Coleman, 1960b). Although it does not kill all the bacteria associated with *E. caudatum*, it is an essential component of many of the antibiotic mixtures used to render protozoal suspensions bacteria-free (Bonhomme et al., 1982b, 1982c; Coleman, 1962; Hino and Kametaka, 1977; Jarvis, 1974; Onodera et al., 1977c). In more recent studies, penicillin has been replaced by ampicillin which is also active against Gram negative bacteria (Bonhomme et al., 1982a; Coleman, 1978c).

One of the first attempts to prepare bacteria-free entodiniomorphs was that of Oxford (1958) who used comparatively low concentrations of penicillin (40 μg/ml) and neomycin (30 μg/ml) and failed although the medium was made free of bacteria. Subsequently Abou Akkada and Howard (1960) used 50 μg chloramphenicol/ml and claimed to have obtained bacteria-free entodinia although it seems likely that their technique for detecting bacteria was inadequate. Coleman (1962) treated suspensions of *Entodinium caudatum* with a mixture of penicillin G, neomycin, streptomycin, and dihydrostreptomycin and obtained preparations that contained less than one bacterium per 10 protozoa but unfortunately the protozoa died in 3–4 days. Hino and Kametaka (1977) successfully used a mixture of carbenicillin, aminopenicillin, cephaloridine (500 μg each ml), and chloramphenicol and leucomycin (100 μg each/ml) in the presence of *Escherichia coli* to produce suspensions of *Entodinium caudatum* containing only this bacterium. Once the protozoon was growing well under these conditions, the bacteria were removed by treatment with the first three antibiotics mentioned above. These axenic protozoa remained alive for 22 days which is the longest period yet achieved. More recently Bonhomme et al. (1982a, 1982b) used alternate treatments with a carbenicillin-aminobenzylpenicillin-cepharolidine mixture and a chloramphenicol-spiramycin mixture to

Table 4.2 Conditions for the isolation and cultivation of entodiniomorphid protozoa

	Isolation method	Preferred source of starch[a]	Growth on grass alone in absence of starch[b]	Growth on straw alone in absence of starch[c]
Diplodinium monacanthum	Single cell	WF 0.05	+	ND
Diplodinium pentacanthum	Single cell	(WF 0.05)	++	++
Diploplastron affine	Single cell	WF 0.05	(++)	–
Enoploplastron triloricatum	Single cell	(WF 0.05)	++	ND
Entodinium bursa	Single cell	WF 0.03	–	–
Entodinium caudatum	Enrichment	RS 0.15	–	–
Entodinium furca dilobum	Single cell	WF 0.05	–	–
Entodinium longinucleatum	Single cell	WF 0.1	–	–
Entodinium simplex	Differential centrifugation	WF 0.02	–	–
Epidinium caudatum (bovine)	Single cell	WF 0.1	ND	–
Epidinium caudatum (ovine)	Single cell	WF 0.1	+	–
Epidinium tricaudatum	Single cell	WF 0.1	ND	–
Eremoplastron dilobum	Single cell	WF 0.05	++	ND
Eremoplastron bovis	Single cell	WF 0.05	++	ND
Eudiplodinium maggii	Single cell	(WF 0.03) (RS 0.03)	++	+
Metadinium medium	Single cell	(WF 0.03)	++	–
Ophryoscolex caudatus	Single cell	WF 0.25	–	–
Ostracodinium dilobum	Single cell	WF 0.05 (RS 0.05)	++	++
Polyplastron multivesiculatum (+)	Single cell	WF 0.05		–
Polyplastron multivesiculatum (–)	Single cell	WF 0.05 RS 0.05	(++)	+

[a]WF, wholemeal flour; RS, rice starch; the figures are the optimum amount of the material (mg/ml) to be added each day; parentheses indicate that, although the protozoon will grow for a limited time on starch, it is not the preferred substrate.
[b]+, protozoon will grow poorly on grass alone; ++, protozoon will grow well on grass alone; ND, not determined. Symbols in parentheses do not apply to all strains.
[c]+, protozoon will grow for a limited time on straw alone; ++, protozoon will grow well for over one year on straw alone; ND, not determined.

produce an *Entodinium* sp. with a reduced bacterial flora and axenic *Polyplastron multivesiculatum* which survived for five days.

Methods for Determination of Bacteria Inside Protozoa Although it has not been widely done, it is essential in the authors' opinion to disrupt the protozoa by some gentle treatment before determination of the number of intracellular bacteria. If this is not done, all the bacteria in one protozoon appear just as one colony after the protozoon has disintegrated in the agar of the bacterial growth media. Suitable methods for the disruption of protozoa include a Potter homogenizer (Potter and Elvehjem, 1936) and mild

Stimulation of growth of established cultures by rumen fluid[d]	Salts medium preferred[e]	Method of increasing culture	Requirement for other protozoa[f]	Maximum population density (Sprotozoa/ml)		Maximum life of culture
				Starch and grass	Grass alone	
–	S	Dilution	–			13 months
(FRF)	C or S	Dilution	–	400	800	14 months
(FRF)	C	Dilution	–	100	200	27 months
–	S	Dilution	–		250	24 months
FRF	C	Dilution	+	150		12 months
FRF, ARF	C + CAP	Replace medium/ dilution	–	32,000		>27 years
FRF	S	Dilution	–			3 months
FRF, ARF	C	Dilution	–	2,100		15 months
FRF	S	Dilution	–	27,000		8 years
–	C or S	Dilution	–	500		15 months
–	C or S	Dilution	–	1,200	20	21 months
–	C or S	Dilution	–	1,000		18 months
–	S	Dilution	–			>10 months
(FRF)	C	Dilution	–		1,000	15 months
–	C	Dilution	–	300	350	2 years
–	C	Dilution	–		2	6 months
FRF	C	Dilution	–	120		29 months
(FRF)	S	Dilution	–	200	100	>2 years
(FRF)	C	Dilution	(+)	50		3 years
(FRF)	(C) H	Dilution	–	100	20	12 months

[d] FRF, prepared fresh rumen fluid; ARF, autoclaved rumen fluid. Abbreviations in parentheses do not apply to all strains.
[f] Symbols in parentheses do not apply to all strains.
[e] C, caudatum buffer; S, simplex buffer; + CAP, added chloramphenicol at 48 μg/ml; H, Hungate buffer.

treatment with ultrasound (Coleman, 1978c). The presence of bacteria should then be tested for using at least three different media, one of which should be a non-selective rumen fluid-containing medium such as that of Bryant and Small (1960). One of the others should be a rich organic broth (Bonhomme et al., 1982b; Coleman, 1962; Hino and Kametaka, 1977).

Sterilization of Particulate Food Materials As the conventional method of sterilizing things by dry heat, namely 160°C for 1 h, produced charring of starch and dried grass, it is necessary to use other methods. These have included dry heat at 120°C for 24 h (Coleman, 1960a) and irradiation with 5–6 Mrad (^{60}Co source; dose rate 640 rad/min) (Bonhomme et al., 1982b,

1982c; Hino and Kametaka, 1977) for starch and conventional autoclaving in 1% cysteine solution under N_2 for plant material (Hino and Kametaka, 1977).

Media for Survival of Axenic Cultures No medium has yet been devised that will keep axenic cultures of rumen ciliates alive for more than three weeks. Hino and Kametaka (1977) achieved greatest success (22 days survival) using a medium containing filtered rumen fluid, rice starch, α-amylase, peptone, yeast extract, β-sitosterol, lecithin, and B-vitamins. Bonhomme et al. (1982b, 1982c) obtained the survival of *Polyplastron multivesiculatum* for five days on a medium containing peptone, dried plant material, starch, and freeze-dried bacteria. In contrast, Coleman (1962) found that none of a range of possible stimulatory compounds had any effect on the survival of *Entodinium caudatum* suspended in a salts medium containing 14% Seitz-filtered rumen fluid. However, survival was improved by incubation under 100% CO_2.

More dramatic results on the effect of additives have been obtained using protozoa with a "reduced" level of bacterial contamination. Onodera and Henderson (1980) found, in experiments where there were four bacteria per protozoon free in the medium, that activated charcoal, heat-treated yeast, β-sitosterol-coated rice starch, and cell-free extract of rumen bacteria stimulated the protozoa to divide each day for four days to reach a population density of 50000/ml. Bonhomme et al. (1982a) also used bacteria-reduced entodinia to show that vitamin B_{12} prolonged the survival of protozoa incubated in salt solution containing prepared fresh rumen fluid.

Monoxenic Culture Hino and Kametaka (1977) successfully prepared *Entodinium caudatum* cultures containing only *Escherichia coli* by incubation of the protozoa with the bacteria and five antibiotics. The *E. coli* could be replaced by other bacteria including *Streptococcus bovis*. Such cultures survived, with dilution of the culture 2–3 times weekly with an equal volume of fresh medium, for over two months at population densites of 0.5–4.0×10^4/ml.

4.4 Cultivation of Holotrich Ciliates In vitro

Although some of the entodiniomorphid ciliates have been cultivated for extended periods in vitro (Coleman, 1978a; see Section 4.3) the only holotrichs that have been maintained successfully in laboratory culture are *Dasytricha ruminantium* and *Isotricha* spp. Both genera will survive as part of a mixed protozoal population in continuous culture and artificial rumen systems (see Section 4.5) They have also been grown separately in in vitro, small-scale, batch culture systems. The majority of the studies on the cultivation of the individual holotrich genera were undertaken in the 1950 and

1960 era although successful methods for the cultivation of holotrich-containing rumen ciliate populations had been described some time earlier (e. g., Margolin, 1930).

Cultures of individual holotrich ciliates were initially established in supplemented, anaerobic salts solutions and the protozoal cells were used in biochemical studies (Gutierrez, 1955, 1958; Gutierrez and Hungate, 1957). These cultures were maintained for relatively short periods of up to 14 days in duration. Culture longevity has been improved as a result of several studies that examined the specific cultural requirements of the individual species. The media used typically contained rumen fluid and other complex extracts. Thus, supplementation of an anaerobic isotonic buffered salts solution with one or more of the following components has been shown to be beneficial and extend the cultivation period: rumen fluid (Purser and Tompkin, 1965), protozoal extract (Clarke and Hungate, 1966; Kubo et al., 1980), bacteria (Gutierrez, 1958; Gutierrez and Hungate, 1957), antibiotics (Purser and Weiser, 1963), soluble carbohydrates (Bonhomme-Florentin and Hempel-Zawitowska, 1977; Gutierrez, 1955; Sugden and Oxford, 1952), or plant extract (Genskow et al., 1969). The optimal environmental conditions recommended by Quin et al. (1962) were a temperature of 37–38°C, pH 6.8–7.0, redox potential −200 mV to −260 mV and an osmotic pressure of 260 mEq of electrolytes per liter. The basal salts solution typically used included sodium chloride, sodium bicarbonate, potassium phosphate, magnesium sulphate, and calcium chloride. The nutritional requirements have not been characterized for either genus, although both require soluble carbohydrates and are able to utilize amino acids (Quin et al., 1962; Sugden and Oxford, 1952). Differing views have been expressed on the value of added rumen liquor (Clarke and Hungate, 1966; Genskow et al., 1969; Purser and Tompkin, 1965) and the presence of bacteria (Gutierrez, 1958; Quin et al., 1962).

It has, however, been recognized that the holotrich ciliates can only be grown successfully if the numbers of bacteria present in the culture are controlled and the availability of soluble carbohydrates limited. Although bacteria are ingested by the holotrichs, Purser and Weiser (1963) demonstrated that the life of the protozoal culture could be extended when the bacterial numbers were limited by the presence of the antibiotic streptomycin. Competition between the bacteria and protozoa for the available substrates and the detrimental accumulation of bacterial metabolites is thus limited. The availability of soluble sugars should also be controlled as holotrich cell division is apparently impaired in cells containing excess reserve polysaccharide. Excessive amylopectin deposition and bacterial overgrowth are both effectively limited by the specific practice of restricting the amount and frequency of sugar addition and by daily transfer of the protozoa to fresh media (Clarke and Hungate, 1966; Kubo et al., 1980; Purser and Weiser, 1963). The cultivation of the individual genera will be considered briefly in the following sections.

Cultivation of Dasytricha ruminantium The initial investigations on the in vitro cultivation of *D. ruminantium* examined aspects of the soluble carbohydrate, trace element, and nitrogen requirements of the organism (Gutierrez and Hungate, 1957; Sugden and Oxford, 1952). The cultivation period was extended by the presence of rumen liquor, grass extract, and certain soluble carbohydrates, trace elements, and bacteria. Sugden and Oxford (1952) were able to establish and maintain a dividing cell population for up to 1 month.

The successful long term cultivation of *D. ruminantium* was described by Clarke and Hungate (1966). The cultures were established in an anaerobic salts solution containing 30% clarified rumen fluid under an oxygen-free carbon dioxide atmosphere. Sucrose and holotrich extract were added daily; some 2–4 hours after these additions the protozoa were transferred to an unsupplemented medium. These cultures, 10 ml total volume in test tubes, were maintained for up to 60 days reaching a population size of 1300 cells/ml from an inoculum of 200 cells. Under these conditions the cells divided once approximately every 48 hours.

Kubo et al. (1980) described a cultivation procedure in which the generation time of the ciliate was 1.3 days and population densities, in a 100ml culture volume, of up to 10^4 cells/ml were attained. The basal salts solution was similar in composition to that used by Clarke and Hungate (1966); it was supplemented with trace elements but not rumen fluid. The daily procedures adopted were also similar; some 30 minutes after a sucrose addition half of the culture supernatant was removed and replaced by fresh salts solution. The culture was then supplemented with a solid feed mixture (500 mg) containing powdered clover and concentrates. Cultures were maintained for periods of 1–6 months in the presence of *Entodinium caudatum, Ent. simplex,* or an extract of *Ent. caudatum.*

In both studies *D. ruminantium* exhibited an obligate requirement for a factor, or factors, of protozoal origin. These essential factors were removed on dialysis but could be replaced by bovine serum; various rumen bacteria and extracts of rumen bacteria or plant material, however, did not replace these factors (Clarke and Hungate, 1966). Bacteria did not stimulate protozoal growth in the absence of the protozoal extract. Clarke and Hungate (1966) also concluded that rumen fluid contained essential dialysable growth factors, and the protozoa did respond positively to a mixture of menadione and vitamin B_{12}. The source of the rumen fluid can markedly influence the survival of *D. ruminantium* in cultivation and care must be exercised in the selection of donor animals and their diets (Clarke and Hungate, 1966; Purser and Tompkin, 1965).

Cultivation of Isotricha spp. The original studies on the in vitro cultivation of *Isotricha prostoma* (Gutierrez, 1958) and mixed *Isotricha* spp. (Clarke and Hungate, 1966; Gutierrez, 1955; Sugden and Oxford, 1952) indicated that their requirements in culture were similar to those of *D. rumi-*

nantium. *Isotricha* spp. were successfully cultured for periods up to 3 weeks by Gutierrez (1955) and appeared to divide once every 48 hours; the cultures of Clarke and Hungate (1966) survived for 50 days.

Quin et al. (1962) developed a chemically defined culture medium suitable for the continous cultivation of *Isotricha* spp. in the absence of bacteria or complex extractive supplements. The initial *Isotricha* population of 2×10^4 cells/ml was maintained for several days before declining. The growth medium had over 60 components but further studies (Quin et al., 1968) confirmed that certain of the short-chain volatile fatty acids, some mono- and disaccharides, and several vitamins, including the principal B vitamins, were not essential to support the growth of *Isotricha* spp. The value of bacteria and rumen fluid is equivocal. The observations made by Gutierrez (1958) on bacterial feeding by *I. prostoma* indicated the the ciliates were able to derive some of their nitrogenous nutrients from ingested bacteria, whereas the studies of Quin et al., (1962) would indicate that ruminal bacteria were unnecessary for the growth of *Isotricha* provided that the growth conditions were rigorously controlled. These latter authors and Genskow et al. (1969) were also able to grow the ciliates in the absence of rumen fluid while Gutierrez (1955) and Clarke and Hungate (1966) advocated the inclusion of up to 30% (v/v) rumen fluid. However, in the absence of rumen fluid, growth factors appear to be associated with the chloroplast in exogenously supplied plant material (Genskow et al., 1969).

Progress has thus been made in defining the growth requirements of both *D. ruminantium* and *Isotricha* spp. However, the procedures for their routine maintenance in the laboratory are labor-intensive and not realistically suited for the long-term maintenance of the organism when access to the host ruminant is not readily available. In addition, the procedures have not been adapted for the large-scale generation of actively growing cells from defined conditions and therefore are not practicable for routine biochemical studies. Although there are advantages to be gained from controlled laboratory cultivation it must be stressed that such cells may respond differently. Cells isolated directly from the rumen have not been subjected to the imposed nutritional constraints of in vitro culture which may directly or indirectly affect the metabolic capabilities and responses of the protozoa.

4.5 Continuous Culture Techniques

Although it is possible to maintain some of the rumen ciliates in in vitro culture for extended periods, the methods described are principally small-scale batch culture techniques. It is now well-established that a limited ciliate population can be maintained in certain types of continuous flow artificial rumen systems. However, there have been relatively few attempts to utilize continuous fermentation systems for the production of either monocultures, or mixed cultures, of protozoa under controlled conditions for biochemical or other studies. The use and further development of continuous culture systems for protozoal cultivations will be considered in this section.

An artificial rumen represents a model in vitro system for studying rumen function under controlled conditions. Many designs have been described and used with varying degrees of success to simulate the conditions occurring in the rumen ecosystem. They have also been used to maintain an active and representative rumen microbial population. The range and complexity of the systems used has been extensively reviewed by Czerkawski (1986), and therefore further considerations will be restricted to examples of systems that have been shown to maintain a varied protozoal population.

It is possible to cultivate some types of nonrumen protozoa using classical continuous culture techniques and equipment (e. g., Curds et al., 1978). However, chemostats modified for the continuous in vitro culture of mixed rumen microorganisms (e. g., Quin, 1962; Rufener et al., 1963; Slyter et al., 1964; Stewart et al., 1961) have not proven to be satisfactory for the maintenance of the rumen protozoal population. Although apparent steady states are reached, as assessed by metabolite concentrations, the protozoal population frequently declines rapidly and may be lost completely from the fermentation. Protozoal populations, consisting principally of small *Entodinium* spp. have been maintained at a level of 10^3–10^5/ml for periods of up to 3 weeks (Itabashi and Kurihara, 1972; Slyter et al., 1964). The holotrich and large entodiniomorphid ciliates are not effectively retained in this type of fermentation system. The retention of the entodiniomorphid ciliates, however, was improved in a modified permeable continuous culture apparatus that incorporated a continuous dialysis system (Abe and Kumeno, 1973).

The systems discussed above are, in reality, modifications of laboratory fermenters developed for bacterial cultivation and as such contain no specific design features to aid protozoal retention. The rumen protozoa do not grow well in conventional continuous culture apparatus but can be grown successfully in systems that provide a suitable matrix for sequestration. The rate of liquid turnover is thus higher than the protozoal outflow rate. The solid feed is not, therefore, dispersed evenly by stirring but is enclosed in nylon bags that are suspended and mechanically agitated in the fermentation liquor (Aafjes and Nijhof, 1967; Czerkawski and Breckenridge, 1977; Nakamura and Kurihara, 1978a; Weller and Pilgrim, 1974). The protozoa will also sequester when inert matrices are provided (Abe and Kurihara, 1984; Czerkawski and Breckenridge, 1979). An alternative, albeit successful, approach is to add the solid at either predetermined intervals (Gijzen et al., 1986, 1988) or continuously to the fermenter (Fuchigami et al., 1989; Hoover et al., 1976; Merry et al., 1987; Teather and Sauer, 1988); both the solid and liquid outflow rates are controlled.

These more complex artificial rumens support a varied protozoal

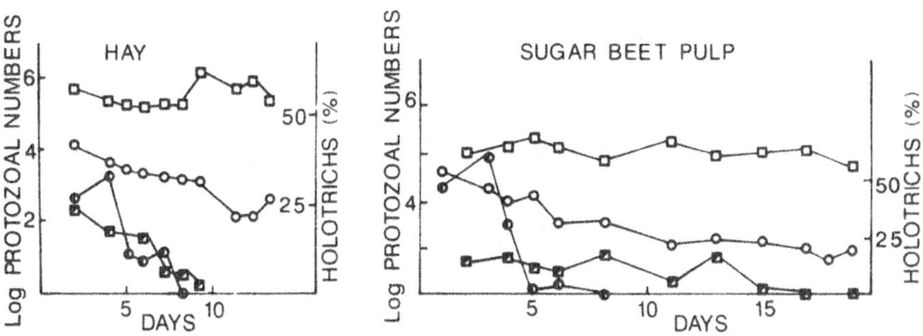

Fig. 4.6 Variation in the numbers (ml^{-1} or g DM^{-1}) of protozoa associated with the matrix solids (□) and in the liquid phase (○) of an *in vitro* artificial rumen system (RUSITEC) with a matrix (i.e., feed material) of hay or sugar beet pulp. The proportion of holotrichs in the liquid (◐) and matrix-associated populations (◧) is indicated. Data are from Hillman (1987).

population for periods ranging from 10–64 days. The types of protozoa present included holotrichs, large entodiniomorphid ciliates, and some smaller *Entodinium* spp. Ciliate populations approximating at 10^6/ml were present in the liquid associated with the solid substrate but were consistently lower in the liquid effluent (Abe and Kurihara, 1984; Czerkawski and Breckenridge, 1977, 1979; Gijzen, 1987; Nakamura and Kurihara, 1978a; Weller and Pilgrim, 1974). The distribution and population size of individual protozoal genera in two of the artificial rumen systems are shown in Figs. 4.6 and 4.7, and these data exemplify the close association of the protozoa with the available matrices.

Artificial rumen systems have only been used infrequently in investigations on the protozoa. The projects undertaken have primarily been designed to examine factors which could influence protozoal survival in the rumen or fermentation system. The simulators have thus been suitable for use in studies in which the objective was either to evaluate the effects of nutritional factors on protozoal survival (Bonhomme-Florentin and Hempel-Zawitkowska, 1977; Imai et al., 1979b; Nakamura and Kurihara, 1978b; Spinsi et al., 1980) or to examine the effect of feed components on the size and generic composition of the ciliate population (Abe and Iriki, 1978; Michalowski, 1975b, Muszynski and Michalowski, 1979). Artificial rumens are particularly useful for assessing the effects of solid and liquid flow rates on protozoal retention (Crawford et al., 1980; Hoover et al., 1976). The further development of these continuous in vitro systems for protozoal cultivation under strictly controlled conditions warrants consideration.

The modification of artificial rumens to function as anaerobic fermentation systems for the continuous in vitro cultivation of monocultures or mixed preparations of rumen ciliate protozoa is desireable. Michalowski (1979) described a simple system for the long-term cultivation of ciliates.

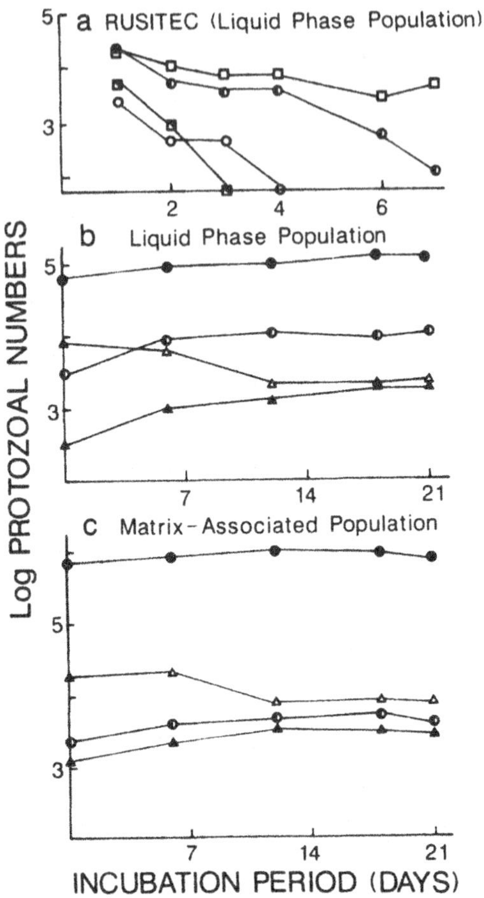

Fig. 4.7 Variations in the numbers (ml^{-1} or gDM^{-1}) of *Dasytricha ruminantium* (◐), *Isotricha* spp. (○), large entodinimorphid ciliates (◪), small entodiniomorphid ciliates (□), *Epidinium* spp. (▲), *Diplodinium* spp. (△), and *Entodinium* spp. (●) are shown for: (a) the liquid phase of the artificial rumen system (Rusitec); and for the (b) liquid and (c) solid matrix of the culture system described by Abe and Kurihara (1984). (a) Data are from Hillman (1987). (b) Data are from Abe and Kurihara (1984).

The culture vessel which did not contain a dead space for sequestration was continuously infused with buffer. Anaerobiosis and intermittent mixing were achieved by gassing with carbon dioxide. Holotrich and entodiniomorphid ciliates were maintained in cultures of up to 100 days duration, although population densities were only of the order of 10^2–10^3 cells/ml (Michalowski et al., 1979). A more complex fermentation system was described and evaluated by Abe and Kurihara (1984). The culture vessel included an inert sponge matrix for protozoal sequestration; the buffer infusion, gassing, and product removal by dialysis were continuous and cell

Table 4.3 Some factors that may influence protozoal growth in continuous in vitro cultivation

Environmental factors	Nutritional factors
Dilution rate	Buffer composition
Rate and type of mixing	"Feed" composition
pH	Physical form of supplied food
Redox potential	Matrix type
Temperature	Number of bacteria present
Anaerobic gas mixture (level of anaerobiosis)	Type of bacteria present (i.e., mono- or mixed culture)
Rate and frequency of gassing	
Availability of matrix for sequestration	
Selective protozoal retention by use of outflow filters (e.g., Hoover et al., 1976)	

Table 4.4 Protozoal numbers in the incubation liquor of the artificial rumen system RUSITEC with different matrix materials

	Protozoal numbers[a]			
Matrix	Day 1	Day 3	Day 5	Day 7
No matrix	7.5(13)	ND	ND	ND
Hay	50(40)	NA	4.2(12)	2.5(13)
Pre-incubated hay	28(16)	32(0.5)	11(ND)	9.2(ND)
Hay + sugar beet pulp	50(48)	24(60)	15(3.3)	7.0(ND)[b]
Nylon string	12(45)	18(19)	4.5(4.0)	NA
Nylon pan scourers	30(31)	28(17)	2.0(ND)	0.3(ND)
Sponge	10(31)	1.5(33)	0.2(ND)	ND
Glass wool	15(25)	3.5(21)	NA	1.8(ND)[c]

[a] The data are expressed as $10^{-3} \times$ total ciliate population/ml. The holotrich values expressed as a percentage of the total population are given in parentheses. ND, not detected; NA, not assessed.
[b] Day 8 values
[c] Day 6 values
Based on data from Hillman, 1987.

numbers were comparable to those found in vivo. However, cultivation conditions have not been optimized for any of the protozoa in either system, and although it is possible to control and hence vary some of the more important environmental factors (see Table 4.3), the effects of these parameters on the survival, growth, or metabolic activity of rumen ciliates in long-term continuous in vitro cultures are largely unknown. Nakamura and Kurihara (1978a) indicated that the physical form and quantity of the substrate supplied and the frequency and degree of agitation affected the stability of the protozoal population. Additional studies on the survival of single holotrich species and mixed ciliate populations in the artificial rumen system RUSITEC (Czerkawski and Breckenridge, 1977) confirmed that protozoal survival was influenced by the substrate available, the buffer infusion, and the type of matrix (Hillman, 1987). Protozoal retention was highest at pH 6.5 to pH 7.0 with a dilution rate of 0.65 per day and niacin supplementation of the infusion buffer (c.f., Dennis et al., 1982; Spinsi et al., 1980). Barley straw and predigested ryegrass hay were useful matrices (Czerkawski and Breckenridge, 1979); the inert matrices evaluated were not particularly effective and holotrich numbers declined in mixed populations (see Figs. 4.6, 4.7, and Table 4.4).

Batch culture studies have shown that the ciliates have complex nutritional requirements and may only be grown axenically for limited periods following extensive washing of the protozoal inoculum in the presence of antibiotics (e.g., Bonhomme et al., 1982a, 1982b; Jarvis, 1974; Onodera and Henderson, 1980; Onodera et al., 1977c). However, for certain investigations agnotobiotic culture of the ciliates with one or a limited number of selected bacterial strains with known biochemical characteristics may be practicable (Hino and Kametaka, 1977). Likewise the development of a fully chemically defined medium may not be necessary. However, for detailed biochemical studies it is advantageous for the cells to be maintained for several generations under controlled steady-state growth conditions. The initial success of Quin et al. (1962) in cultivating *Isotricha* spp. in a continuous culture for 5 days in a defined medium and in the absence of viable bacteria has not, as yet, been repeated in other fermentation systems or with other rumen ciliates.

5

Association of Bacteria and Fungi with Rumen Ciliates

All rumen ciliates whether grown in vivo or in vitro contain bacteria in vesicles in their endoplasm and attached to the outside of the pellicle. However studies in the electron microscope show that the numbers vary considerably between species and with growth conditions. Fig. 5.1, which is a section through *Entodinium caudatum* grown in vitro, shows that this protozoon contained few bacteria in the endoplasm and apparently none attached to the pellicle. In contrast, Figs. 5.2 and 5.3 show bacteria attached to the pellicles of, respectively, an *Entodinium* sp. isolated from the rumen and an *Epidinium caudatum* grown in vivo. Fig. 5.4 shows that protozoa (species unknown) can contain very many bacteria in their cytoplasm.

5.1 Intracellular Bacteria

There is relatively little information on the number or species of bacteria found inside rumen ciliates except for *Entodinium caudatum* grown in vitro. White (1969) found *Klebsiella aerogenes* and *Proteus mirabilis* in the proportions of approximately 2:1 in this protozoon, and the exact proportions depended on the growth conditions. The total numbers decreased with time since the protozoa were last fed; there were 37, 10, and 2 present in each organism last fed 18, 42, or 66 h before, but the bacteria never disappeared completely even on prolonged incubation. In addition approximately one *Butyrivibrio* sp. and one *Bacteroides* sp. were present in each protozoon. *Kelbsiella aerogenes* and *Proteus mirabilis* were the predominant bacteria in the growth medium and are presumed to have colonized the

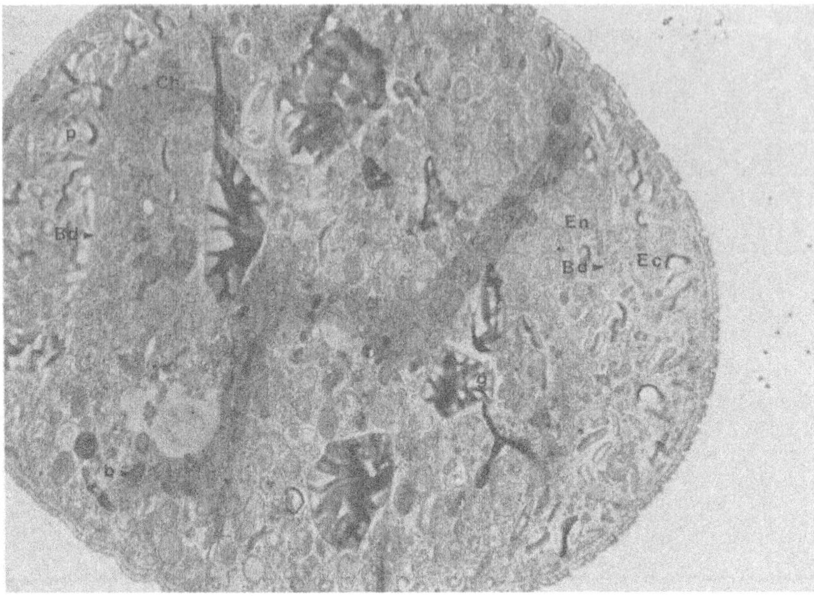

Fig. 5.1 An electron micrograph of a section of cultured *Entodinium caudatum* prepared immediately after harvesting from the growth medium. The endoplasm contains rice starch grains (g) and chloroplasts (Ch) which were engulfed from the medium before harvesting. Note that there are only a few bacteria (b) in vesicles in the endoplasm and none attached to the pellicle. Bd, boundary between endoplasm and ectoplasm; Ec, ectoplasm; En, endoplasm; p, protozoal polysaccharide granule.

protozoa. In contrast Hino and Kametaka (1977), also using *Entodinium caudatum*, found seven bacterial types in the growth medium and six in the protozoa. The bacteria found included Corynebacteria, Bacilli, *Streptococcus bovis* (?), an Entobacterium, and a *Bacteroides* sp. These differences between authors probably reflect differences in the culture medium rather than differences between the protozoa as it is agreed that the bacteria inside the protozoa are the same as those in the growth medium. When Coleman and White (1970) put cultured *Entodinium caudatum* back into the rumen of a defaunated sheep, the total number of intracellular bacteria remained unchanged, but the species changed to those typical of the rumen.

Fig. 5.3 Electron micrograph of a section through a cultured *Epidinium caudatum* showing the two bacterial types (A and B) attached to the outside of the pellicle (p). Note that this micrograph is also an autoradiogram through a protozoon that had metabolized 6-^{3}H-glucose for 4 h before fixation. The density of the silver grains (s) is much higher over the protozoal polysaccharide than the bacteria indicating that the bacteria probably utilize glucose poorly.

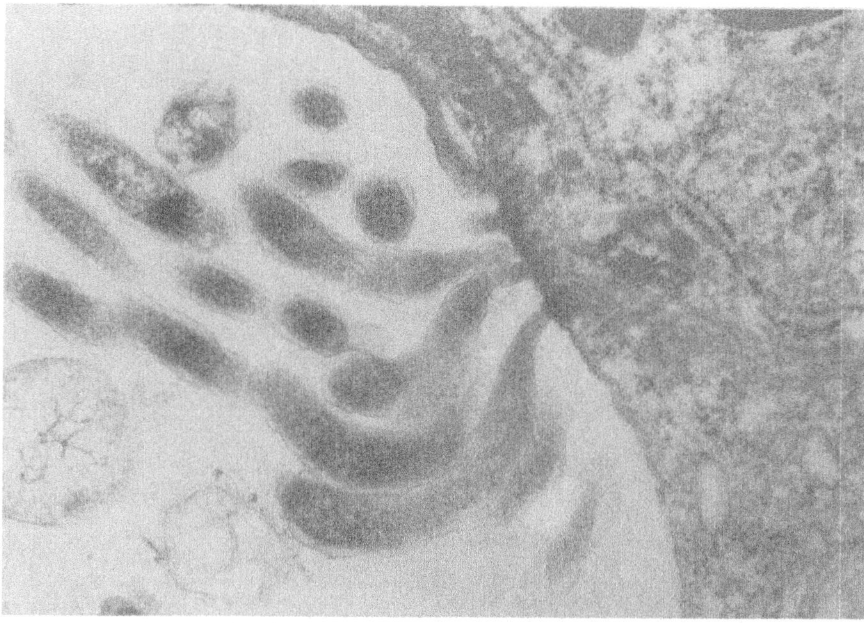

Fig. 5.2 An electron micrograph of a section through an *Entodinium* sp. isolated from the rumen. Note the bacteria attached to the outside of the pellicle.

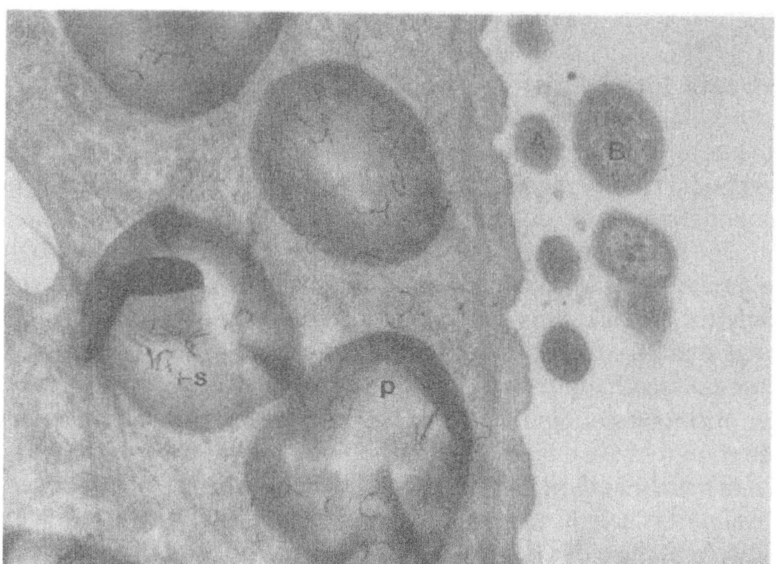

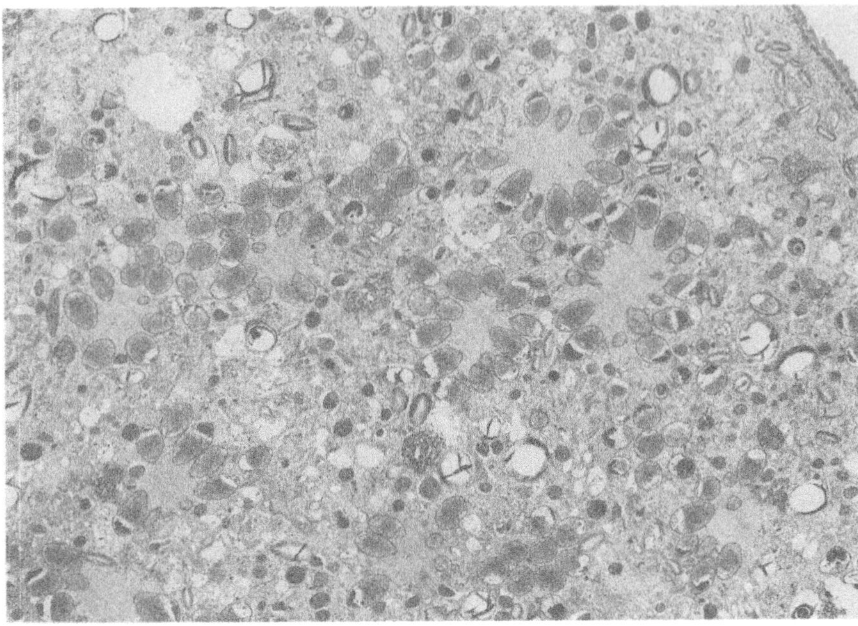

Fig. 5.4 Electron micrograph of a section through an unknown ciliate protozoon isolated from the rumen. Note the presence of many bacteria in the endoplasm, some of which appear to be being taken up from vesicles into the cytoplasm.

Survival of Intracellular Bacteria When ^{14}C-labelled washed suspensions of *Klebsiella aerogenes* and *Proteus mirabilis* (Bacteria 31 and 7G2 of Coleman, 1967a) (isolated from *Entodinium caudatum* and grown aerobically in conventional culture media), are incubated with suspensions of this protozoon, both bacteria are engulfed. However after 2.5 h of continuous engulfment, 35% of the former and 82% of the latter are still viable. Although this shows that *Proteus mirabilis* could well survive inside protozoa, this seemed unlikely for *Klebsiella aerogenes*. Subsequent experiments (Coleman, 1975b) showed that resistance to digestion is associated with the presence of a glucose-polysaccharide capsule which the bacterium synthesizes only under anaerobic conditions. It has been suggested that *Klebsiella aerogenes* inside the protozoa utilizes maltose and glucose (derived from the breakdown of starch) in the vesicles to synthesise this capsule, and, also that the survival of this baterium depends on the relative rates at which the bacterium can form the capsule and the rate at which the protozoon can digest it. The number of bacteria will therefore tend to fall on starvation of the protozoa when intracellular sugars become less plentiful. Further support for the hypothesis comes from autoradiography in the electron microscope of sections of protozoa that have taken up [6-^{3}H]-glucose for 4 h. Dead bacteria labelled with ^3H can be seen in the endoplasm suggesting that living bacteria have taken up glucose and then been killed by the protozoa (Coleman, 1975b).

Holotrich Protozoa There is less information on bacteria inside holotrich protozoa. Stern et al. (1977) observed unidentified bacteria in food vacuoles of an *Isotricha* sp. and Nouzarede (1978) found "endosymbiotic" bacillus-like bacteria, which formed spores, in the perinuclear region of *Isotricha prostoma*. Both these authors used protozoa isolated from the rumen and, since holotrich protozoa have been observed to engulf bacteria (see Chapter 8), it is uncertain if any bacteria survive the engulfment and digestion process and colonize the protozoa.

5.2 Bacteria Attached to the Pellicle

Many, but not all, Entodiniomorphid protozoa have bacteria attached to the outside of their pellicles, but the numbers and distribution are much more variable in general than with the intracellular bacteria. There have been reports of several different bacteria being found, namely *Streptococcus bovis* and *Ruminococcus albus* (Imai and Ogimoto, 1978), Bacteria A and B (Fig. 5.3; Coleman and Hall, 1974), and, on the basis of fluorescence studies, methanogenic bacteria (Vogels et al., 1980). Although they have shown that the bacteria adhering to 11 species of rumen ciliates are all methanogens, the number of bacteria attached to each protozoon is quite variable. In one sheep, Vogels et al. (1980) found that 80% of *Entodinium longinucleatum*, >50% of *Eudiplodinium maggii*, and all of the *Entodinium bursa* had surface methanogens. They also found that in different rumens the proportion of *Eremoplastron bovis* with surface methanogens varied from 10–100%.

The proportion of ciliates with methanogens also depends on the nutritional status of the host and decreased in one sheep from 65% before feeding to 25% after feeding (Stumm et al., 1982). Evidence was also presented, on the basis of flushing the rumen with nitrogen or hydrogen, that the methanogens attach themselves to the protozoa as a source of hydrogen. It was argued that they become detached when hydrogen is pumped into the rumen or becomes freely available after the host animal has been fed. However, when protozoal cultures are grown under 95% H_2 + 5% CO_2, but not when grown under 95% N_2 + 5% CO_2, there is a massive association of methanogens with entodiniomorphid protozoa to the extent that they can be completely coccooned in a mass of bacteria. Although this observation does not preclude the explanation of Stumm et al. (1982), it does suggest that the methanogens can attach themselves to protozoa for other reasons and use them possibly just as a surface. Whether or not the methanogens play any major role in the removal of hydrogen

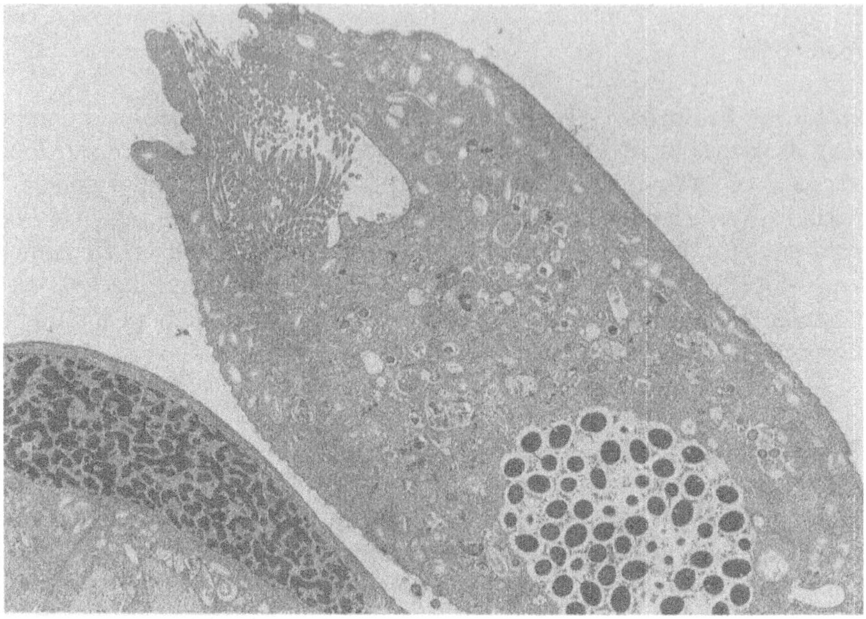

Fig. 5.5 Electron micrograph of an *Entodinium* sp. isolated from the rumen showing bacteria contained in an unusual thick-walled vesicle in the endoplasm.

from the protozoa must be debatable, although evidence has been produced that when the number of methanogens in a culture of *Entodinium* spp. increases, the rate of protozoal fermentation also increases (Hino, 1983). Hydrogen and acetate production increase as does ATP yield. In contrast, when the partial pressure of free hydrogen in the culture was increased, protozoal fermentation was depressed.

Comparatively little is known about these methanogens attached to the pellicle although, on the basis of autoradiographic studies in the electron microscope, it has been shown that there are two different species attached to cultured *Epidinium caudatum*. One (designated B) is sensitive to ampicillin and takes up ^{3}H-glycine more rapidly than the other (designated A) (Coleman and Hall, 1974; Coleman, 1978c).

Holotrich Protozoa Bacteria have been observed living between the cilia and round the body of holotrich protozoa. Imai and Ogimoto (1978) estimated densities of 40–50/200 μm^2 of surface area and suggested that the bacteria might have been trapped between the cilia as the protozoa were not washed before fixation. There is no evidence that the bacteria are methanogens (Vogels et al., 1980).

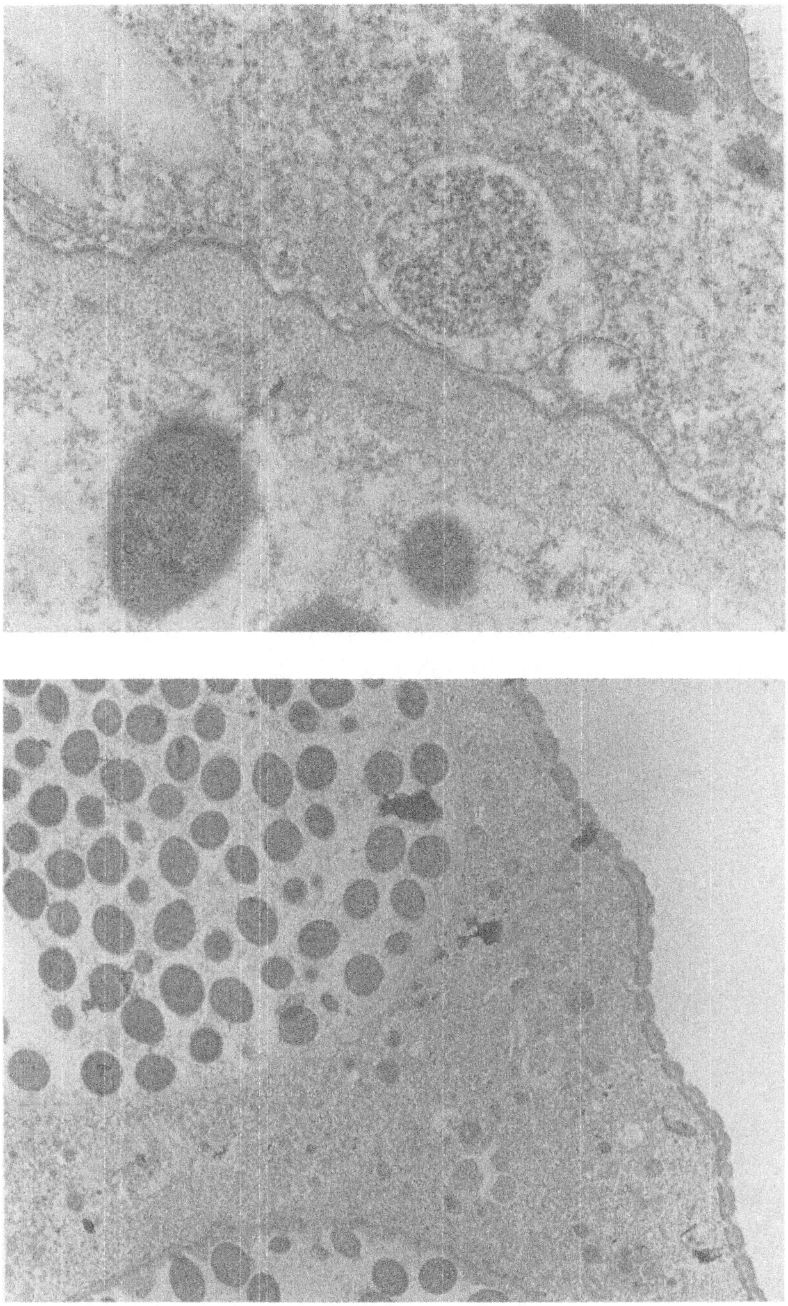

Fig. 5.6 Higher magnification micrographs of parts of the section shown in Fig. 5.5, showing the thick wall of the vesick.

5.3 Intracellular "Parasites" of Rumen Ciliates

From time to time those studying rumen ciliates in the light or electron microscopes come across protozoa that contain intracellular organisms that are apparently "parasitizing" the host. For example, Figs. 5.5 and 5.6 show bacteria contained in an unusual thick-walled vesicle in the endoplasm. It is unknown what effect, if any, these bacteria have on the host protozoon.

Intracellular Parasitic Fungi Lubinsky (1955a) described *Sphaerita hoari,* a chytrid fungus, in a number of entodiniomorphid protozoa, principally *Eremoplastron bovis,* from domesticated ruminants in India. Although present in 19% of *E. bovis* examined, the fungus appeared to do little harm to the host as at least some of the protozoa continued to divide. The fungus consists of a plasmodial thallus usually situated in the endoplasm near the macronucleus of the host. On maturity the thallus develops into a sporangium in which a large number of spores form. Jirovec (1933) also described *Sphaerita* spp. from *Entodinium simplex*. Lubinsky (1955b) described *Sagittospora cameroni* and closely related species parasiting *Eudiplodinium maggii* and other entodiniomorphid protozoa from a goat in India. The parasite invariably killed the host protozoon but was found in only 1% of the rumens examined and then only in 2% of the protozoa. The development of the fungus was similar to that described for *Sphaerita hoari*.

6

Metabolism of Entodiniomorphid Protozoa

Information on the methods used in the preparation of suspensions of protozoa for metabolic experiments is given in Section 4.1 and on incubation conditions and disruption of protozoa in the Appendix to this chapter.

6.1 Metabolism of Starch

As far as is known all the entodiniomorphid protozoa engulf starch grains although the smaller protozoa can only take up the smaller grains such as those found in rice. The grains are taken up through the esophagus or cytostomal invagination into vesicles in the endoplasm. The mechanism probably varies between species (Chapter 7). All the entodiniomorphid protozoa we examine here utilize starch grains for growth, although only *Entodinium* spp. have an absolute requirement for starch. Starch is replacable by cellulose (as dried grass) for the larger ciliates (Coleman, 1987a and Chapter 4).

We will consider the metabolism of starch by species with any results that have only been obtained with mixed protozoa being considered under *Entodinium caudatum*. The comparative rates of breakdown of starch, cellulose, polygalacturonic acid, and some intermediates is shown in Table 6.1.

Role of Bacteria Associated with the Protozoa Most entodiniomorphid protozoa have bacteria attached to their pellicles (Coleman, 1978c; Coleman and Hall, 1974; Vogels et al., 1980) and contained in vesicles in their endoplasm (Coleman, 1975b, 1978c; White, 1969; Coleman and White, 1970;

Table 6.1 Carbohydrate-degrading enzyme levels in the cytoplasmic fraction of various entodiniomorphid protozoa[a]

Species	Source of protozoa[b]	Relative volume	Carboxymethyl cellulase	β-Glucosidase
Diplodinium pentacanthum	C, GW			
	C, G			
	R		(6)	
Diploplastron affine	C, GW	5.3	54	223
	C, G		43	
	R		31	162
Entodinium bursa	C, GW	3.9	0	
	R	2.0		
Entodinium caudatum	C, GRS	1.0	0	390
Spined	R	1.8		
Spineless	R	1.0	2	365
Entodinium longinucleatum	C, GW	1.7	0	
Entodinium simplex	C, GW	0.3	0	88
	R			
Epidinium caudatum	C, GW	5.5	41	22
	C, G			
	R		100	100
Eremoplastron bovis	C, GW		16	377
	C, G		27	
	R	2.8	43	530
Eremoplastron dilobum	C, GW		32	110
Eudiplodinium maggii	C, GW	10.3	6	
	C, G		76	
	R		64	385
Ostracodinium dilobum	C, GW	6.0	88	
	C, G		143	
	R		89	420
Ophryoscolex caudatus	C, GW	11.9	72	104
	R		116	40
Polyplastron multivesiculatum	C − Epi, GW	13.8	55	228
	C − Epi, G			
	C + Epi, GW	37.8		
Mixed small protozoa (entodinia)	R		11	186
Mixed large protozoa	R		40	225
Natural mixed A-type protozoa	R		60	
Natural mixed B-type protozoa	R		200	

[a]The values given are relative to those found in *Epidinium caudatum* (= 100).
[b]C, cultured in vitro; GW, grown on wholemeal flour and dried grass; G, grown on dried grass alone; R,

Coleman and Hall, 1974), although only with *Entodinium caudatum* has a detailed study been made (White, 1969; see also Chapter 5). In any study on the metabolism of these protozoa, it is essential to remember that both the bacteria and the protozoa may contribute to the final result. In the case of intracellular bacteria, any added substrate has to be taken up by the protozoa and then pass into the vesicles containing the bacteria which utilize any such substrate as a source of energy and as a source of carbon for growth. In addition to utilizing these substrates the bacteria are also being killed and digested by the lytic enzymes produced by the protozoa. It is

Amylase maximum velocity	α-Glucosidase	Hemicellulase B	Xylanase	Endopectate lyase maximum velocity
100				
18				
(127)				
716	15		225	460
175				
228	12	176	184	
55	14		5	21
				32
95	14	8	4	0
				0
145	14	6	4	
155				
321	8	5	3	
215				
246	36			98
233				
100	100	100	100	100
640	23			(71)
207				
212	13	93	121	1600
1280	19			
164				240
79				
216	9	61	81	265
50				980
81				
71	16	142	148	530
475	32			360
214	55	116	146	200
287		65	71	1450
125				
490	17	105	140	(6)
	17	48	25	
	4			15
101				
109				

grown as single species in the rumen of a sheep fed on hay (800 g) and oats (100 g); GRS, grown on rice starch and dried grass; − Epi, + Epi, grown in absence or presence of epidinia.

possible, for example, for glucose in the medium to be taken up by the protozoa and metabolized by the bacteria with the incorporation of glucose carbon into bacterial protein. After the death of the bacterium, the bacterial protein is digested by the protozoon. The constituent amino acids are then incorporated into protozoal protein giving the impression that protozoa incorporate glucose carbon directly into their protein.

As antibiotics penetrate slowly into the protozoa (Coleman, 1962; Hino and Kametaka, 1977), incubation of the ciliates with antibiotics does not necessarily stop all bacterial activity.

Entodinium caudatum

Entodinium caudatum takes up starch grains from the medium very rapidly. If a suspension of rice starch is added to starved protozoa incubated at 39°C, many of the ciliates are full of starch by the time the material can be examined microscopically. The protozoa must exert considerable suction because when they attempt to engulf a "large" starch grain, it will disappear more rapidly than the eye can discern into the organism. The protozoa are apparently stimulated by the presence of starch in the medium because after its addition they move around much more rapidly and attempt to engulf more starch even when apparently full (Coleman, unpublished observations).

The first detailed study on the metabolism of starch by *E. caudatum* was made by Abou Akkada and Howard (1960) who used starch-containing protozoa isolated from the rumen. This endogenous starch was fermented with the production of acetic and butyric acids (together 80–90% of the total volatile fatty acids) with lesser amounts of formic and propionic acids and some CO_2 and H_2. Further added starch did not increase the rate of fermentation, and soluble sugars were apparently not metabolized as determined by substrate disappearance or increase in acid or gas production. However Coleman (1962, 1969b) showed that the life of starved "bacteria-free" protozoa could be prolonged for short periods by starch, maltose, and glucose but not by a number of other polysaccharides and sugars.

Amylase *E. caudatum* contains in its cytoplasm an α-amylase with an optimum pH of 6.0 (Abou Akkada and Howard, 1960; Bailey and Clarke, 1963a; Coleman, 1986b) which produces maltose (Abou Akkada and Howard, 1960; Coleman, 1986b). A crude cytoplasmic fraction of *E. caudatum* degrades amylose and amylopectin at the same rate, intact starch grains at 20% of this rate, and β-limit dextrin at 170–180% of this rate (Coleman, 1986b). The specific activity of *E. caudatum* amylase in the cytoplasm is higher in protozoa grown in vivo than in those grown in vitro on rice starch as the sole added source of starch. *E. caudatum* also contains amylase in its particulate fraction (after disruption of the protozoa). The activity is 150–170% of that in the cytoplasm with a specific activity 130–210% of that in the cytoplasmic fraction in rumen-grown protozoa. However it is unclear if this is protozoal enzyme absorbed onto starch grains or bacterial enzyme (Coleman, 1986b).

Maltase *E. caudatum* contains a specific maltase (optimum pH 5.0–6.1) which cannot hydrolyse α-methylglucoside, αα-trehalose, turanose, sucrose, or maltitol (Bailey and Howard, 1963; Coleman, 1969b).

Metabolism of Starch by Intact Protozoa Engulfed starch grains are degraded in the endoplasm and in nongrowing protozoa 50% of the starch is degraded to products that appear in the medium (maltose 35% and glucose 65%). In freshly harvested cultured protozoa, the concentrations of sugars in the pool decreased over 4 h, in the presence or absence of added starch, by 33% and 31% respectively for maltose and 37% and 42% for glucose, showing that provided some starch is present, the levels of sugars in the pool is independent of the amount. The concentrations of maltose and glucose in the pool of freshly harvested protozoa are 13.2 mM and 14.8 mM respectively. Both amylase and maltase in the protozoa are subject to product inhibition (Coleman, 1969b) and it is likely that it is this phenomenon which maintains the concentrations of these sugars in the pool relatively constant. Increase in the concentration of maltose or glucose for any reason tends to decrease the rate at which these sugars are formed from starch and maltose respectively. The concentration of glucose and hence the rate of production of fermentation products tends to remain constant.

Although *E. caudatum* contains α-glucan phosphorylase (Coleman, 1969b; Hoshino et al., 1982) and phosphoglucomutase, they are comparatively inactive. Starch is utilized by degradation to maltose and glucose followed by phosphorylation of the latter to glucose 6-phosphate. This is then utilized for the production of energy via glycolysis (Coleman, 1964a; 1981; Hoshino et al., 1982) or for the synthesis of the protozoon's own intracellular polysaccharide granules. In *E. caudatum* this is an amylopectin-like polysaccharide with an average chain length of 19 glucose units (Eadie et al., 1963). Mixed entodinia contain a glycogen synthase which exists in two forms, one glucose 6-phosphate dependent and the other glucose 6-phosphate independent. The concentration of glycogen synthase increases when the protozoa are starved and the amount of intracellular polysaccharide drops. There is also conversion of the dependent form into the independent as the polysaccharide content of the protozoa falls (Hoshino et al., 1982). Under the conditions of harvesting from the rumen this polysaccharide amounts to 6–7% of the dry weight of the protozoon (Eadie et al., 1963). The polysaccharide content of mixed entodinia increases for 2 h after the host ruminant is fed and then declines (Wakita and Hoshino, 1979). However, in mixed protozoa isolated from the rumen of an animal fed a starch-rich ration, intracellular carbohydrate (both ingested and synthesized) can amount to 37.7% of the dry weight (Jouany and Thivend, 1972).

Although it is difficult to be certain of the exact extent, *E. caudatum* has only very limited ability to synthesize amino acids from starch (Coleman and Sandford, 1980).

Metabolism of Maltose and Glucose Studies with ^{3}H- or ^{14}C-labelled sugars showed that maltose and glucose are taken up by *E. caudatum* principally into the cell pool, protozoal polysaccharide, and intracellular bacteria (Coleman, 1969b; Coleman and Hall, 1974). There is a small incorporation of ^{14}C from [U-^{14}C]glucose into the protein of the cytoplasmic and particulate fractions obtained after disruption of the organisms. Incorporation into

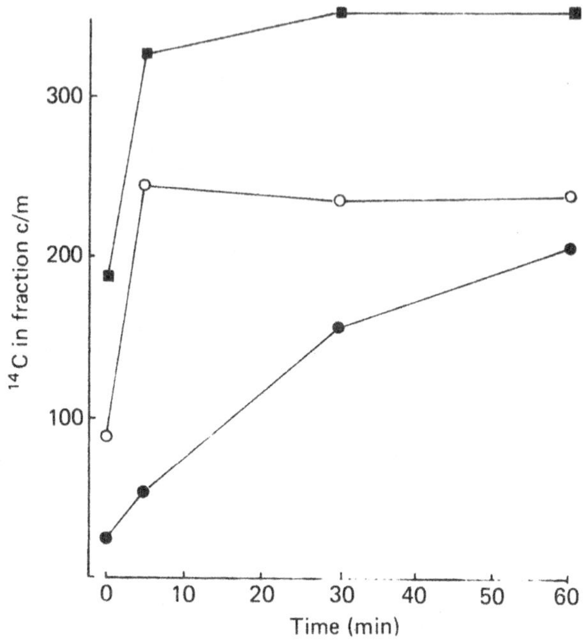

Fig. 6.1 Incorporation of ^{14}C from [U-^{14}C]maltose into various fractions in the pool of *Entodinium caudatum*. ■, into glucose; ○, into maltose; ●, into hexose phosphate.

the latter probably represents bacterial protein synthesis as it is sensitive to antibiotics and all amino acids are labelled. In the cytoplasmic fraction ^{14}C is found principally in aspartic acid and alanine and this may represent a very limited ability to synthesize protein (Coleman and Sandford, 1980).

All soluble compounds are taken up by the protozoa by an active process (predominant at low external concentrations. They are also taken up by a passive process (predominant at high external concentrations) that is probably the trapping inside the cell of medium containing the substrate under investigation. For details see Section 7.12. Maltose is taken up actively at 225% the rate of glucose but is rapidly broken down to glucose. The incorporation of ^{14}C from [U-^{14}C]maltose into free glucose and maltose in the protozoal pool reaches a steady level after five min, although incorporation into hexose phosphate increases steadily (Fig. 6.1). Maltose carbon is incorporated at the same rate as glucose carbon into protozoal polysaccharide but more rapidly into bacterial polysaccharide. The reason for this difference is that one of the principal intracellular bacteria (*Klebsiella aerogenes*) incorporates maltose carbon at three times the rate of glucose carbon when the sugars are present at equimolar concentrations (Coleman, 1969b). During the uptake of [U-^{14}C]glucose (80 mM) by cultured *E. caudatum* the concentration of free glucose in the protozoal pool declined over 6 h while the

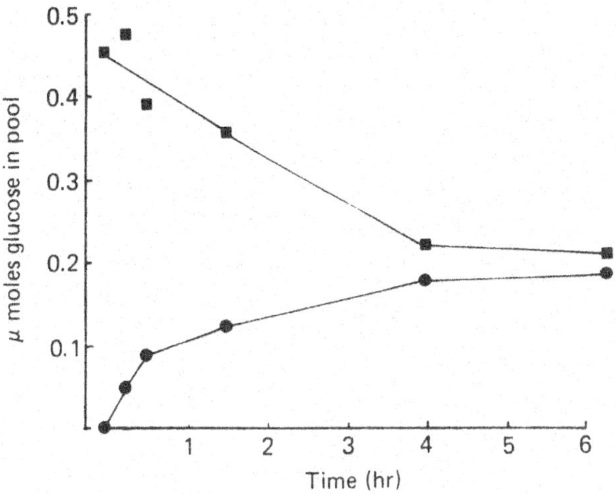

Fig. 6.2 Effect of incubation in the presence of 80 mM [U-^{14}C]glucose on the concentration of glucose in the pool of *Entodinium caudatum*. ■, total glucose in pool as determined by the glucose oxidase method; ●, [^{14}C]glucose incorporated into pool as calculated from ^{14}C found in pool and known specific activity of glucose added initially.

amount of ^{14}C in this glucose steadily increased. This suggested that the formation of glucose from intracellular starch and maltose had decreased due to negative feedback inhibition of the maltase and amylase and had been replaced by glucose from the medium (Fig. 6.2).

With rumen-grown protozoa, the spined form of *E. caudatum* (grown in the presence of *E. bursa*) incorporates glucose at 60 times the rate found with the spineless form (Coleman and Sandford, 1980).

There is evidence (see Section 7.12) that the ectoplasm of *E. caudatum* is freely permeable to low-molecular-weight compounds and that the permeability barrier is the boundary between the ectoplasm and endoplasm.

It is of interest that entodinia are common in the rumen of ruminants fed sugar-based rations that contain little starch and that these entodinia contain no intracellular polysaccharide. This suggests that the protozoa can utilize sugars as a source of energy but not for the synthesis of polysaccharide, possibly because they are taken up too slowly to ever accumulate inside the cell and be available for polysaccharide synthesis (Coleman, 1979a).

Subcellular Distribution of Enzymes Experiments on the subcellular distribution of enzymes in protozoa disrupted in mannitol/EDTA showed that most of the activity was in fractions sedimenting at 1000 g for 10 min, 10,000 g for 10 min, and 20,000 g for 20 min. Only 20–40% of the polysaccharide-degrading enzymes and glycoside hydrolase enzymes were soluble inside the cell (Williams et al., 1986).

Entodinium simplex There is much less evidence available on the carbohydrate metabolism of *E. simplex*. As with *E. caudatum*, starch grains are engulfed although it has been wrongly stated (Abou Akkada and Howard, 1960) that they are not. In the absence of antibiotics, a small (0.67%) proportion of the engulfed starch is converted into amino acids but, as this is decreased to 0.17% in the presence of antibiotics, it was probably due to bacterial protein synthesis (Coleman, 1972).

As with *E. caudatum*, glucose is taken up principally into the pool, intracellular bacteria, and protozoal polysaccharide with very limited amounts in protein. Disruption of the protozoa before incubation with glucose increases the incorporation of glucose carbon into the particulate fraction that contains any intracellular bacteria by 85% showing that the intracellular bacteria are not normally freely in contact with the external medium. Evidence from experiments with inhibitors suggests that bacteria only utilize sugars previously taken up by the protozoa (Coleman, 1972). Sugars are taken up both by an active and passive processes.

The reserve polysaccharide is an amylopectin-like glucose polymer of average chain length 22–25 units. In rumen-grown protozoa it accounts for 7.6% of the dry weight of the protozoa (Wakita and Hoshino, 1980).

Entodinium bursa There is only limited information available on *E. bursa*, although it engulfs starch grains and utilizes starch grains from engulfed *E. caudatum*. Free glucose is taken up principally into the cell pool, intracellular bacteria, and protozoal polysaccharide granules and only to a limited extent into protein (Coleman and Hall, 1984).

Entodinium longinucleatum As with the other entodinia, starch grains are engulfed and glucose is taken up but neither are used for significantly important protein synthesis by *E. longinucleatum* (Owen and Coleman, 1977). This protozoon contains an active amylase whether grown of starch grains in vitro (Coleman, 1986b) or on a starch-free ration in vivo (Bailey and Clarke, 1963a). A mixture of entodinia including *E. longinucleatum, E. ovinum, E. indicum, E. biconcavum*, and *E. bicarnitum* grown on a starch-free ration in vivo contained enzymes for the degradation of xylan, arabinoxylan, laminarin, and cellodextrins (Bailey and Clarke, 1963a).

Epidinium caudatum *Epidinium*, like the other entodiniomorphid protozoa, engulf starch grains which are degraded by α-amylase(s). The α-amylase concentration is lowest in protozoa isolated from the rumen of a sheep fed on hay and oats, higher in protozoa cultured on grass alone, and highest in those cultured on wholemeal flour and grass in the approximate proportions (1:2.5:5.0) (Bailey, 1958; Coleman, 1986b). Rumen-grown protozoa, at least, contain a specific maltase (Bailey and Howard, 1963).

On incubation of intact protozoa with ^{14}C-soluble starch, there is rapid uptake of ^{14}C that appears sequentially in maltose, glucose 6-phosphate, and glucose. Incorporation into protozoal polysaccharide is slow. The specific activity of the glucose 6-phosphate is often higher than that of the free sugars and glucose 1-phosphate is also labelled; this suggests that at least some of the glucose 6-phosphate is formed from starch by a mechanism not involving hydrolysis to free sugars. The final metabolic products of starch metabolism are acetic and butyric acids, H_2, and CO_2 with traces of formic, propionic, and lactic acids (Gutierrez and Davis. 1962). In the presence of antibiotic (ampicillin), ^{14}C from ^{14}C-starch grains is incorporated in appreciable amounts (8% of starch engulfed) into protozoal protein as principally serine, glutamic acid, lysine, and alanine, and it is believed that this represents protozoal protein synthesis. There is also appreciable incorporation of ^{14}C into the bacteria-containing particulate fraction, but it is uncertain what proportion, if any, is of protozoal origin (Coleman and Laurie, 1974a).

Metabolism of Glucose On incubation of *Epidinium caudatum* with low concentrations of [U-^{14}C]glucose, ^{14}C first appeared in the protozoa as glucose-6-phosphate and then in maltose and protozoal polysaccharide. The incorporation into glucose-1-phosphate was always small. ^{14}C appeared in cellular glucose after it had appeared in all other compounds. From the shape of the uptake curves glucose 6-phosphate and glucose appeared to be intermediates and polysaccharide and maltose final products (Coleman and Laurie, 1976). From a study of the effects of inhibitors and of lowering the incubation temperature, it is considered that glucose was taken up initially unchanged into pool 1 in a reaction inhibited by 2-deoxyglucose but energy independent. The glucose is then transferred to pool 2 and at the same time phosphorylated to glucose 6-phosphate, some of which is then dephosphorylated to glucose. This second pool contains all the free sugars in the cell and it is from this that they are removed for glycolysis or polysaccharide synthesis and into which starch degradation products are released. The maltose from the ^{14}C-glucose is labelled solely in the reducing glucose moiety and this suggests that high specific activity glucose in the first pool must exchange with maltose before passing into the second pool. The synthesis of protozoal polysaccharide depends on the square of the glucose concentration, suggesting that two molecules are involved in the synthesis, but the reason for this is obscure (Coleman and Laurie, 1976). Autoradiography in the electron microscope showed that ^3H from ^3H-glucose appeared in polysaccharide granules in both the skeletal plate and the cytoplasm (Coleman and Hall, 1974). Glucose carbon is incorporated, especially at low external concentrations, into intracellular bacteria and those attached to the pellicle (Coleman and Laurie, 1974a), although autoradiographic studies showed that not all bacteria in either situation take up ^3H from ^3H-glucose (Coleman and Hall, 1974) (Fig. 5.3). *Epidinium caudatum* takes up glucose at over 400 times the rate found with *Entodinium caudatum*.

Epidinium takes up ^{14}C from ^{14}C-maltose at only 20–30% of the rate found with glucose and ^{14}C appeared initially in maltose and glucose 6-phosphate and later in glucose. Unlike the results found with glucose, the proportion of the ^{14}C appearing in protozoal polysaccharide did not increase with increasing maltose concentration.

In a study of the intracellular distribution of polysaccharide-degrading enzymes, 60% of all the activity in protozoa disrupted in mannitol/EDTA was sedimentable at 100,000 g for 60 min and 30–40% of this sedimented at 10,000 g for 10 min. The distribution was similar to but not identical to that of acid phosphatase (Williams et al., 1986).

Eudiplodinium maggii Although *E. maggii* is a cellulolytic protozoon, the survival of starved antibiotic-treated (ampicillin) protozoa is prolonged for 4 days by rice starch compared with $5\frac{1}{2}$ days by cellulose and 2 days in the absence of added food. In the rumen of a hay-fed sheep the addition of 100 g of oats per day to the ration increases the population density from approximately 2000 to 20,000/ml when *E. maggii* was the only ciliate protozoon present (Coleman, unpublished observation). Both of these observations suggest that starch plays an important role in the metabolism of this protozoon.

Suspensions of *E. maggii* grown in vitro were incubated with [U-^{14}C]-starch grains in the presence of chloramphenicol to inhibit bacterial protein synthesis. The results showed that 17% of the ^{14}C appeared in volatile fatty acids (acetic acid 71%, butyric acid 21%, propionic acid 8%), 26% as free sugars in the medium (glucose 60%, maltose 40%), and 8% in the protozoa. The remainder of the ^{14}C was assumed to be in CO_2. Of the ^{14}C in the protozoa, 6% was in cytoplasmic protein suggesting that there had been some synthesis of amino acids from glucose (Coleman, 1978b). The rate of disappearance of engulfed starch from rumen-grown *E. maggii* is 220 pg/ng protozoal protein/h.

Glucose Metabolism ^{14}C from [U-^{14}C]glucose was incorporated by *E. maggii* in short-term experiments in the absence of antibiotic, principally into the cell pool (30%) and polysaccharide fractions (34%) (presumably both protozoal and bacterial), with smaller amounts into cytoplasmic (2.8%) and particulate fraction protein (20%). The incorporation into protein fractions is sensitive to chloramphenicol, being decreased by 87% (cytoplasmic and 44% (particulate), suggesting that much of the protein was bacterial or derived from bacteria. Aspartic acid, glutamic acid, and alanine were the most heavily labelled amino acids (Coleman, 1978b).

Polyplastron multivesiculatum *P. multivesiculatum* contains, when grown on wholemeal flour and grass, one of the highest concentrations of α-amylase found in any entodiniomorphid protozoon (Table 6.1; Coleman, 1986b). However, when the protozoon is grown on grass alone, the concentration is decreased to 25–43%. The enzyme has a lower optimum pH than the other protozoal amylases and maltose is the principal product (Abou Akkada et al., 1963; Coleman 1986b). There is, however, some confusion about the further breakdown of maltose. Abou Akkada et al. (1963) reported the presence of a comparatively inactive maltase that did not hydrolyse α-methyl glucoside whereas Williams et al., (1984) found α-glucosidase activity in cultured and rumen-grown protozoa. Evidence obtained from short-term (2 min) incubations of *P. multivesiculatum* with ^{14}C-soluble starch showed that phosphorolysis of starch, and possibly maltose, to glucose 1-phosphate was an important pathway of starch breakdown in this protozoon (Coleman and Laurie, 1977). After disruption of the protozoa in mannitol/EDTA, approximately 60% of the polysaccharide depolymerase and glycoside hydrolase enzymes are sedimentable by centrifugation and are found in the 1000 g for 10 min and 10,000 g for 10 min fractions (Williams et al., 1986; Williams and Ellis, 1985).

To determine if *P. multivesiculatum* could synthesize amino acids from starch, ^{14}C-labelled starch grains were incubated with protozoa in the absence of antibiotics and with protozoa that had been preincubated with ampicillin in an attempt to kill any intracellular bacteria. The results showed that the protozoa had a very limited ability to synthesize alanine, aspartic acid, glutamic acid, and serine, at least, and that the variety of amino acids increased in the absence of antibiotics (Coleman and Laurie, 1977).

Metabolism of Glucose On incubcation with [U-^{14}C]glucose, ^{14}C appeared sequentially in glucose 6-phosphate, maltose, glucose, and glucose 1-phosphate with the highest specific activity in glucose 6-phosphate, suggesting that this was the primary product of glucose uptake (Coleman and Laurie, 1977). In longer experiments (4 h) in the presence of chloramphenicol, ^{14}C was found principally in the pool (40%), bacterial polysaccharide (19%), protozoal polysaccharide (8%), cytoplasmic protein (11%), and particulate protein (5%). This suggests that the protozoa may have limited ability to synthesize amino acids from glucose.

Eremoplastron bovis This protozoon contains an active α-amylase when grown in vitro in the presence of starch but the activity is decreased to 30–35% in protozoa isolated from the rumen or grown in vitro in the absence of starch (Bailey and Clarke, 1963b; Coleman, 1986b; Table 6.1). This protozoon also contains a range of glycoside hydrolase and polysaccharide depolymerase enzymes, all of which are contained in vesicles inside the cell (Williams et al., 1984, 1986).

Ophryoscolex spp. *Ophryoscolex caudatus*, which grows in vitro on starch and dried grass but not on dried grass alone (Coleman and Reynolds, 1982a), contains an active α-amylase when grown under these conditions. However the concentration is decreased to 25–45% when the protozoon is grown as the only ciliate of an animal fed on hay and oats

(Table 6.1; Coleman, 1986b). An α-glucosidase is present under both conditions (Williams et al., 1984). Starch is fermented by *O. caudatus* (Williams et al., 1961) and *O. purkynjei* (Mah and Hungate, 1965) to acetic, lactic, and butyric acids plus H_2 and CO_2.

Glucose Metabolism *O. caudatus* grown in vivo takes up glucose principally into protozoal polysaccharide with a small incorporation into cytoplasmic protein. Similar results are obtained with protozoa grown in vitro except that there is more incorporation of glucose carbon into the protozoal protein fractions and there is evidence that this might have been due to the activities of intracellular bacteria (Coleman and Reynolds, 1982b).

Metadinium medium The only study of this protozoon has been made by Naga and El-Shazly (1968) who called the organism *Eudiplodinium medium*. Intact protozoa degrade starch, amylose, and amylopectin with release of soluble sugars into the medium and the production of acetic and butyric acids and unusually large amounts (25% of total) of formic acid.

Entodinium furca, Ostracodinium spp., Diplodinium spp., Eremoplastron bilobum, and Diploplastron affine These all contain α-amylase and α-glucosidase (Williams et al., 1984; Coleman, 1986b), but nothing else is known except that *Diploplastron affine* takes up glucose (Coleman and Reynolds, 1982b).

6.2 Metabolism of Cellulose and Cellobiose

Cellulase Determination There are three major problems in the determination of the total cellulase activity in rumen ciliates: 1) a smaller or larger proportion of the "cellulase" (and there are probably several in each protozoon—Bonhomme et al., 1986; Coleman, 1983b) is bound to particulate matter and is not easily released; 2) there is a non-linear relationship between time or enzyme concentration and all products of cellulose breakdown however measured; and 3) the variety of different methods available, all of which measure different facets of the action of an enzyme on a large and complex molecule. It is therefore difficult to compare different authors' results, especially if, as often occurs, no regard has been taken of the non-linear enzyme-product relationship.

Methods of Cellulase Assay The following methods have been used by the authors (Coleman, 1985a; 1986a) and others.

1. *Drop in viscosity of carboxymethylcellulose or hydroxyethylcellulose.* The authors have used the reciprocal of the time taken for the

viscosity to drop to 50% of the original value as a measure of activity whereas Bonhomme et al. (1986) have measured relative viscosity and then used predetermined formulae.

2. *Nephelometric assay using finely ground cellulose as substrate*. Nummi et al. (1981) found a linear relationship between rates of reaction and enzyme concentration. Coleman (1985a) found this difficult to measure accurately and used the method described below.

Both these methods give high activities at low enzyme concentrations because a few breaks in the cellulose strand produce shorter chains with lower viscosities or turbidity but with the production of few extra reducing end groups. For example, when the viscosity of carboxymethylcellulose has been decreased to 50% by an enzyme preparation, less than 0.5% of the reducing groups have been liberated (Coleman 1985a).

3. *Release of dye from cellulose azure*. Hotten et al. (1983) used a formula to obtain a linear response whereas Coleman (1985a) used the method described below. The dye release method depends on how the dye is bound to the cellulose. The authors have found that even after prolonged incubation with protozoal preparations only 36% of the substrate is degraded.
4. *Release of reducing compounds from carboxymethylcellulose*. Krishnamurti and Kitts (1969) found that a linear relationship was obtained if a dialysed enzyme preparation was used whereas Groleau and Forsberg (1981) assumed that there was a linear relationship at low enzyme concentrations.
5. *Release of reducing compounds from, or loss of weight of, microcrystalline cellulose as powder or cotton threads*. The rate of breakdown of microcrystalline cellulose powder (e.g., Sigmacell) is very slow (0.1% of the rate) compared with that of carboxymethylcellulose and therefore prolonged incubation (up to 10 days) is needed. In the authors' experience native cotton fibers are resistant to attack by soluble protozoal enzymes, although the method has been widely used in the rumen by, for example, Stewart (1977) and Mould and Orskov (1983/84).

Calculation of Activities In the authors' experience the most satisfactory way of overcoming the problem of nonlinearity of product formed with enzyme concentration or time is to run, at the same time, a standard curve using different concentrations of a standard enzyme preparation. Using this curve the amount of product formed with the enzyme under test can be converted into a volume of the standard enzyme preparation. Depending on the experiment, the actual or specific activity of the test preparation can be expressed relative to that of the standard preparation taken as 100 (Coleman, 1985a).

Cellulase Activity of Individual Protozoal Species During growth in vitro the various protozoal species utilize either starch (fed as wholemeal flour) plus a little dried grass or powdered dried grass alone, although some species will grow on both food sources. Table 6.1 shows the relative activities of the cytoplasmic fractions of protozoa against carboxymethylcellulose when they are grown in the rumen or on one or both of the two substrates in vitro.

Entodinium caudatum and other Entodinium spp. There appears to be disagreement as to the ability of *Entodinium* spp. to degrade cellulose, although the disagreement may be only in the interpretation of the results. Hungate (1943) first stated that entodinia probably do not degrade cellulose. However Bonhomme-Florentin (1967; 1968; 1975a) showed that cell-free filtrates of mixed entodinia isolated from the rumen degrade both native cellulose and carboxymethylcellulose at rates of 0.038 and 0.142 μmole glucose/mg protein/h (calculated by the present authors from the results of Bonhomme-Florentin, 1975a). This author also has evidence that intact entodinia, although unable to grow on cellulose alone, digest cellulose when grown on starch. There is good evidence (Coleman and Hall, 1969) that *E. caudatum* at least engulfs plant material such as chloroplasts, but there is also evidence that these may be voided via has anus (Coleman and Hall, 1984). Coleman (1985a) found that *E. caudatum* cultured in vitro contains almost undetectable amounts of cellulase, although mixed entodinia isolated from the rumen occasionally had carboxymethylcellulase activities corresponding to the lower levels of those found by Bonhomme-Florentin 1975a) (Coleman, 1985a; 1986a). The levels were equivalent at best to only about 5% of the activities found in an actively cellulolytic protozoon such as *Eudiplodinium maggii*. Approximately half of the cellulase in *Entodinium caudatum* was present in the particulate fraction after disruption of the organisms and could therefore be bacterial (Coleman, 1986a).

Epidinium spp. Early experiments on the carbohydrases of *Epidinium* spp. (Bailey, 1958; Bailey and Gaillard, 1965; Bailey et al., 1962) suggested that these protozoa did not contain cellulase. This agreed with the observation of Coleman et al. (1976) that epidinia did not grow well in vitro on dried grass as the sole added source of food. However, the concentration of cellulase in the cytoplasm of *Epidinium caudatum* is among the highest in the protozoa tested (Coleman, 1985a; Table 6.1) although there is also some activity (up to 25% of that in the cytoplasm) in the particulate fraction after disruption of the protozoa (Coleman, 1986a). Epidinia grown in the rumen of a sheep fed on hay and oats tended to contain more cellulase (up to 2.5 times) than those grown in vitro on wholemeal flour and dried grass. Evidence from the differential rates of loss of viscosity and appearance of compounds from carboxymethylcellulose suggested that the cellulase(s) of epidinia, and other rumen ciliates, are endoenzymes, i.e., they attack cellulose molecules in the middle.

E. caudatum and *E. ecaudatum* also contain enzymes that hydrolyze cellopentose and cellobiose, but the activity tends to be lower (up to 90%) of that found in other entodiniomorphid protozoa (Bailey and Gaillard, 1965; Williams et al., 1984).

Attack on Plant Tissues Epidinia are unique among the rumen ciliates in that they actively invade plant tissues and break them down. Bauchop and Clarke (1976) first showed that in New Zealand sheep, where epidinia are the predominant protozoa, these attached themselves to cut leaf surfaces of plant feed material and concentrated beneath the epidermis. These observations were followed up by those of Amos and Akin (1978) and Akin and Amos (1979) who showed that in cattle epidinia preferentially attacked cool rather than warm season forage leaves. During protozoal attack the cell walls of mesophyll, parenchyma bundle sheath, and epidermis cells became swollen and frayed to reveal microfibrillar networks. The protozoa then engulfed plant cells and cell walls. It was suggested that the protozoa secrete extracellular enzymes which degrade plant tissues. As the experiments were carried out in the presence of high concentrations of streptomycin, it is unlikely that bacteria were responsible for the results. With the cool season forages over 10% of the dry matter disappeared in 22 h as the result of protozoal action. The results of Bauchop (1979) confirmed that epidinia are probably the primary invaders of plant tissue and that bacteria are not involved. More recently Grain and Senaud (1984; 1985) have shown that the epidinia attach themselves to plant tissue by an anterior pleated zone, covered with a thick glycoprotein coat, that is distinct from the normal mouth area. The engulfment of plant particles involves enlargement of the mouth or vestibulum and formation of "pseudopods" (phagoplasmic tongue) from nonciliated parts of the vestibulum. The pseudopods surround isolated cells or extend along large plant particles and bring about engulfment of the particles. The proximal ends of particles too large to be engulfed are digested in channel-shaped vacuoles until pieces break off and are taken into secondary vesicles where digestion is completed. The ultrastructural studies of Furness and Butler (1983, 1988) are consistent with this mechanism. Bacteria are believed to play no role in this process although Delfosse-Debusscher et al. (1979a) and Thines-Sempoux et al. (1980) state that bacteria play a major role in the breakdown of plant material by protozoa (see below).

Eudiplodinium maggii This protozoon, which will grow on starch (Coleman, 1987a) or dried grass, contains an active cellulase, the activity of which depends on the growth conditions. It is highest in protozoa grown in vitro on dried grass alone, lower (83% of this) in protozoa taken from the rumen of a sheep fed on hay and oats, and lowest (8%) in protozoa grown in vitro on starch and dried grass. Almost all the protozoal cellulase is soluble; only about 10% is associated with particulate matter (Coleman, 1985a). The concentration of β-glucosidase is high in rumen-grown protozoa, the only condition under which it has been tested (Williams et al., 1984).

In experiments on the survival of starved antibiotic-treated *E. maggii* grown originally in vitro on dried grass, cellulose kept the protozoa alive for 5 days compared with 4 days by starch and 2 days with no added food. This suggests that cellulose is the preferred food source.

Although Hungate (1943) showed that cellulose digestion products were used by *E. maggii* for the synthesis of amylopectin-like material, it was only recently that Coleman (1986d) showed that 2.6 ng per protozoon per hour (39 pg/mg protein/h) was synthesized by the protozoa from cellulose.

The Role of Bacteria in Protozoal Cellulolysis Ever since Hungate (1942, 1943) demonstrated that the larger entodiniomorphid protozoa digested cellulose and utilized the products for polysaccharide synthesis, there has been debate on whether the enzymes responsible for cellulose breakdown were of bacterial or protozoal origin. Hungate believed that as the particulate matter obtained after disruption of the protozoa did not produce reducing material in the presence of cellulose, bacteria were not involved. More recently (Coleman, 1978b) it has been shown that the cellulase content of the cytoplasmic fraction of *E. maggii* decreased during incubation in salt solutions for 22 h but that this was not further decreased in the presence of chloramphenicol or ampicillin. In contrast the activity of the particulate fraction decreased by 35%. These results suggested that the cytoplasmic enzyme was of protozoal origin and that the much smaller activity in the particulate fraction was of bacterial origin. Studies using the electron microscope showed that there were few bacteria associated with ingested plant fibers. Bonhomme et al. (1986) reached a similar conclusion when they showed that after incubation of *Polyplastron multivesiculatum* with seven antibiotics to kill all the cellulolytic bacteria, appreciable cellulase activity remained inside the protozoa. Grain and Senaud (1984) also believe that the cellulase is of protozoal origin although Delfosse-Debusscher et al. (1979a) and Thines-Sempoux et al. (1980), in an electron microscope study of unnamed protozoa taken from the rumen, obtained evidence that the cellulase is produced by intracellular bacteria. The enzyme is believed to be concentrated by the protozoa and then released into vesicles containing plant fibers. Unfortunately this question will remain unresolved until the cellulolytic protozoa can be grown axenically.

Rate of Cellulose Utilization It is difficult to measure the rate of cellulose utilization by rumen ciliates because of the difficulty in separating cellulose from intact protozoa or from protozoal debris and because of appreciable endogenous production of volatile fatty acids by the protozoa. However, by the use of starved *E. maggii* incubated with cellulose for 24 h in the presence of ampicillin, it could be shown that the following acids were produced (in pmoles/protozoon/h with the production in the absence of cellulose in parentheses): acetic 7.4 (1.28), propionic 0.6 (0.17), and butyric 2.60 (0.25) (Coleman, 1978b). Experiments with ^{14}C-cellulose suggested that *E. maggii* can synthesize limited amounts of amino acids from cellulose although some was of bacterial origin. The maximum rate of this protein synthesis was 2.9 pg/protozoon/h which, if this were the sole sole source of protein, would allow division every 8,300 h.

Polyplastron multivesiculatum Hungate (1943) first reported the presence of cellulase in *P. (Diplodinium) multivesiculatum* and later Abou Akkada et al. (1963) found that protozoal extracts slowly attacked cellulose.

Although some strains will only grow in vitro in the presence of starch, others grow well on dried grass as sole added source of food (Coleman, 1987a). The concentration of cellulase in the protozoa is approximately half that in *Epidinium* spp. and is independent of the presence of starch in the medium (Coleman, 1985a). There are possibly three endo 1,4-β-glucanases (based on activity against carboxymethylcellulose) in antibiotic-treated protozoa (Bonhomme et al., 1986) and these are probably located in vesicles inside the protozoa (Williams and Ellis, 1985). The protozoa also contain at least one β-glucosidase (Williams et al., 1984; Bonhomme et al., 1986).

Ophryoscolex caudatus Williams et al. (1961) first reported that, on the basis of gas production, *O. caudatus* did not metabolize cellulose and this was supported by the observation of Coleman and Reynolds (1982a) that *O. caudatus* would not grow in vitro on dried grass alone. However studies on extracts of *O. caudatus* showed the presence of comparatively high concentrations of cellulases in organisms grown both in vivo and in vitro (Coleman, 1985a; Table 6.1). In contrast the concentration of β-glucosidase is comparatively low (Williams et al., 1984).

Eremoplastron bovis Hungate (1943) first reported that this protozoon, which he called *Eudiplodinium neglectum*, contained cellulase, although Bailey and Clarke (1963b) were unable to confirm this. Coleman (1985a) found that the cellulase concentration was very dependent on the growth conditions of the protozoa. If the content of protozoa isolated from a sheep fed on hay and oats is 100, then that in protozoa grown in vitro on grass alone is 63 and in protozoa grown on starch and grass 37. *E. bovis* also contains a high concentration of β-glucosidase especially when isolated from the rumen (Williams et al., 1984).

Ostracodinium dilobum This protozoon, which grows readily in vitro on dried grass alone or on powdered straw alone (Coleman, 1987a), contains one of the highest concentrations of cellulase in all the rumen ciliates

(Table 6.1). The concentrations in protozoa grown under different conditions was as follows, relative to that in protozoa taken from a sheep fed on hay and oats taken as 100: in vitro on dried grass alone, 160–270; in vitro on powdered straw, 110; in vitro on wholemeal flour and dried grass, 100. It is noteworthy that of all the protozoal species listed in Table 6.1, *O. dilobum* has the lowest ratio of amylase to cellulase. This means that this protozoon could be relatively more important in cellulose than starch digestion than any other species. *O. dilobum* also contains one of the highest concentrations of β-glucosidase (Williams et al., 1984).

Diploplastron affine This protozoon grows in vitro on dried grass as sole added source of food, but with some strains the population density is increased in the presence of starch (Coleman et al., 1976; Coleman, 1987a). *D. affine* contains only about half the concentration of cellulase found in epidinia and this is relatively unaffected by the growth conditions or the presence of starch in the medium (Coleman, 1985a; Table 6.1). An active β-glucosidase is present especially in cultured protozoa (Williams et al., 1984).

Diplodinium spp. These will all grow on dried grass as the sole added source of food in vitro but the only quantitative investigation showed that *D. pentacanthum* contained a low concentration of cellulase (Coleman, 1985a).

Enoploplastron triloricatum This protozoon grows well in vitro on dried grass alone and contains cellulase, although the concentration is unknown (Coleman et al., 1976).

6.3 Metabolism of Pectin, Hemicelluloses, and Related Compounds

Epidinium spp. Bailey et al. (1962) first showed that rumen-grown *E. ecaudatum* hydrolyzed wheat flour xylan with the initial production of arabinose followed by xylobiose and finally xylose, glucose, galactose, and uronic acid. The hydrolysis involved cleavage of xylobiose from the ends of the chains. The optimum pH was 5.8–6.2. *Epidinium* extracts were also shown to be active against clover hemicellulose, wheat straw hemicellulose, and commercial xylan. This work was extended by Bailey and Gaillard (1965) who showed that epidinial extracts hydrolyzed three hemicellulose fractions of pasture plants at different rates, but with eventual complete breakdown of these polymers to their constituent monosaccharides. Pentoses were liberated more rapidly than galactose or uronic acids. These authors isolated an arabinofuranosidase which removes arabinose from xylan chains and two xylanases with different specificities. These epidinial extracts would also hydrolyze laminarin, lichenin, and laminaribose.

More recently Williams and Coleman (1985) compared the rates of degradation of *Lolium perenne* hemicellulose B and oat spelt xylan by extracts of rumen ciliates (Table 6.1) and showed that extracts of epidinia had comparable activity to that in the other large entodiniomorphid protozoa. Arabinose was released more rapidly than xylan. A hemicellulase with broad specificity has been purified from epidinia from the goat rumen (Rao et al., 1977).

A detailed study of the glycosidases present showed that epidinia contained a wide range of enzymes required for the final degradation of hemicelluloses to their constituent monosaccharides. There is, however, no evidence that entodiniomorphid protozoa can utilize pentoses or uronic acids as a source of energy or for growth (Oxford, 1958, Coleman et al., 1980b).

Pectin Wright (1960) first showed that a mixture of protozoa rich in epidinia contained pectin esterase and polygalacturonase activity. However, as the result of later experiments, Wright (1961) believed that it was the bacteria rather than the protozoa that were responsible for gas production from pectin: there was no gas production from polygalacturonic acid. Subsequently Coleman et al. (1980b) showed that the enzyme in epidinia and all the other rumen ciliates tested was a polypectate lyase with an optimum pH of 8.5 which differs from values of 6.5–7.2 found by Wright (1960).

Entodinium spp. Abou Akkada and Howard (1961) examined extracts of *Entodinium caudatum* taken from the rumen of sheep fed on hay and concentrates and found no activity against wheat flour pentosan. Subsequently Bailey and Clarke (1963a) found high activity against xylan and arabinoxylan in mixed entodinia (from which *E. caudatum* was missing?) taken from a sheep fed rye grass that was very poor is starch. Recently (Williams and Coleman, 1985) these findings have been quantified and it has been shown that *Entodinium* spp. breakdown hemicellulose B and oat spelt xylan at 2–25% of the rate found with epidinia. Mixed *Entodinium* spp. grown in vivo tended to be more active than individual species grown in vitro. The concentration of glycosidases such as α-arabinofuranosidase are also much lower in the entodinia than the larger entodiniomorphs.

Pectin *Entodinium* spp. do not possess endopectate lyase, polygalacturonase, or pectin esterase activity (Bailey and Clarke, 1963a; Abou Akkada and Howard, 1961; Coleman et al., 1980b).

Polyplastron multivesiculatum Abou Akkada et al. (1963) first showed that extracts of *P. multivesiculatum* grown in vivo hydrolyzed wheat flour pentosan with an optimum pH of 6.3. These results were confirmed by Williams and Coleman (1985) who showed that *P. multivesiculatum* had comparable activity to the other large entodiniomorphs towards hemicellulose B and xylan, and also that with the latter, arabinose was produced more rapidly than xylose. This protozoon also contains high concentrations of enzymes such as α-arabinofuranosidase and α- and β-xylosidase (Williams et al., 1984).

Pectin Abou Akkada et al. (1963) found that extracts of rumen-grown *P. multivesiculatum* hydrolyzed polygalacturonic acid and contained pectin esterase activity. The former finding was confirmed by Williams et al. (1986) who found that the polygalacturonase activity was almost the highest found in any entodiniomorphid protozoon. It is likely, however, that this enzyme is an endopectate lyase (Coleman et al., 1980b) and there is evidence that the activity in protozoa grown in vitro is much less than that found in rumen-grown organisms (Coleman et al., 1980b).

Eudiplodinium maggii and Eremoplastron bovis Bailey and Clarke (1963b) first showed the presence of xylanase in *E. bovis* and Williams and Coleman (1985) found that the activity against hemicellulose B and xylan is higher in *Erem. bovis* than in *Eudip. maggii*. The same was true with respect to endopectate lyase (Coleman et al., 1980b). Both species contained enzymes active against intermediates in hemicellulose breakdown (Williams et al., 1984). Polygalacturonic acid may prolong the life of these protozoa to a limited extent (Coleman et al., 1980b).

Metadinium medium Naga and El-Shazly (1968) investigated the metabolic activities of this organism, which they called *Eudiplodininum medium*, and found that it degraded hemicellulose, xylan, and araban, although only with the first and last was there any increase in volatile fatty acid production. A xylanase was purified almost thirty times.

Diploplastron affine, Ophryoscolex caudatus and Ostracodinium dilobum These protozoa all contain high concentrations of enzymes active against hemicellulose B, oat spelt xylan, polygalacturonic acid, and the intermediates in the breakdown of these compounds (Coleman et al., 1980b; Williams et al., 1986; Williams and Coleman, 1985).

6.4 Engulfment and Utilization of Bacteria

Bacteria, after engulfment, probably provide the entodiniomorphid protozoa with most of the nitrogenous compounds required for growth. Fig. 6.3 is an electron micrograph of a section through *Entodinium caudatum* that was actively engulfing *Escherichia coli* and vesicles containing 5–7 bacteria are clearly visible.

In this section there will be consideration of the factors affecting the rate of bacterial uptake and the fate of bacterial constituents inside the protozoa. Description of the fine structural changes associated with bacterial

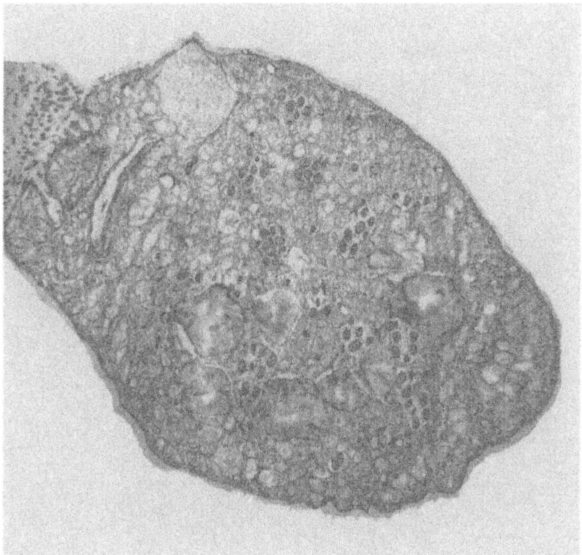

Fig. 6.3 Electron micrograph of a section through an *Entodinium caudatum* cell that had been incubated for 5 min with *Escherichia coli*. Note the many large vesicles containing five to seven bacteria.

engulfment will be found below. The importance of protozoa-bacterial interrelationships in the rumen will be discussed in Chapter 10.

Factors Affecting the Rate of Uptake of Bacteria

Bacterial Population Density The number of bacteria engulfed by *Entodinium caudatum* and other protozoa increase with increasing bacteria population density until a saturating density is reached. To measure this accurately double reciprocal plots of bacterial uptake against bacterial population density have been prepared (Fig. 6.5) and the maximum bacterial uptake measured by extrapolation to the ordinate. As not all bacteria are the same size, this figure must be multiplied by the relative volume of the bacterium under investigation in order to be able to compare the volume of bacterial material engulfed. This has been done for *Entodinium simplex* in Table 6.2. With *Entodinium caudatum* engulfing *Escherichia coli* this maximum uptake varied from 4×10^3 to 11×10^3 (average 6.3×10^3) bacteria/protozoon. As the volume occupied by each bacterium in a close-packed pad was 2 μm^3, the average volume available for bacteria inside *Entodinium caudatum* was 12.6×10^3 μm^3 (Coleman, 1964b). As the volume of each *E. caudatum* was 40×10^3 μm^3 (Coleman, 1967a), this means that 31% of the protozoal volume could be occupied by engulfed bacteria. When comparison is made of the rates of uptake of different bacterial species, it is found to be easier in

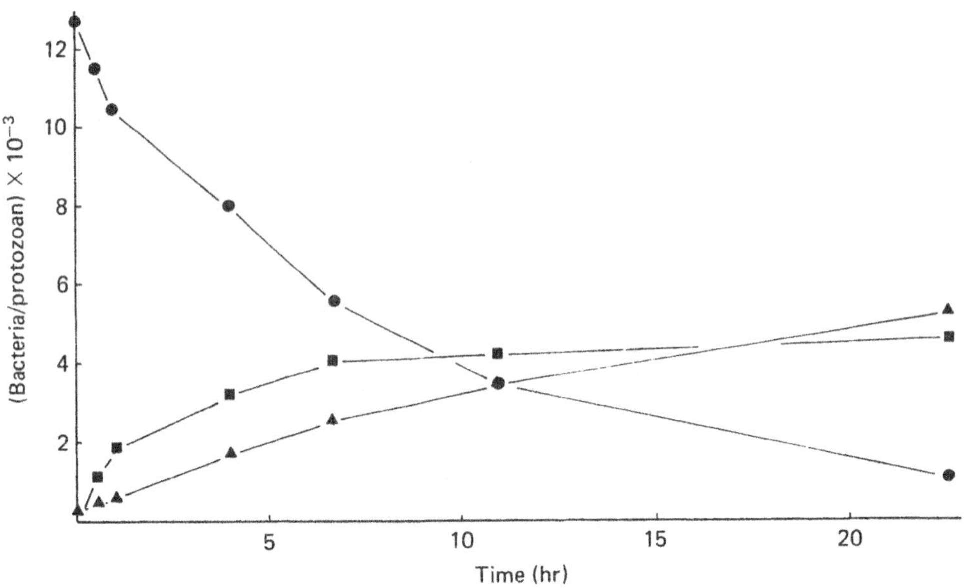

Fig. 6.4 Metabolism of [U-^{14}C] *Escherichia coli* (0.9×10^9 organism/ml) by 73×10^3 *Entodinium caudatum*/ml. ■, ^{14}C in protozoa; ●, ^{14}C in free bacteria; ▲, ^{14}C free in medium.

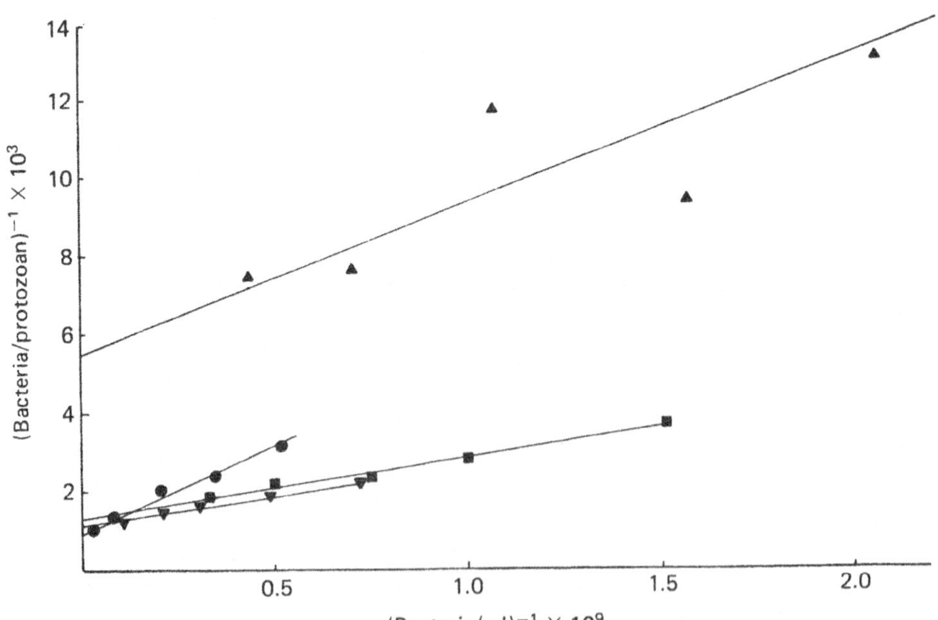

Fig. 6.5 *Entodinium caudatum* was incubated with ^{14}C-labelled bacteria for 3 h and then the number of bacteria engulfed was estimated as described in the text and in Coleman (1972). ▲, *Bacillus megaterium;* ▼, *Escherichia coli;* ●, *Proteus vulgaris;* ■, *Streptococcus faecalis*.

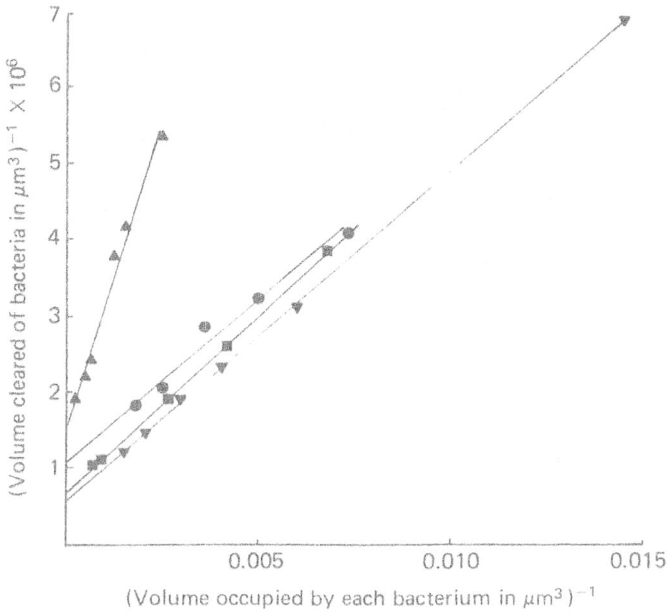

Fig. 6.6 *Entodinium caudatum* was incubated with ^{14}C-labelled bacteria for 3 h, and then the volume of medium cleared of bacteria was estimated as described in the text and in Coleman (1972). ▲, *Bacillus megaterium;* ▼, *Escherichia coli;* ●, *Proteus vulgaris;* ■, *Streptococcus faecalis*.

Table 6.2 The engulfment of various bacterial species by cultured *Entodinium simplex*

Bacterial species	Relative volume[a]	Maximum number of bacteria/h/ protozoon[b]	Maximum volume of bacteria/h/ protozoon[c]	Maximum volume of medium cleared of bacteria/h/protozoon ($\mu m^3 \times 10^{-6}$)[d]
Bacillus megaterium	10.8	105	1130	0.83
Butyrivibrio fibrisolvens	1.0	2800	2800	2.34
Escherichia coli	1.65	900	1530	0.72
Klebsiella aerogenes	1.8	4100	7400	1.72
Proteus mirabilis	1.8	3700	6700	1.61
Streptococcus bovis	2.0	700	1400	0.77
Streptococcus faecalis	1.0	2100	2100	1.8

[a] Expressed relative to that of *Streptococcus faecalis* and determined from the number of bacteria in a packed cell pad.
[b] Number of bacteria taken up from an infinitely dense suspension.
[c] Value in previous column multiplied by relative bacterial volume.
[d] Volume of medium cleared of bacteria from an infinitely dilute suspension.

practice to compare the volumes of bacteria to that of *Butyrivibrio fibrisolvens* taken as 1.0 (Table 6.3).

The uptake of bacteria can also be measured in terms of the volume of medium cleared of bacteria by a protozoon. For example, in a suspension of

Table 6.3 Comparison of the uptake and digestion of three bacterial species by five species of entodiniomorphid protozoa grown in the rumen

	Entodinium caudatum	*Epidinium caudatum*	*Eremoplastron bovis*	*Eudiplodinium maggii*	*Ostracodinium dilobum*
Relative volume[a]	1.0	5.5	2.8	10.3	6.0
Butyrivibrio fibrisolvens (relative volume = 1.0)					
Bacteria inside protozoa	142	1430	305	1700	500
Relative digestion[b]	15	12	4.4	0.6	2.4
Actual uptake after allowing for digestion[c]	2270	18600	1650	2720	1700
Actual uptake/unit protozoal volume[d]	2270	3380	590	260	283
Selenomonas ruminantium (relative volume = 1.6)[a]					
Bacteria inside protozoa	95	2900	580	143	80
Relative digestion[b]	9.5	3.8	0	0	2.3
Actual uptake after allowing for digestion[c]	1000	13920	580	143	264
Actual uptake/unit protozoal volume[d]	1000	2530	210	14	44
Uptake/protozoal vol/ unit bacterial vol[e]	623	1580	130	9	28
Proteus mirabilis (relative volume = 1.8)[a]					
Bacteria inside protozoa	85	690	490	260	310
Relative digestion[b]	0	14	2.4	12	2.5
Actual uptake after allowing for digestion[c]	85	10350	1667	3380	1085
Actual uptake/unit protozoal volume[d]	85	1880	595	328	180
Uptake/protozoal vol/ unit bacterial vol[e]	47	1045	330	182	100

[a]Relative protozoal and bacterial volumes were determined from the number of organisms in a packed cell pad and expressed relative to *Entodinium caudatum* and *Butyrivibrio fibrisolvens* respectively.
[b]Digested bacteria in medium divided by bacteria in protozoa after 4 h.
[c]Number of bacteria inside protoza × (1 + relative digestion) = A.[d]
[d]A ÷ (relative protozoal volume) = B.
[e]B ÷ (relative bacterial volume).

10^9 bacteria/ml, each bacterium occupies 10^3 μm^3, and an uptake of 6.3×10^3 bacteria means that 6.3×10^6 μm^3 (6.3×10^{-6} ml) of medium has been cleared of bacteria.

Two parameters have therefore been used to compare the uptake of various bacterial species by protozoa. The first is the number of bacteria taken up per hour from an infinitely dense bacterial suspension. This parameter probably measures the ability of a protozoon to engulf bacteria and may depend on the size and shape of the bacterium. The second parameter is the volume of medium cleared of bacteria from an infinitely dilute bacterial suspension. The parameter probably measures a protozoon's ability to seek, find, and catch bacteria and may depend on the attraction between the bacterium and the protozoon. The first parameter is obviously more relevant to condition in the rumen. As the bacterial population density in the rumen is of the order of 10^9/ml, the values have also been calculated under these conditions where the significance of bacterial uptake and digestion on rumen metabolism is being considered (Tables 6.5 and 6.11).

The effects of bacterial population density on the rate of uptake of bacteria and on the volume of medium cleared of bacteria are shown for three bacterial of similar size (*Escherichia coli, Proteus vulgaris* and *Streptococcus faecalis*) and for a larger (6.5 ×) bacterium (*Bacillus megaterium*) in Figs 6.5 and 6.6. It is apparent that the three small bacteria behave similarly whichever parameter is considered. If the larger size of *Bacillus megaterium* is taken into consideration, then the uptake of bacterial mass is the same as with the smaller bacteria. However, the volume cleared of *B. megaterium* is appreciably less than that with the other bacteria, suggesting that *Entodinium caudatum* has difficulty in finding, catching, and engulfing a large bacterium present at low density (Coleman, 1975a). With a small protozoon such as *Entodinium simplex*, the uptake of *Bacillus megaterium* mass is appreciably less than that of other bacteria (Table 6.2).

Length of Incubation and Digestion of Bacteria When most protozoal species are incubated with most bacterial species, the latter are engulfed by the former and are often then digested with release of the digestion products into the medium (Fig. 6.4) (Coleman, 1964b; Onodera and Kandatsu, 1969b; Coleman, 1975b). It is apparent that with *Entodinium caudatum* and *Escherichia coli*, for example, the amount of bacterial carbon in the protozoa reaches a constant level after 8–10 h associated with a steady increase in the amount of soluble bacterial digestion products (measured as carbon) free in the medium. These results show that after 10 h the rate of engulfment of bacteria and the rate of bacterial digestion in the protozoon and release into the medium are constant. However, each *E. coli* is not completely digested because the outer membranes remain almost intact (Coleman and Hall, 1972). These empty shells must take up space inside the protozoa and prevent, to some extent, further bacterial uptake.

Although the digestion of engulfed bacteria is a very important phenomenon as such, it is also important in the calculation of the number of bacteria engulfed. The number of bacteria found inside the protozoa in the experiments quoted in Table 6.3 have therefore been corrected for 1) the number of bacteria engulfed and digested and 2) the different sizes of the protozoa relative to that of *Entodinium caudatum*. For example the uptake of 305 *Butyrivibrio fibrisolvens*/h by *Eremoplastron bovis* has been multiplied by

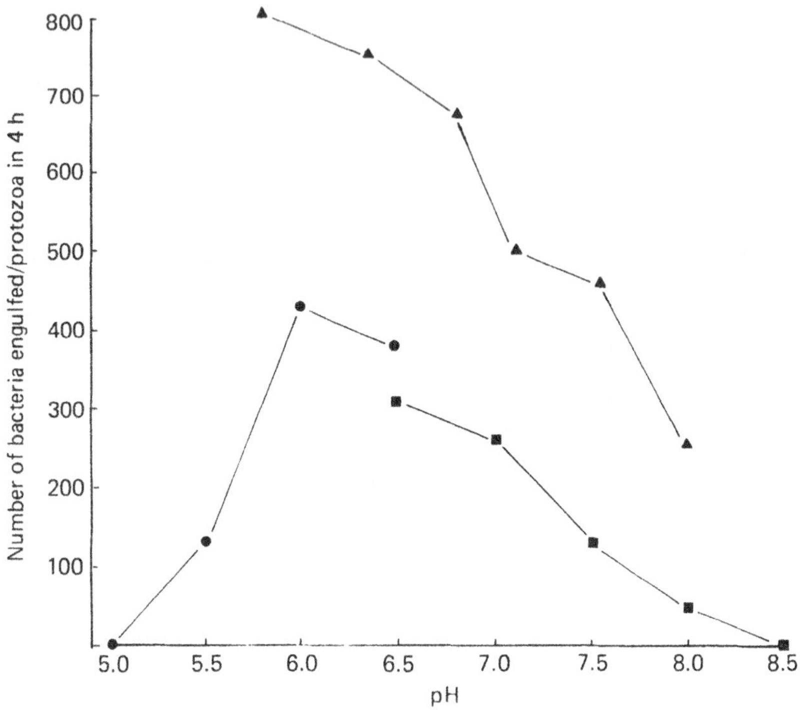

Fig. 6.7 Effect of pH value on the uptake of mixed rumen bacteria labelled with ^{35}S (2.2×10^9/ml) by *Entodinium* spp. (143×10^3/ml) grown in vivo. ▲, salt solution D (Coleman, 1972); ■, 33 mM TES buffer: ●, $\beta\beta'$-dimethylglutarate buffer.

5.4 to give the total number of bacteria engulfed and then divided by 2.8 (volume of the protozoon relative to *Entodinium caudatum* to give the actual uptake per unit protozoal volume of 590. The results in Table 6.3 show that there is considerable variation in the rate of bacterial uptake per unit protozoal volume with the three bacterial and five protozoal species used.

pH The optimum pH for the uptake of bacteria is 6–7 and the uptake declines to zero and pH 5.0 and 8.5 (Fig. 6.7) (Coleman, 1967b; Coleman and Sandford, 1979a). The number of bacteria engulfed also depends upon the buffer used (Fig. 6.7). It is likely that the effects of alteration of pH are on the general well-being of the protozoa rather than on the specific engulfment mechanism.

Salt Concentration The rate of uptake of bacteria is fairly constant over a range of 65–140% of standard salt concentration (Coleman and Sandford, 1979a) but declines rapidly at lower concentrations. The rate of release of soluble bacterial digestion products is also more rapid, at least initially, at 65–100% of standard salt concentration than at higher values (Fig. 6.8).

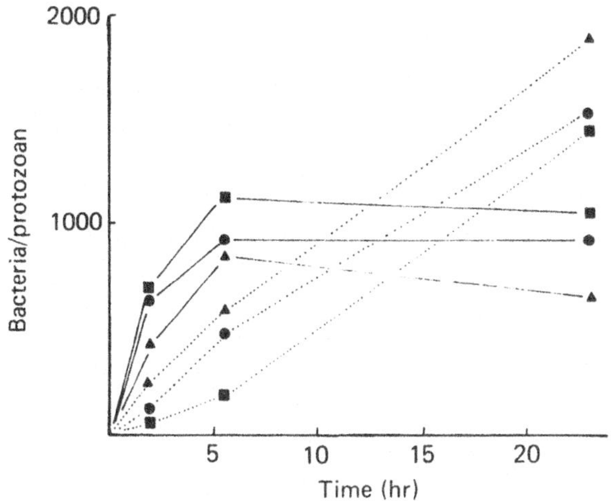

Fig. 6.8 Effect of salt concentration and of added [^{12}C]L-isoleucine on the uptake and digestion of mixed rumen bacteria labelled with [U-^{14}C]isoleucine (1.52×10^9/ml) by *Entodinium* spp. (143×10^3/ml) grown in vivo. Solid lines, bacteria in protozoa; dotted lines, bacterial digestion products released into the medium; ■, standard salt concentration (salt solution D); ●, low salt concentration (62% of standard); ▲, low salt concentration + 7.7 mM [^{12}C]L-isoleucine.

Free Amino Acids If the uptake and digestion of a bacterium is being followed by the use of bacteria specifically labelled with ^{14}C in one amino acid rather than with uniformally ^{14}C-labelled bacteria, the same results are obtained unless the ^{12}C-form of that amino acid is free in the medium. Under these conditions the ^{14}C-form in the bacterial protein appears to exchange with the ^{12}C-form in the medium and less of the ^{14}C-form is incorporated by the protozoa. This reaction is sensitive to changes in the salt concentration in the medium and occurs much more rapidly at low external salt concentrations (Fig. 6.8) (Coleman, 1967b; Coleman and Sandford, 1979a).

Physical Form of the Bacteria The physical form in which bacterial material is presented to the protozoa is also important at least with epidinia where it has been investigated (Coleman and Laurie, 1974a). Sonication of [U-^{14}C]-labelled *Escherichia coli* for 0.5, 1.0, 2.0, 3.0, 5.0, or 10.0 min, to produce after 3 min an almost completely clear suspension increased the rate of uptake of ^{14}C by the epidinia by 2.7, 3.9, 4.8, 4.9, 4.9, and 5.2 times respectively over that of untreated suspensions. In contrast, sonication of ^{14}C-labelled *Proteus mirabilis*, which is engulfed about nine times more rapidly than *E. coli* when both are intact, decreased the rate of uptake of ^{14}C by 97%. As sonicated ^{12}C-*E. coli* did not increase the rate of uptake of ^{14}C-*E. coli* or *P. mirabilis*, this suggests that sonicated *E. coli* did not generally stimulate the protozoal feeding process and that therefore epidinia prefer small pieces of *E. coli* to intact bacteria.

Starvation and Effect of Other Particulate Matter *Entodinium caudatum*, at least, takes up all particles that are small enough to be engulfed. In addition to bacteria, these include starch grains, olive oil droplets, polystyrene latex beads, and particles of palladium black (Coleman and Hall, 1969). The pattern of uptake of starch, for example, is quite different from that of the bacteria because the former are taken up to apparently completely fill the protozoa in just a few seconds whereas, as shown above, bacterial engulfment is a slow, steady process lasting many hours. On incubation of cultured *Entodinium caudatum* with 10, 52, 260, or 1670 rice starch grains per protozoon, the uptake of *Escherichia coli* was depressed by 0, 0, 18, and 68% respectively in the first hour compared with uptake in the absence of added starch. At no starch concentration was bacterial uptake completely abolished. The rate of uptake of bacteria does not increase when the protozoa are starved for 18 or 42 h but when the protozoa have been starved for 66 h, the rate of bacterial uptake is 70% higher in the first hour (Coleman, 1975a). Starvation also increased the rate of release of bacterial digestion products into the medium. A slightly different result was obtained with *Entodinium caudatum* grown in vivo; protozoa taken from a sheep last fed 23 h before engulfed mixed rumen bacteria more slowly than protozoa taken from a recently fed animal. In this case the rate of uptake of bacteria by the starved protozoa was increased by the addition of starch grains. These starved protozoa did not release bacterial digestion products into the medium (Coleman and Sandford, 1979a). These results are thought to show that, provided sufficient energy is present, the less starch in the protozoa, the faster they will engulf bacteria and digest them. However, once the protozoon's energy reserves fall below a certain level, both these processes decline and eventually cease.

Metabolism of Individual Bacteria by Individual Protozoal Species Information of the engulfment of bacteria is summarized in Table 6.4.

Entodinium caudatum Cultured protozoa take up bacteria in the proportion in which they are present in the medium (Coleman, 1964b). Although there is some variation, all bacterial and yeast species are taken up at the same rate by protozoa grown in vivo, except for *Butyrivibrio fibrisolvens* which is engulfed and digested rapidly by the spined form of the protozoon (Coleman and Sandford, 1980).

E. caudatum is the only species where a detailed study has been made of the intracellular bacteria. On harvesting from the growth medium, cultured protozoa contain 16–20 *Klebsiella aerogenes*, 8–10 *Proteus mirabilis*, 1 *Butyrivibrio* sp., and 1 *Bacteroides* sp. (White, 1969). All these bacteria are found in the growth medium. However, the numbers of bacteria in the protozoa are not constant and in protozoa last fed 18, 42, or 66 h before, the total number of viable bacteria was 37, 10, and 2 respectively (Coleman, 1975a). Of the two principal bacteria found in the protozoa, *Proteus mirabilis*

Table 6.4 The single bacterial species not engulfed and poorly or preferentially engulfed by entodiniomorphid protozoa[a]

Protozoal species	Growth condition[b]	Bacterial species: Not engulfed	Poorly engulfed	Preferentially engulfed	Bacterial species preferentially digested – relative digestion: > 10	3–10	2.3 (about)
Diploplastron affine	C	Br, Ec, Ka, Pm	0	Me, Rf	Bf, Sr		
Entodinium bursa	C	0	0	0		Ec, Pm	
	R	Ka, Pm	0	Sr, Sb	0		
Entodinium caudatum	C	0	0	0			
Spined	R	0	0	Bf		Bf, Sr	
Spineless	R	0	0	0		Rf, Sr	
Entodinium longinucleatum	C	Br	Ec	Pm, Ka	0		
Entodinium simplex	C	0	0	Ka, Pm			
Epidinium caudatum	C	0	Ka, Sb	Pm	0		
	R	Ec	Ka, Sb	Bf, Sr, Pm	Bf, Pm	Sr	
Eremoplastron bovis	R	Ec, Ka	0	Bf, Pm	0	Bf	Pm
Eudiplodinium maggii	C	Br, Ka	0	Bf, Sr	Bf, Sr		
	R	0	Ka, Sr	Pm, Bf	Pm		
Ostracodinium dilobum	C	0	Pm	0			
	R	Ec, Me	0	0			
Ophryoscolex caudatus	C	Ec, Ka	Me, Sr	Bf, Sb		Me, Sb	Bf, Sr, Pm
	R	Euc, Me, Rf	Ec, Ka, Pm, Sr	0		Sr	
Polyplastron multivesiculatum	C + Epi	0	0	Pm		Pm	
	C – Epi	0	Bf	Pm			

[a] 0, none; Bf, *Butyrivibrio fibrisolvens*; Br, *Bacteroides ruminicola*; Ec, *Escherichia coli*; Euc, *Eubacterium cellulosolvens*; Ka, *Klebsiella aerogenes*; Me, *Megasphaera elsdenii*; Pm, *Proteus mirabilis*; Rf, *Ruminococcus flavefaciens*; Sb, *Streptococcus bovis*; Sr, *Selenomonas ruminantium*.
[b] C, cultured *in vitro*; R, grown as a single species in the rumen.

survives because the cell envelope is comparatively resistant to digestion and *Klebsiella aerogenes* survives because it synthesizes a glucose-based polysaccharide capsule which is also comparatively resistant to digestion. This capsule is produced from maltose and glucose released during the breakdown of starch grains inside the protozoon. In starved protozoa the concentration of these free sugars falls, less capsule is formed, and the bacteria become less resistant to attack by protozoal lytic enzymes; the number of bacteria therefore falls (Coleman, 1975b). On reintroduction of cultured *Entodinium caudatum* into the rumen of a defaunated sheep, the total number of intracellular bacteria remains the same, but the *Proteus mirabilis* and *Klebsiella aerogenes* disappear and are replaced by typical rumen bacteria (Coleman and White, 1970).

Entodinium simplex *Entodinium simplex*, which is the smallest protozoon to be investigated, also engulfs all bacterial species readily, showing a slight preference for *Proteus mirabilis* and *Klebsiella aerogenes* which are both present in the growth medium. The large *Bacillus megaterium* is taken up comparatively slowly presumably because of the difficulty experienced by a small protozoon in engulfing a large bacterium (Table 6.2). Of the bacteria engulfed over 1 h, 90% of *Escherichia coli, Klebsiella aerogenes, Staphylococcus aureus, Bacillus subtilis*, and *Streptococcus faecalis* are killed whereas *Serratia marcescens* and *Proteus* spp. are comparatively resistant. All the bacteria that were killed rapidly were digested with release of digestion products into the medium (Coleman, 1972).

Entodinium longinucleatum *Entodinium longinucleatum* engulfs all bacteria except for *Bacteroides ruminicola* and a *Pseudomonas* sp. Of the others, *Escherichia coli* and *Bacillus megaterium* are taken up comparatively slowly whereas *Proteus mirabilis* and *Klebsiella aerogenes* are engulfed preferentially. The maximum volume of medium cleared of bacteria is very similar for all species except the last. Only *Proteus mirabilis, Klebsiella aerogenes*, and *Bacillus subtilis* are digested with release of digestion products into the medium (Owen and Coleman, 1977).

Entodinium bursa *Entodinium bursa* grown in vivo does not engulf *Proteus mirabilis* or *Klebsiella aerogenes* but rapidly takes up *Bacillus megaterium, Selenomonas ruminantium, Torulopsis glabrata*, and *Streptococcus bovis*, although only the last is digested with release of soluble material into the medium. Protozoa grown in vitro engulfed each of the bacteria tested, taking up *Megasphaera elsdenii* and *Proteus mirabilis* most rapidly. Individual bacterial species tend to be engulfed more rapidly (up to 20 times) by protozoa grown in vivo than those grown in vitro (Coleman and Hall, 1984).

Epidinium spp. Although *Epidinium caudataum* has five times the volume of *Entodinium caudatum*, it only takes up most bacteria at the same rate when grown in vitro. These cultured protozoa engulf *Klebsiella aerogenes* and *Streptococcus bovis* comparatively slowly and engulf *Proteus mirabilis* and possibly the yeast *Saccharomyces fragilis* preferentially. Of the bacteria tested, only *Bacillus megaterium* and other lysozyme-sensitive bacteria are digested to give soluble digestion products in the medium, and this occurs in the medium by the action of a lysozyme-like enzyme released by the protozoa. The digestion does not occur in vesicles in the cytoplasm (Coleman and Laurie, 1974a, 1974b). *Epidinium caudatum* grown in vivo engulfs individual species at a rate per unit volume comparable to other protozoal species although it does not take up *Escherichia coli* and only engulfs *Klebsiella aerogenes* and *Streptococcus bovis* slowly. *Butyrivibrio fibrisolvens, Selenomonas ruminantium,* and *Proteus mirabilis* are all taken up preferentially and digested rapidly (Coleman and Sandford, 1979a).

Eudiplodinium maggii *Eudiplodinium maggii* engulfs all bacteria tested except for *Bacteroides ruminicola* and *Klebsiella aerogenes,* and it takes up *Butyrivibrio fibrisolvens* and probably *Selenomonas ruminatium* rapidly when grown in vitro. These last two species are also digested with release of soluble products into the medium. Although these species are taken up preferentially, the maximum volume of medium cleared of bacteria from an infinitely dilute suspension is similar with all species (Coleman and Sandford, 1979b). Protozoa grown in vivo behave similarly except that they also engulf *Proteus mirabilis* rapidly and digest it with release of products into the medium (Coleman and Sandford, 1979a).

Eremoplastron bovis *Eremoplastron bovis* grown in vivo does not engulf *Escherichia coli* or *Klebsiella aerogenese* and takes up and digests *Butyrivibrio fibrisolvens* and *Proteus mirabilis* rapidly (Coleman and Sandford, 1979a).

Polyplastron multivesiculatum *Polyplastron multivesiculatum,* which is the largest ciliate investigated, engulfs all bacteria tested, showing a slight preference for *Proteus mirabilis*. There is, however, considerable variation in the number and mass of bacteria engulfed between the different species, although there is less variation in the maximum volume of medium cleared of bacteria. As would be expected, the smaller organisms grown in the absence of other protozoa engulfed fewer bacteria than the larger protozoa grown in the presence of epidinia. *Proteus mirabilis* is the only bacterium that is digested with release of soluble products into the medium. *P. multivesiculatum* is unusual in than it takes up the small particles obtained after the disruption of mixed rumen bacteria more rapidly than intact bacteria. The protozoa also take up the products of the spontaneous lysis of many bacterial species (Coleman and Laurie, 1977).

Ophryoscolex caudatus *Ophryoscolex caudatus* grown in vivo engulfs bacteria at only approximately 10% of the rate found with those grown in vitro.

Those grown in vitro do not engulf *Escherichia coli* and *Klebsiella aerogenes*, and they take up other bacteria at very different rates. Only *Bacillus megaterium*, *Megasphaera elsdenii*, *Proteus mirabilis*, and *Streptococcus* bovis are digested with release of soluble compounds into the medium. Protozoa grown in vivo do not engulf *Eubacterium cellulosolvens*, *Megasphaera elsdenii*, or *Ruminococcus flavefaciens* and only digest *Proteus mirabilis*, *Selenomonas ruminantium*, and mixed rumen bacteria (Table 6.4; Coleman and Reynolds, 1982b).

Diploplastron affine *Diploplastron affine* grown in vitro did not engulf *Escherichia coli Klebsiella aerogenes*, *Bacteroides ruminicola* or, unusually for most rumen ciliates, *Proteus mirabilis*. Of the other bacteria *Megasphaera elsdenii* and *Ruminococcus flavefaciens* are taken up a little more rapidly than the others. Only *Bacillus megaterium*, *Butyrivibrio fibrisolvens*, *Selenomonas ruminantium* and mixed rumen bacteria are digested with release of soluble products (Coleman and Reynolds, 1982b).

Summary of Uptake and Digestion of Individual Bacterial Species With all the protozoal species tested, *Bacteroides ruminicola* and *Escherichia coli* tend to be taken up poorly or not at all whereas *Butyrivibrio fibrisolvens* and *Proteus mirabilis* tend to be engulfed preferentially. *Klebsiella aerogenes* is apparently either actively liked or disliked by the protozoa. Experiments on bacterial digestion showed that *Bacillus megaterium*, *Butyrivibrio fibrisolvens*, and *Selenomonas ruminantium* were digested rapidly by many protozoal species whereas *Escherichia coli*, *Streptococcus bovis*, *Torulopsis glabrata*, and *Megasphaera elsdenii* were digested only slowly or only by a few species. *Proteus mirabilis* is digested rapidly by some species but not by others.

These results mean that on faunation of a rumen with a mixed protozoal population, the population densities of *Butyrivibrio fibrisolvens* and *Selenomonas ruminatium* will tend to selectively decrease. It is not clear at present how relevent these results are to those of Coleman (1986a) and Kurihara et al. (1978) who showed that entodinia tend to encourage the growth of cellulolytic bacteria in the rumen whereas the larger cellulolytic protozoa have the opposite effect.

Metabolism of Mixed Rumen Bacteria by Individual Protozoal Species Although it is valuable to know which bacteria are selectively engulfed by the protozoa, it is also important to know whether or not some protozoal species take up all bacteria faster than others. The method used to follow uptake and digestion involves labelling the bacteria by incubation of a crude bacterial fraction from the rumen with ^{14}C- or ^{35}S-labelled compounds and following the fate of the label during incubation with the protozoa (for details of method see Coleman and Sandford, 1979a). Although the labelling of mixed bacteria is much less exact than that of single bacterial species, the results are believed to be valid because the same values for uptake and digestion are obtained after labelling with such different compounds as [^{14}C]isoleucine, [^{14}C]guanine, and [^{35}S]sulfate. Results obtained with [^{14}C]glucose showed lower uptakes and higher digestion, suggesting that the glucose carbon was incorporated into surface structures (Coleman and Sandford, 1979a). Values for the maximum bacterial uptakes and for maximum volumes cleared of bacteria are shown in Table 6.5 together with these values normalized with respect to protozoal volume. Also shown is the number of bacteria digested relative to those found in the protozoa. The results show that all protozoal species engulf mixed rumen bacteria and that those grown in vitro engulf them faster (1.5 to 17.6 times) than those grown in vivo. The same is also true, with one exception, to the maximum volume of medium cleared of bacteria. With protozoa grown in vitro the variation in maximum uptake per unit protozoal volume is from 1750 (with *Ostracodinium dilobum*) to 9235 (with *Entodinium longinucleatum*) and, with protozoa grown in vivo, from 390 (with *Ophryoscolex caudatus*) to 3930 (with *Eremoplastron bovis*).

Fine Structural Studies on the Digestion of Bacteria Unfortunately there have been no studies on the digestion of true rumen bacteria by rumen ciliates. The only information comes from experiments with cultured *Entodinium caudatum* and bacteria originally isolated from other habitats. There is no evidence for any fixed circulation of food vesicles because vesicles containing bacteria in different stages of digestion appear next to each other in sections (Fig. 6.9). Protozoa that have engulfed bacteria for a period and are then allowed to metabolize these bacteria for a further period still contain bacteria in all stages of digestion, so the sequence of events can only be determined by logic (Coleman and Hall, 1972).

When *E. coli* is engulfed there is a progressive loss of cell contents until only part of the apparently intact cell membrane remains. This remainder consists of two unit membranes 9.4 nm wide which are composed of two electron dense lines 2.7 nm across and a lighter central area, and are separated by a space 13.4 nm wide (Coleman and Hall, 1972). It is rare to see multilayered lamellar systems or myelin figures as found in slime moulds (Hohl, 1965). Studies on the digestion of U-^{14}C-cell envelopes by extracts of *Entodinium caudatum* showed that 30% of the ^{14}C was rendered soluble and subsequent investigation in an electron microscope revealed that it was a dense layer between the two unit membranes that had been removed. Experiments using *Escherichia coli* specifically labelled with galactose in lipopolysaccharide component of the cell envelopes showed that *Entodinium caudatum* was unable to digest the lipopolysaccharide of *Escherichia coli* (Coleman and Hall, 1972).

Investigations into the metabolism of other Gram negative bacteria such as *Proteus mirabilis* and *Serratia marcescens*, which were known to be comparatively resistant to digestion by *Entodinium caudatum*, showed that

Table 6.5 Engulfment of mixed rumen bacteria by entodiniomorphid protozoa

	Growth condition[a]	Relative volume[b]	Bacteria engulfed/h/protozoon	
			Maximum[c]	At 10^9 bacteria/ml
Diploplastron affine	C	5.3	21600	2200
Entodinium bursa	C	3.9	21000	1060
	R	2.0	7500	1080
Entodinium caudatum	C	1.0	2700	950
Spined	R	1.8	4070	340
Spineless	R	1.0	1070	130
Entodinium longinucleatum	C	1.7	15700	2920
Entodinium simplex	R	0.3	1030	133
Epidinium caudatum	C	5.5	10100	3600
	R		13000	770
Eremoplastron bovis	R	2.8	11000	1040
Eudiplodinium maggii	C	10.3	68000	6480
Ostracodinium dilobum	C	6.0	10500	2200
	R		5500	120
Ophryoscolex caudatus	C	11.9	100000	21200
	R	14.5	5700	2030
Polyplastron multivesiculatum	C + Epi	37.8	83000	11900
	C − Epi	13.8	97000	8900

[a] C, cultured in vitro; R, grown as a single species in the rumen.
[b] Volume relative to that of cultured *Entodinium caudatum*.
[c] Number of bacteria taken up from an infinitely dense suspension.
[d] Number of bacteria taken up per protozoon per hour divided by the relative protozoal volume.

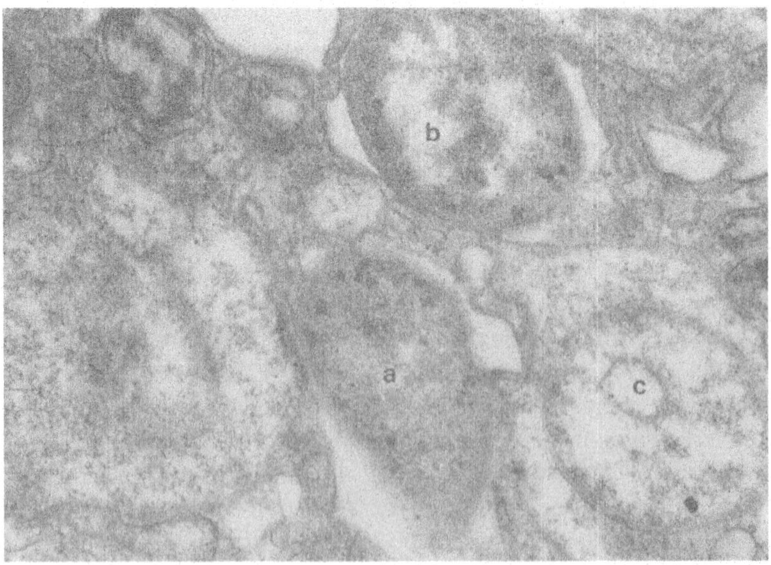

Fig. 6.9 Electron micrograph of a section through an *Entodinium caudatum* cell that had been incubated for 60 min after it ceased to engulf cells of *Escherichia coli*. The *E. coli* cells labelled a, b, and c are at different stages of digestion.

Rate of bacterial uptake per unit protozoal volume[d]		Volume of medium cleared of bacteria/h/protozoon		
Maximum	At 10^9 bacteria/ml	Maximum[e]	Maximum/protozoal vol[f]	Relative rate of bacterial digestion[g]
4075	415			4.6
5380	272	0.46	0.118	1.2
3750	540	0.80	0.40	0
2700	950	1.5	1.5	1.0
2260	190	0.4	0.22	0
1070	130	0.18	0.18	0–1.5
9235	718	1.43	0.84	0
3430	446			1.8
1840	654	8.7	1.58	6.8
2360	140	3.2	0.58	0
3930	370	0.34	0.12	0
6600	600	6.9	0.67	0–6.0
1750	365	2.2	0.37	
920	20	0.48	0.08	
8400	1780	3.9	0.33	0.3
390	140	1.5	0.10	6.3
2190	315	17.0	0.45	0
7030	645	15.0	1.09	0

[e] Volume of medium ($\mu m^3 \times 10^{-6}$) cleared of bacteria from an infinitely dilute suspension.
[f] Values in the previous column divided by the relative protozoal volume.
[g] Number of digested bacteria in the medium divided by the number inside the protozoa after a 4-h incubation.

the cell envelopes of these bacteria were very resistant to digestion and that this was probably why they survived well inside the protozoa (Coleman and Hall, 1972).

Bacillus megaterium is killed and digested more rapidly than any of the Gram-negative bacteria tested, and after a period of engulfment by *Entodinium caudatum* followed by immediate fixation, no intact bacteria were visible in sections. At this time most of the bacterial cell wall had disappeared leaving a layer of material under where the wall would have been. The remainder of the bacteria gradually disappeared over 3–6 h. Similar results were obtained with *Micrococcus lysodeikticus* and *Bacillus subtilis*. Experiments with U-^{14}C-*Bacillus megaterium* cell walls showed that 85–95% of the walls were digested by the protozoa (Coleman and Hall, 1972). The release of diaminopimelic acid, which is found only as a component of the peptidoglycan of bacterial cell walls, has been used by Denholm and Ling (1984) and Masson and Ling (1986) to follow the metabolism of *Bacillus megaterium* cell walls. In experiments on the digestion of U-^{14}C-labelled cell walls of *Bacillus megaterium*, the protozoal extracts with the highest specific activities are those from *Diploplastron affine* (rumen-grown), *Eremoplastron bovis* (rumen-grown), *Entodinium simplex* (cultured), and the lowest with *Ophryoscolex caudatus* (rumen-grown and cultured) (Coleman, unpublished results).

Other Gram-positive bacteria such as *Streptococcus faecalis*, *Staphylococcus aureus*, *Bacillus cereus*, and *Clostridium welchii* which are more resistant to lysozyme are digested in a different way. Typically the cell contents are digested first leaving apparently intact cell walls which subsequently slowly are broken down into small pieces (Coleman and Hall, 1972).

Fate of Bacterial Constituents after Engulfment by Protozoa

Protein Amino Acids Although *Escherichia coli* is not found in appreciable numbers in the rumen, it has been used as a test bacterium to determine the fate of bacterial constituents because strain B, at least, can be grown aerobically on glucose and ammonia as sole sources of carbon and nitrogen. It is also possible by the addition to the growth medium of the appropriate ^{14}C-amino acid and the appropriate ^{12}C-amino acids to prepare bacteria labelled with just one ^{14}C-amino acid. For details see Roberts et al. (1955). On incubation of these bacteria with washed suspensions of entodiniomorphid protozoa, it is possible to follow the fate of the ^{14}C using the techniques described in Chapter 6 Appendix.

Table 6.6 gives the quantities of CO_2, volatile acid, and nonvolatile materials liberated into the medium per 100 moles of carbon incorporated by the protozoa. The chemical nature of these compounds and of those found in the protozoal pool is also given if known. There is evidence for limited decarboxylation of some amino acids especially those with hydrocarbon side chains. For example leucine is decarboxylated and deaminated to give isovaleric acid. The nonvolatile compounds in both the medium and pool are similar and are either the free amino acid or the N-acetyl or N-formyl derivatives. The evidence available suggests that the pool constituents are not intermediates between bacterial and protozoal protein (Coleman, 1967b).

All the bacterial amino acids tested (alanine, aspartic acid, glycine, glutamic acid, leucine, isoleucine, proline serine, and valine) are incorporated into protozoal protein and there is no evidence for any interconversion of even closely related amino acids, such as glutamic acid and proline, by resting suspensions or growing cultures of *Entodinium caudatum* (Coleman, 1967b) or *E. simplex* (Coleman, 1972).

The amino acids can be divided into two groups depending on the effect of the ^{12}C-form free in the medium on the incorporation into the protozoa of ^{14}C from the ^{14}C-form present in bacterial protein. Group 1 amino acids (alanine, arginine, aspartic acid, glutamic acid, and serine) had no effect or a slightly stimulatory effect. Those in group 2 (isoleucine, leucine, methionine, phenylalanine, proline, and valine) decreased the uptake of ^{14}C, especially at low salt concentrations, due principally to the loss of soluble compounds from the pool into the medium. When low-molecular-weight compounds are retained in the pool, 39–44% of the protozoal ^{14}C is in protozoal protein (as distinct from possibly bacterial protein

Table 6.6 The products of the metabolism of the amino acids in *Escherichia coli* by *Entodinium caudatum*[a]

	Products of amino acid metabolism (moles/100 moles of carbon)			
			Non-volatile material	
Amino acid labelled	Volatile acid	CO_2	In medium	In protozoal pool
[U-^{14}C]alanine	79 (acetic 80; formic 20)	18	29 (alanine 85; N-formylalanine 15)	(alanine 65; N-acetylalanine 28; formylalanine 7)
[1-^{14}C]alanine	24	62	28	
[U-^{14}C]glutamic acid	6	5	26 (glutamate 60; N-formylglutamate 40)	(glutamate 35; N-formylglutamate 65)
[U-^{14}C]-glycine	8	1	26 (glycine 65; N-acetylglycine 35)	(N-acetylglycine 100)
[U-^{14}C]isoleucine	38 (α-methylbutyric)	6	28 (isoleucine 100)	(isoleucine 100)
[U-^{14}C]leucine	39 (isovaleric)	7	57 (leucine)	(leucine)
[1-^{14}C]leucine	32	133	31	
[U-^{14}C]lysine	3	1	23	(lysine, others)
[U-^{14}C]phenylalanine	15	3	45	
[U-^{14}C]proline	21	3	86	
[U-^{14}C]valine	74 (isobutyric)	9	40	(valine)

[a] Protozoa were incubated anaerobically for 4–5 h in the presence of *Escherichia* coli labelled with ^{14}C-amino acids as shown. The results are expressed in moles/100 moles of carbon incorporated into the protozoa. Where known, the nature of the compounds in each fraction, as well as the percentage of that compound present in the fraction, is shown in parentheses.

in particulate material in the protozoa). Where some of the pool constituents are lost, 56–70% is in protozoal protein. Most of the remainder is present in the pool fraction. The metabolism of free amino acids is described in Section 6.6.

Nucleic Acid Constituents Unfortunately it is not possible to prepare *Escherichia coli* labelled with ^{14}C in just one purine or pyrimidine. However, on incubation of *Entodinium caudatum* with *Escherichia coli* labelled with ^{14}C-uracil and ^{14}C-cytosine, 70% of the ^{14}C was released as low-molecular-weight compounds into the medium, but 70% of that incorporated was present in protozoal nucleic acid as uracil and cytosine. The ratio of ^{14}C in cytosine to that in uracil was lower in the protozoa compared to the bacteria (0.77 compared with 1.35). Free [^{12}C]-cytosine or uracil and their corresponding nucleosides and nucleotides are effective competitors for the incorporation of the corresponding base into protozoal nucleic acid. Similar results are obtained with *Escherichia coli* labelled with adenine and guanine except that 85% of the protozoal ^{14}C was present in the pool as xanthine and hypoxanthine (Coleman, 1968).

Experiments in which measurements were made of the ratios of ^{14}C in pyrimidine bases to ribose to ^{32}P in associated phosphate suggested that the transfer of nucleic acid bases from bacteria to protozoa occurred at the level of nucleotides (Coleman, 1968).

Cell Wall Constituents The only constituent the metabolism of which has been extensively studied is diaminopimelic acid. Onodera et al. (1974) first showed that mixed rumen ciliates converted 22% of the diaminopimelic acid in *Escherichia coli* cell envelopes into lysine and pipecolic acid and that 72% of this was lysine. More recently, Denholm and Ling (1984) and Masson and Ling (1986) have repeated this work using [^{3}H]-diaminopimelic acid bound in the cell walls of *Bacillus megaterium* and have showed that lysine was the major intracellular product and that free diaminopimelic acid, a number of diaminopimelic acid-containing peptides, their derivatives, and a small amount of lysine were released into the medium. The metabolism of free diaminopimelic is described on page 217.

6.5 Engulfment and Utilization of Ovals and Fungi

There is comparatively little information on the engulfment of "ovals" and fungi by entodiniomorphid protozoa. Much of the evidence comes from observations in the light microscope on predation by these protozoa. This has been observed by Orpin (1983/84—*Entodinium* spp. on *Neocallimastix* sp.) and by Mackie (1987—*Enoploplastron confluens* and *Entodinium lobosospinosum* on fungal zoospores). However, there is some electron micrographic evidence and Fig. 6.10 shows an *Entodinium* sp. apparently en-

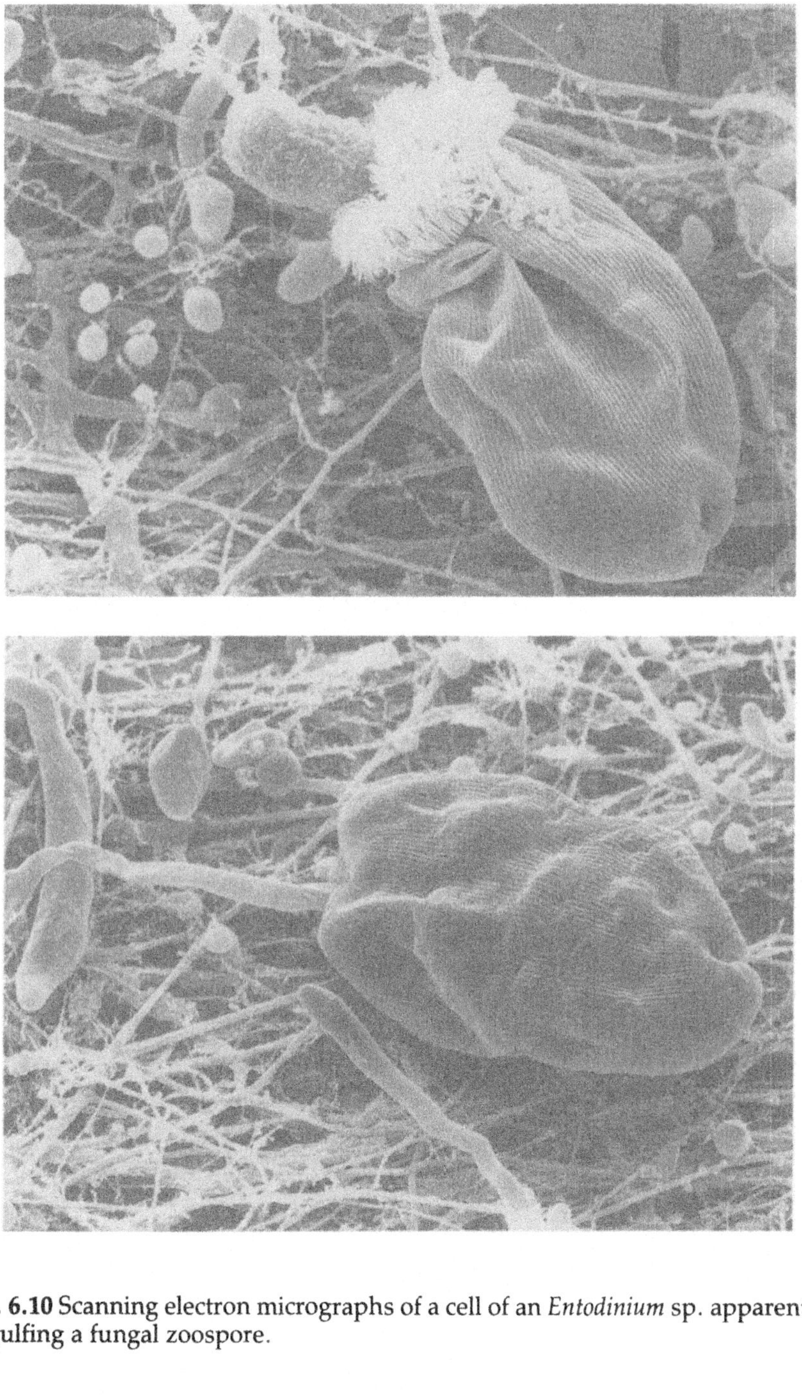

Fig. 6.10 Scanning electron micrographs of a cell of an *Entodinium* sp. apparently engulfing a fungal zoospore.

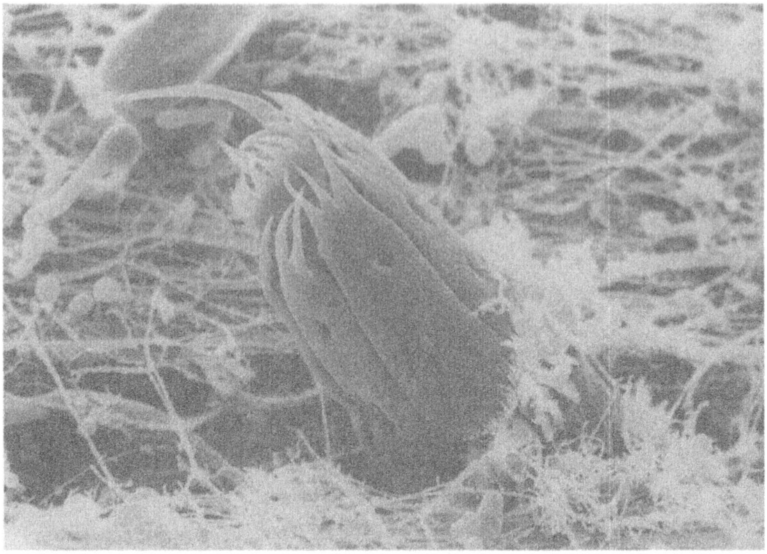

Fig. 6.11 Scanning electron micrograph of a cell of an *Ophryoscolex* sp. (*O. caudatus*?) burying itself in the mycelium of a rumen Phycomycete fungus.

gulfing a fungal zoospore while Fig. 6.11 shows an Ophryoscolex sp. apparently burying itself in fungal mycelium. Recently Newbold and Hillman (1990) have shown that fungal populations are slightly higher (37%) in defaunated than normally faunated sheep.

6.6 Metabolism of Free Amino Acids

In all experiments on the uptake of free amino acids by rumen ciliates, it must be remembered that these organisms contain bacteria in vesicles in the endoplasm and attached to the outside of the pellicle. The addition of antibiotics does not render these bacteria instantly metabolically inactive. Autoradiography in the electron microscope has shown that, after incubation of *Entodinium caudatum* and *Epidinium caudatum* with [^{3}H]-glycine, tracer is present not only in the cytoplasm but also in some, but not all, of the bacteria in the endoplasm and with the epidinia some, but not all, of the bacteria attached to the pellicle (Coleman and Hall, 1974; Coleman, 1978c).

On incubation of *Entodinium caudatum* with ^{14}C-amino acids in the presence or absence of antibiotics, there is a progressive uptake of ^{14}C for 6–10 h although in the presence of e.g. ampicillin or chloramphenicol the uptake is only 25–57% of that in its absence. In the absence of antibiotic, 60% of the ^{14}C is found associated with particulate matter, suggesting that this might be incorporation into bacteria. This is confirmed if the effect of e.g., ampicillin on the uptake of e.g., serine by intact and disrupted *Epidinium caudatum* is investigated (Coleman, 1978c). In the absence of antibiotic the incorporation of ^{14}C from [^{14}C]-serine into the particulate fraction is markedly increased (3.5 times) by disruption of the protozoa (to liberate the intracellular bacteria) prior to the incubation. This uptake is almost completely abolished in the presence of antibiotic, suggesting that this uptake is entirely bacterial and that in intact protozoa these bacteria are not freely in contact with the medium. Although in much of the authors' work to be described below it has been assumed that any ^{14}C found in the soluble material (cytoplasmic fraction) inside the protozoa must be the result of protozoal metabolism, it is possible for external amino acids to be taken up by intracellular bacteria which are then digested by protozoal enzymes. Amino acids in the medium could then be incorporated into protozoal protein after incorporation into bacterial protein.

In some early investigations, attempts were made to measure amino acid uptake from the rate of disappearance of α-amino nitrogen from the medium (Onodera and Kandatsu, 1968). Unfortunately these were frustrated by the spontaneous liberation into the medium by the protozoa of ammonia, peptides, and amino acids during the incubation (Onodera and Kandatsu, 1970c). It has been estimated that 1% of the cellular nitrogen is excreted per hour in this way by *Entodinium caudatum* (Abou Akkada and Howard, 1962).

Another approach to the problems posed by the presence of bacteria is that of Onodera (e.g., Onodera et al., 1983; Onodera and Migita, 1985) who incubated washed mixed ciliates with antibiotics for 12–24 h in the absence of food in order to obtain starved bacteria-free ciliates for metabolic experiments. Although this approach overcomes the problems of allowing for bacterial activity, these authors noted that the activity of the preparations was variable and it is possible that antibiotic-treated protozoa are abnormal and dying, as was found with *Entodinium caudatum* treated with antibiotics for longer periods by Coleman (1962).

Effect of Amino Acid Concentration Amino acid uptake increases with increasing amino acid concentration and on a double reciprocal plot of uptake against amino acid concentration there is a discontinuity in the curve at, for phenylalanine, 3.0 mM (Fig. 6.12; Coleman, 1967a). This discontinuity occurs at 1.2–3.0 mM with isoleucine, leucine, phenylalanine, proline, and valine (group 2 amino acids—Coleman, 1967a) and at 0.09–0.22 mM with alanine, arginine, aspartic acid, glutamic acid, and serine (group 1 amino acids). There is no discontinuity with glycine (Coleman, 1963, 1964a). The slope of the left-hand part of the curve shown in Fig. 6.12 is similar for all amino acids and the curve always passes through or near the origin. This means that at high external amino acid concentrations (nearest the origin on a double reciprocal plot) the uptake is proportional to the external concentration; doubling the concentration doubles the uptake.

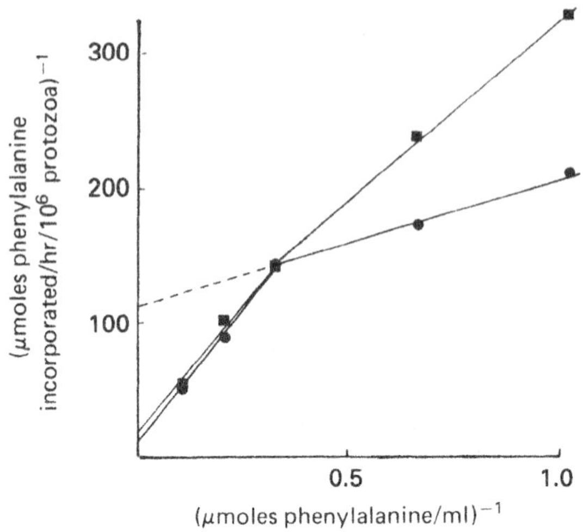

Fig. 6.12 Effect of substrate concentration on the incorporation of [U-^{14}C]L-phenylalanine by *Entodinium caudatum* incubated in the presence of 1000 units penicillin G + 400 μg neomycin sulfate per ml and in the presence (■) or absence (●) of 10 mM-DL-*p*-fluorophenylalanine.

This could be due to medium being trapped inside the protozoa after harvesting from the incubation medium. Such uptake has been termed "passive." At low external concentrations, shown in the right-hand part of the curve in Fig. 6.12, the uptake is higher than the "passive" uptake and has been termed "active." This "active" uptake can be abolished by e.g., amino acid analogues and, as shown in Fig. 6.12, the active uptake of phenylalanine is abolished by p-fluoro-phenylalanine while the passive uptake is unaffected. The maximum active uptake can be determined by extrapolation of the right-hand part of the curve to the ordinate. The maximum active uptake of amino acids by cultured *Entodinium caudatum* varies from 0.6–4.2 nmoles/h/10^6 protozoa with group 1 amino acids to 10.8–27.5 nmole/h/10^6 protozoa with group 2 amino acids (Coleman, 1967a). Glycine is taken up much more rapidly at 200 nmole/h/10^6 protozoa (Coleman, 1963). Details of rates of uptake of individual amino acids by *Entodinium caudatum* are given in Table 6.7. As it is of importance in the determination of the sources of amino acids for protozoal growth in vivo, the rates of uptake of six selected amino acids by all the protozoa investigated are given in Table 6.8. The uptakes have been measured at 0.1 mM amino acid concentration which is approximately that found in the rumen (Wright and Hungate, 1967).

Effect of Salt Concentration and pH Fig. 6.13 shows that within the limits of salt concentration the protozoa will tolerate without dying, and at low amino acid concentration there was little effect on the rate of uptake of group 1 amino acids (Table 6.7), but the incorporation of group 2 amino acids was markedly increased at low salt concentrations. The uptake of group 2 amino acids was insensitive to pH change at high salt concentration

Table 6.7 The incorporation rate for amino acids by cultured *Entodinium caudatum*

Amino acid	Group	Maximum velocity (nmoles/h/10^6 protozoa)[a]
L-Alanine	1	2.35
L-Arginine	1	4.20
L-Aspartic acid	1	0.67
L-Glutamic acid	1	0.61
Glycine		200
L-Isoleucine	2	27.5
L-Leucine	2	22.0
L-Lysine		0.82
L-Phenylalanine	2	8.8
L-Proline	2	10.8
L-Serine	1	3.33
L-Valine	2	14.5

[a]Calculated by extrapolation to the ordinate of the right-hand part of the curve (see Fig. 6.12) obtained on a double reciprocal plot of uptake against amino acid concentration.

(within limits tolerated by the protozoa), but sensitive with a clear optimum at 6.5–7.2 at 50% of standard salt concentration (Coleman, 1967a).

Extracellular Products of Amino Acid Metabolism Although the catabolism of amino acids by entodiniomorphid protozoa has not been exhaustively studied, there is no evidence that these organisms breakdown amino acids either rapidly or extensively. As shown below and in Table 6.6, the products are often N-acetyl or N-formyl derivatives or the result of decarboxylations or deaminations. More detailed information on some amino acids is given below.

Glycine One of the most rapid reactions recorded is that for the N-acetylation of glycine by *Entodinium caudatum* (Coleman, 1964a). On incubation of [^{14}C]-glycine with the protozoa, there is a steady uptake of ^{14}C for 5 h until a maximum is reached when the rate of uptake of glycine is balanced by the rate of release of N-acetyl glycine.

Threonine and Methionine Experiments with starved mixed protozoal suspensions that had been rendered bacteria-free by incubation with antibiotics showed that threonine is deaminated to produce 2-oxobutanoic acid which was then coverted into 2-aminobutyric acid and propionic acid. The relative proportions of the two depends on the nutritional state of the protozoa. The 2-aminobutyric acid tends to be produced by starved cells and propionic acid by better fed organisms (Onodera and Ushijima, 1982; Onodera and Migita, 1985).

The first step in the breakdown of methionine is dethiomethylation to produce methanethiol (Merricks and Salsbury, 1973, 1974, 1976) and an unknown compound which gives rise to 2-oxobutanoic acid (Onodera and Migita, 1985) which is then further metabolized as described under threonine.

Table 6.8 Uptake rate of free amino acids by entodiniomorphid protozoa

Protozoon	Growth condition	Amino acid uptake rate[a]					
		Alanine	Glutamate	Glycine	Isoleucine	Methionine	Phenylalanine
Diplodinium pentacanthum	C	7.2	14.8	3.5	5.4	13.5	8.0
	R	0.98	1.98	0.59	0.54	0.53	0.73
Diploplastron affine	C	2.14	1.03	4.87	1.83	0.41	9.85
Entodinium bursa	C	9.52	9.4	2.7	6.8	ND	6.9
	R	5.25	7.3	5.2	ND	5.5	8.9
Entodinium caudatum	C	0.074	0.153	0.487	0.419	0.327	0.347
Spined	R	0.79	0.73	0.64	0.85	0.92	0.84
Spineless	R	0.071	0.147	0.43	ND	ND	ND
Entodinium longinucleatum	C	ND	1.34	1.47	ND	ND	ND
Entodinium simplex	C	0.89	3.53	3.00	ND	1.79	ND
Epidinium caudatum ovine	C	1.90	1.00	5.88	ND	1.01	1.35
bovine	C	12.5	14.8	14.4	ND	15.0	10.7
Eremoplastron bovis	C	8.24	9.13	10.5	7.64	8.63	5.77
	R	1.61	3.17	3.51	2.24	ND	18.0
Eudiplodinium maggii	C	74.8	36.7	70.9	17.0	13.4	42.9
Ophryoscolex caudatus	C	91.7	33.8	67.5	45.9	50.7	75.9
	R	25.8	23.5	32.2	15.7	49.2	110.5
Ostracodinium dilobum	C	5.42	5.88	3.07	ND	ND	3.63
Polyplastron multivesiculatum	C + Epi	659	1175	195	ND	ND	ND
	C − Epi	72	126	68	62	71	86

[a]Uptake rate expressed as pg amino acid taken up/protozoon in one hour in the absence of antibiotics from a 0.1 mM solution. ND, not determined.

Arginine, Citrulline, Ornithine, and Proline Arginine is degraded by starved, mixed, bacteria-free protozoal suspensions firstly to ammonia and citrulline which is then converted into ornithine, NH_3, and CO_2. The ornithine is then converted into proline with the formation of ammonia. The proline is then reduced to δ-aminovaleric acid which appears to be the end product. Evidence has also been obtained for the incorporation of ^{14}C from [^{14}C]-citrulline into arginine and from [^{14}C]-ornithine into citrulline. No evidence has been obtained for the involvement of urea in any of the reactions (Onodera et al., 1977b; 1983).

Leucine, Isoleucine, and Valine Leucine is apparently deaminated and decarboxylated to give isovaleric acid, although the formation of NH_3 and CO_2 have not been demonstrated. The rate of degradation is about 38% of the rate of uptake by the protozoa (Coleman, 1967a). Although no experiments have been made with free isoleucine and valine, it is assumed that they are metabolized in the same way as bound amino acids to give α-methylbutyric acid and isobutyric acids respectively. No evidence was obtained for any N-acetyl or N-formyl derivatives.

Alanine For every 100 μmoles of alanine carbon incorporated by the protozoa, 46 μmoles of acetic acid, 740 μmoles N-acetylalanine, and 100 μmoles N-formylalanine were produced (Coleman, 1967a).

Diaminopimelic Acid On incubation of [1,7-^{14}C]diaminopimelic acid with mixed rumen ciliates for 12 h, up to 28% is converted into lysine which is found free in the medium. About 10% of the lysine is further metabolized to pipecolic acid (Onodera and Kandatsu, 1974).

Glutamic Acid Even after prolonged incubation of ^{14}C-glutamic acid with suspensions of *Entodinium caudatum* there is only very limited (5%) formation of acid volatile products, N-formylglutamate and α-N-formylglutamine (Coleman, 1967a).

Glutamine and Asparagine Although there is no direct evidence, Abou Akkada and Howard (1962) reported that ammonia was produced during the breakdown of casein by *Entodinium caudatum* and that this ammonia came from amide groups rather than from the deamination of amino acids.

Lysine Mixed rumen ciliates metabolize lysine to pipecolic acid although the rate of reaction is slow (Onodera and Kandatsu, 1974).

Phenylalanine *Entodinium caudatum* breaks down phenylalanine slowly to phenylacetic acid (Coleman, 1967a).

Intracellular Products of Amino Acid Metabolism After incubation for 5 h of cultured *Entodinium caudatum* with [U-^{14}C]phenylalanine in the presence of penicillin and neomycin, 38% of the ^{14}C was in the cytoplasmic fraction and of this 77% was in protein and only 9% in the pool. Of the ^{14}C in the particulate fraction, 82% was in protein (Coleman, 1967a). Similar results were obtained with other amino acids and other protozoa, although the proportions of ^{14}C is variable between the fractions.

Interconversion of amino acids There is no evidence for any interconversion of amino acids with *Entodinium caudatum* (Coleman, 1967a), *E. longinucleatum* (Owen and Coleman, 1977), or, except minimally, with *Polyplastron multivesiculatum*. Limited interconversions were found with the following species; 1) *Epidinium caudatum*: serine to alanine, cysteine, glycine, glutamic acid principally; aspartic acid to nine other amino acids (Coleman and Laurie, 1974a); 2) *Eudiplodinium maggii*: aspartic acid to alanine and glutamic acid plus lesser conversions to other amino acids (Coleman and Sandford, 1979b); 3) *Ophryoscolex caudatus*: no metabolism of lysine but limited conversion to related amino acids of glutamic acid, aspartic acid, serine, and glycine (Coleman and Reynolds, 1982b); and 4) *Entodinium bursa*: limited conversion of glutamic acid, arginine and proline and extensive conversion of aspartic acid into all amino acids (Coleman and Hall, 1984).

Little information is available on the constituents of the protozoal pool during the uptake of amino acids except that during alanine uptake the pool contained alanine and N-acetylalanine and during the uptake of glutamic acid it contained glutamic acid and N-formylglutamine (Coleman, 1967a).

6.7 Metabolism of Protein

There is little information on the utilization by rumen ciliates of proteins apart from the protein of intact bacteria (see above). Bacterial protein that has been precipitated with trichloroacetic acid and then broken up to form a fine suspension is taken up by *Entodinium caudatum* and digested in the same way as the protein in intact bacteria (Coleman, 1964b). Abou Akkada and Howard (1962) showed that soluble casein was rapidly degraded by *E. caudatum*, but Onodera and Kandatsu (1970b) disagreed, finding that mixed entodiniomorphid protozoa only digested insoluble casein. Naga and El Shazly (1968) working with *Metadinium* (*Eudiplodinium* in their classification) *medium* found that it digested a number of proteins and that the rate was proportional to the solubility of the protein. Although the rate of breakdown was increased by amylopectin, the overall rate was less than 20% of that found with *Entodinium caudatum* (Abou Akkada and Howard, 1962). Evidence on the role of protozoa in the digestion of protein in the rumen will be considered in Chapter 10.

There is, however, much more information on the proteolytic activities of rumen ciliates, although there is little agreement between the various workers in the field on the optimum pH, optimum temperature, or the measured activity (Table 6.9). Forsberg et al. (1984) and Coleman (1983a)

Table 6.9 Proteolytic activity of rumen ciliates as described by various authors

Authors	Protozoal preparation	Substrate	Substrate concentration (mg/ml)	Optimum pH	Optimum temperature	Activity[a] at 39°C, optimum pH	Stimulation by reducing agents
Shinchi and Kandatsu (1981)	Mixed protozoa cytoplasmic fraction	Casein	1.2	5–6	70–75°C	0.044	–
	Extracellular	Casein	1.2	5.5–6.0	55–60°C	4.3	+
Brock et al. (1982)	Mixed protozoa sonicate	Azocasein	10.0	ND (6.8)[b]	ND	0.028	Inhibition
Coleman (1983a)	*Ent. caudatum* cytoplasmic fraction	Fraction 1 leaf protein	1.0	3.2	ND		Variable
		Casein	1.0	3.2			
Forsberg et al. (1984)	Mixed protozoa autolysate	Azocasein	10	6–9	ND	0.020	Variable
		Endogenous		5.8	ND	0.89	Variable
Lockwood et al. (1988)	*Ent. simplex* supt after lysis by Triton × 100	Azocasein		4.0–4.5	ND	0.72	Variable

[a] Activity expressed as mg substrate hydrolyzed/h/mg protein.
[b] The value in parentheses is the pH actually used; no optimum was determined.

found little activity of protozoal enzymes against casein at neutral pH, whereas others (e.g., Abou Akkada and Howard, 1962; Brock et al., 1982) experienced no difficulty in measuring activity under these conditions. Coleman (1983a) found an optimum pH of 3.2 compared with 5.5 (Shinchi and Kandatsu, 1981) and 6.0–9.0 (Forsberg et al., 1984) for the activity of protozoal enzymes against casein. However these findings require qualification as Shinchi and Kandatsu's experiments were made at 70°C because they found no activity below 60°C. All other workers have used 38–39°C. Although Forsberg et al. (1984) found no sharp optimum pH for casein breakdown, there was an obvious optimum at 5.8 for the breakdown of endogenous proteins.

In the only comparative study on the proteolytic activity of rumen ciliates, it was shown that the highest activities were present in *Entodinium caudatum* and *E. simplex* and the lowest with cellulolytic species grown in vivo (Table 6.10; Coleman, 1983a). Activities were higher when Fraction 1 leaf protein (the enzyme ribulose 1,5-bisphosphate carboxylase which is found in chloroplasts) was used as substrate rather then casein.

Table 6.10 Breakdown of Fraction 1 leaf protein and casein by the cytoplasmic fraction from entodiniomorphid protozoa

Protozoal species	Growth condition[a]	Enzyme activity[b] against Fraction 1 protein: Maximum	At 1 mg/ml	Enzyme activity[b] against casein: Maximum	At 1 mg/ml
Diplodinium anacanthum	C	1.8	0.17		
Diplodinium pentacanthum	R	5.7	0.28		
Diplodinium tricanthum	C	1.8	0.32		
Diploplastron affine	C	1.24	0.35		
Entodinium bursa	C	0.38	0.26	0.22[c]	0.085
Entodinium caudatum	C	3.8	0.51	0.78	0.48
	R	1.15[c]	0.47		
Entodinium simplex	C	6.4	4.22	1.42[c]	1.21
Epidinium caudatum	C	1.00	0.047	0.13 at 2.5mg/ml	
	R	0.22	0.050		
Eremoplastron bovis	C	0.55	0.41		
	R	0.16	0.04		
Eremoplastron dilobum	C	3.7	0.27		
Eudiplodinium maggii	C	1.6	0.68		
	R	0.19	0.093	0.14	0.071
Ophryoscolex caudatus	C	0.53	0.16[c]	0.22[c]	0.15
	R	1.65	0.11[c]		
Ostracodinium dilobum	C	1.37	0.60		
	R	0.2	0.12	0.053	0.041
Polyplastron multivesiculatum	C	4.3	0.92		
Mixed small protozoa	R	1.13	0.080	0.26	0.18
Mixed large protozoa	R	0.18	0.085	0.11	0.092

[a] C, grown in culture; R, grown as single protozoon in the rumen.
[b] Expressed as μmol acid-soluble tyrosine formed/h/mg protein.
[c] Subject to substrate inhibition.

Although the results tend to be ambiguous, the balance of evidence suggests that the proteolytic enzymes in entodiniomorphid protozoa are of the "thiol" and "carboxyl" types (Shinchi and Kandatsu, 1981; Coleman, 1983a; Forsberg et al., 1984).

The significance of these findings in terms of overall protozoal metabolism is difficult to assess because the conditions in food vesicles in the protozoal endoplasm are probably pH 6 (Weineck, 1934) and temperature 39°C. Although under these conditions the proteolytic activity of protozoal extracts is minimal, there must be appreciable activity present because of the rate at which bacterial proteins are degraded.

All ciliates contain a high concentration of leucine aminopeptidase (Forsberg et al., 1984; Prins et al., 1983).

Extracellular Enzymes Ciliates isolated from the rumen and incubated in vitro release nitrogenous compounds into the medium. Although Abou Akkada and Howard (1962) and Onodera and Kandatsu (1970d) showed that these were principally ammonia and amino acids, Shinchi and Kandatsu (1981, 1983) found that proteolytic enzymes were also released by mixed rumen ciliates. This release is not due to autolysis but it could be the result of the trauma experienced by the protozoa during the washing procedures.

6.8 Sources of Amino Acids for Protozoal Growth

As none of the entodiniomorphid protozoa have been cultured axenically, it is necessary to determine the sources of amino acids for protozoal growth by indirect means. As none of the protozoal species tested, with the possible exception of *Ophryoscolex caudatus*, synthesize quantitatively important amounts of amino acids from starch or glucose, all the amino acids must be obtained preformed. The possible sources are 1) bacteria by engulfment and digestion, 2) plant protein, associated with chloroplasts and other plant material, after engulfment and digestion, 3) engulfment and digestion of other protozoa, and 4) the uptake of free amino acids. Measurements are made of the rates at which free amino acids, amino acids in bacterial protein, and the amino acids in engulfed protozoa are incorporated into protozoal cytoplasmic protein. Then from a knowledge of the amount of cytoplasmic protein present in each protozoon, calculations can be made of the length of time it would take for this to double if one or all of these sources of amino acids were available. Details of the calculations are given in Coleman and Reynolds (1982b) and Coleman and Hall (1984).

Table 6.11 gives information on the rates of uptake of 18 amino acids (assuming on the basis of the evidence available that all are taken up independently) from a 0.1 mM solution (0.1mM is the concentration in the rumen). The rate of uptake of mixed rumen bacteria from a population density of 10^9 bacteria/ml, which is the density of viable bacteria in the rumen, is also given along with the rate of incorporation of bacterial protein into

Table 6.11 Relative importance of various sources of amino acids for protozoal protein synthesis

	Entodinium bursa	*E. caudatum* (spined)	*E. caudatum* (spineless)	*E. longi-nucleatum*	*Diploplastron affine*
Grown in culture (C) or in rumen (R)	C	R	C	C	C
Minimum generation time (h)	6	(5.5)	33	31	13
Protein in cytoplasmic fraction (ng/protozoon)	17.0	4.1	2.3	3.9	8.3
Bacteria engulfed/protozoon/h at 10^9 bacteria/ml	1600	340	1150	1220	2200
Bacterial protein incorporated into protozoal cytoplasmic protein (ng/protozoon/h)	0.158	0.031	0.135	0.132	0.21
Minimum generation time from bacterial protein (h)	108	132	17	30	39
Amino acid incorporation into protozoal cytoplasmic protein (ng/protozoon/h)	0.063	0.011	0.0017	0.008	0.015
Minimum generation time from free amino acids (h)	271	386	1370	482	553
Minimum generation time from bacteria and amino acids (h)	77	98	17	28	37
Other protozoa engulfed/h (max)	2.2				
Other protozoal protein taken up (ng/protozoon/h)	5.1				
Minimum generation time from other protozoal protein (h)	3.3				

protozoal protein. The results show that cultured *Entodinium caudatum, E. longinucleatum, Epidinium caudatum* (bovine), and *Ophryoscolex caudatus* engulf bacteria at a rate which would allow division about every 24 h. If both bacteria and free amino acids are considered, then cultured *Polyplastron multivesiculatum* would also divide every 24 h. Table 6.11 also shows that where another ciliate is engulfed (e.g., by *Polyplastron multivesiculatum* or *Entodinium bursa*) this is a very important source of protein. This is reflected by the low generation time found in these cultures. With some protozoal species, e.g., the spined form of *Entodinium caudatum* and rumen grown *Ostracodinium dilobum,* the rate of bacterial engulfment is highly dependent on bacterial population density and increases from 340 to 4070 and from 320 to 1220 bacteria/protozoon/h respectively on an increasing bacterial population density from 10^9 to 10^{10}/ml. However, with those protozoal species where neither amino acids nor bacteria provide amino acids sufficiently rapidly for the protozoa to divide at the observed rate, it is assumed that plant protein must make up the shortfall. In all species the observed proteolytic activity is high enough for this to occur, even with the cellulolytic protozoa.

Epidinium caudatum (ovine)	*E. caudatum* (bovine)	*Eudiplodinium maggii*	*Ophryoscolex caudatus*	*O. caudatus*	*Ostracodinium dilobum*	*Polyplastron multivesiculatum*
C	C	C	C	R	C	C
26	?	?	?	?	?	14
14.2	14.2	24	28	28	9.2	69
860	5000	6200	21200	2000	320	12000
0.105	0.504	0.41	1.34	0.19	0.020	0.72
135	28	59	21	147	460	96
0.050	0.20	0.36	0.184	0.092	0.016	3.6
284	70	70	152	304	575	19
91	20	31	18	99	255	16
						0.4
						5.7
						12.1

6.9 Metabolism of Free Constituents of Nucleic Acids

Purines *Entodinium caudatum* takes up ^{14}C from [8-^{14}C]adenine and guanine principally into the cell pool (85%) and nucleic acid (14%). ^{14}C from[^{14}C]adenine is found in the pool as adenine, adenosine, adenylic acid, xanthine and hypoxanthine and in the medium as xanthine and hypoxanthine. In the nucleic acid 96% is present as adenine and 4% as guanine. ^{14}C from [^{14}C]guanine is found in the pool as adenine (15%), guanine (25%), and xanthine (60%) and in the medium principally as xanthine. In the nucleic acid 84% was present as guanine and 16% as adenine (Coleman, 1968). From these results it appears that adenine and guanine are metabolised by adenase and guanase to hypoxanthine and xanthine respectively and that the hypoxanthine is then oxidized even under anaerobic conditions to xanthine. Similar results were obtained with *Eudiplodinium maggii* (Coleman and Sandford, 1979b).

Pyrimidines *Entodinium caudatum* takes up uracil more rapidly than thymine but metabolizes the latter rapidly to dihydrothymine which is the only product found in the medium or the pool. The significance of this rapid reduction will be considered in section 7.13. Little or no thymine is incorporated into nucleic acid. Uracil is reduced more slowly to dihydrouracil and is incorporated into nucleic acid as uracil (70%) and cytosine (30%) (Coleman, 1968). With *Eudiplodinium maggii* [U-^{14}C]uridine 5′ monophosphate is taken up twice as rapidly as [2-^{14}C]uracil but the uracil from uridine monophosphate is taken up twelve times as fast as free uracil into protozoal nucleic acid. [^{14}C]uridine monophosphate is metabolized in the medium to uridine, uracil, and dihydrouracil in that order suggesting that the first reaction was a splitting off of phosphate: no free ribose or ribose 5′ phosphate were detected (Coleman and Sandford, 1979b).

Ribose [^{14}C]ribose is taken up by *Entodinium caudatum* and *Eudiplodinium maggii* and is converted into glucose by either the protozoa or the intracellular bacteria: none is incorporated into the nucleic acid (Coleman, 1969b; Coleman and Sandford, 1979b).

Phosphate Free inorganic phosphate is taken up by *Entodinium caudatum* into nucleic acid and phospholipids.

6.10 Metabolism of Lipids

In contrast to the studies with carbohydrates, proteins, amino acids, and phospholipids, there have been almost no investigations into the metabolism of lipids by pure suspensions of rumen entodiniomorphid protozoa. *Entodinium caudatum,* at least, engulfs particulate matter including oil droplets (Coleman and Hall, 1969) and chloroplasts (Hall et al., 1974) and both in culture and in vivo receives lipids only in particulate form. Almost all the linoleic acid in grass is present in chloroplasts.

Uptake of Fatty Acids Washed suspensions of mixed entodiniomorphid protozoa take up linoleic acid when the latter is dispersed in buffer (Chalupa and Kutches, 1968), washed suspensions of *Entodinium simplex* take up stearic, oleic, linoleic, and palmitic acids (Gutierrez et al., 1962), and suspensions of *Entodinium caudatum* take up stearic, oleic, linoleic, and palmitic acids although they only incorporate the last three into phospholipid (Broad and Dawson, 1975).

Some information on the lipid composition of rumen protozoa is given by Williams et al. (1963) and Williams and Dinusson (1973a).

Hydrogenation Mixed entodiniomorphid protozoa incubated in the presence of penicillin and streptomycin hydrogenated 95% of the added free linoleic acid to octadecanoic acid and stearic acid, which is found both in the protozoa and medium. The rate of hydrogenation was decreased to

55–60% when the antibiotics were omitted from the medium. The reverse was found when a bacterial preparation was used (Chalupa and Kutches, 1968). It was suggested that the linoleic acid was first converted to the *cis* monoene which was isomerised to the *trans* monoene before hydrogenation to stearic acid. The proportion of monoene had to exceed that of linoleic acid before the reaction could proceed. These authors noted that in the presence of antibiotics, 82% of the monoene was *trans* whereas it was only 50% in the absence of antibiotic. The significance of this last finding is unknown. No direct information on the hydrogen donor was obtained, although incubation under 100% CO_2 rather than 95% N_2 + 5% CO_2 inhibited hydrogenation. It was suggested that hydrogen might be the ultimate donor acting via a carrier.

Wright (1959), using a protozoal preparation rich in epidinia, showed that triene in chloroplasts was rapidly reduced to diene and that the reduction of monoene to saturated acid was also rapid. However, the conversion of diene to monoene was slow. Abaza et al. (1975) confirmed that entodiniomorphs from the rumen reduced oleic acid to stearic acid in the presence of chloramphenicol and that the reaction was stimulated by the addition of iron and manganese salts. The rate of hydrogenation was also increased in the presence of starch, cellulose, urea, and sodium formate whereas glucose, casein, ammonium sulphate, L-cysteine, and n-valeric acid had no effect. It is not obvious why these compounds should have the effects observed. On incubation of these protozoa in the absence of substrate the endogenous fatty acids became progressively more unsaturated.

Although these results and the effects of antibiotics suggest that it is the protozoa rather than the bacteria which are important in the hydrogenation of fatty acids, the matter cannot be taken as proven. The effect of the presence or absence of protozoa in the rumen on biohydrogenation in vivo will be considered in Chapter 10.

Hydrolysis of Triglycerides The evidence that entodiniomorphid protozoa can hydrolyze lipids is weak and contradictory. Prins (1977) quoted a statement of Warner that *Entodinium caudatum* can hydrolyze olive oil droplets. However, Wright (1961) found that extracts of a mixed protozoal preparation rich in epidinia hydrolyzed tributyrin rapidly and trihexanoin slowly but had no activity against olive oil, linseed oil, or triolein. This observation was supported by the finding that the addition of penicillin to the rumen markedly decreased the rate of linseed oil hydrolysis. Latham et al. (1972) estimated that 30% of the lipolytic activity in the rumen is associated with the protozoa.

Synthesis of Fatty Acids and Triglycerides Unfortunately all the studies on the synthesis of fatty acids by rumen ciliates have been made with mixed rumen microorganisms, mixed ciliates, or holotrich protozoa only. It is therefore possible to give only preliminary and possibly erroneous information on the biosynthesis of fatty acids and triglycerides by entodiniomorphid protozoa.

Little (1.2%) of the glucose-carbon taken up by *Entodinium caudatum* is incorporated into lipid (Coleman, 1969b) and from the studies of Saini et al. (1977) it is likely that this in the glycerol moiety of phospholipids rather than neutral lipids.

^{14}C from [2-^{14}C] acetate is incorporated principally into polar lipids and the proportion increases with time. The amount in neutral lipid decreases from 17.7% after 30 min to 6.8% after 90 min. Of this 19–34% was in diglyceride and Saini et al. (1977) suggested that this was subsequently incorporated into phospholipid. The other constituents of the neutral lipid fraction were sterol (24–34%) and triglyceride (14–19%). It is presumed that the ^{14}C in the glycerides was in the fatty acid moiety. In contrast to these results, ^{14}C from [1-^{14}C]butyrate was not incorporated into protozoal lipids.

Emmanuel (1974) obtained somewhat different results using mixed rumen ciliates. ^{14}C from [1-^{14}C]butyrate was incorporated principally into 16 carbon saturated fatty acid, that from [1-^{14}C]propionate into 15 and 17 carbon saturated acids, and that from [1-^{14}C]acetyl CoA and [2-^{14}C]malonate into 16 carbon saturated acid. These results were interpreted as showing that acetate, propionate, and butyrate were used as chain initiators and malonate as a chain extender, as in *Escherichia coli* fatty acid synthesis (Bressler and Wakil, 1961). Emmanuel (1974) also obtained evidence for α- and β-oxidation of long-chain fatty acids and suggested that odd-numbered fatty acids which contained labelling from [1-^{14}C]acetyl CoA could have been labelled by α-oxidation of palmitic acid (C_{16}) to yield pentadecanoic acid (C_{15}) followed by two carbon addition to give C_{17} acid. Evidence was also obtained for incorporation of ^{14}C from [6-^{14}C]isoleucine into branched chain acids.

However, both Emmanuel (1974) and Demeyer et al. (1978), on the basis of very rough estimates, produced evidence that rumen microorganisms, including protozoa, prefer to utilize performed fatty acids rather than synthesize them de nove.

6.11 Metabolism of Phospholipids

Metabolism of Choline $(CH_3)_3 \cdot N^+ \cdot CH_2 \cdot CH_2OH$ Free [methyl-^{14}C]choline is rapidly taken up by *Entodinium caudatum* at a rate comparable to that for the reduction of thymine. The choline taken up is rapidly converted into phosphorylcholine by the action of choline kinase, which is widely distributed in the cell. No free choline has ever been found in the protozoa even if uptake was measured at 10°C. The uptake was not affected by the presence of betaine or ethanolamine at 400 times choline concentration or by the simultaneous ingestion of latex beads which might be expected to stimulate membrane and hence phospholipid biosynthesis. In contrast, the choline transport inhibitor hemicholinium-3 is a highly effective inhibitor giving 100% inhibition at 10 mM. The uptake mechanism is substrate saturable and the apparent Km for uptake (25 μM) was the same as that for the phosphorylation of choline by cell-free preparations. It has therefore been suggested that even if choline kinase is not part of the transport mechanism, it does provide a highly efficient trapping mechanism for the uptake of choline as the cell is not permeable to phosphorylcholine (Broad and Dawson, 1975; Bygrave and Dawson, 1976).

The phosphorylcholine in the protozoa is converted into cytidine diphosphorylcholine (CDP-choline) by the action of choline phosphate cytidylyltransferase in the soluble fraction of the cell. During the uptake of [^{14}C]choline, ^{14}-C labelled CDP-choline is not normally found as it is rapidly converted into phosphatidylcholine by CDP-choline 1,2 diacylglycerol choline phosphotransferase found principally in the membrane fraction. Choline phosphate cytidylyltransferase is probably the rate-limiting step in the conversion of choline into phosphatidylcholine. Disrupted *Entodinium caudatum* preparations incubated with appropriate additions such as ATP, CTP, and Mg^{++} will also convert [^{14}C]choline into phosphatidylcholine and with ATP alone these preparations have been used to prepare phosphorylcholine and CDP-choline from free choline (Broad and Dawson, 1975; Bygrave and Dawson, 1976).

Choline is essential for the growth of *Entodinium caudatum*, at least (Broad and Dawson, 1976), and, in cultures, it is taken up associated with the starch grains on which the protozoon feeds. The choline cannot be replaced by ethanolamine which is consistent with the finding that *Entodinium caudatum* cannot methylate ethanolamine or its mono- and dimethyl derivatives (Broad and Dawson, 1975). In this respect *Entodinium caudatum* resembles *Plasmodium knowlesi* (Rock, 1971) and differs from *Tetrahymena pyriformis* (Smith and Law, 1970) and *Euglena gracilis* (Tipton and Swords, 1966) where methylation of ethanolamine is an important source of choline.

As choline is taken up specifically by rumen ciliates and not by the bacteria, it has been used as a label for ciliate protozoa in studies in vivo (Coleman et al., 1980a; Leng, 1982).

Metabolism of Ethanolamine Free [2-^{14}C]ethanolamine is taken up by *Entodinium caudatum* and incorporated into CDP-ethanolamine and phosphatidylethanolamine but the rate is much slower (7%) than that found with choline (Broad and Dawson, 1975). After prolonged incubation the highest specific activity was in phosphatidylethanolamine followed by ceramide phosphorylethanolamine and N-(1-carboxyethyl) phosphatidylethanolamine in that order. In each compound only the ethanolamine moiety was labelled. The phosphonolipids, diglyceride aminoethylphosphonate, ceramide aminoethylphosphonate, and phosphatidylcholine were all unlabelled (Broad and Dawson, 1975). Further experiments by Broad and

Dawson (1975) and Coleman et al. (1971) showed that [^{32}P]phosphatidyl [^{3}H]ethanolamine was converted unchanged by *Entodinium caudatum* into ceremide phosphorylethanolamine. It was also hydrolyzed to lysophosphatidylethanolamine, then glycerylphosphorylethanolamine, and finally glycerophosphate and phosphate. The ceremide phosphorylethanolamine once formed was also degraded to phosphatidylethanolamine (Broad and Dawson, 1973, 1975; Coleman et al., 1971).

Metabolism of Aminoethylphosphonic Acid $NH_2.CH_2.CH_2.P(OH)_2$ This compound is characterized by the presence of a carbon-phosphorus bond and occurs as a plasmalogen or diglyceride derivative. It was first isolated by Horiguchi and Kandatsu (1959). It accounts for 20% of the phospholipid phosphorus in *Entodinium caudatum* (Broad and Dawson, 1975). Little is known about the biosynthesis of this compound except that it is derived from phosphatidylethanolamine (Coleman et al., 1971; Broad and Dawson, 1975). However, on incubation of *Entodinium caudatum* with [^{14}C]starch, pyruvate, or lactate, there is appreciable incorporation of ^{14}C into the aminoethyl moiety of aminoethylphosphonate diglyceride, suggesting that a glycolytic intermediate such as phospho-enolpyruvate is involved (Trebst and Geike, 1967; Warren. 1968; Liang and Rosenberg, 1968; Coleman et al., 1971).

Aminoethylphosphonate has been used as a protozoal marker because theoretically it should only be found in the rumen in the ciliate protozoa. There was considerable disagreement in the earlier literature about the usefulness of aminoethylphosphonate, but this could have been due to difficulties in assaying the compound. Dufva et al. (1982) found it in feed materials and Whitelaw et al. (1983) found that the concentration in ciliates was variable. Recently Ankrah et al. (1989) have developed an improved assay and found that aminoethylphosphonate is present in mixed rumen bacteria at one third of the concentration (1.34 compared with 4.3 mg/g dry matter in mixed rumen protozoa and in feed at 0.7–6.1% of that found in the protozoa). Aminoethylphosphonate was also found in rumen and duodenal fluid of defaunated sheep. It must therefore be concluded that aminoethylphosphonate is of no value as a marker of protozoal passage out of the rumen.

Metabolism of N(2-Hydroxyethyl)alanine $HO \cdot CH_2 \cdot CH_2 \cdot NH \cdot CH(CH_3) \cdot COOH$ This compound was first isolated bound to phosphatidic acid in the phospholipids of *Entodinium caudatum* by Kemp and Dawson (1969a, 1969b) and was shown to have the L configuration by Coleman et al. (1971). This phosphoglyceride only occurs in anaerobic protozoa where it appears to replace phosphatidylserine.

There are various pieces of evidence on the origin of hydroxyethyl alanine. As mentioned above, ^{32}P is rapidly taken up by *Entodinium caudatum* and it is incorporated into phosphatidic acid and other phospholipids but not hydroxyethylalanine phosphoglyceride. If *E. caudatum* is allowed to incorporate [^{14}C]ethanolamine until the phosphatidylethanolamine is labelled and then an excess of [^{12}C]ethanolamine is added, incorporation of ^{14}C into hydroxyethylalanine continues, suggesting that the synthesis of the latter proceeds via a labelled pool of an intermediate derived indirectly from ethanolamine. Incubation of ^{32}P and ^{14}C-labelled phosphatidylethanolamine with *Entodinium caudatum* produces labelled hydoxyethylalanine phospholipid in which the ^{32}P/^{14}C ratio is the same as in the protozoal phosphatidylethanolamine. This suggests again that ethanolamine phospholipids are the precursors of hydroxyethylalanine phospholipids. The limited evidence available suggests that a three-carbon, glycolytic intermediate is the probable precursor of the carboxyethyl grouping and that this presumably substitutes on the amino group of the ethanolamine (Coleman et al., 1971; Broad & Dawson, 1975).

Metabolism of Phosphate *Entodinium caudatum* takes up inorganic phosphate rapidly and incorporates 10% of this uptake into phospholipids. When the quantity of phosphate taken up is considered, the highest incorporation is into phosphatidylethanolamine + phosphonolipid followed by phosphatidylcholine, phosphatidylinositol, ceramide phosphorylethanolamine + phosphonolipid, and phosphatidylhydroxyethylalanine in that order. However, when the specific activity is considered, that of phosphatidylinositol was higher (35%) than that of phosphatidylcholine and much higher than that of phosphatidylethanolamine. Pulse-chase experiments confirmed that the turnover of phosphatidylinositol was much greater than that of other phospholipids (Coleman et al., 1971; Broad and Dawson, 1975). Mixed rumen ciliates also take up free phosphate principally into phosphoinositol (Saini et al., 1977).

Metabolism of Fatty Acids *Entodinium caudatum* takes up free long-chain fatty acids slowly with palmitic acid taken up most rapidly. Most of the fatty acids remain unesterified with no mono-, di-, or triglycerides detected. However, there is progressive incorporation of oleic acid, and to a lesser extent, palmitic and linoleic acids into phospholipids. Stearic acid, although taken up by the protozoa, is not incorporated into neutral or phospholipid. Palmitic acid is found principally in phosphatidylcholine and phosphatidylethanolamine (Emmanuel, 1974; Broad and Dawson, 1975).

6.12 Metabolism of Lactic Acid

Entodiniomorphid protozoa (probably all species) isolated from the rumen metabolize D- and L- lactic acids and the D isomer is attacked at 51–88% of the rate with the L. The rate of metabolism per g protein is also dependent on the ration on which the host animal had been fed, with protozoa from silage-fed animals showing the highest rate (1.12 g lactate disappearing/g protozoal protein/h). It is likely that the major products of lactate metabolism are propionic and butyric acids but it is difficult to determine these accurately because of the high rates of production from endogenous reserves.

Of the lactate carbon that was taken up by the protozoa, 60% was present as soluble compounds inside the cell, 17% as hot trichloroacetic acid-soluble material (possibly as starch), and 18% in protein (Newbold et al., 1987).

6.13 Uptake and Metabolism of Other Entodiniomorphid Protozoa

There have been many examples quoted by various authors of the predation of one rumen ciliate by another, and these are listed in Table 6.12. However, of these only the predation of *Polyplastron multivesiculatum* and of *Entodinium bursa* have been studied in any detail.

Polyplastron multivesiculatum Some strains of *Polyplastron multivesiculatum* will only grow well in vitro in the presence of epidinia which are engulfed and apparently used as a source of food (Figs. 6.14 and 6.15;

Table 6.12 Predation by entodiniomorphid protozoa

Predator	Prey	Reference
Diplodinium costatum	*Entodinium* spp.	Dogiel (1927)
Diplodinium rangiferi	*Entodinium* spp.	Dogiel (1927)
Elytoplastron bubali	*Entodinium* spp.	Lubinsky (1957d)
	Entodinium bifidum	Lubinsky (1957d)
	Metadinium rotundatum	Lubinsky (1957d)
Enoploplastron confluens	*Entodinium* spp.	Lubinsky (1957d)
Entodinium bursa	*Entodinium caudatum*	Poljansky and Strelkow (1938); Coleman et al. (1977)
Entodinium longinucleatum	*Entodinium* spp.	Ogimoto and Imai (1981)
Eudiplodinium maggii	*Entodinium* spp.	Lubinsky (1957d)
Metadinium medium	*Eudiplodinium maggii*	Lubinsky (1957d)
Metadinium tauricum	*Ostracodinium trivesiculatum*	Lubinsky (1957d)
Ostracodinium trivesiculatum	*Entodinium aculeatum*	Lubinsky (1957d)
Polyplastron multivesiculatum	*Diplodinium monacanthum*	Coleman et al. (1977)
	Epidinium caudatum	Coleman et al. (1972); Imai et al. (1979a)
	Epidinium tricaudatum[a]	Abou Akkada et al. (1969)
	Entodinium spp.	Imai et al. (1979a)
	Eudiplodinium maggii	Eadie (1967); Coleman et al. (1977)
	Metadinium medium	Dogiel and Fedorowa (1925b)
	Polyplastron multivesiculatum	Eadie (1967)

[a] This species is specifically not engulfed.

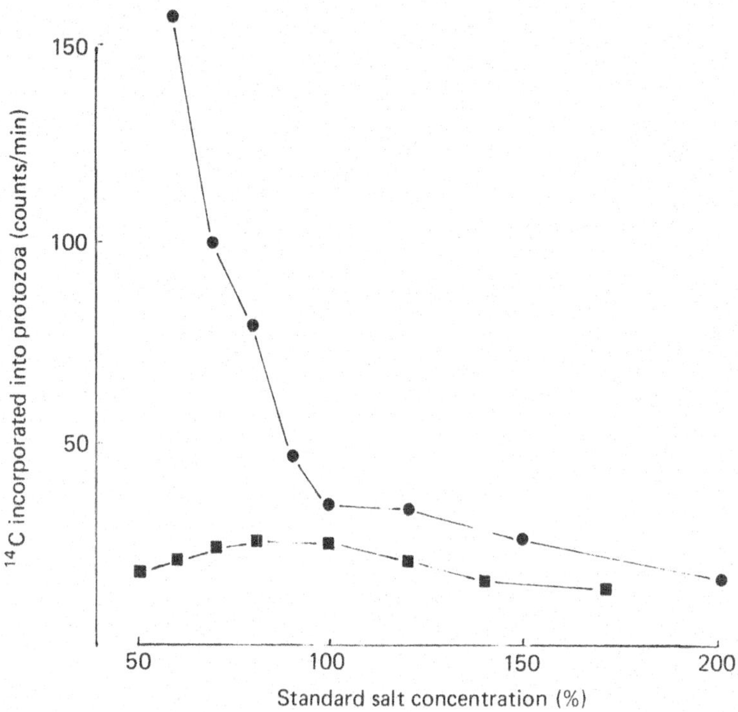

Fig. 6.13 Effect of salt concentration on the incorporation of ^{14}C from ^{14}C-labelled group 1 (■) or group 2 (●) amino acids (see text and Table 6.7) by *Entodinium caudatum* incubated in the presence of 1000 units penicillin G + 400 μg neomycin sulfate per ml.

Coleman et al., 1972; Coleman, 1987a). If the *P. multivesiculatum* has not previously been grown in the presence of epidinia, there is a lag of 3–4 days before engulfment begins. The average generation time of *P. multivesiculatum* under these conditions is 0.6 days and each engulfs 10 epidinia/day. The size of *P. multivesiculatum* varies considerably with the growth conditions being 123 × 98 μm when grown in vitro in the absence of other protozoa, 175 × 127 μm when grown in an A-type population in vivo, and 205 × 123 μm when actively engulfing epidinia in vivo or in vitro. Calculations to show the possible relative importance of bacteria, free amino acids, and epidinia in the nutrition of *P. multivesiculatum* are shown in Table 6.11.

Ingested epidinia are initially digested only slowly with the gradual formation of holes in the epiplasm. Thereafter the protozoa disintegrate rapidly until only a large almost empty vesicle containing fragments of membrane remains in the *P. multivesiculatum*. The cytoplasm of *P. multivesiculatum* grown under these conditions is unique among the rumen ciliates in containing irregularly shaped, heavily staining bodies, larger than hydrogenosomes, which may have been pieces of cytoplasm from engulfed epidinia (Coleman and Hall, 1978).

Experiments on the specificity of the requirement of *P. multivesiculatum* for epidinia showed that *Epidinium caudatum* (ovine and bovine) and *Epi. tricaudatum* all supported equally good growth, that *Eudiplodinium maggii* and *Diplodinium monacanthum* supported limited growth for short periods, and that *Entodinium* spp. and *Diploplastron affine* had no activity. Eadie (1967) reported cannibalism between *P. multivesiculatum* organisms in vivo but this has not been observed in vitro.

Entodinium bursa *Entodinium bursa* only grows in vitro in the presence of the spineless form of *Entodinium caudatum* which is engulfed at the rate of 2.2 *E. caudatum/E. bursa* per hour. Under these conditions the mean generation time is 6 h (Coleman et al., 1977). As the *E. bursa* can eliminate all the *E. caudatum* in a culture, more have to be added each day. It is not advisable to add an excess of *E. caudatum* as, under these conditions, they gradually (over several weeks) develop the caudal spines characteristic of the species. These spined protozoa are either not engulfed or are engulfed only slowly because the *E. bursa* engulfs the *E. caudatum* posterior end first and a long caudal spine is presumably an impediment to this process. The requirement for *E. caudatum* is absolute, as all other *Entodinium* spp. are inactive. A photomicrograph of *E. bursa* containing an engulfed *E. caudatum* is shown in Fig. 6.16. Photomicrographs of the spined and spineless forms of *E. caudatum* are shown in Figs. 2.2, 2.3, and 2.4.

Entodinium bursa and *E. caudatum* will also grow together as the only protozoa in an otherwise ciliate-free rumen with the numbers of *E. caudatum* fluctuating cyclically from 1000 to 150,000/ml with a peak-to-peak time of eight days. The population density of the *E. bursa* also varies cyclically from 5000 to 20,000/ml but there is no consistent connection between the two cycles (Coleman and Hall, 1984).

After engulfment into a vesicle, *Entodinium caudatum* becomes rounded and the first stage in the digestive process is an attack on the surface layer of the pellicle (glycocalyx) often at the bottom of the corrugations present. The underlying membrane and epiplasm are then digested at these places. These initial stages occur relatively slowly but once a hole is made, the *E. caudatum* is rapidly digested. Amino acids in the *E. caudatum* proteins and purines, and presumably the pyrimidines, in the nucleic acid are incorporated into the *E. bursa* protein and nucleic acid although appreciable amounts (up to 50%) are lost into the medium. Some, at least, of the intracellular bacteria and polysaccharide granules in the *E. caudatum* are transferred intact into the *E. bursa* (Coleman and Hall, 1984).

Fig. 6.15 A photomicrograph of a *Polyplastron multivesiculatum* containing an engulfed *Epidinium caudatum* cell.

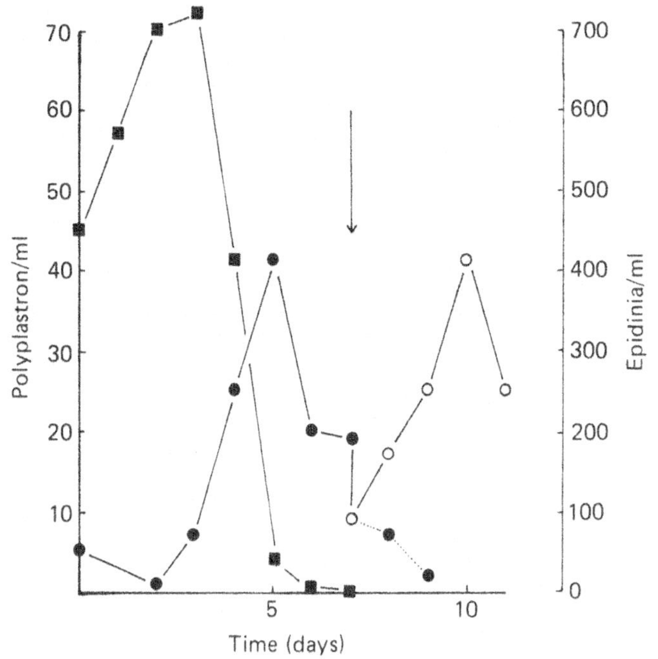

Fig. 6.14 The disappearance of *Epidinium caudatum* (■) during growth of *Polyplastron multivesiculatum* (●). At the arrow, after all the epidinia had disappeared, the culture was diluted with an equal volume of fresh medium and divided between two tubes, to one of which was added 330 epidinia/ml. Number of *Polyplastron* cells: ●---●, without added epidinia; ○——○, with added epidinia.

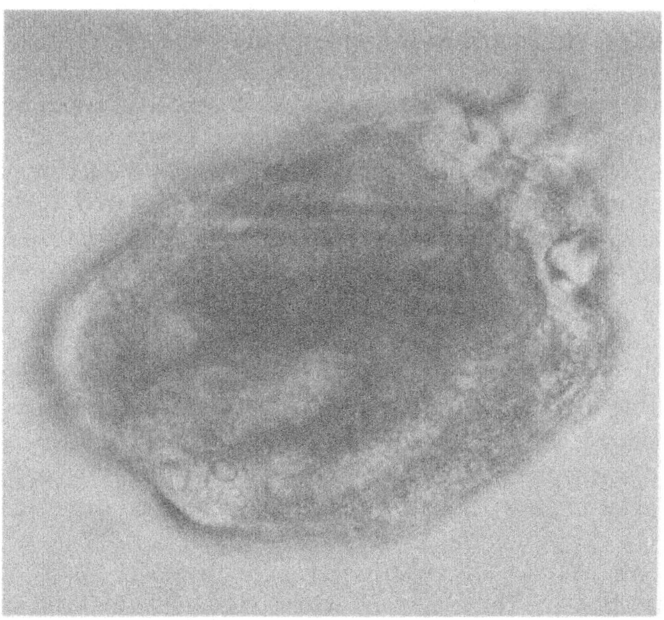

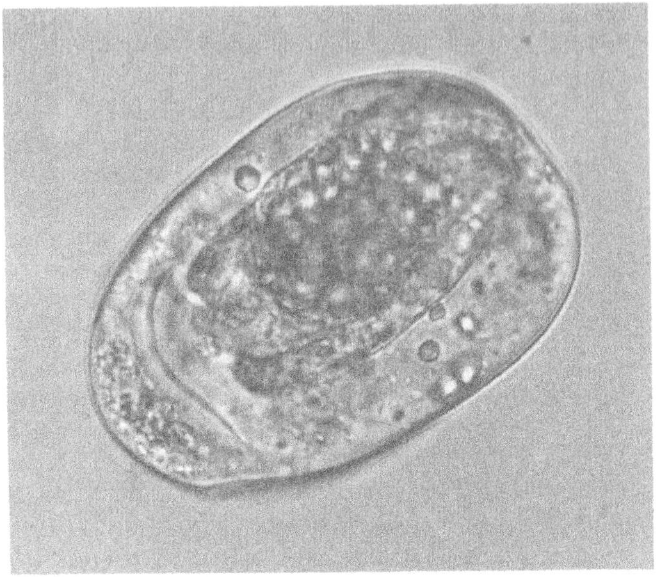

Fig. 6.16 A photomicrograph of an *Entodinium bursa* cell containing an engulfed *Entodinium caudatum* cell.

Appendix: Techniques Used in Experiments on the Metabolism of Entodiniomorphid Protozoa

Information on the techniques available for the preparation of clean suspensions of entodiniomorphid protozoa is given in Section 4.1. Such protozoa still have bacteria in vesicles in the endoplasm and attached to the outside of the pellicle.

In the following section there is a description of the techniques available for studies on the metaboblic activity of these protozoa. As the organisms are all strict anaerobes, it is essential that the appropriate methods are used, as it is very easy to kill all the protozoa if the incubations are not made in reduced media in the absence of air. Entodiniomorphid protozoa will tolerate traces of oxygen in the suspension media at room temperature but will be killed if the suspension is incubated at 39°C. Traces of oxygen introduced during the washing procedures used to prepare clean suspensions do not therefore kill the protozoa. However once these suspensions are prepared it is essential to bubble an O_2-free gas mixture, such as $N_2 + CO_2$ (95:5, v/v), for at least 1.5 min through the suspensions before they are used.

Suspension Media The most satisfactory basal medium is salt solution D (Coleman, 1972) which contains (g/l): K_2HPO_4, 6.3; KH_2PO_4, 5.0; NaCl,

0.65; dried $CaCl_2$, 0.045; $MgSO_4.7H_2O$, 0.09; pH6.8. This is used as such for washing the protozoa but should be supplemented with 0.2% (w/v) L-cysteine hydrochloride (neutralized just before use) or, if it is undesirable to use an amino acid, 0.1% (w/v) ascorbic acid (neutralized before use). The pH is not changed significantly by bubbling N_2 + CO_2 (95:5) through this solution and this partial pressure of CO_2 is satisfactory for those protozoal species grown under 5% or 100% CO_2.

Incubation Conditions Incubations should be made in thick-walled centrifuge tubes (to facilitate subsequent manipulations) that contain 0.03–1.00 ml of ^{14}C-labelled bacteria, amino acid, sugar, or other compound, 0–0.1 ml antibiotic (if required), 0–1.0 ml protozoal suspension (500 large protozoa/ml to 250,000 small protozoa/ml), and 0–1.47 ml salt solution D containing cysteine or ascorbic acid (see above) to give a final volume of 1.5 ml. The protozoal suspension should be the last item added and should be used immediately after it has been prepared and bubbled with gas. Protozoal suspensions deteriorate rapidly when standing at room temperature. After inoculation N_2 + CO_2 (95:5) should be bubbled through the contents of each tube for 15 sec at room temperature and the tube then sealed with a rubber bung and incubated at 39°C. At the end of the incubation the contents of all tubes should be examined microscopically to ensure that all the protozoa are actively motile.

Harvesting of Protozoa After an Incubation After incubation the protozoa are centrifuged (on a bucket-head centrifuge for 20 sec from rest, maximum speed equivalent to 200 g) and washed three times by centrifugation in salt solution D at room temperature. The washed protozoa are then suspended in 2 ml water and disrupted with ultrasound. The crude homogenate is then centrifuged (5500 g for 20 min at 4°C) to give a supernatant fluid which is the cytoplasmic fraction of the protozoa and a pellet, which after washing once in water contains any starch grains, bacteria, pieces of protozoal pellicle, etc., present in the protozoa. These two fractions can then be analyzed as appropriate.

Disruption of Protozoa with Ultrasound The tube containing the washed protozoa should be immersed (to the depth of the liquid in the tube) in an ultrasonic cleaning bath (average output 80 kHz) until no protozoa can be detected microscopically. It is essential to use thick-walled tubes as thin-walled tubes shatter and to use tubes free of scratches as the process takes much longer in such tubes. The length of sonication necessary varies considerably with protozoal species. Entodinia disintegrate very quickly whereas *Eudiplodinium maggii* requires long periods of sonication. Even when the contents have been dispersed, the pellicles of *E. maggii* remain intact.

7

Structure of Entodiniomorphid Protozoa and its Relationship to Function

7.1 General Description of Structure

Fig. 7.1 is a schematic representation of the structure of an *Entodinium* sp. and shows that the organism is divided into two regions, the ectoplasm and endoplasm, which are separated by a boundary layer. In living organisms the two can be distinguished because the endoplasm and the food materials in it are slowly, but continually moving and appear to be revolving while the ectoplasm does not move. An electron micrograph of a section through *Entodinium caudatum* is shown in Fig. 7.2.

At the anterior end of the protozoa is a U-shaped extrudable peristome. In resting protozoa this is retracted and the inner and outer lips cover the vestibulum to apparently seal the organism off from the external medium. In swimming and feeding protozoa the peristome is extruded and the cilia both propel the organism along and waft food particles down the cytostomal invagination to the cytostome where the particles are enclosed in food vacuoles (Coleman and Hall, 1969). Studies in the electron microscope have shown that the endoplasm contains food materials such as bacteria, starch grains, and choloroplasts contained in vacuoles (Figs. 6.4, 6.9, and 7.3) and also a variety of other vesicles and vacuoles which are empty or filled with amorphous material (Coleman and Hall, 1969; Hall et al., 1974). Both the endoplasm and ectoplasm contain small polysaccharide granules of similar size, although it is uncertain if they are of similar origin (Fig. 7.3). Those in the endoplasm could be partially digested food starch grains whereas those in the ectoplasm are almost certainly grains synthesized by the protozoon from food digestion products (Coleman and Hall, 1969).

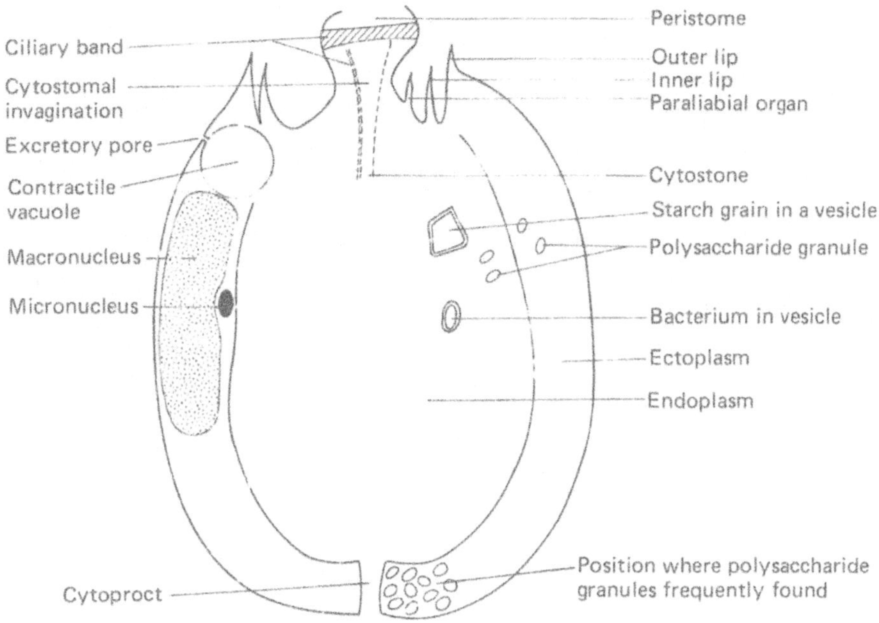

Fig. 7.1 A schematic longitudinal section through an *Entodinium* sp. showing the position of the various organelles and subcellular inclusions.

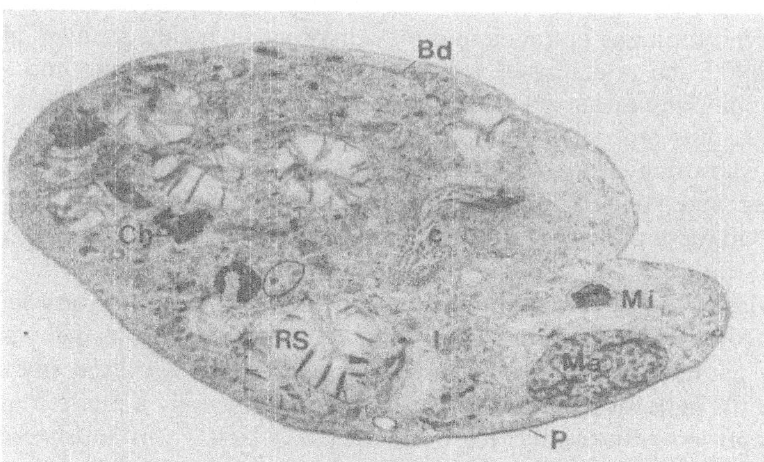

Fig. 7.2 An oblique transverse section of an *Entodinium caudatum* cell prepared immediately after harvesting the organism from the growth medium. Bd, boundary layer between ectoplasm and endoplasm; Ch, ingested chloroplast; C, cilia in cytostomal invagination (esophagus); Ma macronucleus; Mi, micronucleus; P, pellicle; RS, ingested rice starch grain.

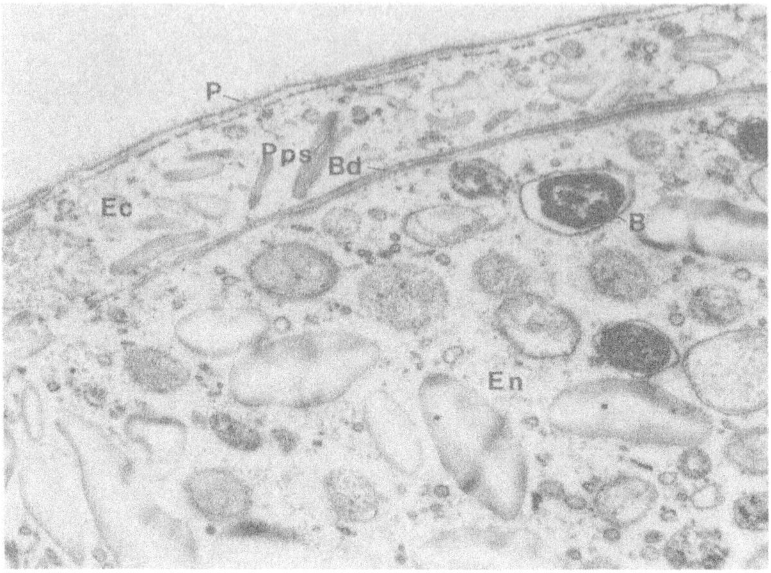

Fig. 7.3 Electron micrograph of a section through the outer part of an *Entodinium caudatum* cell that had been allowed to engulf *Escherichia coli* for 5 min. B, bacterium in vesicle; Bd, boundary layer between ectoplasm and endoplasm; Ec, ectoplasm; En, endoplasm; P, pellicle; Pps, protozoal polysaccharide granule.

Although mitochondria have been described in entodiniomorphs by some authors (Noirot-Timothée, 1960), it is now agreed that they do not occur. However, the larger entodiniomorphs contain small, highly granular microbodies (0.5 μm in diameter) which resemble hydrogenososmes and have similar biochemical capabilities (Snyers et al., 1982; Yarlett et al., 1984). The endoplasm is probably also the source of the membrane-bound, pleomorphic, lysososme-like vesicles (0.1–1.0 μm in diameter) that contain phosphatase and polysaccharide depolymerase and glycoside hydrolase enzymes in these protozoa (Williams and Ellis, 1985; Williams et al., 1986).

Ectoplasm The ectoplasm contains the two nuclei. The macronucleus is a comparatively large, often rod-shaped body, which contains many chromatin granules of approximately the same size. It is usually easily seen under the light microscope in starved protozoa especially if phase contrast optics are used. In contrast the micronucleus, which is very much smaller and which is situated close to the macronucleus, is very difficult to see. The micronucleus is located between the macronucleus and the endoplasm in the entodinia and between the macronucleus and the pellicle in the diplodinia and the more complex entodiniomorphs. In the convention adopted in this book, the macronucleus is, by definition, on the dorsal side on the protozoon. One or more contractile vacuoles are also present in the ectoplasm. In the entodinia there is only one which is usually situated near the anterior end whereas in the more complex protozoa there is often a row of them near the macronucleus. Each vacuole has an excretory pore to the exterior (Imai and Tsunoda, 1972). Information on the functioning of these vacuoles is given below.

In those entodiniomorphs that have "skeletal plates" these are also present in the ectoplasm. They are built up of an ordered array of "amylopectin-like" granules that fit together to form the plate, although in epidinia, at least, the characteristics of the granules show a gradual anterior-posterior change (Furness and Butler, 1983). When protozoa are progressively starved, the storage polysaccharide in the other parts of the cell disappears first so that when such partially starved protozoa are stained with iodine the skeletal plates are clearly visible. As the protozoa are starved further, the amylopectin in the plates also disappears until the plates and their platelets are visible only in outline. Although, as the name suggests, these plates were originally thought to have a supporting role, this is not proven. Furness and Butler (1983) have evidence that, in epidinia, the anterior end of the plates may have a structural function, and Ogimoto and Imai (1981) state that fibrillar structures extend from the plates to the inside of the body. Whether or not the presence or absence of amylopectin affects such a role is not known.

As mentioned above, amorphous polysaccharide granules are also present throughout the ectoplasm. However, they tend to be concentrated at the anterior and posterior parts of the organisms and especially in caudal lobes if these are present.

7.2 Structure of Vestibular and Cytopharyngeal Regions

Entodinium caudatum The peristome, when extruded, has a band of cilia around most of its circumference below the tip. One end of this band runs down the side of the cytostomal invagination to the cytostome where particles are taken up (Fig. 7.4; Coleman and Hall, 1971; Furness and Butler, 1985b). The nonciliated peristomal surface bears crests which arise anteriorly from the inner edge of the ciliary band and pass obiquely over the tip to the concave side. They also run transversely around the inner surface of the cytostomal invagination. From these crests microtubular ribbons, of which four types have been described, run into the body of the peristome. A semicircular palisade of nematodesmata run parallel to the long axis of the peristome. Furness and Butler (1985b) believe that the force necessary for the engulfment of the small particles taken up by entodinia could be produced by membrane movement and cytoplasmic flow across the cytopharyngeal microtubular ribbons. Entodinia, unlike eudiplodinia and epidinia, also have myoneme-like elements in the peristome and these could be involved in the extension and retraction of the peristome. There is

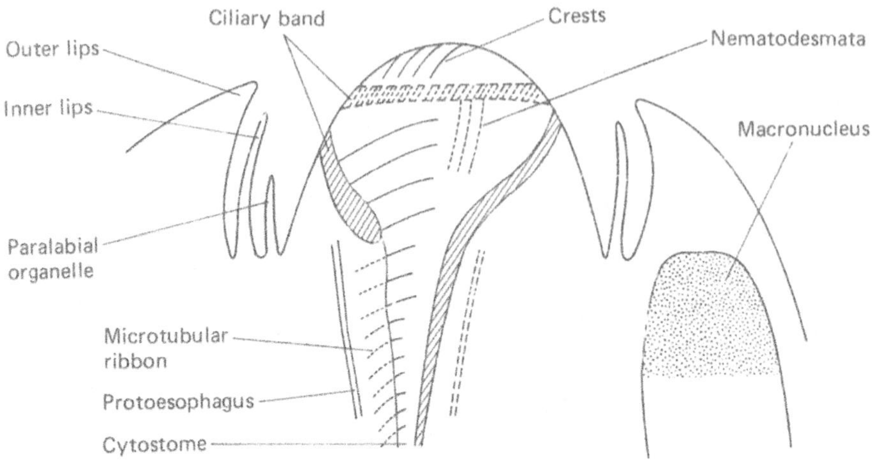

Fig. 7.4 A schematic longitudinal section through the anterior and of an *Entodinium* sp.

a protoesophagus, composed of microtubular and fibrous material, which forms a rough half-cylinder along the inside of the peristome adjacent to the cytoplasm. For more detailed information see Furness and Butler (1985b).

Eudiplodinium maggii *E. maggii* (Fig. 7.5) differs from *Entodinium* spp. in not only being larger but also in having a convoluted esophageal sheath which is a fibrous/microtubular structure that extends over half the length of the protozoon. Near the macronucleus it appears to be derived from the ectoplasm/endoplasm boundary layer. It surrounds material which is continuous with the peristomal phagoplasm. Furness and Butler (1985a) report that the peristome of *E. maggii* is cup-like, but this may only be the form of the retracted organelle which may resemble that of entodinia when extruded. The peristome contains up to five concentric rows of longitudinal nematodesmata in a vertical crescent palisade. Two types of cytopharyngeal ribbons are situated under the peristomal membranes. It has been suggested that, as these microtubule ribbons run longitudinally rather than transversely as in the entodinia, they are better suited for guiding elongated plant fragments to the cytostome. It has also been suggested that the well-developed esophagus is the main force generator for engulfment. The absence of vesicles in the phagoplasm in the esophagus suggests that no digestion takes place there (Fig. 7.5; Furness and Butler, 1985a, 1988).

Epidinium caudatum The structure of epidinia is more complex than that of the previous two species (Fig. 7.6). The peristome is covered with a U-shaped band of cilia with a phagocytic border running longitudinally

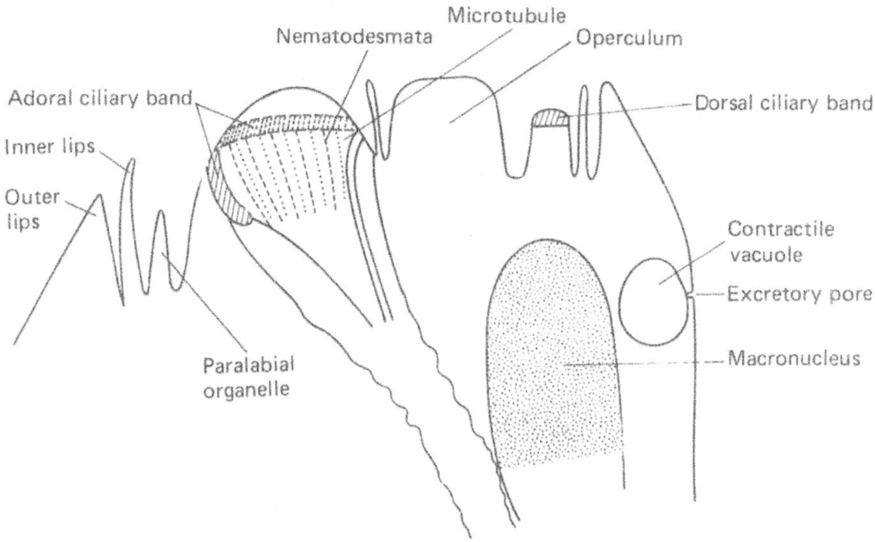

Fig. 7.5 A schematic longitudinal section through the anterior end of an *Eudiplodinium maggii*.

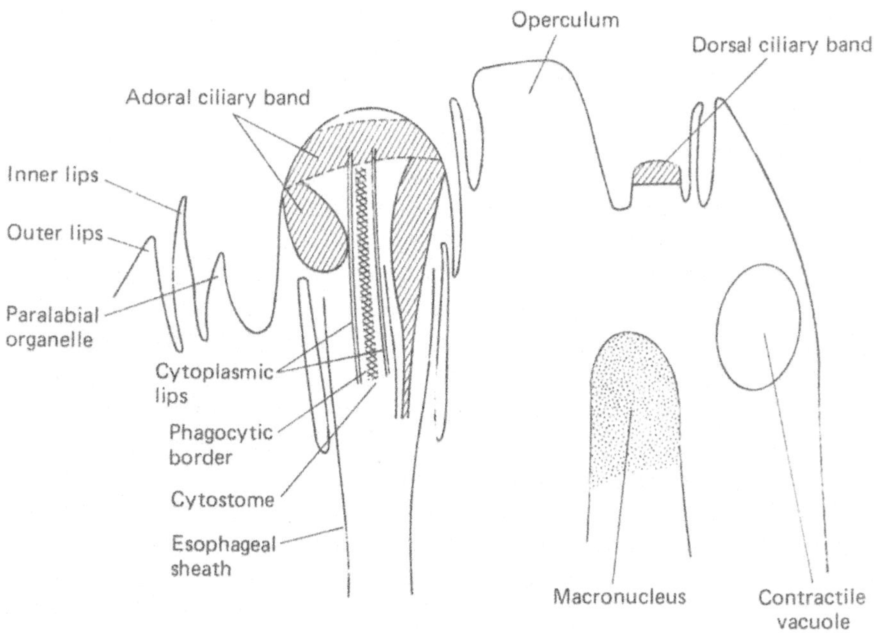

Fig. 7.6 A schematic longitudinal section through the anterior end of an *Entodinium* sp.

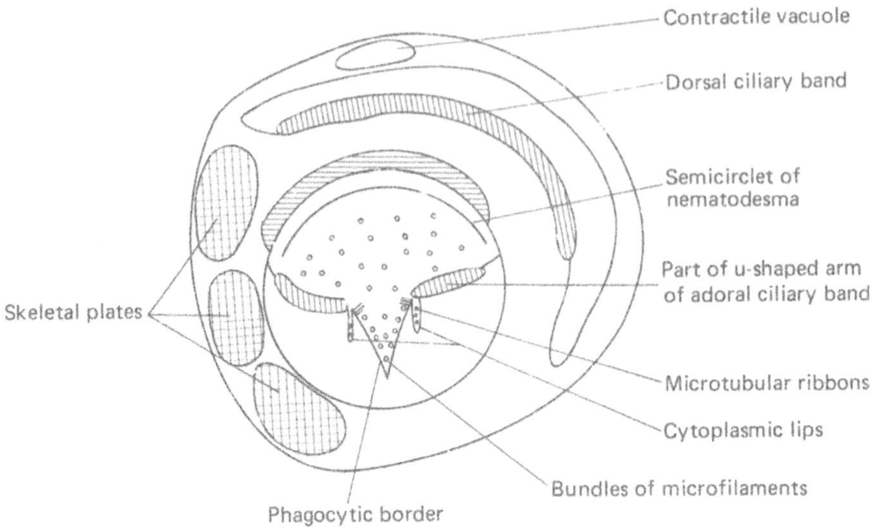

Fig. 7.7 A schematic transverse section through the anterior end of an *Entodinium* sp. near the base of the peristome.

down one side. This border, which may produce pseudopodia, is flanked posteriorly by a pair of cytoplasmic lips. Down one side of the phagocytic border runs the cytostomal tract, associated with which is one end of the ciliary band. The peristomal phagoplasm continues posteriorly to the cytostome as the highly filamentous tube-like esophagus which extends three-quarters of the length of the body. The esophageal cytoplasm is surrounded by a convoluted microtubular- and microfilament-containing wall and, at the level of the cytostome, by a fibrous esophageal sheath. Along the peristomal edge of the phagocytic border are rows of nematodesmata and peristomal microtubular ribbons in the cytopharynx. These extend back into the peristome as a funneled semicircular palisade of nematodesmata which anteriorly contain the highly filamentous phagoplasm. Unlike the arrangement in the other two species mentioned above, the longitudinal microtubule ribbons are within the peristome rather than at the surface (Fig. 7.7; Furness and Butler, 1983, 1988; Grain and Senaud, 1985). Furness and Butler (1983) suggested that ingestion forces could be derived from microtubule ribbons and/or the esophagus and its fibrillar components, but both would involve the highly filamentous phagoplasm. Epidinia, unlike the eudiplodinia, have many vesicles in the phagoplasm. These could have a local lubricative or digestive function or they could be the source of the enzymes secreted uniquely by epidinia (Coleman and Laurie, 1974b; Bauchop and Clarke, 1976).

Ophryoscolex purkynjei From the little information available, the structure appears to be similar to that of the epidinia (Schrenk and Bardele, 1987; Furness and Butler, 1988).

7.3 The Cilia and Infraciliature

The entodinia possess only one ciliary band, the adoral, which is situated on the peristome (Fig. 7.1). There has been much discussion in the past over how these cilia were arranged. Dogiel (1927) assumed that they were organized into membranelles and Bretschneider (1960a, 1960b) drew groups of cilia surrounded by membranes. However, subsequent work by Coleman and Hall (1971b), Imai et al. (1983), and Furness and Butler (1985b) produced no evidence for an enveloping membrane and it has been suggested that the term "syncilia" or "synciliary tuft" (Gelei and Sebestyen, 1932; Corliss, 1979) should be used for closely packed groups of cilia. More recently, Imai et al. (1983) have shown that the adoral ciliary zone could be divided into two parts. The outer part is composed of "membranelle-like" structures rather than tufts while the inner part is composed of scattered aggregates of cilia.

All the more complex entodiniomorphs also have a dorsal zone of cilia which is composed mainly of syncilia and which increases in extent as the genera become larger and more complex (Furness and Butler (1988).

The infraciliature of the dorsal and adoral ciliary zones is identical and the kinetids of all species so far investigated are similar except for those of *Epidinium* spp. which have additional interkinetal microtubules. The structure of a cilium and its basal structures is shown in Fig. 7.8. The kinetosomes and the kinetodesma are joined together in rows called "kineties" (sing. kinety) which run transversely around the edge of the peristome. Under each kinetosome is the subkinetosomal plate, rows of which are joined together to form the subkinetosomal rod. On mild treatment of entodinia with ultrasound, sheets of kinetosomes and subkinetosomal rods are released and on more prolonged treatment single rods are found (Coleman and Hall, 1971b). As mentioned above the units in a kinety are also linked together by bifurcated kinetodesma (Furness and Butler, 1983). Running at right angles to the subkinetosomal rods are interkinetal microtubules, transverse microtubules, and fibers.

Attached to the base of certain subkinetosomal plates are nematodesmata (retrociliary fiber of Noirot-Timothée, 1960; myoneme of Bretschneider, 1960a, 1960b) which extend longitudinally to form the cytopharynx in epidinia and eudiplodinia. The nematodesmatal palisade is less extensive in entodinia.

Noirot-Timothée (1960) stained rumen ciliates by the silver-impregnation method and described from studies in the light microscope a com-

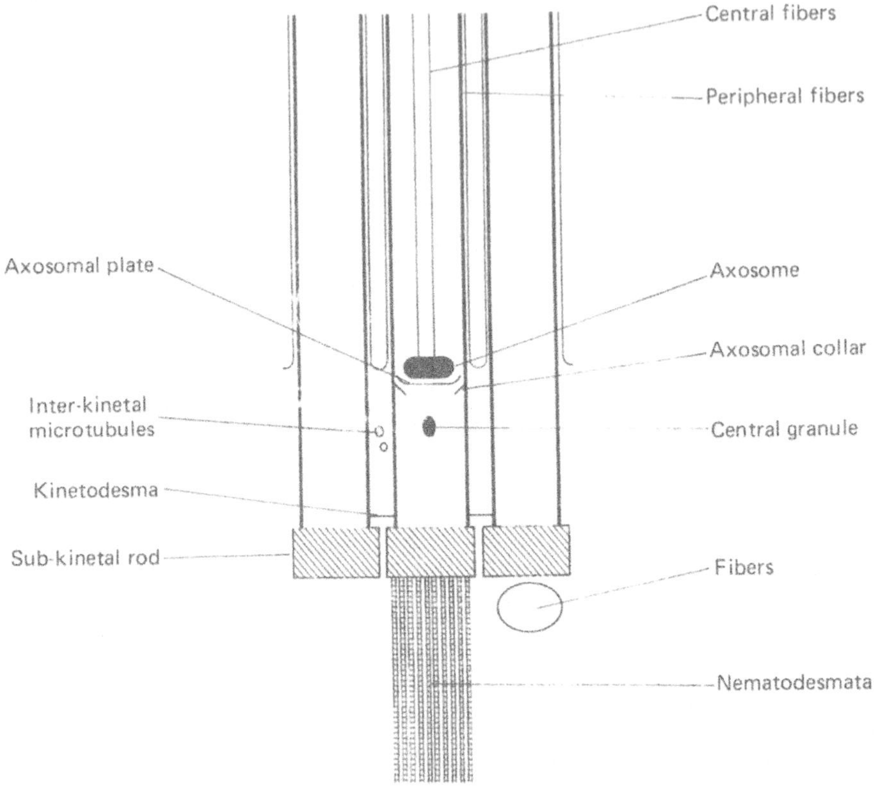

Fig. 7.8 A schematic longitudinal section through a cilium and its infrastructure.

plex system of argentophilic fibres at the anterior end of several species of entodiniomorphid protozoa. However she was unable to correlate these fibers with structures visible under the electron microscope. Although Ogimoto and Imai (1981) state that the method stains kinetosomes, Coleman and Hall (1971b) showed that the cilia, an area under the subkinetosomal sheet, and the lining of the cytopharynx (esophagus) are heavily stained by silver. The fibers shown by Noirot-Timothée (1960) in protozoa with retracted peristomes could therefore be cilia and the "peri-esophageal ring" could be silver deposited on the lining of the cytopharynx.

7.4 The Paralabial Organelle

All entodiniomorphia protozoa so far investigated have a group of modified cilia, the paralabial organelle, situated in a narrow crypt ventral to the peristome and between this and the inner oral lips (Bretschneider, 1962; Roth and Shigenaka, 1964; Coleman and Hall, 1971a; Furness and Butler, 1985a, 1985b; Schrenk and Bardele, 1987). In *Ophryoscolex purkynjei* it is 15 μm long, 4 μm wide, and 10 μm high and consists of alternate rows of short modified cilia and thin pellicular folds which form ridge-like structures. Some kineties bear cilia with 9 + 2 axonemes which are 2 μm long whereas others have cilia with a 9 + 0 axoneme and are only 0.5–0.7 μm long. The fine detail is described by Schrenk and Bardele (1987). Bretschneider (1962) first suggested that the organ could have a sensory function as, in its extended state, it would be too short to contribute to the swimming of the protozoa or the ingestion of food. Although Roth and Shigenaka (1964) suggested that it was a tuft of not yet fully developed cilia, all other authors believe that it has a sensory function especially as these protozoa demonstrate a response to compounds, such as sugars, in the medium and there must be a sensory organelle somewhere in the ciliary area.

7.5 The Contractile Vacuoles

The structure, function, and operation of the contractile vacuole(s) have been relatively little studied in entodiniomorphid protozoa although all have one or more vacuoles. The only extensive studies on the pulsatory cycle were made by MacLennan (1933) and Noirot-Timothée (1960) and are therefore over thirty years old, although there have been more recent studies in the electron microscope on vacuoles in randomly fixed protozoa.

Entodinium spp. The contractile vacuole of entodinia is located toward the anterior of the organism and probably has a single discharge canal (Kofoid and MacLennan (1930), although Imai and Tsunoda (1972) in studies in the scanning electron microscope could find no excretory pore in *Entodinium* spp. they examined. Furness and Butler (1985b) state that the spongiosome consists of 50–100 nm diameter canals continuous with larger irregular vesicles. There is no evidence on the existence of a pulsatory cycle.

Eudiplodinium maggii *Eu. maggii* has two contractile vacuoles (25–30 × 10–12 μm) which are surrounded by small vesicles. Each vacuole has a permanent discharge canal (2–4 μm in diameter and 10–15 μm long) which opens by a pore (3–8 μm long and 2–5 μm wide) onto a depression on the surface of the protozoon. The vacuole is isolated from the exterior by a vacuolar membrane. In the canal the epiplasm of the pellicle (cortex) is retained but the glycocalyx is reduced. Under the epiplasm are closely packed, helically wound microtubules. Microfilaments and somatic barren kinetosomes occur around the canal (MacLennan, 1933; Noirot-Timothée, 1960; Furness and Butler, 1985a). MacLennan (1933) described the pulsatory cycle as follows. In preparation for systole the resting vacuole rounds up for 5–8 sec and when this has occurred, there is a sudden convulsion with systole lasting for 2–10 sec. The residual vacuole pulls away from the excretory pore which is closed by a membrane. Accessory vacuoles form even before systole is complete but do not fuse until after systole is complete. This process of small vesicles appearing and fusing occurs for 30–120 sec and then stops when a narrow layer of granular material surrounds the vacuole. The main vacuole continues to enlarge without the formation of accessory vacuoles and the roughened wall gradually becomes smoother until the resting vacuole is produced. This then remains quiescent for from 15 sec to an indefinite period. MacLennan (1933) and Noirot-Timothée (1960) obtained evidence that the accessory vacuoles are formed from osmiophilic granules.

Epidinium spp. Epidinia have two contractile vacuoles, one anterior (12–22 × 8–13 μm) and one posterior which is 2–4 μm larger in all dimensions. Just before systole the vacuole rounds up and then with a convulsion of the vacuole the membrane over the canal ruptures and the contents of the vacuole expelled in about 4 sec. During systole small vacuoles are carried into the space occupied by the old vacuole and fuse with its remains. This process takes about a minute and then the vacuole increases in size without the addition of more vacuoles. The entire cycle lasts 1–60 min. In *Epidinium* and *Ophryoscolex* spp. the accessory vacuoles arise from the entire dorsal ectoplasm whereas in *Ostracodinium* spp. this only occurs over a small area. Osmiophilic granules are associated at various stages with both the accessory and main vacuoles.

Polyplastron multivesiculatum *P. multivesiculatum* usually has up to 10 contractile vacuoles, but the number can vary from 7–11 depending on the activity of the protozoon. The fully expanded vacuoles are comparatively small varying from 12–30 × 8–15 μm. When the number of vacuoles is low, the "missing" ones exist just as the excretory canal and pore. Just before systole the vacuole rounds up for 1–3 sec and then empties for 1.5–3.0 sec and completely disappears. After systole is complete, accessory vacuoles fuse to form a new contractile vacuole which gradually increases in size. The total cycle time is 50 sec to 10 min with an average of 2 min.

7.6 The Cytoproct (Rectum)

The cytoproct, which is the site for the expulsion of solid waste material, is situated at the posterior end of the protozoon (Fig. 7.1). A single naked membrane separates the endoplasm from the external medium. Where the tract passes through the ectoplasm it is lined by modified pellicle (*Entodinium* spp.) or the boundary layer (*Eudiplodinium maggii*) (Furness and Butler, 1985a, 1985b).

Kofoid and MacLennan (1930) described the structure, as seen in the light microscope, of the cytoproct of many entodinia in detail. The simplest cytoproct, found in *Entodinium brevispinum* is just a narrow cylinder bounded by a membrane whereas in *E. laterospinum* there is a circular fibril at the anterior end of the cylinder which could act as a sphincter. In *E. rostratum* there are two additional fibrils which run anteriorly from the circular fibril along the boundary layer. *E. biconcavum* has many circular fibrils around the cytoproct which Kofoid and MacLennan (1930) observed to expand considerably during the expulsion of material. They believed that the fibrils act as a sphincter. They also stated that entodinia that engulf large food particles have cytoprocts with heavy walls whereas those that ingest small food particles have cytoprocts with thin walls. *E. indicum* is unusual in that that opening of the cytoproct is slit-like and situated at the base of a caudal spine. Limited information on the cytoprocts of more complex protozoa is given by Kofoid and MacLennan (1932) and Furness and Butler (1985a) who also believe that complex cytoprocts would be required to expel bulky food residues.

There is little evidence as to the nature of the material expelled through the cytoproct, although *Entodinium bursa*, at least, voids a wide range of materials including polysaccharide granules, chloroplasts, and membranes (Coleman and Hall, 1984). *Entodinium caudatum* also expels polystyrene latex beads when these are present in the medium (Coleman, 1975a). It remains to be shown whether or not these ciliates can selectively void undigestible material.

7.7 The Pellicle (Cortex)

The pellicles of all entodiniomorphod protozoa have the same basic five-layered structure although the thickness of each increases with the size of the organism (Fig. 7.9).

The surface layer is the glycocalyx which consists of granular or filamentous material overlying two (in entodinia) or three (in other genera) membranes, the outermost of which stains densely when prepared for and examined in the electron microscope. Under these membranes is a granular epiplasm which fills in any corrugations on the surface. The next layer is electron-transparent and contains one or two longitudinal rows of separate groups of up to seven microtubules separated by septa. The basal layer is composed of microfilaments organized into large bundles running at right angles to the microtubules (Noirot-Timothée, 1960; Furness and Butler, 1983; 1985a, 1985b; Coleman and Hall, 1984). In some species, e.g., epidinia (Furness and Butler, 1983), there are surface pores at the bottom of the surface corrugations. There is a barren kinetosome near each pore. Imai and Tsunoda (1972) studied the surface striations of entodiniomorphid protozoa under the scanning electron microscope and found that these were characteristic for individual species.

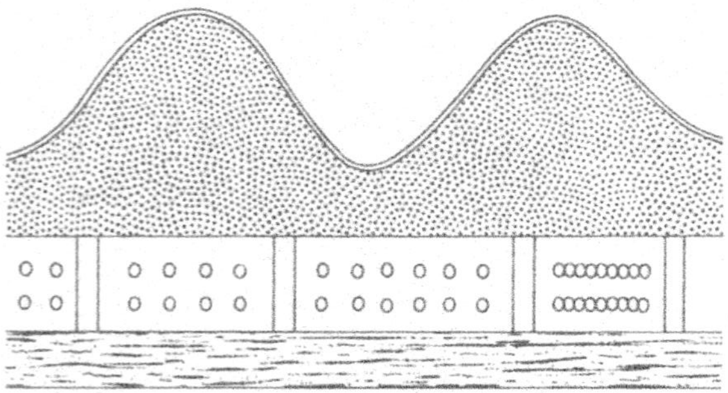

	Entodinium spp.	*Eudiplodinium maggii*	*Epidinium caudatum*
Glycocalyx	40-50 nm	200 nm	< 55 nm
Membranes	2	3	3
Epiplasm	45-80 nm	125-400 nm	60-280 nm
Clear zone containing microtubules	65-110 nm	140 nm	150 nm
Microfilament layer	40-70 nm	220 nm	100-200 nm

Fig. 7.9 (a) A schematic section through the pellicle of an idealized entodiniomorphid protozoon.

7.8 The Boundary Between Ectoplasm and Endoplasm

This boundary, which is also known as the ectoplasmic-endoplasmic boundary, the boundary layer, and the *tela corticalis*, separates these two areas. It is said to be poorly defined in *Entodinium* spp. (Furness and Butler,1985b) although it is clearly visible in Fig. 7.3. Kofoid and MacLennan (1930) also found that it was easily seen in most entodinia but was, for example, almost invisible in *E. indicum*. However, it was clearly visible in sections of cultured *E. caudatum* where it appeared as a "membrane" 110 nm wide made up of 2–4 components each of which was 14 nm wide and composed of two electron-dense lines 3 nm across and a lighter central area. The boundary layer is clearly visible in the more complex entodiniomorphs and appears to be fibrous rather then membranous with two layers of fibers running at right angles to each other (Noirot-Timothée, 1960). Furness and Butler (1983) believe that the layer is elastic and can be deformed without damage.

7.9 Division of Entodiniomorphid Protozoa

The doubling (division) times of various protozoal species grown in vivo and in vitro are given in Table 7.1 and vary considerably with species and conditions. In his studies on the division of entodinia in sheep fed once daily, Warner (1962b) found that the minimum doubling time was $5\frac{1}{2}$ h and that the time from the beginning of the first signs of the division process to

Table 7.1 Minimum generation times of rumen entodiniomorphid protozoa grown *in vivo* and *in vitro*

	Growth *in vitro* (h)	Growth *in vivo* (h)
Entodinium bursa	6 (1)	
Entodinium caudatum	22 (3), 23 (2)	
Entodinium longinucleatum	31 (4)	
Entodinium simplex	38 (5)	
Entodinium spp.		5.5 (6), 6.8 (7)
Entodinium spp.—re-establishment in defaunated rumen—from numbers/ml		28 (8)
Epidinium caudatum (ovine)	26 (9)	
Diplodinium pentacanthum	8.5 (10)	
Diploplastron affine	13 (10)	
Metadinium medium	2* (11)	
Polyplastron multivesiculatum	14 (9)	14 (12)

*Limited time
(1) Coleman et al. (1977); (2) Coleman (1960a); (3) Hino and Kametaka (1977); (4) Owen and Coleman (1976); (5) Coleman (1969*a*); (6) Warner (1962b); (7) Potter and Dehority (1973); (8) Hino and Kametaka (1974); (9) Coleman *et al.* (1972); (10) Coleman *et al.* (1976); (11) Naga and El-Shazly (1968); (12) Senaud *et al.* (1973)

its completion was only 15 min with no change at different times of day. Warner (1962b) used the following formula of Scherbaum (1957) to calculate the doubling time of protozoa in the rumen:

$$t_d = \frac{\log_e 2 \cdot T}{\log_e (1 + 2I)/(1 + I)}$$

where t_d is the doubling time, T is the time needed to complete division and I is the division index, i.e. the proportion of dividing organisms.

Morphological studies on the changes that occur during divison have been described in some detail for many species by Noirot-Timothée (1960) and in more detail for *Entodinium dubardi* by Zielyk (1961), for *Eremoplastron* (but called *Eudiplodinium*) *neglectum* by Kraschenninnikow (1955), and for *Eudiplodinium maggii* (principally the infraciliature) by Furness and Butler (1986). The following is a broad outline of the process and the reader is referred to the above papers for fine detail.

1. The first stage is marked by an elongation of the micronucleus which becomes 8-shaped. At the same time the cilia of the adoral and dorsal zones (except in entodinia, which do not have a dorsal zone) begin to form in a subequatorial anlagen. The adoral cilia arise in two areas which converge and the dorsal cilia in only one.
2. The two daughter micronuclei move apart but can still be joined by an elongated intermediary spindle, although this has not been observed in entodinia (Noirot-Timothée, 1960; Zielyk, 1961). A constriction begins to form around the middle of the protozoon. In protozoa with skeletal plates, very indistinct traces of an anlagen of the plate can be seen in the posterior daughter.
3. The spindle between the two daughter nuclei disappears as these nuclei migrate toward the ends of an elongated macronucleus which develops an indentation on the dorsal side.
4. The macronucleus gradually divides in two in association with the progressive constriction that develops around the middle of the protozoal body. The skeletal plate in the posterior daughter still tends to be poorly developed. A new contractile vacuole and cytopharynx develops in the posterior daughter.
5. All the various organelles continue to develop in the posterior daughter which is joined by an ever-decreasing band to the anterior daughter. The macronucleus which tended to be spherical immediately after division gradually assumes its normal shape.

In *Eudiplodinium maggii*, at least, the new ciliary zones of the posterior daughter are derived from subcotical barren kinetosomes which replicate and migrate to the subequatorial anlagen. Evidence was presented by Furness and Butler (1986) that a "pattern generator" was present in the barren kinetosomes that formed the vestigial somatic ciliature in the parent organism.

7.10 Conjugation

Conjugation, which is a form of sexual reproduction, occurs when two individuals join together by their anterior ends and exchange genetic material. It occurs infrequently and erratically in the rumen although, for no obvious reason, many conjugating individuals are occasionally seen in one animal. The process has been most frequently observed in African antelopes (Dogiel, 1923b, 1925b; Noirot-Timothée, 1960). Conjugation has also been observed in cultures of *Entodinium caudatum* (Williams and Coleman, 1988) and of *Ophryoscolex purkynjei* (Mah, 1962). According to Noirot-Timothée (1960) the micronucleus of a protozoon that is about to conjugate undergoes a pregamic division to give two nuclei that are much larger than normal. The protozoon then divides. One daughter then becomes paired with another individual of the same species with the two protozoa becoming joined by their anterior ends. In each conjugant the micronucleus passes from the ectoplasm into the endoplasm where it divides to form two presumably haploid nuclei, one of which degenerates while the other divides to form two gametic nuclei (one male and the other female?). Gametic nuclei are then exchanged between the two conjugants via a bridge of endoplasm. The gametic nuclei fuse in each conjugant and then divide. One becomes the new micronucleus and the other become surrounded by a large vesicle, increases in size to become the new macronucleus, and replaces the old ones which have degenerated and disintegrated while the conjugation process was proceeding. The importance of the conjugation process is not known but, as it has been observed in cultures that apparently grow indefinitely (>27 years) but not in those which die after 2–3 years, it may be a 'rejuvenating' process.

7.11 Protozoal Chemotaxis and Feeding

Bauchop and Clark (1976) and Bauchop (1979) first showed that epidinia attach themselves to plant tissue. The protozoa invade damaged plants and preferentially attack mesophyll cells and portions of the parenchyma bundle sheath. The protozoa readily attack cool-season grasses (orchard grass) but are much less active against warm seasons grasses such as bermuda grass which have a resistant epidermis (Akin and Amos, 1979). The epidinia produce enzymes which break up both the cellular network and individual cells so that the protozoa can engulf both whole cells and bits of cells.

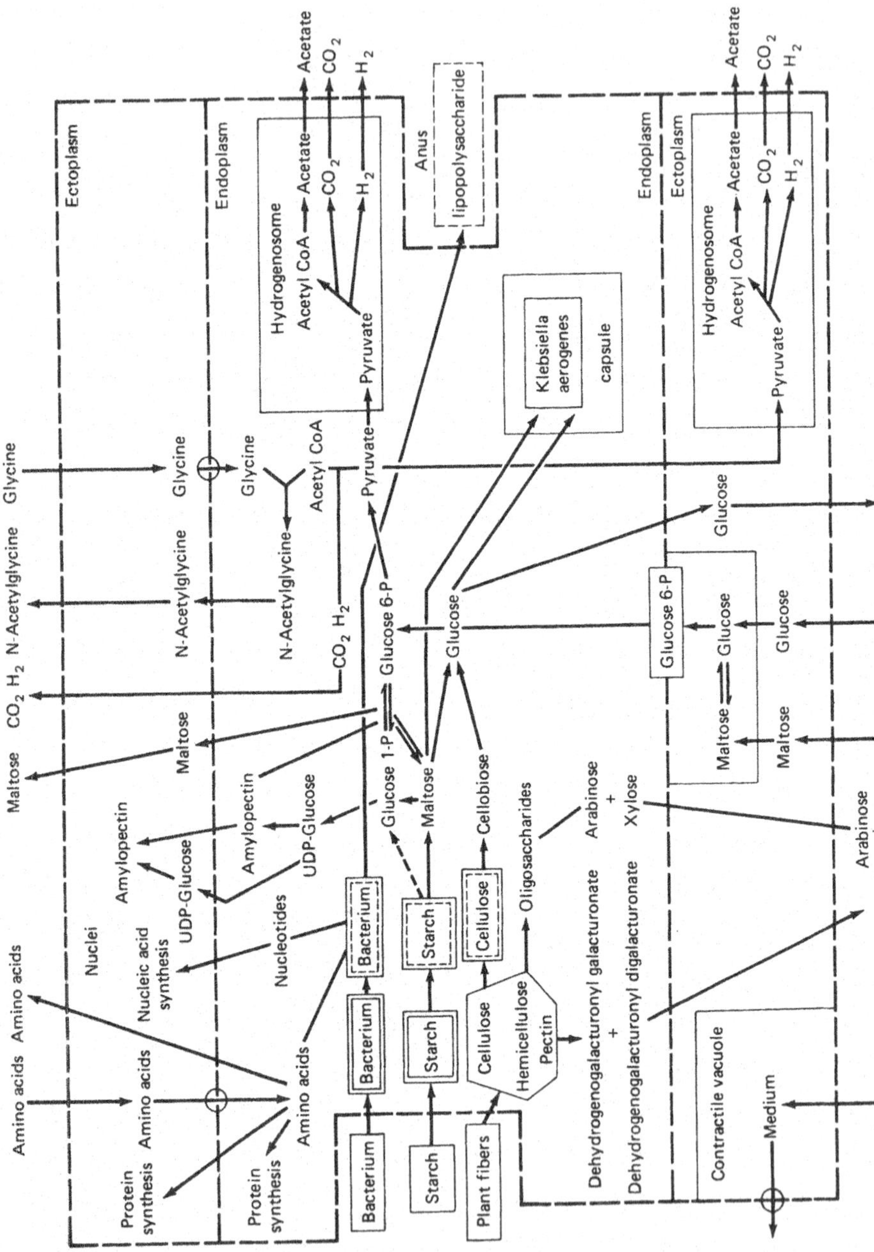

Fig. 7.10 A diagramatic representation of the principal metabolic reactions known to occur in entodiniomorphid protozoa. ○, active transport.

Orpin (1985) extended this work and investigated the attraction of eight protozoal species to fresh plant material and hay. Holotrich protozoa were attracted most rapidly and to the greatest extent (85–95% in 5 min) whereas only 10–20% of the entodinia were attracted after 15 min. Attractions with the larger entodiniomorphs were *Epidinium* spp. 49–76%, *Eudiplodinium maggii* 20–69%, *Ophryoscolex caudatus* 10–96%, and *Polyplastron multivesiculatum* 10–52% with the results increasing with increases in the amount of plant material and the degree of fineness of the material. The association of the protozoa with the plant material only lasted for 1–2 h and was not observed with material that had been extracted with water, suggesting that the protozoa were attracted to soluble compounds in the plant material. Although starch-rich material such as ground wheat were not used in these studies, the observations of Weller and Pilgrim (1974), Nakamura and Kurihara (1978a, 1978b), and the present authors suggest that entodinia are attracted to starch grains. From the evidence presented above and because epidinia are unique among the entodiniomorphs in secreting enzymes, the epidinia are the primary invaders of plant tissue in the rumen. However, since many ruminants do not contain epidinia but still thrive, their role must be taken over by other species in these circumstances.

7.12 Protozoal Metabolic Compartments

The volume occupied by each cultured *Entodinium caudatum* in a packed cell pad is 4.7×10^4 μm^3 (Coleman, 1967a). Of the volume of such a pad, 86% is impermeable to *Escherichia coli* at 20°C, suggesting that 14% of the pad is interstitial water. The actual volume of each protozoon is therefore 4.04×10^4 μm^3. MacLennan (1933) also showed that "vital" dyes penetrate very slowly into resting protozoa. Of the actual protozoal volume 15–20% is starch and about 10% other solid material. The amount of solid material therefore varies from 10% in starved organisms to 30% in protozoa which are full of starch grains. On short term (5 min) incubation of the protozoa at 20°C with low-molecular-weight compounds such as sugars and amino acids, 52–73% of the protozoal volume is imprenetrable by these compounds (Coleman, 1967a, 1968, 1969b). If this variation is due to variation in the amount of solid material, then $42 \div 0.70$ or $0.90 = 47$ or 60% of the liquid volume in impenetrable and 53 or 40% penetrable by such compounds. The magnitude of these values suggests that the compartments must be the endoplasm and ectoplasm. On the basis of the differential affects of inhibitors and because antibiotics do not readily kill bacteria in the endoplasm, Coleman (1969b) suggested that it was the ectoplasm which was freely permeable to low-molecular-weight compounds in *Entodinium caudatum* and that the permeability barrier was the boundary between the endoplasm and ectoplasm. If this is true, then the "active" uptake of, e.g., amino acids (Coleman, 1967a) must be into the endoplasm (see Chapter 6) assuming that protozoa at 39°C behave in the same way as resting organisms at 20°C. A freely permeable ectoplasm would appear to present difficulties to the organism in terms of synthesizing polysaccharide granules from sugars produced in the endoplasm and transmitting messages to and from the nuclei, but it has long been known that approximately 75% of the cell volume of *Escherichia coli* is also freely permeable to low-molecular weight compounds (Roberts et al., 1955). Presumably in both cases the external membrane is impermeable to compounds such as UDP-glucose and AMP-amino acids.

There is also evidence on the existence of another metabolic compartment in *Entodinium caudatum*. If the "passive" uptake of an amino acid, where uptake is proportional to external concentration, is regarded as the trapping of external medium inside the cell during incubation and subsequent washing, then the volume occupied would be 8–15% of the protozoal volume. In other experiments (Coleman, 1969b), it was shown that iodoacetate abolished both ciliary activity and the active uptake of glucose, but increased the "passive" uptake six times. It was suggested that during a normal uninhibited incubation at 39°C, external medium plus dissolved solutes penetrate into the contractile vacuole. This would contract periodically and expel its contents. If this occurred during washing, any external medium contained therein would be lost. However, in iodoacetate-treated organisms, the vacuole would not contract but remain full of medium plus solute. If this is true, then calculation shows that the vacuole occupied 11% of the cell volume (12–16% of the liquid volume). Although it is far from proved, there is therefore evidence for another metabolic compartment which could be the contractile vacuole. Although it is morphologically part of the ectoplasm, it behaves as part of the endoplasm.

A diagramatic representation of the sites of the principal metabolic reactions known to occur in entodiniomorphid protozoa is shown in Fig. 7.10.

7.13 Rate of Clearance of Soluble and Particulate Materials

The soluble compounds metabolized most rapidly by *Entodinium caudatum* are glycine (converted into N-acetyl glycine—Coleman, 1964a), thymine (converted into dihydrothymine—Coleman, 1968), and choline (taken up—Broad and Dawson, 1975), and all the substrate in 8.7×10^4 μm^3, 11×10^4 μm^3, and 8.6×10^4 μm^3, respectively, is attacked by one protozoon each hour. If it is assumed that all the substrate in a volume of medium is metabolized in one "passage" through a protozoon, then these very similar volumes (approximately 10×10^4 μm^3) must represent the rate at which medium passes in and out of the protozoa. If this is the maximum rate, it is therefore not unexpected that the rate is not increased by, for example, increased use of choline for membrane synthesis. It is not known where the reactions referred to above take place but, with glycine, for example, N-acetyl glycine appears and increases in concentration inside the protozoa before being released into the medium. If the ectoplasm is freely permeable at 39°C and if medium must "pass" into the endoplasm before its solutes are metabolized, then $10 \times 10^4 \div$ (maximum volume of endoplasm, $.4 \times 4.04 \times 10^4$) = 6.2 endoplasm volumes (or 2.5 protozoal volumes) of medium are metabolized each hour.

These rates of metabolism are slow compared with those at which bacteria are cleared from the medium; 2.9×10^6 μm^3 and 1.8×10^6 μm^3 of medium are cleared hourly of *Escherichia coli* from an infinitely dilute suspension and a population of 10^9 bacterial/ml respectively. These rates correspond to 72 and 45 protozoal body volumes per hour, respectively, i.e., 18–29 faster than the rate of metabolism of soluble compounds. As bacteria are engulfed via the cytopharynx and are removed from a stream of medium apparently passing down one side and up the other side (propelled by the cilia down one side only), at least this volume of medium must pass in and out of the cytopharynx each hour. If soluble compounds are taken up from this stream of medium, then only about 5% of the solute would be removed during each passage of the medium. If, however, the uptake of soluble compounds involves diffusion of medium and/or solutes into and out of the protozoa, then this would be a much slower process.

8

Metabolism of Holotrich Protozoa

The rumen entodiniomorphid and holotrich ciliates depend upon the plant material, ingested by the host ruminant, for their principal supply of food; both protozoal groups also digest bacteria. However, they do not compete directly for the available nutrients in that the holotrichs primarily use soluble carbohydrates whereas the entodiniomorphid ciliates ingest and utilize particulate material (Williams, 1989, Williams and Coleman, 1988). Although their nutritional niches are well-defined, the rumen ciliates are not as well-characterized biochemically as the rumen bacteria and the studies undertaken to date have been restricted to relatively few of the protozoal species. Biochemical studies on the holotrich ciliates are in fact restricted to *Dasytricha ruminantium* and the two species of *Isotricha;* there have been no metabolic studies on other members of the Isotrichidae or on any of the rumen Buetschliidae, Blepharocorythidae, or Paraisotrichidae, and the metabolic potential of these latter organisms in the rumen ecosystem is unknown. This chapter, therefore, reviews the biochemical and metabolic studies that have been undertaken with the three principal isotrichid holotrich species. The biochemistry of the rumen entodiniomorphid ciliates is detailed in Chapter 6.

8.1 Carbohydrate Metabolism

Carbohydrate Utilization The range of carbohydrates utilized by the holotrich ciliates has been determined manometrically (Gutierez, 1955; Heald and Oxford, 1953; Howard, 1957, 1959a; Williams and Harfoot,

1976), and through the capability of the carbohydrate to either stimulate storage polysaccharide deposition (Kubo and Kandatsu, 1968; Masson and Oxford, 1951; Oxford, 1951; Prins and Van Hoven, 1977; Van Hoven and Prins, 1977) or prolong the life of the protozoa in culture (Sugden and Oxford, 1952). The common concensus from these surveys is that the holotrich ciliates are involved principally in the utilization of soluble carbohydrates and non-structural storage polysaccharides of plant origin. The holotrichs are, however, only able to utilize the monosaccharides D-glucose, D-fructose, and D-galactose and certain soluble oligosaccharides and polysaccharides comprising one or more of these sugars (Table 8.1). The range of oligosaccharides metabolized is genus-dependent as maltose, cellobiose, and lactose are not effective substrates for *Isotricha* spp. (Gutierrez, 1955; Howard, 1959a, 1963; Prins and Van Hoven, 1977). *Isotricha* spp., unlike *D. ruminantium*, will, however, ingest and ferment starch grains of an appropriate size (Gutierrez, 1955; Howard, 1957, 1959a; Yoshida and Katsuki, 1980), and is also able to utilize melibiose (Prins and Van Hoven 1977; Van Hoven and Prins, 1977). Surveys of *Isotricha* spp. are however limited in that apart from the studies of Prins and Van Hoven (1977) with *I. prostoma* all other investigations of carbohydrate metabolism in the genus have used mixed preparations of *I. intestinalis* and *I. prostoma*; the range of carbohydrates utilized by *I. intestinalis* has not been ascertained.

Fructose and fructose-containing saccharides (e.g., sucrose, raffinose, inulin) are rapidly metabolized by both genera. Some glucose- and galactose-containing oligomers are also readily utilized although the rate of carbohydrate utilization and the proportions of the metabolites formed are substrate-dependent. Although some of the earlier studies were unable to demonstrate the utilization of either galactose or galactose-containing oligosaccharides (see Table 8.1), it became evident in later studies (Prins and Van Hoven, 1977; van Hoven and Prins, 1977; Williams, 1979a; Williams and Harfoot, 1976) that galactose and some of the galactose-containing oligomers were metabolized, albeit at a slower rate than that of glucose. The relative rates of carbohydrate utilization as determined by the incorporation of ^{14}C-labelled sugars are sucrose > fructose > glucose > galactose. No comparative data have been published on the relative rates of individual sugar disappearance from culture supernatants of either genus. Pentoses, hexuronic acids, poly- and oligouronides, cellulose, and hemicellulosic polysaccharides are however not metabolized (Table 8.1) while mannose and certain hexosamines are toxic to the holotrichs. In cultural studies mannose, glucosamine, and galactosamine were shown to be toxic to both genera of the holotrich ciliates (Sugden and Oxford, 1952). The rate of mannose uptake by *D. ruminantium* is low and extensive cell lysis occurs in incubation systems containing this monosaccharide (Howard, 1959a, Williams and Harfoot, 1976). The synthesis rates of amylopectin and other fermentation products are reduced to below the endogenous rates even in the presence of readily assimilated carbohydrates (Williams, 1979a). *Isotricha* spp. are also

Table 8.1 References for the presence or absence of carbohydrate fermentation ability by various holotrich ciliates as assessed by growth stimulation, manometry, or reserve polymer deposition

	Reference number[a]					
	Dasytricha ruminantium		*Isotricha* spp.		Mixed holotrichs	
Carbohydrate	+	−	+	−	+	−
Esculin	5		5			
Amygdalin	5		5			
Arabinan		10				
Arabinose		8, 9, 10		7, 8		6
Arbutin	5		5			
Carboxymethyl cellulose	10					
Cellobiose	3, 4, 5, 8, 9, 10		4, 8	3, 5, 7	6	
Cellulose		8		8		
Fructose	3, 4, 8, 9, 10		3, 4, 7, 8		6, 12	
Fucose		8		8		
Galactan		10				
Galactosamine		8, 10		8		
Galactose	5, 8, 9, 10	4	7, 8	4, 5	12	6
Galacturonic acid		10		7, 8		6
Glucosamine		8, 10		8		6
D-Glucose	3, 4, 8, 9, 10		3, 4, 7, 8		6, 12	
Glucose-1-phosphate		8, 10		8		
Glucose-6-phosphate		10				
L-Glucose		10				
Glucuronic acid		4, 8		4, 8		
Hexitols		8		8		
Inulin	3, 4, 8, 10		3, 4, 8		6	
Lactose	10	4, 8, 9		4, 7, 8	12	6

[a]The references are as follows: 1, Abou Akkada et al. (1961); 2, Eadie (1957); 3, Gutierrez (1955); 4, Heald and Oxford (1953); 5, Howard (1957, 1959a); 6, Masson and Oxford (1951), Oxford (1951); 7, Prins and Van Hoven (1977); 8, Sugden and Oxford (1952); 9, Van Hoven and Prins

adversely affected by mannose and are not able to effectively utilize glucose in the presence of mannose (Howard, 1959a; Prins and Van Hoven, 1977; Sugden and Oxford, 1952).

In vitro, *D. ruminantium* will take up approximately 20 μmol glucose per mg cell protein per hour; the uptake rate of *Isotricha* spp. is some twofold higher (Williams and Yarlett, 1982). The rate of carbohydrate utilization by *D. ruminantium* is affected by the nature and concentration of the available substrate and by the environmental temperature and pH (Williams and Harfoot, 1976). Sugar uptake was not reduced by the presence of storage polysaccharide within the cell and consequently uptake rates were constant in cells isolated from the rumen at different times during the

Table 8.1 *Continued*

	Reference number[a]					
	Dasytricha ruminantium		*Isotricha* spp.		Mixed holotrichs	
Carbohydrate	+	−	+	−	+	−
Laminarin					2	
Levan (bacterial)	8		8		6	
Levan (ryegrass)	4		4			
Maltose	5, 9, 10	4, 8		4, 5, 7, 8	12	6
Maltotriose	10					
Mannose		4, 8, 10		4, 8		6
Melezitose		4, 10		4, 8		
Melibiose		4, 9	7	4		
Methylcellulose		10				
Pectin	1, 9, 10		1, 7			
Polygalacturonic acid		10		7		
Raffinose	3, 4, 5, 8, 9, 10		3, 4, 5, 7, 8		12	6
Rhamnose		8, 10		8		6
Ribose		8, 10		8		6
Salicin	3, 5		5	3		
Sorbose		8		8		6
Starch		3, 4, 5, 8, 11	3, 5, 8	4	12	6
Sucrose	3, 4, 5, 8, 9, 10		3, 4, 5, 7, 8		6, 12	
Trehalose		8		8		
Turanose		4, 8		4, 8		
Xylan		10				
Xylose		4, 5, 8, 9, 10		4, 5, 7, 8		6

(1977); 10, Williams and Harfoot (1976), Williams (1979a); 11, Yoshida and Katsuki (1980); 12, Zenbayashi et al. (1969).

postprandial period. Glucose uptake and metabolism is however markedly inhibited by lactic acid (Williams and Morrison, 1982) and influenced by the presence and level of other soluble carbohydrates. There is no evidence to indicate diauxic sugar utilization by *D. ruminantium* as the protozoon does not regulate sugar entry by preferential or sequential utilization and a concomitant uptake of soluble carbohydrates occurs irrespective of their effectiveness as substrates for the cell (Williams, 1979a).

The effects of alternative potential substrates on the rate of sugar uptake by *D. ruminatium* are shown in Table 8.2. Although the rate of uptake of an individual sugar is decreased by an alternative substrate, the total rate of sugar uptake is maintained, as is exemplified by the co-utilization of

Table 8.2 Sugar incorporation by *Dasytricha ruminantium* in the presence of an alternative carbohydrate substrate

	Relative rate of incorporation		
Alternative carbohydrate	Glucose	Sucrose	Galactose
Glucose	100	65.5	118.5
Sucrose	57.8	100	113.3
Galactose	82.4	91.2	100
Fructose	53.1	55.3	118.5
Cellobiose	72.9	79.6	122.2
Lactose	90.0	100	91.5
Maltose	85.7	98.4	91.7
Inulin	82.9	76.3	127.8
Xylose	111.7	106.5	95.0
Mannose	5.3	12.3	29.6

Data compiled from Williams (1979a).

glucose and fructose (Fig. 8.1). The individual rates of glucose and fructose uptake were 3.52 and 3.90 mg per mg cell per hour although in combination uptake rates of >4.6 mg per mg protein per hour occurred. The presence of a readily assimilated substrate also results in enhanced galactose uptake by *D. ruminantium* (Table 8.2).

The mechanism of sugar uptake by the holotrichs has not been determined but the evidence indicates that several sugars may be transported by the same route as uptake rates decrease when more than one substrate is available. The increased rate of galactose uptake in the presence of a readily assimilated substrate indicates that this sugar is probably taken up by a different, albeit energy-dependent, uptake mechanism. It has not been established whether disaccharides and higher oligomers are taken up undegraded or as their component monosaccharides as exocellular carbohydrases are formed by *D. ruminantium*.

Degradation of Plant Polysaccharides

Association with Plant Material The rumen holotrich protozoa *Isotricha intestinalis* and *I. prostoma* exhibit chemotaxis to sucrose, glucose, and fructose and will attach themselves by means of an attachment organelle on the anterior cell surface to particulate sources of these carbohydrates (Orpin and Letcher, 1978). The attachment organelle was described in Chapter 3 and is evident in Fig. 3.7. *Isotricha intestinalis* and *I. prostoma* will also attach to plant fragments in the rumen, the association being more pronounced with fresh plant particles in which the content of water soluble components is greater. Studies in vitro by Orpin (1985) demonstrated that the holotrich ciliate populations consistently showed the highest degree of association with some 85–90% of the cells becoming attached; maximum association

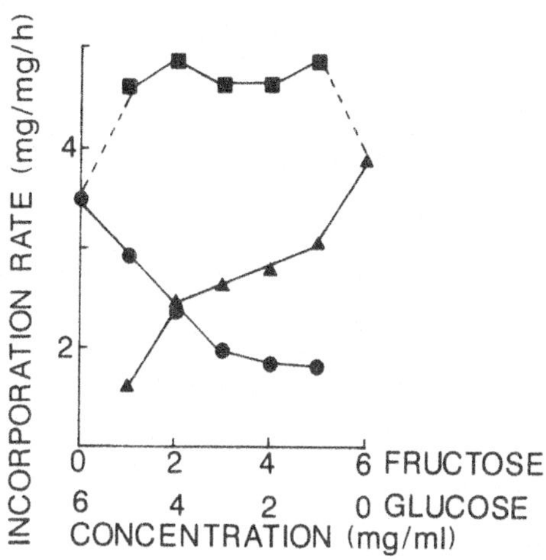

Fig 8.1 Carbohydrate substrate co-utilization by *Dasytricha ruminantium*. The rate of incorporation of glucose (●) and fructose (▲) in incubations containing varying proportions of the two substrates is shown to exemplify concomitant sugar utilization. Total sugar incorporation during the experiment (■) is also shown. Data compiled from Williams (1979a).

was achieved within 5 minutes of incubation. It has been estimated by Orpin and Letcher (1978) that in animals fed once daily up to 80% of the *Isotricha* population will become attached to plant particles 2.5 hours after feeding.

Dasytricha ruminantium has also been observed to rapidly colonize lucerne stem fragments in the rumen (Bauchop, 1980, 1989); the protozoon will also attach to grass particles (Orpin, 1985) and glass surfaces (Gutierrez, 1955) in vitro, although a specific attachment organelle has yet to be described. The colonization of the lucerne stem was specifically in the central area of the pith, and although the plant tissues were not futher degraded following colonization, the protozoon was able to penetrate deeply within damaged parenchymatous cells. Sequestration by the holotrich protozoa within the particulate phase of the rumen digesta will enable the organisms to take full advantage of the nutrients available in the ingested plant tissues. Orpin (1985) showed that the protozoa which had the closest association with the plant material were those species that had the ability to degrade the polysaccharides or were dependent upon maintaining a close proximity to plant tissue sources of soluble carbohydrates.

The rumen protozoal population contains a range of enzymes that are active against both plant structural and storage polysaccharides (Williams and Strachan, 1984), and Demeyer (1981) estimated that 30–40% of the ruminal fiber degradation was protozoal. Electron microscopic studies of the degradation of plant material (Akin and Amos, 1979; Amos and Akin, 1978; Bauchop, 1979, 1980; Bauchop and Clarke, 1976; Senaud et al., 1986) and enzymic surveys (Coleman, 1985a; Williams and Coleman, 1985) confirmed the role of the larger entodiniomorphid species in the degradative process. However, the holotrich ciliates may not be entirely passive as they also form an array of muralytic enzymes. In the following section these activities will be discussed.

Storage Polysaccharide Degradation The two principal storage polysaccharides present in plant material are starch and inulin. Amylolytic enzymes are produced by both *Dasytricha* and *Isotricha* spp. (Bailey and Howard, 1962; Mould and Thomas, 1957, 1958; Williams, 1979b; Williams et al., 1984, 1986). The amylases of each genus are inhibited by maltotriose, but have different pH optima at pH 4.8 (*Isotricha* spp.) and the dual values of pH 5 and pH 6 (*D. ruminantium*) (Mould and Thomas, 1957, 1958). Athough the amylase activity of the two genera is similar, the α-glucosidase of *Dasytricha* is the more active. The maltase of *D. ruminantium* is a specific α-glucosidase which has maximum activity at pH 5.5 and 50°C (Bailey and Howard, 1963). The enzyme has some activity against maltotriose and has transferase properties at higher maltose concentrations forming the trisaccharides maltotriose and panose [O-α-D-glucopyranosyl-(1 $\rightarrow$ 6)-O-α-D-glucopyranosyl-(1 $\rightarrow$ 4)-D-glucopyranose]. Some α,1$\rightarrow$6 glucosidase activity is also present in the holotrich ciliates as isomaltose was degraded to glucose by an enzyme preparation obtained from a mixed suspension of *Isotricha* and *Dasytricha* spp. (Williams, 1979b).

Fructosans are also produced as reserve carbohydrates in certain plants. A major component of the soluble carbohydrate of perennial ryegrass (*Lolium perenne*) is a fructosan in which the fructose moieties are linked 2$\rightarrow$6, whereas in inulin, a reserve polymer found in the tubers of the *Compositae* family, the linkage is 2$\rightarrow$1; the polyfructose chains in both types of polymer are terminated by an α-D-glucopyranose residue (Aspinall and Hirst, 1965; Tomasic et al., 1978). The plant fructosans and fructose-containing oligosaccharides are readily fermented by the holotrichs (Table 8.1). Grass fructosan is hydrolyzed four to five times more rapidly than inulin by the holotrichs with *Dasytricha* exhibiting higher activities against both substrates (Thomas, 1960). It was believed that invertase was responsible for the hydrolysis of all fructose-containing saccharides (Punj et al., 1970; Thomas, 1960). However, the purified enzyme of *I. prostoma*, although fully active on sucrose, displayed only 30% and 1% of this activity against raffinose and inulin respectively (Dauvrin and Thines-Sempoux, 1989c).

Several studies have confirmed the presence of invertase in the holotrich ciliates (Table 8.1, 8.3). The Km values for sucrose for *D. ruminantium* and *Isotricha* spp. are 4.17 and 2.9 mM respectively with pH and temperature optima for both genera occurring at pH 5–5.5 and 50°C (Dauvrin and Thines-Sempoux, 1989c; Thomas, 1960; Williams, 1979b). The enzyme has transglycosylase activity transferring fructosyl moieties at higher sucrose concentrations; the principal products of the transfructosylation are trisac-

Table 8.3 Properties of some holotrich enzymes

Enzyme	Protozoal genus	Characteristics and optima pH	K_m (mM)	Temperature (°C)	Reference[a]
Polygalacturonase	*Isotricha*	8.7–9			1
Pectin esterase	*Isotricha*	8.6–8.8			1
Amylase[b]	*Isotricha*	4.8			4, 5
Amylase	*Dasytricha*	5, 6			4, 5
Maltase	*Dasytricha*	5.5		50	2
Invertase					
Intracellular	*Isotricha*	5–5.5	2.9	50	3, 6
Extracellular	*Isotricha*	5–5.5	2.9	50	6
Intracellular	*Dasytricha*	5–5.5	4.17	50	6
Extracellular	*Dasytricha*	5–5.5	4.25	50	6
β-Glucosidase					
Intracellular	*Dasytricha*	5.5–6	2.17	>60	6
Extracellular	*Dasytricha*	5.5–6	2.17	>60	6

[a] The references are as follows: 1, Abou Akkada and Howard (1961); 2, Bailey and Howard (1963); 3, Dauvrin and Thines-Sempoux (1977c); 4, Mould and Thomas (1957); 5, Mould and Thomas (1958); 6, Williams (1979b).
[b] The isoelectric point for this enzyme is 5.4.

charides (Carnie and Porteous, 1959; Christie and Porteous, 1957; Howard, 1959b). The purified invertase of *I. prostoma* comprises at least two distinct polypetides one of which is glycosylated. The purified enzyme is a tetramer with an apparent molecular weight of 350,000 (Dauvrin and Thines-Sempoux, 1989a, 1989b, 1989c). Subcellular fractionation studies indicated that the invertase occurred as an insoluble aggregate and was not associated with vesicles or organelles (Dauvrin and Thines-Sempoux, 1989a). These aggregates, termed invertosomes, contained two distinct forms of invertase (Mr 10^7 and $> 4 \times 10^7$).

Invertase is also the most active of the extracellular carbohydrases formed by the holotrichs and although the activity optima (Table 8.3) do not coincide with rumen conditions (i.e., 39–40.8°C, pH 5.8–6.8), the properties of the invertase present in rumen liquor are very similar to those of the holotrich exocellular enzymes. The holotrich ciliates would, therefore, appear to be an important source of free invertase in rumen contents (Thomas, 1960; Williams, 1979b). The invertase of *I. prostoma* was associated with an extracellular mucus secretion (Dauvrin and Thines-Sempoux, 1989a; Thines-Sempoux and Dauvrin, 1989).

Degradation of Plant Cell Wall Structural Polysaccharides The isotrichid ciliates can associate with plant material in the rumen and are able to effectively utilize the principal plant reserve polysaccharides starch and inulin. Their ability to degrade plant cell wall structural polymers cellulose, hemicellulose, and pectin is more limited and will be considered in the next section.

Isolated pectin is utilized by *D. ruminantium* (van Hoven and Prins, 1977; Williams and Harfoot, 1976) and *Isotricha* spp. (Abou Akkada and Howard, 1961; Prins and Van Hoven, 1977) stimulating gas, acidic metabolite, and storage polymer formation. However, polygalacturonic acid, oligo-uronides, and galacturonic acid are not fermented by either genus. Various neutral sugars are covalently linked to pectin and it is their release by glycosidase action and subsequent utilization that is likely to stimulate metabolite formation. Holotrich cell extracts do, however, possess both pectin esterase and polygalacturonase activities; products of both polygalacturonic acid lyase and polygalacturonic acid hydrolase action have been detected (Abou Akkada and Howard, 1961; Coleman et al., 1980b). Hydrolytic enzyme activities were higher in the *Isotricha* spp. preparation which had pH optima in the range pH 8.5–9.0 (Table 8.3). The principal products of the enzymic degradation of pectin in both genera are oligo-uronides and methanol; these products are not metabolized further by the protozoa. Pectolytic enzyme activity has also been detected in cell-free incubation supernatants of the holotrichs (Abou Akkada and Howard, 1961; Williams, 1979b).

Cellulose is a major structural component of the plant cell wall. It is a homopolysaccharide composed of β,1→4 linked glucose residues. The rumen protozoal population is actively cellulolytic (Williams and Strachan, 1984) and a survey by Coleman (1985a) confirmed the widespread occurrence of cellulase in the entodiniomorphid ciliates. The data relating to the holotrichs are, however, more equivocal. Although metabolite formation by *D. ruminatium* is not enhanced by carboxymethylcellulose (CMC) there is an initial transient increase in gas formation (Van Hoven and Prins, 1977; Williams and Harfoot, 1976) and cell-free extracts of both *D. ruminatium* and *Isotricha* spp. will release reducing sugars from CMC (Williams et al., 1986). The activity was low and the presence of discreet cellulase activity has yet to be confirmed, as both genera have β-glucosidase and β-cellobiosidase activities (Williams et al., 1984). Similarly, the holotrichs are able to degrade the β, 1-3 linked disaccharide laminaribiose (Howard, 1959b), although their activity against polysaccharides containing β, 1-3 linkages (e.g., laminarin, lichenan) is not known. The entodiniomorphid ciliate *Epidinium ecaudatum* degrades both β, 1-3 and β, 1-4 glucans (Bailey and Gaillard, 1965) and holotrich ciliates are present in lichen-consuming Arctic ruminants (Dehority, 1986c) and seaweed-eating sheep (Orpin et al., 1985). Holotrichs from seaweed-eating sheep were found to store amylopectin when the rumen liquor was incubated with laminarin (Eadie, 1957), although it was not established whether the holotrichs were primary degraders or scavenged the lower saccharides (e.g., glucose, laminaribiose) released by rumen bacterial or fungal enzymes.

Although the data accumulated for the presence of holotrich β-glucanase activity is inconclusive, it has now been established that both genera have certain hemicellulolytic activities. Ryegrass hemicellulose and oat xylan are degraded to ethanol-soluble oligomers and the component monosaccharides, xylose, and arabinose, by endo-acting β-D-xylanase(s), β-D-xylosidase, and α-L-arabinofuranosidase (Williams and Coleman, 1985; Williams et al., 1984). Prior to these studies very low levels of hemicellulase activity were detected in partially purified holotrich extracts by colorimetric determination of reducing sugar formation (Hayer et al., 1976; Rao et al., 1977). Initial studies using manometric and paper chromatographic techniques failed to demonstrate xylan or arabinan utilization by holotrich preparations or cell extracts (Bailey et al., 1962; Prins and Van Hoven, 1977; Van Hoven and Prins, 1977; Williams and Harfoot, 1976). The depolymerization products are not metabolized further and this explains the failure of manometric techniques to detect hemicellulolytic activity.

Hemicellulose-degrading enzymes are formed by many of the rumen ciliates (Williams and Coleman, 1985). The highest activities were present in the cellulolytic entodiniomorphid ciliates; the holotrich activities were approximately 5–10 fold lower. The specific activity of the *Isotricha* spp. preparation was 3–7 times greater than that of the *Dasytricha* preparations although their modes of action were apparently similar. The holotrichs thus have the enzymic capability to depolymerize isolated pectic and hemicellulosic polysaccharides, although the products of these degradative activities are not utilized. The ability of the protozoa to degrade and remove these polymers from intact plant cell walls has not been established. However, both genera associate with plant material in the rumen and it is possible that these fibrolytic activities are localized by the attachment organelle at the point of contact with the plant cell wall and thereby disrupt the structure sufficiently to permit the release and uptake of the more readily fermentable intracellular carbohydrates.

Glycoside Hydrolase Enzymes Effective polysaccharide breakdown requires the interaction of polysaccharide depolymerase and glycoside hydrolase enzymes. Glycosidase activity is essential if complete polysaccharide degradation is to occur as these enzymes remove side chains from the main polymer backbone (e.g., Bailey et al., 1962; Bailey and Gaillard, 1965, 1969) and hydrolyze the oligosaccharidic fragments generated by the polysaccharidases. In addition various glycosides are present in the plant tissues and seeds ingested by the host ruminant. A wide range of α- and β-glycoside hydrolases have been detected in the rumen ciliate population (Delfosse-Debusscher et al., 1979; Williams and Strachan, 1984). There have been several studies on the range of glycosidases formed by *D. ruminantium* and *Isotricha* spp. and these are summarized in Table 8.4. These surveys using both natural carbohydrate and chromogenic substrates have confirmed that both genera have activity against glycosides of pentoses, hexoses, hexosamines,and hexuronic acids. Although there are differences in the range of disaccharides fermented by *Dasytricha* and *Isotricha*, the enzyme profiles of both genera are the same. Thus, although maltose and cellobiose are not effective substrates for *Isotricha* spp., α-glucosidase and

Table 8.4 References for the glycoside hydrolase (glycosidase) enzymes detected in the holotrich ciliates[a]

	Dasytricha ruminantium		*Isotricha* spp.	
Enzyme substrate	Detected	Not detected	Detected	Not detected
Maltose	1, 5, 6, 8		5, 6	
Maltotriose	1, 6, 8		6	
Isomaltose	8			
Trehalose	5	1, 4		4, 5
p-Nitrophenyl-α -D-glucoside	8, 9, 10, 11, 13		9, 10, 12	
Cellobiose	1, 4, 5, 8		5	4
Laminaribiose	4, 5		4, 5	
Amygdalin	5		5	
Esculin	5		5	
Arbutin	5		5	
Salicin	5		5	
p-Nitrophenyl-β -D-glucoside	4, 5, 8, 9, 10, 11, 13		4, 5, 9, 10, 12	
p-Nitrophenyl-β -D-cellobioside	9, 10		9, 10	
Sucrose	1, 2, 3, 4, 5, 7, 8		2, 3, 4, 5, 7, 8	
Raffinose	4, 5, 8		4, 5	
Melezitose		5		5
Lactose	8	4, 5		4, 5
Melibiose	5, 8	4		4, 5
p-Nitrophenyl-α -D-galactoside	8, 9, 10, 11, 13		9, 10	
p-Nitrophenyl-β -D-galactoside	8, 9, 10		9, 10	
Xylobiose		4, 5		4, 5

[a] The references are as follows: 1, Bailey and Howard (1963); 2, Carnie and Porteous (1959); 3, Christie and Porteous (1957); 4, Howard (1957); 5, Howard (1959b); 6, Mould and Thomas (1958), 7, Thomas (1960); 8,

β-glucosidase activities are present in cell extracts. There are other apparent anomalous observations as certain of the glycosidase activities present (e.g., xylosidase, arabinofuranosidase, uronidase, mannosidase, hexosaminidase) will release monosaccharides that are not utilized or are even toxic to the protozoa. The inability of the holotrich ciliates to metabolize these monomers may be due to the evolutionary loss of one or more key enzymes from a metabolic pathway.

Of the glycoside hydrolases formed only the invertase, α-glucosidase, and β-glucosidase activities have been examined in any detail (Table 8.3). Invertase and α-glucosidase were discussed in "Storage Polysaccharide Degradation," above. The β-glucosidase activity of *D. ruminantium* is approximately five- to tenfold greater than that present in *Isotricha* preparations (Howard, 1957; Williams et al., 1984, 1986). However, both β, 1-3 (laminaribiase) and β, 1-4 (cellobiase) activities are formed by both genera and the β-glucosides esculin, amygdalin, arbutin, and salicin are degraded by *D.*

Table 8.4 *Continued*

	Dasytricha ruminantium		*Isotricha* spp.	
Enzyme substrate	Detected	Not detected	Detected	Not detected
p-Nitrophenyl-α-D-xyloside		9, 10	10	9
p-Nitrophenyl-β-D-xyloside	9, 10		9, 10	
p-Nitrophenyl-α-L-arabinofuranoside	8, 9, 10, 11, 13		9, 10	
p-Nitrophenyl-α-D-mannoside	8, 10		10	
p-Nitrophenyl-β-D-mannoside		9, 10	10	9
p-Nitrophenyl-α-L-fucoside		10		10
p-Nitrophenyl-β-L-fucoside		10		10
p-Nitrophenyl-β-D-fucoside	9, 10		9, 10	
p-Nitrophenyl-β-D-glucuronide	9, 10, 11, 13		9, 10	
p-Nitrophenyl-β-D-galacturonide	9, 10		9, 10	
p-Nitrophenyl-N acetyl-β-D-galactosaminide	10		10	
p-Nitrophenyl-N acetyl-β-D-glucosaminide	9, 10		9, 10	

Williams (1979b); 9, Williams et al. (1986); 10, Williams et al (1984); 11, Yarlett et al. (1981); 12, Yarlett et al. (1983a); 13, Yarlett et al. (1985).

ruminantium and *Isotricha* spp. enzyme preparations (Howard, 1957, 1959b). The β-glucosidase of *D. ruminantium* exhibits increasing activity to 60°C and has a pH optimum at pH 5.5–6.0; identical pH and temperature characteristics were exhibited with either cellobiose or p-nitrophenyl-β-D-glucopyranoside as the substrate (Williams, 1979b). The K_m of the *Dasytricha* enzyme for the latter substrate was 2.17 mM.

In addition to the pectolytic enzymes and invertase, several other holotrich glycosidases and p-nitrophenyl (aryl) glycoside hydrolases have been detected extracellularly (Williams, 1979b). The activity of many of the extracellular enzymes was low representing approximately 2–3% of the total activity of that enzyme in the protozoon. Incubation of pre-feed rumen contents with a range of carbohydrates did not appear to induce exoenzyme formation. Intracellular activity against ineffective substrates was increased by pre-incubation with a readily utilized carbohydrate (e.g., glucose, sucrose, inulin) whereas the level of activity of the exocellular enzymes associated with starved cells was increased by pre-incubation in the absence of added carbohydrate or when the substrate was only poorly utilized (e.g., maltose, lactose, melibiose, etc.). It was suggested by Williams (1979b) that these enzymes are actively secreted when nutrients are not readily available to enhance the exocellular breakdown of polymeric carbohydrates. The attachment organelle may have an important role in this process. In addition to fixing the protozoal cell to the plant material, the organelle may also function both to secrete and localize extracellular enzymes as well as participating in the absorption of the released nutrients (Orpin and Letcher, 1978). These functions would confer obvious advantages to the protozoa in an ecosystem as competitive and complex as the rumen.

There are many unresolved aspects of holotrich exoenzyme formation and it has yet to be determined whether the extracellular activity is due to active secretion or exocytosis (i.e., excretion) following the completion of intracellular vacuolar digestion. The properties of the intracellular and extracellular forms of the enzymes are similar (Table 8.3), and approximately 70–80% of the acid hydrolase activity is sedimentable by centrifugation and located in membrane-bound subcellular vesicles (Fig. 8.2). In *Dasytricha*

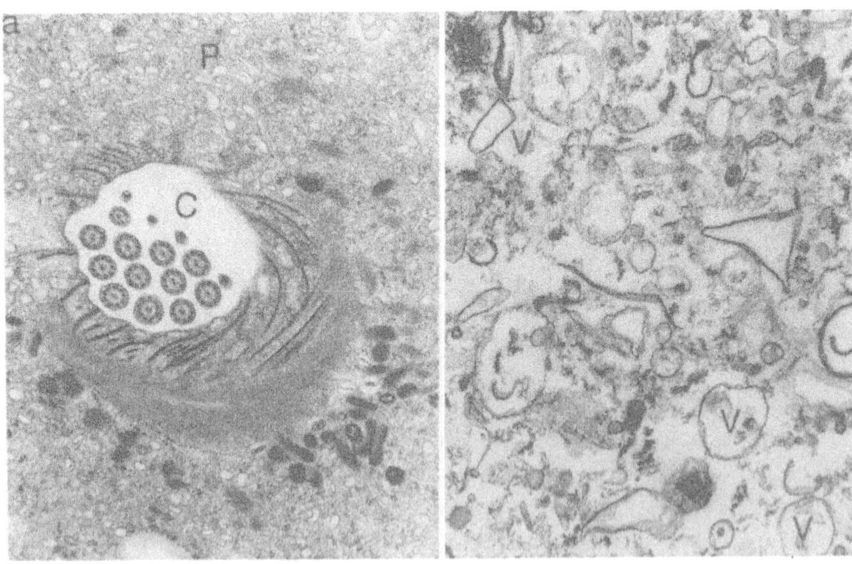

Fig. 8.2 (a) The phagoplasm (P) of *D. ruminantium* is characterized by the presence of numerous membrane-bound vesicles. The cytostome (C) and hydrogenosomes (H) are indicated in the section. Subcellular fractions enriched with these glycoside hydrolase-containing vesicles can be prepared by differential centrifugation. (b) The enzyme-containing vesicles (V) were sedimented from cell-free homogenates by centrifugation at for 10^5 g·min with precentrifugation of 10^4 g·min.

ruminantium, two distinct populations of acid hydrolase-containing organelles would appear to be present as a large particle population containing α-L-arabinofuranosidase and α-D-galactosidase sedimented on centrifugation for 10^4 g·min, whereas α- and β-D-glucosidase and the lysosomal marker enzyme (acid phosphatase) were characteristic of a smaller particle population that sedimented at 4×10^5 g·min (Yarlett et al., 1981). Very small particles, enriched in β-D-glucuronidase, that sedimented at 6×10^6 g·min may have been derived from Golgi vesicles. There is as yet, however, no definitive evidence confirming the existence of a Golgi apparatus in holotrich ciliates. Acid phosphatase and the glucosidase activity of *Isotricha* spp. also co-sediment in a 4×10^5 g·min subcellular fraction (Williams et al., 1986; Yarlett et al., 1983a). The glucosidase- and acid phosphatase-containing organelle of *D. ruminantium* has a mean equilibrium density on sucrose isopycnic gradient centrifugation of 1.13 g/ml (Yarlett et al., 1985). The acid hydrolases present in the rumen entodiniomorphid ciliates are also located in distinct populations of membrane-bound vesicles 0.1–0.8 μm in size (Hellings et al., 1981, 1982; Paul et al, 1989b; Williams et al., 1986). The subcellular structures of *Polyplastron multivesiculatum* contain acid phosphatase and an array of other glycoside hydrolases. After isopycnic centrifugation in sucrose density gradients, the acid hydrolases and polysaccharide depolymerases co-sedimented in vesicles with a mean equilibrium density of 1.17 (1.15–1.19) g/ml (Williams and Ellis, 1985). Approximately 50–60% of the hemicellulose and carboxymethylcellulose-degrading enzymes of the holotrichs were sedimentable after differential centrifugation of cell homogenates (Williams et al., 1986). The location of these enzymes within the cell ultrastructure still has to be established.

Metabolic Products of Carbohydrate Fermentation

Acidic Metabolites The metabolic end products formed during carbohydrate utilization by *Dasytricha* and *Isotricha* spp. are lactic acid, acetic acid, butyric acid, hydrogen, carbon dioxide, and storage polysaccharide. Formic acid and propionic acid have also been detected as fermentation products (Table 8.5). The metabolic products are the same with all substrates. The proportions of the metabolites formed and their rates of synthesis are, however, affected by the concentration and type of carbohydrate available (Table 8.6) (Prins and Van Hoven, 1977; Van Hoven and Prins, 1977; Williams, 1979a; Williams and Harfoot, 1976). In addition environmental pH and temperature (Williams and Harfoot, 1976), metabolic interactions with methanogens (Hillman, 1987; Hillman et al., 1988), the presence of low levels of oxygen (Hillman et al, 1985a; Prins and Prast, 1973; Yarlett et al., 1983b) and the ionophore antibiotic monensin (Hino, 1981) can affect the proportions of the metabolites produced. Although non-metabolizable substrates do not impede the uptake and metabolism of effective substrates,

Table 8.5 Products of endogenous and carbohydrate fermentation by the holotrich ciliates; a chronological survey

Protozoal preparation	Metabolite[a] Amylopectin	Acetate		Butyrate		Lactate		Propionate		Formate		Carbon dioxide		Hydrogen		Reference
	S	E	S	E	S	E	S	E	S	E	S	E	S	E	S	
Holotrichs	+															Oxford (1951)
Holotrichs	+															Masson and Oxford (1951)
Holotrichs	+	+	+	+	+	+	+	ND	tr			+	+	+	+	Heald and Oxford (1953)
D. ruminantium	+															Eadie and Oxford (1955)
Isotricha spp.	+															Eadie and Oxford (1955)
D. ruminantium	+	+	+	+	+	+	+	ND	tr			+	+	+	+	Gutierrez (1955)
Isotricha spp.	+	+	+	+	+	+	+	ND	tr			+	+	+	+	Gutierrez (1955)
Isotricha spp.	+		+				+						+			Howard (1957)
Holotrichs	+		+		+		+						+			Howard (1959a)
Holotrichs	+		+		+		+		tr		tr		+		+	Howard (1963)
Holotrichs			+		+								+			Zenbayashi et al. (1969)
D. ruminantium	+	+	+	+	+	+	+		tr			+	+	+	+	Williams and Harfoot (1976)
D. ruminantium	+	+	+	+	+	+	+			ND	+	+	+	+	+	Van Hoven and Prins (1977)
I. prostoma	+	+	+	+	+	+	+					+	+	+	+	Prins and Van Hoven (1977)
D. ruminantium	+		+		+		+						+		+	Williams (1979a)

E, endogenous fermentation; S, fermentation of a carbohydrate substrate; ND, not determined; tr, trace only.

Table 8.6 Relative rates of gas, amylopectin, and acidic metabolite formation by *I. prostoma* and *D. ruminantium* from various soluble carbohydrates

	Relative rate of formation[a] (Reference)[b]								
	For *I. prostoma*		For *D. ruminantium*						
Carbohydrate	Hydrogen (2)	Amylopectin (2)	Gas (1)	Hydrogen (3)	Gas (4)	Acetate (4)	Butyrate (4)	Lactate (4)	Amylopectin (2)
D-Glucose	100	100	100	100	100	100	100	100	100
Endogenous	NA	NA	NA	NA	37	26	45	18	NA
Galactose	33	26	72	108	81	59	55	28	35
Fructose	225	188	NA	99	115	97	92	102	139
Sucrose	117	115	100	86	110	89	106	100	144
Cellobiose	15	8	100	143	103	69	54	57	90
Lactose	3	2	NA	0	59	63	53	20	3
Mlibiose	32	31	NA	0	NA	NA	NA	NA	0
Maltose	0	0	33	80	89	49	62	38	19
Maltotriose	NA	NA	NA	NA	78	58	51	35	NA
Raffinose	86	67	100	105	104	104	97	85	95
Inulin	NA	NA	NA	NA	110	90	80	106	NA
Mannose	NA	NA	NA	NA	26	13	41	16	NA

[a] All rates are expressed relative to the glucose value, which was designated as 100%. NA, Not Assayed.
[b] The data are compiled from (1) Howard (1959a); (2) Prins and Van Hoven (1977); (3) Van Hoven and Prins (1977); (4) Williams and Harfoot (1976).

maximum rates of metabolite synthesis are not maintained in the presence of more slowly utilized substrates (Williams, 1979a).

L-lactic acid is the principal acidic metabolite of *D. ruminantium* (lactate > butyrate > acetate > formate), although some D-lactic acid and formic acid are produced at higher sugar concentrations. Maximum rates of acid formation occur at pH 7–7.5 and 39–40°C; glucose uptake and product formation rates are similar in cells isolated at various stages throughout the diurnal cycle of the rumen (Williams and Harfoot, 1976). *Dasytricha ruminantium* does not exhibit preferential or sequential substrate utilization and the presence of readily assimilated substrates does not preclude the uptake and effects of less effective, non- metabolizable, or toxic carbohydrates (Williams, 1979a). The rates at which the principal metabolites are formed is reduced when toxic (e.g., mannose) or poorly utilized substrates (e.g., galactose) are available with readily assimilated substrates. The presence of sugars that are not metabolized (e.g., arabinose, xylose) does not affect metabolite formation rates (Van Hoven and Prins, 1977; Williams, 1979a). Alanine has recently been identified using [^{13}C]nmr as a minor product of 1-[^{13}C]-D-glucose catabolism by *D. ruminantium* (Ellis et al., 1991; in press).

The carbohydrate fermentation products of mixed preparations of *I. prostoma* and *I. intestinalis* are the same as those formed by *D. ruminantium* (Table 8.5). However, of the two species of *Isotricha* only *I. prostoma* has been examined and the metabolic capabilities of *I. intestinalis* are unknown. At high rates of fermentation L-lactic acid is the principal acidic metabolite of *I. prostoma* although butyric acid and acetic acid predominate at lower rates (Prins and Van Hoven, 1977). The fermentation rates of *I. prostoma* are higher than those measured with *D. ruminantium* (Table 8.7); with the exception of butyric acid the rates of product formation by both genera are dependent upon the concentration of the available carbohydrate (Fig. 8.3; Prins and Van Hoven, 1977; Van Hoven and Prins, 1977; Williams and Harfoot, 1976).

Formation and Degradation of Amylopectin An important feature of *Dasytricha ruminantium* and *Isotricha* spp. is their ability to synthesize and store an intracellular reserve polysaccharide during the periods when sugars are available in the rumen. This evolutionary adaptation enables the protozoa to compete effectively for limited substrates as the reserve polymer is utilized during periods when exogenous substrates are not available. Dehority (1970), however, was not able to demonstrate the accumulation of an iodophilic polysaccharide in *Buetschlia parva*.

The level of soluble carbohydrates in the rumen ecosystem is elevated in the immediate post-feed period, but rapidly declines to pre-feed concentrations (see Williams and Harfoot, 1976). The holotrichs are well-adapted to maximize their usage of the available carbohydrates. Both genera (*Dasytricha* and *Isotricha*) are attracted by chemotaxis to sources of soluble carbohy-

Table 8.7 Metabolite formation rates in *D. ruminantium* and *I. prostoma*

Protozoon	Substrate	Metabolite formation rate (pmol/cell/h)[a]						Reference[e]
		Acetate	Butyrate	Lactate	Formate[d]	Amylopectin	Gas	
D. ruminantium	Endogenous	5.4	4.9	3.4	ND	−11.0	30.2	2
	Endogenous[b]	3.2	2.2	2.8	NA	NA	10.0	3
	Glucose, 5 mM	11.0	8.0	65	1	105	28[c]	2
	Glucose, 24 mM	19.0	5.0	105	10	115	18[c]	2
	Glucose, 28 mM[b]	6.0	5.3	32.2	NA	57.4	39.7	3
I. prostoma	Endogenous	27	26	14.5	NA	−52	142[c]	1
	Glucose 5 mM	90	86	101	NA	390	471[c]	1
	Glucose 11 mM	175	145	186	NA	611	869[c]	1

[a]Cells were maintained in vitro for 20 h in presence of antibiotics prior to assay. A negative value indicates utilization rather than formation.
[b]Calculated on the basis of 6.6 ng protein per cell (Gutierrez, 1955); the data are mean values for a 24-h diurnal cycle.
[c]Hydrogen data only.
[d]ND, not detected; NA, not assayed.
[e]References: 1, Prins and Van Hoven (1977); 2, Van Hoven and Prins (1977); 3, Williams and Harfoot (1976).

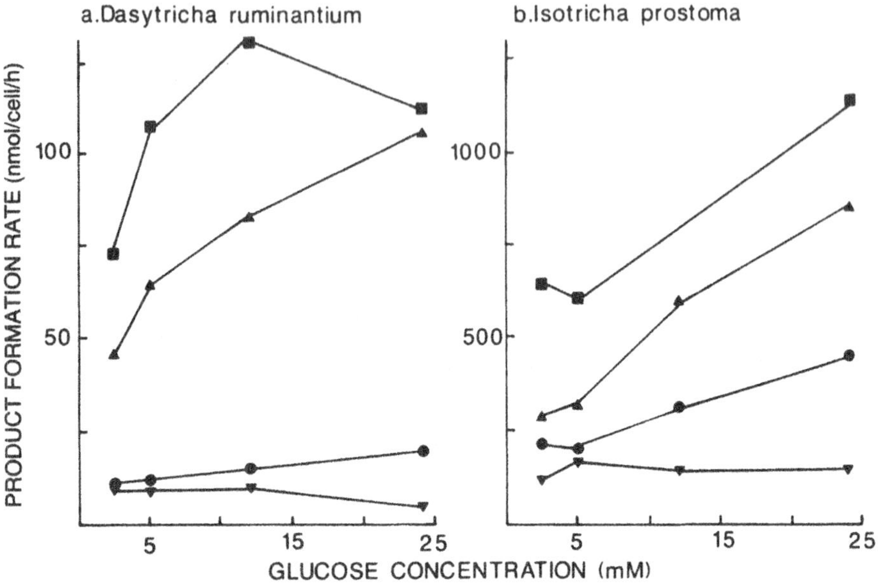

Fig. 8.3 Effect of glucose concentration on the rates of formation of amylopectin (■), lactic acid (●), and butyric acid (▼) by (a) *Dasytricha ruminantium* and (b) *Isotricha prostoma*. Compiled from Prins and Van Hoven (1977) and Van Hoven and Prins (1977).

drates and will nonselectively assimilate the sugars as soon as they are present. The absence of selectivity in sugar uptake enables a concomitant uptake and co-utilization of carbohydrate substrates (Williams, 1979a). When soluble sugars are not readily available the holotrichs may secrete enzymes through their attachment organelle to facilitate further substrate release.

Approximately 75–80% of the sugar taken up by the holotrichs may be converted to the reserve polymer, while over 90% of the cell-associated ^{14}C-labelled carbon derived from [U-^{14}C]glucose in a 60 minute incubation was found in the reserve polymer (Williams and Harfoot, 1976). Intracellular polysaccharide levels in the rumen are maximal some 2–4 hours after feeding; thereafter the levels decline (Oxford, 1951; Williams and Harfoot, 1976). This subsequent fermentation of the stored material allows the protozoa to have a constant supply of energy throughout the day, although the initial carbohydrates may have only been available for a very limited period. The protozoa are thus able to compete effectively with other saccharolytic microorganisms in the rumen and also exert some control on the overall rate at which short-chain volatile fatty acids are formed in the rumen. Coleman (1979a) estimated that approximately one-third of the sugars consumed by the host ruminant were converted by the holotrichs into storage polysaccharide. Holotrich ingestion of soluble sugars and starch grains

(Heald and Oxford, 1953; Mackie et al., 1978; Oxford, 1955a) is of benefit to the host animal as the alternative bacterial fermentation, which could lead to an accumulation of lactate in the rumen and lowering of rumen pH, is prevented. The holotrich protozoa may, therefore, help to alleviate the onset of lactic acid acidosis by rapidly assimilating soluble sugars; the subsequent fermentation of the reserve material, and ingested starch, over a prolonged period thus modulates the rate of volatile fatty acid formation and stabilizes the fermentation of ingested carbohydrates.

The reserve polysaccharide is a high molecular weight α, 1-4-1inked homoglucan which has a unit chain length of 22 glucose residues that are cross-linked by α, 1–6 linkages. The amylose content of the polysaccharide is negligible and the highly branched molecular structure is closely similar to that of amylopectin (Forsyth and Hirst, 1953). It is synthesized by the energy consuming phosphorylase mechanism common to plants and animals (Mould and Thomas, 1958). Approximately 90% of the adenosine-5′-triphosphate (ATP) formed in the fermentation of glucose by *D. ruminantium* is used by the ciliate in amylopectin synthesis whereas only 30% of the energy available through galactose fermentation was used for synthesis (Van Hoven and Prins, 1977). A wide range of carbohydrates are converted into the polymer (Table 8.1) and its deposition was used initially as a criterion for assessing substrate utilization (Masson and Oxford, 1951; Oxford, 1951). All substrates that stimulate the production of hydrogen by *I. prostoma* are precursors for amylopectin formation (Prins and Van Hoven, 1977). The rate of polysaccharide deposition is affected by both the substrate and its concentration; for example the rate of amylopectin deposition by *D. ruminantium* increases up to a glucose concentration of 13 mM but is reduced at 24 mM (Fig. 8.3). The maximum rates of deposition by *D. ruminantium* and *I. prostoma* are 130 and 1300 pmol of amylopectin-hexose/cell/hour from glucose, whereas with galactose, which is a less effective substrate for amylopectin synthesis in both genera, the rates of amylopectin storage are only 7 and 120 pmol/cell/hour respectively (Prins and Van Hoven, 1977; Van Hoven and Prins, 1977). These latter two investigators calculated that, with glucose as substrate, the maximum amount of amylopectin synthesized by *I. prostoma* was approximately 250 ng per cell. Polymer deposition in vitro is also affected by temperature (Eadie and Oxford, 1955).

Simultaneous synthesis and utilization of the amylopectin storage polysaccharide occurs and there are threshold concentrations (e.g., glucose, galactose 20 mM; fructose 2–5 mM) at which there is no net loss of amylopectin from the protozoal cell. The protozoa are, however, able to control the total amount of amylopectin deposited. It had been proposed, on the basis of earlier observations, that the holotrichs were unable to control the intracellular deposition of amylopectin, and in the presence of excess substrate polysaccharide synthesis would continue until the over accumulation was sufficient for cellular degeneration to ensue. Cellular deterioration and bursting in vivo (Clarke, 1965a,b) and in vitro (Oxford, 1955b; Sugden and

Oxford, 1952) was ascribed to over-synthesis of amylopectin. Holotrich degeneration in the presence of excess carbohydrate was used to explain the observed post-feed decline in holotrich numbers, and as the cell contents of burst holotrichs have foam-stabilizing properties which could influence the foaming properties of rumen digesta, bursting in vivo was suggested as a possible cause of the potentially fatal ruminant disorder bloat (Clarke, 1965b). However, the observed lysis of the holotrichs is not, as first proposed, due merely to an overaccumulation of storage polysaccharide. Polymer deposition is controlled but lysis occurs as the protozoa do not selectively control the entry of sugars into the cells; the subsequent metabolism of the carbohydrates does not lead to an uncontrolled synthesis of amylopectin but to a detrimental intracellular build-up of acidic fermentation products, in particular lactic acid (Van Hoven and Prins, 1977; Williams, 1979a; Williams and Harfoot, 1976).

Amylopectin Degradation The metabolites produced from the endogenous fermentation of reserve amylopectin are approximately equal proportions of acetic acid and butyric acid and equimolar amount of hydrogen and carbon dioxide. *Dasytricha ruminantium* forms only L-lactate from amylopectin whereas *I. prostoma* produces a racemic mixture of both isomers during the endogenous fermentation (Tables 8.5, 8.7); the proportion of lactic acid formed is more variable (Prins and Van Hoven, 1977; Van Hoven and Prins, 1977; Williams and Harfoot, 1976). The endogenous rates of metabolite formation are some 2- (acetate, butyrate) to 10-fold (lactate, gas) lower than the rates determined with a rapidly assimilated sugar. The endogenous rates of acid formation by *I. prostoma* are approximately five times higher than those of *D. ruminantium*, and the maximum endogenous rates of amylopectin breakdown are by *D. ruminantium* and *I. prostoma* are 12 and 85 pmol/cell/hour respectively (Prins and Van Hoven, 1977; Van Hoven and Prins, 1977).

In the absence of an exogenous substrate, a constant utilization of the stored amylopectin occurs, although endogenous fermentative activity is inhibited by lactic acid and the toxic monosaccharides glucosamine and mannose. The endogenous activity of *D. ruminantium* preparations remained constant in cells recovered from the rumen during a nine hour postprandial period. The carbohydrate content of the cells had not declined markedly during this post-feed period; however, by 24 hours after feeding the level of reserve carbohydrate had fallen and the endogenous metabolic activity was also reduced (Williams and Harfoot, 1976). There is also some limited information to indicate that the rate of amylopectin fermentation by *I. prostoma* was influenced by the reserve levels in the cell (Prins and Van Hoven, 1977).

Intermediary Metabolism of Carbohydrates Fermentative metabolism in the rumen holotrich (isotrichid) ciliates *Dasytricha ruminantium, Isot-*

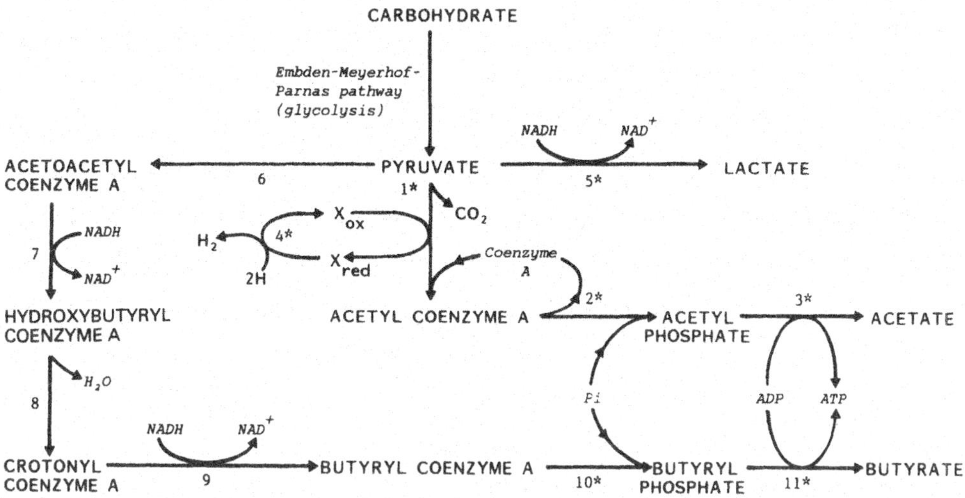

Fig. 8.4 Proposed pathways leading to metabolite formation in *Dasytricha ruminantium* and *Isotricha* spp. The enzymes indicated are: 1, pyruvate synthase (pyruvate : ferredoxin oxidoreductase); 2, phosphoacetyltransferase; 3, acetate kinase; 4, hydrogenase; 5, lactate dehydrogenase; 6, acetyl-CoA : acetyl-CoA acetyltransferase; 7, 3-hydroxybutyryl-CoA dehydrogenase; 8, 3-hydroxyacyl-CoA hydrolase; 9, 3-hydroxyacyl-CoA dehydrogenase; 8, 3-hydroxyacyl-CoA hydrolase; 9, 3-hydroxyacyl-CoA reductase; 10, phosphobutyryl transferase; 11, butryrate kinase. Some or all of the activity of enzymes marked* is present in the hydrogenosomal fraction; the other enzymes are located in the cell cytosol.

richa prostoma and, in all probability, *I. intestinalis* involves the utilization of soluble carbohydrates and the formation of short-chain acidic metabolites. The metabolites are of value in the nutrition of the host, but despite their importance, the intermediary metabolism of the ciliates has not been studied extensively and our understanding of the metabolic pathways and the factors controlling their expression is incomplete.

The enzymes involved in the formation of hydrogen, acetic acid, butyric acid, and lactic acid have been detected in both genera and are located either within the cell cytosol or a microbody-like subcellular organelle the hydrogenosome (Yarlett et al., 1981, 1982, 1983a, 1985). The principal pathways are summarized in Fig. 8.4 and the available evidence would indicate that they are common to both *Dasytricha* and *Isotricha* spp.

Anaerobes with few exceptions transform sugars via glycolysis or by one of three pathways involving 6-phosphogluconic acid, which is formed by the enzyme glucose-6-phosphate dehydrogenase. As this enzyme has not been detected in the holotrichs it is reasonable to presume that glucose is metabolized to pyruvic acid solely via the Embden-Meyerhof-Parnas glycolysis pathway. All the glycolytic enzymes are present in *D. ruminantium* and specific activity measurements indicated that aldolase and triosephosphate isomerase had a role in controlling the rate of pyruvate formation in this protozoon (Yarlett et al., 1985). Subcellular fractionation studies indicated that all the enzymes were predominantly nonsedimentable by centrifugation at 6×10^6 g·minutes, and occurred in the cytosolic (soluble) fraction. There was thus no evidence to indicate the presence of a subcellular organelle such as the glycosome. This organelle has only been described in certain trypanosomid protozoa and is the site at which the key enzymes of glycolysis are localized within the cell (Opperdoes, 1987). Extracts of *I. prostoma* contain an active fructose-2,6-biphosphate-insensitive pyrophosphate:fructose-6-phosphate phosphotransferase. The purified enzyme had an M_r of 95,000 and was comprised of subunits with an M_r of 48,000 (Mertens et al., 1989).

Hydrogen is a common protozoal carbohydrate fermentation product, although mechanisms for its production have been illucidated in relatively few species (e.g., *Trichomonas* sp; Bauchop, 1971). In bacteria two separate pathways of hydrogen formation occur. Both proceed via the intermediate formation of pyruvate which may be converted to acetyl coenzyme A and formate or alternatively to acetyl coenzyme A, carbon dioxide, and hydrogen. The former mechanism is found in the Enterobacteriaceae and the formate produced is further metabolized to hydrogen and carbon dioxide by formate hydrogenlyase. Saccharolytic *Clostridium* spp. utilize the latter reaction sequence in which formate is not an intermediate. Formate dehydrogenase has not been detected in the holotrichs and this would imply that hydrogen is formed from pyruvate by a reaction analagous to that occurring in the saccharolytic clostridia and the trichomonad group of flagellate protozoa. The tricarboxylic acid and glyoxylate cycles do not operate in *D. ruminantium* (Yarlett et al., 1981, 1983a).

The two enzymes, pyruvate synthase (pyruvate:ferredoxin oxidoreductase) and hydrogenase, that are involved in the conversion of pyruvate to acetyl coenzyme A and hydrogen are located within the subcellular organelle the hydrogenosome in both *D. ruminantium* and *Isotricha* spp. The subsequent release of acetate and the regeneration of coenzyme A is also organelle-associated as the enzymes phosphoacetyltransferase and acetate kinase are present both in the hydrogenosome-containing and cytosolic fractions. The final reaction in acetate formation is an energy-generating substrate level phosphorylation and the principal route of ATP synthesis in the holotrichs occurs through glycolysis. The molar yield of ATP to the protozoa is at least 4 from the conversion of glucose into pyruvate and its further transformation to acetate via acetyl phosphate. The pathway for the regeneration of coenzyme A from acetyl coenzyme A in the hydrogenosome-containing trichomonad flagellates is markedly different; in these protozoa regeneration involves succinate thiokinase and succinate-acetate coenzyme A transferase (Muller, 1980, 1982, 1988). These enzymes have not been detected in holotrich cell extracts (Yarlett et al., 1981, 1983a).

Six enzymes (Fig. 8.4) are involved in the ATP-generating conversion of pyruvate into butyrate in both holotrich genera (Prins, 1977; Yarlett et al., 1985). These metabolic reactions proceed predominantly in the cytosol, although the activity of phosphobutyryltransferase and butyrate kinase is partly sedimentable occurring both in the cytosol and in association with the hydrogenosomes. The initial enzymic condensation of acetyl coenzyme A to produce acetoacetyl coenzyme A was the rate-limiting reaction (Yarlett et al., 1985).

It is apparent that the hydrogenosomes have an important role in the conversion of pyruvate into acetyl coenzyme A; it is not unusual for the formation of this coenzyme to occur in specialized organelles. In many eukaryotic cells the β-oxidation of fatty acids to yield acetyl coenzyme A is compartmentalized within peroxisomes, glyoxysomes, or even mitochondria. The range of enzymes associated with the hydrogenosomal fraction of the rumen holotrichs is different to that reported for other protozoa, and the organelles have lower isopycnic densities than have been reported for *Trichomonas* spp. (Muller, 1980) and rumen entodiniomorphid ciliates (Paul et al., 1989b; Snyers et al., 1982). In addition to the enzymes involved in acetate, butyrate, and hydrogen formation that have been discussed in the preceding paragraphs, some lactate dehydrogenase activity is also found in association with the organelle in *Dasytricha ruminantium*. Although the full complement of hydrogenosomal enzymes has still to be determined, there are obvious differences between the holotrich ciliates and trichomonad flagellates. One pronounced difference concerns the location of decarboxylating malate dehydrogenase. The oxygen sensitive enzymes pyruvate synthase and hydrogenase are associated with the organelle in the holotrichs, whereas the oxygen-tolerant malate dehydrogenase is located in the cytosol, suggesting that compartmentation within the hydrogenosome may afford some protection to the enzymes that are particularly oxygen-sensitive. However, in the trichomonads (Muller, 1980) decarboxylating malate dehydrogenase is located within the hydrogenosome. Other rumen microorganisms are the same as the holotrichs, since the enzyme has been shown to be non-sedimentable (i.e., cytosolic) in *Polyplastron multivesiculatum* (Paul et al., 1989b) and several other entodiniomorphid ciliates (Yarlett et al., 1984). However, Snyers et al. (1982) did find some sedimentable activity in the hydrogenosomal fraction prepared from a mixed entodiniomorphid ciliate population, and the enzyme is hydrogenosome-associated in the anaerobic ruminal chytrid fungus *Neocallimastix frontalis*. The characterization of the organelle in organisms from other intestinal and anoxic environments will lead to an understanding of hydrogenosomal evolutionary progression in the anaerobic ciliate and aerotolerant flagellate protozoa (Muller, 1988).

Lactic acid is the principal acidic metabolite formed by both genera of holotrichs (Prins and Van Hoven, 1977; Williams and Harfoot, 1976). It is formed from pyruvate through the action of a nicotinamide adenine dinucleotide-dependent lactate dehydrogenase. Approxmately 25% of the activity of this enzyme in *Dasytricha* is hydrogenosome associated. Some kinetic parameters of the enzyme from *Isotricha prostoma* have been determined (Counotte, 1979; Counotte et al, 1980). The K_m values of the enzyme in crude cell extracts for pyruvate and NADH were 1.33 mM and 15 μM respectively. Activity was not markedly affected by pH in the range pH 5–9 and, unlike the rumen bacterial enzymes studied, was independent of fructose 1,6-diphosphate. Activity was modulated by ATP and the relative proportions of the oxidized and reduced forms of the nicotinamide adenine dinucleotide coenzyme.

Factors Affecting Metabolite Formation There have been several attempts to quantify the holotrich contribution to volatile fatty acid production in the rumen. Gutierrez (1955) calculated that the holotrichs contributed some 10% of the total acids in the rumen while in dairy cattle on rich pasture approximately 3–4% of the acids may originate from each genus (Prins and Van Hoven, 1977; Van Hoven and Prins, 1977). The calculations, however, are based on rate determinations made in vitro with rapidly assimilated substrates. Metabolite formation is affected by many factors and more information is needed to establish the relationships between metabolite formation and the diet of the host.

It has been established in studies in vitro that the rates of metabolite synthesis are not constant and are influenced by various parameters, some of which have been detailed in the appropriate preceding sections. Thus the rates of product formation are affected by the nature of the available substrate (Table 8.6) the amount of sugar present (Fig. 8.3, Table 8.7) and the consequences of concomitant sugar utilization (Fig. 8.5; Williams, 1979a). In addition the environmental conditions influence protozoal activity and rates are maximal at pH7 and 40°C (Fig. 8.5; Williams and Harfoot, 1976). Feed additives included in the diet to manipulate the ruminal fermentation can also affect the protozoa. The ionophore antibiotic monensin, for example, is toxic at concentrations greater than 800 ng per ml, although metabolic changes are evident at concentrations of 100–200 ng per ml. At these lower levels the molar proportions of acetate and lactate formed are enhanced, whereas the production of butyrate, formate, and hydrogen are all decreased (Hino, 1981).

Metabolite output from the protozoa is not constant throughout the postprandial period. The level of storage polysaccharide increases in the period immediately after feeding and in animals receiving a diet of molassed sugar beet pulp and hay is maximal some 2–3 hours after the ration has been consumed by the host ruminant. The amount of carbohydrate in the protozoa is continuously depleted throughout the subsequent post-feed period (Williams and Harfoot, 1976). Metabolite production rates are reduced in cells isolated immediately prior to feeding in animals receiving a sugar beet pulp and hay feed (Williams and Harfoot, 1976) or an all-hay ration (Hillman, 1987). Prins and Van Hoven (1977) concluded that the en-

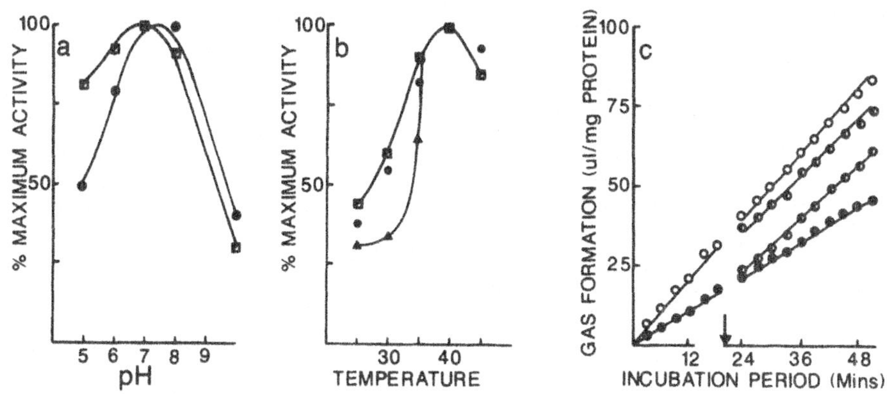

Fig. 8.5 Factors affecting the metabolic activity of *Dasytricha ruminantium*. The effect of (a) pH and (b) temperature on the rate of glucose incorporation (●) and the rates of lactic acid (◩) and butyric acid (▲) formation are indicated, as well as (c) the effects of of substrate availability on gas production. In (c) an alternative substrate (glucose or galactose) was added to an established fermentation at the point indicated by the arrow. The addition of glucose (○) to a galactose fermentation increased gas formation whereas galactose addition to glucose fermentations (●) depressed gas formation. The final rate of gas production during the co-fermentation of glucose and galactose (◐) was the same irrespective of the order of the alternative substrate addition. Data compiled from Williams (1979a) and Williams and and Harfoot (1976). (c) From Williams (1979a).

dogenous rate of fermentation of *I. prostoma* was affected by the amount of amylopectin in the protozoon.

The rates of holotrich endogenous gas and acid formation are also generally lower in protozoa isolated from hay-fed animals as compared with the endogenous metabolic rates in preparations recovered from animals receiving molassed sugar beet pulp-based diets (Table 8.8; Hillman, 1987). The lower metabolic activity of protozoa in the rumen of animals on a diet of hay may well reflect the paucity of soluble metabolizable nutrients in hay as compared to sugar beet. The importance of holotrich protozoa in the rumen fermentation is, therefore, in all probability dependent on the type of feed consumed by the ruminant and upon the length of time that has elapsed following food intake by the host animal.

A further variable associated with different diets is the concentration of dissolved gases in the liquor phase of rumen contents (Hillman et al., 1985b). Although the levels of oxygen and hydrogen were similar (0–3 μM) across the three diets examined, the carbon dioxide concentrations were more diet-dependent (viz. Hay 590–3800 μM; sugar beet pulp 550–1300 μM; grass nuts 160–320 μM); the levels of all three gases fluctuated appreciably in the immediate post-feed period. Each of these gases has been shown to influence metabolite formation by the holotrich ciliates (Hillman, 1987; Hillman et al., 1990; Ellis et al., 1991; in press).

Table 8.8 Endogenous metabolite formation rates of holotrichs isolated from animals receiving different diets[a]

Metabolite	Time post-feed (h)	*Isotricha* spp. SBP	*Isotricha* spp. Hay	*D. ruminantium* SBP	*D. ruminantium* Hay
Acetic acid[b]	2	1.62	0.57	1.14	0.31
	5	1.06	0.52	1.75	0.28
Butyric acid[b]	2	0.60	0.39	0.41	0.22
	5	0.62	0.61	0.31	ND[d]
Lactic acid[b]	2	0.43	0.46	0.55	0.04
	5	0.36	0.24	0.13	0.07
Gas[c]	2	86.2	57.8	104	13.7
	5	27.1	16.2	96.0	16.1

[a] The two diets were: SBP, molasses sugar beet pulp (600 g) and hay (300 g), or hay *ad lib*. Data based on Hillman (1987).
[b] Metabolite formation rates measured in μmol/mg protein/h.
[c] Gas formation rate measured in μl/mg protein/h.
[d] ND, not detected.

The production of acetate, butyrate, and propionate by suspensions of *Dasytricha* and *Isotricha* spp. in the presence of glucose in vitro was modulated by increasing concentrations of carbon dioxide (20–70% vol/vol in nitrogen) (Hillman et al., 1990). A cytosolic flux to butyrate in *D. ruminantium* was observed under a CO_2 partial pressure of 66 kPa (Ellis et al., 1991; in press). Rumen gases contain approximately 60% (vol/vol) carbon dioxide and it is possible that results derived in vitro under a nitrogen atmosphere do not reflect the true in situ rates. Endogenous metabolic rates were also affected by increasing carbon dioxide levels in the headspace gases. However, in the presence of low levels of hydrogen (2.5% and 5% vol/vol), butyrate formation was increased; although some other rates were lower under the hydrogen mixtures than in an all nitrogen atmosphere. Hydrogen, however, is only present in small quantities in the rumen gases (Hillman et al., 1985b, 1985c) and would be unlikely to exert a marked effect on metabolite production by the holotrichs under normal dietary conditions. The metabolic effects of oxygen are, however, more significant and will be considered in the following section.

Although the rumen is generally regarded as being a highly anaerobic environment in situ measurements with an oxygen electrode (Scott et al., 1983) and by mass spectrometry (Hillman et al., 1985b, 1985c) have confirmed that oxygen is present at low concentrations ($<3\ \mu$M) in the liquid phase of the rumen contents of domesticated animals. The levels were higher in the headspace gases of a fistulated cow. The concentration of oxygen in the rumen is not constant and falls to undetectable levels immediately after feeding. The available evidence suggests that oxygen enters the rumen continously and pre-feed levels are re-established relatively quickly (Scott et al., 1983). The physiological levels of oxygen in the rumen, however, have pronounced effects on ruminal methanogenesis (Scott et al., 1983).

Although the holotrichs are able to survive exposure to a gas atmosphere containing up to 1% oxygen (Yarlett et al., 1982) the presence of oxygen does influence the proportions of the metabolites formed by both genera. The rate of glucose consumption and acetate production by *D. ruminantium* is increased at low oxygen tensions of 1–3 μM (Ellis et al., 1991; in press). Prins and Prast (1973) observed that the addition of small amounts of oxygen (approximately 20 μmol) to in vitro *Isotricha* incubations resulted in increased acetate production whereas the amount of butyrate and hydrogen formed was reduced. There was no apparent change in lactate production. Similar results were obtained with *Dasytricha ruminantium* (Hillman, 1987; Hillman et al., 1988), although under 1–3 μM O_2 the ciliate will consume lactate when glucose is depleted (Ellis et al., 1991; in press). The effect of oxygen on hydrogen production has been studied in more detail and quantified by mass spectrometry (Hillman et al., 1985a; Yarlett et al., 1983b).

Hydrogen production by both holotrich genera has been quantified using mass spectrometry. Hydrogen evolution by *Dasytricha* and *Isotricha* spp. is reversibly inhibited at physiological oxygen concentrations. The extent of the inhibition increases almost linearly up to oxygen level of 1.5 μM, although oxygen tensions higher than approximately 3 μM irreversibly inactivate hydrogen production by *D. ruminantium*. The apparent K_i of oxygen for the inhibition of hydrogen formation by *D. ruminantium* and *Isotricha* spp. is 1.11 ± 0.09 and 1.68 ± 0.47 μM respectively. The metabolic contribution, and in particular hydrogen formation, by the holotrich ciliates in the rumen ecosystem will thus be affected by the environmental levels of oxygen.

Studies on protozoal metabolism have always used cell preparations that have either been thoroughly washed to minimize bacterial contamination or alternatively the suspensions have been pretreated for extended periods with antibiotics to suppress the bacterial activity. However, in the rumen ecosystem it is probable that the protozoa will interact with other microbial groups and these interactions will influence the range and proportions of the metabolites formed. Metabolic interaction between the holotrichs and hydrogen-utilizing methanogenic bacteria, for example, has been demonstrated (Hillman, 1987; Hillman et al., 1988).

The major group of hydrogen-utilizing bacteria in the rumen are the methanogens and metabolic interactions between the methanogenic bacteria and hydrogen-producing bacteria (Chen and Wolin, 1977; Scheifinger et al., 1975) and anaerobic fungi (Bauchop and Mountfort, 1981; Mountfort et al., 1982) are already well-described. The methanogens also closely associate with hydrogen-forming entodiniomorphid ciliates (Krumholz et al.,

Table 8.9 Effect of the presence of a methanogen on the rate of metabolite formation by *Isotricha* spp.[a]

Metabolite	*Isotricha* spp.	*Isotricha* spp. + *Methanosarcina barkeri*
Methane	ND	+
Acetic acid	33.3	26.6
Propionic acid	1.0	2.4
Butyric acid	31.8	24.2
Lactic acid	75.4	30.7

[a] The rate of formation is given in nmol product formed by 10^3 protozoa per h. The fermentation substrate was glucose. ND, not detected; +, trace.
Data based on Hillman (1987) and Hillman et al. (1988).

1983; Stumm et al., 1982) and this close physical association may facilitate interspecies transfer of hydrogen to the methanogen. Although no physical association between the rumen holotrichs and methanogens has been observed in the rumen two separate metabolic interactions are know to occur in vitro with *Methanosarcina barkeri* and the holotrich ciliates.

The methanogens are strictly anaerobic and the rate of ruminal methanogenesis is markedly affected by the presence of low levels of oxygen (Scott et al., 1983). Methanogenesis by *M. barkeri* was inhibitied by a gas phase oxygen concentration of approximately 2 μM; in the presence of holotrich protozoa the oxygen concentration could be increased tenfold to 20 μM before inhibition of methanogenesis occurred (Hillman, 1987; Hillman et al., 1988). It would thus appear that the holotrich ciliates are able to scavenge the low levels of dissolved oxygen which protects the oxygen-sensitive methanogen thereby enabling methanogenesis and hydrogen utilization to continue. In addition to their "protective" function, the hydrogen formed by the ciliates is utilized by the methanogen and does not accumulate during the combined incubation of protozoa and methanogen. A further consequence of the inter-species hydrogen transfer is that lactate and butyrate formation by the holotrich ciliates are reduced. The metabolite profile of glucose fermentation by *Isotricha* spp. is thus different in monoculture and in the presence of a methanogen (Table 8.9). The metabolic consequences of interactions between holotrichs and microbial groups capable of utilizing the products of the ciliate fermentation (e.g., lactate fermenter) warrant further examination because, as when the hydrogen utilizers are present, both the rate and the nature of the protozoal fermentation can be altered by the presence of the end-product utilizers. Such interactions undoubtedly occur in the rumen ecosystem since the holotrichs and many other diverse microbial groups are attracted to, and retained within, the digesta solids. These food particles provide a nutrient-rich matrix which aid microbial retention in the rumen and provide the site for many important interactions. Microorganisms having widely differing metabolic roles are able to adhere to the solids and maintain a close proximity to the members of the consortia responsible for generating the necessary substrates.

8.2 Oxygen Utilization

Historically the holotrich ciliates are described as being obligately anaerobic. Oxygen is, however, present in the rumen ecosystem and in animals fed twice daily, detectable levels of oxygen ($<3\ \mu M$) will be present for at least 18 hours during the day (Scott et al., 1983). The environmental oxygen concentrations will affect product formation (Lloyd et al., 1989) (see Section "Factors Affecting Metabolite Formation," above) although the holotrichs do have a limited capability to "metabolize" oxygen.

It is a frequent observation that holotrich protozoa retain motility and viability in wet mount preparations longer than many protozoa and concentrated suspensions will remain viable in the presence of air for short periods (e.g., Hungate, 1966). The protozoa will not, however, survive long periods under air and cannot be cultured unless anaerobiosis is maintained (Clarke and Hungate, 1966; Gaumont and Grain, 1967). The effects of oxygen tension on the viability of *Dasytricha ruminantium* in vitro have been quantified (Yarlett et al., 1982), and generally substantiate these descriptive observations. *Dasytricha ruminantium* had limited viability and survived only for a few minutes at oxygen tensions above 5 kPa; the oxygen tension in air is 21 kPa. However, at lower oxygen tensions the ciliate was able to survive prolonged periods and was successfully maintained for periods in excess of 12 hours at an oxygen tension of 1 kPa (Yarlett et al., 1982). The holotrich ciliates exhibit a respiratory activity at these low oxygen tensions and under such conditions are as equally capable of utilizing oxygen as the aerobic ciliate *Tetrahymena pyriformis* or the aerotolerant flagellate *Tritrichomonas foetus*. The K_m values have been determined using mass spectrometry (Hillman et al., 1985a) and by bacterial bioluminescence measurements (Lloyd et al., 1982) and under low oxygen tensions *Dasytricha* and *Isotricha* spp. have K_m values for oxygen in the range 0.3–2.5 μM (Table 8.10). Both genera are thus capable of scavenging oxygen at the low levels found in situ in the rumen, although *Dasytricha* has the lower apparent K_m and is thus more efficient. The apparent K_m O_2 values of the entodiniomorphid ciliates *Polyplastron multivesiculatum* and *Eudiplodinium maggii* are 1.7 ± 0.8 and $5.2 \pm 0.7\ \mu M$ respectively (Ellis et al., 1989a, 1989b). In view of the ability of the holotrichs to consume (respire) oxygen they should no longer be classed as obligately anaerobic but may be better considered as aerotolerant or oxyduric.

The stoichiometry of oxygen consumption and the resultant decreased hydrogen generation by *D. ruminantium* indicated that water was the major reaction product (Yarlett et al., 1983b). The nature of the terminal oxidase is not known although it has been confirmed that cytochromes are absent

Table 8.10 Oxygen affinities of the holotrich ciliates and some nonrumen protozoa

Protozoon	Apparent K_m for oxygen[a] Mass spectrometric determination	Bioluminescent measurement
Dasytricha ruminantium	0.34 ± 0.20 (39°C)	1.70 ± 0.30 (25°C)
Isotricha spp.	2.33 ± 0.48 (39°C)	ND
Tritrichomonas foetus KV	ND	1.08 ± 0.18 (19°C)
Tetrahymena pyriformis ST	ND	2.43 ± 0.08 (22.5°C)
Acanthamoeba castellanii (Neff strain)	ND	0.42 ± 0.05 (20°C)

[a] The temperature at which the determination was made is shown in parentheses. ND, no data. The data are based on Hillman et al. (1985a) and Lloyd et al. (1982).

(Yarlett et al., 1981). The terminal oxidases confer a high apparent K_m for oxygen and although these affinites, under low oxygen tensions, are similar to those of some aerobic protozoa, the holotrich oxidases are likely to differ from those found in the mitochondria of the aerobic protozoa. Results of studies on the effects of respiratory inhibitors on cell suspensions also showed that the respiration of *D. ruminantium* was unaffected by the mitochondrial inhibitor antimycin A, although sensitivity to quercitin, chloroquine, and other quinoline derivatives would indicate the presence of flavoprotein-mediated electron transport, analagous to that described in the aerotolerant flagellate protozoon *Tritrichomonas foetus* (Ohnishi et al., 1980).

Inhibitor susceptibilities and electron microscopic evidence (Yarlett et al., 1981, 1982, 1983a; Paul et al., 1989a) confirm that the holotrich ciliates do not possess organelles identifiable as mitochondria; the organelles initially described as mitochondria (Gaumont and Grain, 1967; Grain, 1966b) were on the basis of morphology and location hydrogenosomes. Under anaerobic conditions the reactions involved in the conversion of pyruvate to acetate and hydrogen are associated with the organelle. However, the hydrogenosome in the presence of oxygen will act as a respiratory organelle (Yarlett et al., 1982). Respiration by the hydrogenosomal fraction isolated from rumen entodiniomorphid ciliates has also been demonstated (Snyers et al., 1982). The respiratory activity of disrupted hydrogenosomes from *Dasytricha* was highest in the presence of coenzyme A and pyruvate indicating that oxygen uptake accompanies the conversion of pyruvate to acetyl coenzyme A. The respiration of pyruvate was also stimulated by phosphate, but other potential mediators in the release of bound coenzyme A could not be found. Respiratory control by ADP could not be demonstrated in hydrogenosomes from *D. ruminantium*. Oxygen has been shown to affect both hydrogen and acetate formation by the holotrich ciliates and the enzymes involved in these transformations are found in association with the hydrogenosome (Figure 8.4). The organelle has a primary role in carbon and oxygen metabolism in the cell and many of the metabolic effects ascribed to oxygen arise as a consequence of changed fluxes in interacting enzyme systems associated with the hydrogenosome.

The respiratory activity may also have a protective function although other enzyme systems capable of scavenging oxygen are present in the holotrichs, and these include NADH oxidase, NADH perioxidase, and catalase (Prins and Prast, 1973; Yarlett et al., 1981, 1983). Catalase activity was only present at low levels in *Isotricha* spp. and superoxide dismutase was not detectable in either genus. There is uncertainty over the presence of catalase activity (Prins and Prast, 1973) but it is believed that the NADH oxidase and NADH peroxidase present may aid in the protection against the deleterious effects of dioxygen and the products of oxygen reduction. Prins and Prast (1973) confirmed that ATP formation occurred during the oxidation of NADH by *Isotricha* spp., although the P/O ratio was low at 0.1–0.35. Hydrogen peroxide was formed as an intermediate product during the oxidation of NADH but was removed by the rate-limiting peroxidase activity. Both the NADH oxidase and NADH peroxidase are located in the cell cytosol although up to approximately one-third of the total activity may occur in association with the hydrogenosomal fraction. The inhibitor susceptibilities of the enzyme systems from both genera are also similar (Prins and Prast, 1973; Yarlett et al., 1981, 1982, 1983a).

The ability of the holotrich protozoa to take up oxygen may have important consequences in the rumen ecosystem. The ability of the holotrichs to "protect" oxygen sensitive methanogens has already been demonstrated in vitro (Hillman, 1987; Hillman et al., 1988) and their potential importance in helping to maintain the anaerobic conditions should not be neglected. Prins and Prast (1973) concluded that *Isotricha* spp. were responsible for only a very small proportion of the total oxygen uptake by rumen contents. However, these conclusions were based on calculations which only considered oxygen consumption by the cytosolic NADH oxidase. They estimated that an *Isotricha* population of 10^5 cells per ml would consume approximately 0.4 ml oxygen/liter of rumen contents per hour. Uptake by *Dasytricha* would be lower as the maximum oxygen consumption by the cytosolic fraction of this protozoon was shown to be 1.6–2.5 nmol oxygen per minute per mg protein (Yarlett et al., 1982); this corresponds to an approximate uptake rate of 0.6–1.0 pmol oxygen per cell per hour as compared with the value of 3.25 pmol oxygen per cell per minute calculated for *Isotricha* spp. (Prins and Prast, 1973). The holotrich population would thus only appear to be able to consume approximately 0.5 ml of O_2 per liter of rumen content per hour whereas the comparable rate of oxygen uptake by rumen contents was considered to be 32 ml per liter per hour.

However, values based only on cytosolic NADH oxidase activities will underestimate the true capacity of the protozoa. Mass spectrometric measurements showed that at physiological oxygen tensions (i.e., 1–1.5 μM) the rate of oxygen uptake by a cell population of 10^5 cells per ml, of either holotrich genus, was approximately 500 μmol per liter in 1 hour (Hillman et al., 1985a). Using these uptake rates Williams (1986) estimated that the total oxygen uptake by a *Dasytricha* or *Isotricha* population of 10^5 cells/ml in a bovine rumen of 100 liters volume would be almost 27 liters a day, a value that is considerably higher than the 1 liter value derived from NADH oxidase acitivity by Prins and Prast (1973). A smaller holotrich population containing 10^4 cells per ml of each genus would thus be responsible for 7% of the daily oxygen consumption by rumen contents. The contributions of the bacterial and protozoal populations to ruminal oxygen consumption are approximately the same (Ellis et al., 1989a,b)

A further aspect of the involvement of the holotrichs in ruminal oxygen consumption relates to their sequestration on the reticulum wall during the post-feed period (Abe et al., 1981). Oxygen is believed to diffuse across the wall and the epithelial population contains a higher proportion of facultative organisms (Cheng et al., 1979). Williams (1986) speculated that if the holotrichs were able to derive extra ATP from their respiratory activity, the migration and sequestration on the reticulum wall would be beneficial to the protozoa. The observed migratory and sequestration phenomenon would locate the cells in the region where oxygen availability was enhanced. The activities of the protozoa and other oxygen-utilizing microorganisms in the epthelial population would also aid the obligately anaerobic members of the rumen population by minimizing the ingress of oxygen across the wall of the organ and into the ruminal environment.

8.3 Nitrogen Metabolism

Protein Degradation Mixed protozoal preparations containing holotrich ciliates have been used to demonstrate and study protozoal proteolysis (Forsberg et al., 1984; Onodera and Kandatsu, 1970b; Shinchi and Kandatsu, 1980, 1981, 1983). Unfortunately such studies do not provide any information on the proteolytic activities of the individual protozoal genera present in the consortia. Onodera and Kandatsu (1970b) using ^{14}C-labelled casein demonstrated the uptake and degradation of a protein substrate by a holotrich cell suspension, while the observed ingestion and vacuolar digestion of chloroplasts was proposed to provide an amino acid source for incorporation into protozoal protein (Stern et al., 1977). Coleman (1983a) subsequently confirmed that proteolytic enzymes were present in the fourteen species of entodiniomorphid ciliates used in his survey, although it is only recently (Lockwood et al., 1988, 1989) that the presence of proteolytic enzymes in *Dasytricha ruminantium* and *Isotricha* spp. has been confirmed.

Cell lysates of both genera possessed proteinase activity with azocasein as substrate. The specific activity was higher in the *Dasytricha* preparation, although in both genera the proteinase activities were maximal at an acid pH in the range pH 4.0–pH 4.5 and were inhibited by cysteine proteinase inhibitors. Although the inhibitor susceptibilities of the two genera were

Table 8.11 Effect of proteinase inhibitors on the proteinase activities of holotrich ciliates determined with azocasein as substrate[a]

Inhibitor[a]	Percentage change in proteinase activity[b]	
	D. ruminantium	*I. intestinalis/ I. prostoma*
Leupeptin	−66	−92
Iodoacetic acid	−30	−67
1-chloro-4-phenyl-3-tosylamido-L-butane-2-one (TPCK)	−5	−71
N-α-p-tosyl-L-lysine-chloromethylketone (TLCK)	−58	−86
Phenylmethylsulfonyl fluoride (PMSF)	+13	−1
Pepstatin	+23	+14
1,10-phenanthroline	+10	−13
Dithiothreitol	+63	+148

[a] The inhibitor concentration used was 1 mM with the exception of pepstatin and leupeptin, which were tested at a final concentration of 100 μg/ml.
[b] A positive number indicates stimulation of activity and a negative value indicates inhibition. Data based on Lockwood et al. (1988).

similar, the inhibitor profiles and extent of inhibition were genus-dependent (Table 8.11); the proteinase activity of *D. ruminantium* was less susceptible to inhibition by TPCK, iodoacetic acid, or phenanthrolene.

Analysis of the proteolytic enzymes by gelatin-polyacrylamide gel electrophoresis revealed that multiple forms of proteinase were present in lysates of each genus (Lockwood et al., 1988, 1989). Although some similarities were apparent in the proteinase profiles, inter-genus divergence was also marked (Fig. 8.6). Both genera possessed a number of higher-molecular-weight proteinases and a single low-molecular-weight band. However, there were distinct differences in the molecular weights and inhibitor sensitivities of the electrophoretically separated proteinases. The activity bands in the *Isotricha* spp. gel differed in their inhibitor susceptibilities as the higher-molecular-weight enzymes were inhibited by phenanthroline and the low-molecular-weight band by cysteine proteinase inhibitors. However, all the activity bands separated in the *Dasytricha* gel were sensitive to leupeptin, iodoacetate, and phenanthroline. Thus although the metabolic characteristics of the two holotrich genera *Dasytricha* and *Isotricha* are generally regarded as being putatively similar, differences do exist in the activity and range of carbohydrase (see Section 8.1) and proteinase enzymes formed.

Studies with mixed rumen ciliates indicate that products of nitrogen metabolism can accumulate in the culture supernatant during in vitro in-

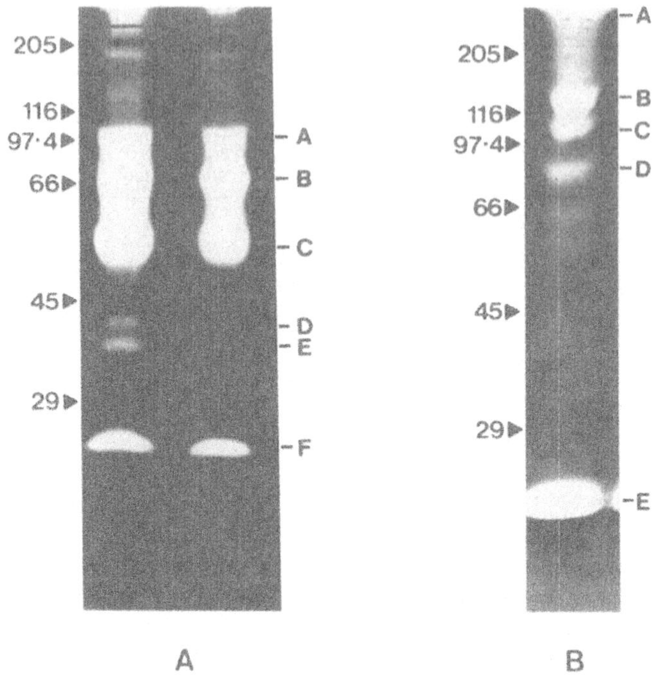

Fig. 8.6 Proteinase profiles of cell lysates of (A) *Isotricha* spp. and (B) *Dasytricha ruminantium* following gelatin-polyacrylamide gel electrophoresis. From Lockwood et al. (1988).

cubations. In addition to ammonia other products that are released from the cells include amino acids, peptides, and specific degradation products such as L-pipecolic acid, δ-aminovaleric acid, and 2-aminobutanoic acid (Onodera and Kandatsu, 1969a; Onodera et al., 1982; Tsutsumi et al., 1975). Onodera and Kandatsu (1970b) observed the specific accumulation of nitrogenous compounds in the culture supernatant of mixed holotrichs incubated with casein. Mixed protozoal suspensions also excrete amino acids (Sarmah and Ahuja, 1979). However, Heald and Oxford (1953) were the first to report the loss of soluble nitrogenous material from the holotrichs during incubations in vitro. Harmeyer (1971b) subsequently showed that *Isotricha* spp. in a 24 hour incubation period excreted one-quarter of the cellular nitrogen, of which some 30% was amino acids. The principal amino acids excreted were alanine, proline, glutamic acid, and aspartic acid. Smaller quantities of citrulline, ornithine, and ammonia were also detected. Some of the nitrogenous material released is active enzymic protein (Abou, Akkada and Howard, 1961; Williams, 1979b). The occurrence of ammonia as a metabolic product could indicate that amino acid dissimilation had taken place. However, there have been no specific studies on the catabolism of amino acids by the holotrich ciliates. Exogenously supplied amino acids are usually directly incorporated into protein without modification by *Isotricha* spp. (Wallis and Coleman, 1967) although Harmeyer (1971a) noted that some degradation of arginine and cystine occurred in incubations with isolated *Isotricha* spp. The amino acids tryptophan (Endo and Kandatsu, 1981), lysine (Onodera and Kandatsu, 1972), threonine, methionine (Merricks and Salsbury, 1976; Onodera and Ushijima, 1982; Patterson and Kung, 1988), arginine, citrulline, ornithine, and proline (Onodera et al., 1977b, 1983) are degraded by mixed rumen ciliate populations. The evidence concerning protozoal ureolytic activity is contradictory and individual protozoal genera have not been assessed for urease activity. There are reports of urea digestion and utilization by ciliate preparations (Bonhomme, 1975b; Ohba et al., 1981), although other studies have not been able to demonstrate ureolysis with washed protozoal suspensions (Onodera et al., 1977a).

Protein Synthesis Rumen ciliates are able to take up free amino acids (Ahuja and Sarmah, 1979; Onodera and Kandatsu, 1968, 1969b). Studies with the holotrich ciliates have been restricted to mixed preparations of *Isotricha prostoma* and *I. intestinalis*. Using ^{14}C-labelled amino acids Wallis and Coleman (1967) were able to confirm amino acid uptake by *Isotricha* spp., though Harmeyer (1971a) had previously been unable to detect any decrease in the extracellular amino acid concentration during the incubation period. The exogenously available ^{14}C-labelled amino acids were taken up singly and from complex mixtures at a constant rate during the 24 hour incubation period although the uptake rates were amino acid dependent (Table 8.12). The amino acids were assimilated without interconversion and no degradation products were detectable in the incubation supernatant (Wallis and Coleman, 1967). Higher amino acid concentrations (52 mM) proved to be toxic to the isolated protozoa in the study of Harmeyer (1971a). *Isotricha* spp. and *D. ruminantium* are also able to take up oligo-

Table 8.12 Amino acid uptake rates of *Isotricha* spp.

Amino acid	Uptake rate[a]
Alanine	0.039
Aspartic acid	0.042
Glutamic acid	0.022
Glycine	3.06
Isoleucine	1.68
Leucine	1.52
Lysine	0.32
Methionine	0.57
Serine	1.76

[a] Uptake rate is expressed as pmol per cell per day.
Data are based on Wallis and Coleman (1967).

peptides with *Dasytricha* being more active (Newbold et al., 1989b; Wallace et al., 1990). The rate of uptake decreased as the chain length of the alanine oligopeptide increased. Dialanine was taken up by both isotrichid genera three times more rapidly than penta-alanine. A variety of amino acids, peptides, polypeptides, and proteins occur in the rumen environment, and further studies are needed to identify the range of nitrogenous materials utilized by the holotrichs and to determine the factors that control their metabolic fate.

The holotrichs also have the capability to synthesize amino acids de novo. Incorporation of ^{14}C-label into the cellular protein fraction of either *Dasytricha* or *Isotricha* spp. has been achieved with ^{14}C-labelled monosaccharides (Williams, 1979a; Williams and Harfoot, 1976), sodium carbonate (Harmeyer, 1965), or sodium acetate (Harmeyer and Hekimoglu, 1968; Williams and Harfoot, 1976). Analysis of the cellular protein fraction of *Isotricha* spp. incubated with ^{14}C-labelled sodium carbonate confirmed that the label was widely distributed in the component amino acids. Approximately 95% of the ^{14}C-label was present in the five amino acids alanine, histidine, threonine, glutamate, and aspartate (Harmeyer, 1965). Protein synthesis by a cell-free polysome-containing extract of mixed rumen protozoa (Chesters, 1968) and *Entodinium* spp. (Hoshino and Sugiyama, 1974) has also been demonstrated. The cell-free protein-synthesizing system examined by Chesters (1968) was prepared from a protozoal suspension that contained over 80% holotrichs (mainly *Isotricha* spp.). The polysomes contained 82S ribosomes that could be dissociated to yield 40S and 55S ribosomes. Amino acid incorporation by the polysome fraction required adenosine triphosphate (ATP), guanosine triphosphate (GTP), and an energy source. Amino acid incorporation was influenced by the concentration of the cations present and was optimal with 5 mM Mg^{2+} and 50 mM K^+ at pH 7–7.5; further stimulation was achieved by adding 0.25 mM manganous chloride or spermidine and although the system was partially inhibited by cylcoheximide (10^{-4}M, 65%; 10^{-6}M, 48%) a chloramphenicol concentration of 100 μg/ml had no effect. The characteristics of the *Entodinium* spp. protein synthesizing system were similar (Hoshino and Sugiyama, 1974).

Ingestion of Bacteria Comparatively little is known about the engulfment and subsequent metabolism of bacteria by the holotrich ciliates, or the obvious ciliate-bacteria associations that have been observed. Imai and Ogimoto (1978) using Gram staining and fluorescent microscopy after staining with calcoflour-white estimated that there were approximately 10–40 bacterial cells per 200 μm^2 of the surface of both *I. prostoma* and *I. intestinalis*; the higher counts were obtained after Gram staining. The numbers of bacteria associated with the individual ciliates varied considerably and it appeared that the bacterial cells were trapped between the numerous cilia rather than adhering to the cell surface. Immunofluorescent staining with a polyvalent

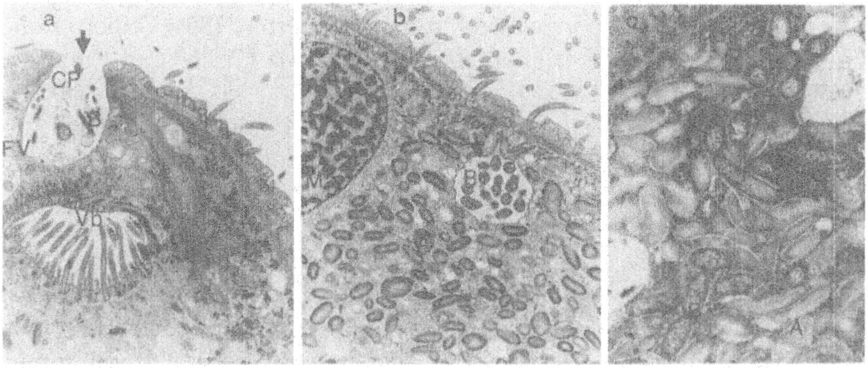

Fig. 8.7 Transmission electron micrographs of sections of *Dasytricha ruminantium* demonstrating the presence of free and vesicle-located bacteria. (a) Excreted bacteria (B) are visible in the cytoproct (CP); the vestibulum (Vb) and food vacuoles (FV) are also indicated. (b and c) Vesicles containing numerous bacteria 0.5–1 μm in length are present in the endoplasm. Other subcellular structures highlighted are: the macronucleus (Ma), hydrogenosomes (H), and amylopectin granules (A).

antiserum revealed that the ciliate surface-associated bacteria were cocci in chains of varying length and which appeared to be primarily *Streptococcus bovis* and *Ruminococcus albus*. However, although the holotrichs are hydrogen-formers, no externally associated methanogens have been reported (Vogels et al., 1980).

Bacteria are also evident within the holotrich cells and viable bacteria have been recovered and cultured following the lysis of *Isotricha prostoma* (Gutierrez, 1958). Intact and partially digested bacteria have been observed enclosed within digestive vacuoles in the holotrichs (Gutierrez, 1958; Gutierrez and Hungate, 1957; Paul et al., 1989a; Stern et al., 1977) and also outwith vacuolar structures and apparently "free" within the endoplasm (Paul, Butler and Williams, unpublished results; Figure 8.7). Nouzarede (1978) reported the division and sporulation by a *Bacillus*-like bacterium in the endoplasm of *I. prostoma,* and *Rickettsia*-like organisms have been observed in the macronucleus of *I. intestinalis* (Bretschneider and Van Vorstenbosch, 1964).

Bacterial predation by the rumen protozoa is the most important cause of bacterial protein turnover in the rumen (Wallace and MacPherson, 1987). However, early studies with the holotrichs demonstrated that they were unable to ingest *Lactobacillus bulgaricus* (Sugden and Oxford, 1952) and it was concluded that intact viable bacteria had no apparent function in the "life processes" of the holotrich ciliates (Appleby et al., 1956). However, it is now believed that bacteria do have an important fuction in holotrich nutrition and the ciliates cannot be successfully cultivated in the absence of viable bacteria (Clarke and Hungate, 1966; Gutierrez and Hungate, 1957). The holotrich ciliates do ingest both rumen (Gutierrez, 1958; Gutierrez and Hungate, 1957) and non-rumen strains of bacteria (Wallis and Coleman, 1967).

Starved cells of *Dasytricha ruminantium* selected and ingested small cocci, 0.5–0.8 μm in diameter, from a diluted preparation of mixed rumen bacteria. The bacteria agglutinated at the mouth and were gradually ingested. Small rod-shaped bacteria, 1.5 × 1.0 μm in size, were also observed within food vacuoles; growth of the ciliate in vitro was stimulated by two rumen bacterial isolates that were morphologically identical to the ingested strains (Gutierrez and Hungate, 1957). *Isotricha prostoma,* however, only selected and ingested certain rod-shaped bacteria from a mixed rumen bacterial preparation (Gutierrez, 1958; Gutierrez and Hungate, 1957). Gutierrez (1958) reported that, as with *D. ruminantium,* the bacteria adhered close to the mouth and were drawn into the cell by the cilia in the region of the cytostome. The bacteria were rapidly ingested into food vacuoles, the larger of which were 20–25 × 12 μm in size. Smaller vacuoles were also formed, some of which coalesced. Vacuole formation took approximately 4–14 minutes, and the contents of the vacuole were gradually digested. Larger bacteria (20 × 0.8 μm) were occasionally observed to be ingested without typical food vacuole formation and these became distributed throughout the endoplasm.

Cultivation studies indicated that *I. prostoma* contained at least 10–100 viable non-cellulolytic bacteria per ciliate (Gutierrez, 1958). Of 18 strains isolated into pure culture, 15 were Gram-negative rods and the remaining three were Gram-negative cocci. The isolates had different colonial and morphological characteristics, but four rod-shaped strains (0.8 × 1.5–4 μm) predominated; three of these when fed to the ciliate were rapidly ingested into food vacuoles by *I. prostoma.* Suspensions of mixed *Isotricha* spp. will also engulf and digest non-rumen bacteria. Wallis and Coleman (1967) demonstrated that the protozoa ingested *Escherichia coli* at an approximately linear rate of up to 3000 bacteria per protozoon per hour during a 24 hour in vitro incubation. The viability of the bacteria declined rapidly following ingestion and after 24 hours all of the bacteria had been rendered non-viable. The bacterial cells are extensively digested and amino acids derived from the bacterial protein are incorporated directly, without modification into protozoal cellular protein. However, the protozoa appeared to preferentially use free amino acids, as amino acid incorporation from the bacterial protein was partially inhibited when the same acid was available exogenously in the surrounding medium. Not all of the bacterial cell components were utilized by the protozoa and some release of soluble digestion products occurred (Wallis and Coleman, 1967). The holotrich ciliates are thus able to derive some of their nitrogenous requirements from the digestion of ingested bacteria, although Coleman (1979b, 1980) calculated that protein synthesis from retained bacterial amino acids would be sufficient to enable cell division once every 120 hours, whereas sufficient protein could be derived from de novo synthesis to allow division every 13–22 hours. In the rumen *I. intestinalis* and *I. prostoma* have been estimated to consume approximately 0.5 g of bacteria per day; starved ciliates may engulf bacteria more rapidly and the total quantity ingested may in fact be greater (Coleman, 1975a).

8.4 Lipid Metabolism

Lipid Composition Analysis of the isolated lipids from mixed holotrich protozoa indicated that 70% of the lipid fraction was phospholipid. The principal phospholipid components have been identified as phosphatidylethanolamine (21%), phosphatidylethanolamine plasmalogen (22%), and phosphatidylcholine (28%) (Katz and Keeney, 1967; Keeney, 1970). All the phospholipid fractions contained significant amounts of branched chain and unsaturated fatty acids located in the $\beta(2)$ position. A more restricted range of fatty acids occurred in the $\alpha(1)$ position; chemical degradation of the plasmalogen indicated that the vinyl ether linkage was in the α position. The unusual amino acid, 2 aminoethylphosphonic acid (AEP) which has been identified as a component of the phospholipids of *Entodinium caudatum* (Dawson and Kemp, 1967), is also present in mixed *Isotricha* spp. preparations (26–223 mg per g total N; Abou Akkada et al., 1968b; Whitelaw et al, 1983) although the identity of the AEP-containing lipid has to be established. Phosphatidyl-N-(2-hydroxyethyl)-alanine may be present in the uncharacterized holotrich phospholipid fraction as this particular phospholipid has been characterized in lipids extracted from holotrich-containing protozoal preparations (Kemp and Dawson, 1969a, 1969b).

The non-phospholipid components that have been identified include sterols (18.5%), unesterified free fatty acids (32%), aliphatic alcohols (6.8%), monoglycerides (4.8%), diglycerides (10.1%), and hydroxyacids (6.8%). The sterols and unesterified fatty acids accounted for approximately half of the fraction. The two principal sterols would appear to be cholesterol and α-spinasterol (Katz and Keeney, 1967). The major fatty acid in all lipids was palmitic acid; lesser amounts of the C14–C18 straight-chain and branched-chain fatty acids were detected (Katz and Keeney, 1967; Williams and Dinusson, 1973a). The fatty acid composition of the lipid fractions isolated from the holotrich ciliates is summarized in Table 8.13.

Metabolism of Dietary Lipids There have been very few studies with the holotrich ciliates and the extent and importance of these protozoa in ruminal lipid metabolism is, therefore, largely speculative. Exogenously supplied C16 and C18 saturated and unsaturated straight-chain fatty acids are taken up and assimilated or metabolized by the holotrich ciliates (Demeyer et al. 1978; Emmanuel, 1974; Girard and Hawke, 1978; Gutierrez et

Table 8.13 Fatty acid composition of holotrich lipids

Lipid component	Fatty acid[a]												
	14:0[b]	14:0br	15:0	15:0br	16:0	16:0br	17:0	17:0br	18:0	18:0br	18:1	18:2	18:3
Total lipid													
Holotrich protozoa	2.8				37.3				9.1		18.2	10.5	4.0
Mixed rumen ciliates	1.2	tr	2.5	3.0	41.0	1.5	1.2	3.2	10.5	ND	15.4	17.0	1.9
Phospholipids													
Phosphatidylcholine													
Position 1(α)-fatty acid	1.1	ND	6.6	ND	74.0	tr	3.0	tr	15.2	ND	tr	ND	ND
Position 2(β)-fatty acid	1.5	0.9	2.2	8.5	8.9	2.0	0.8	3.8	1.4	ND	17.9	34.8	17.1
Phosphatidylethanolamine													
Position 1(α)-fatty acid	tr	tr	6.2	tr	80.3	1.8	2.5	tr	9.1	ND	tr	ND	ND
Position 2(β)-fatty acid	1.1	1.8	1.3	16.9	7.2	10.3	tr	18.6	0.8	ND	9.9	20.1	11.4
Phosphatidylethanolamine plasmalogen													
Position 1(α)-fatty aldehyde	7.3	13.9	6.7	27.5	17.5	2.2	tr	2.6	3.4	ND	10.6	2.9	ND
Position 2(β)-fatty acid	tr	tr	2.1	tr	59.5	4.1	tr	13.5	6.9	ND	8.6	5.3	tr
Other lipid fractions													
Free fatty acids	1.3	tr	6.6	3.3	50.1	2.1	2.3	3.9	8.3	tr	5.9	10.2	6.0
Monoglyceride	1.4	tr	3.0	4.4	27.7	2.4	tr	5.1	10.9	tr	18.3	21.5	5.3
Diglyceride	1.3	0.8	4.4	5.9	40.3	2.1	1.4	4.4	10.5	tr	10.0	13.9	4.7
Unknown glyceride													
Fatty acids	3.1	0.9	4.3	4.9	32.3	2.0	1.1	4.2	7.7	1.7	14.9	15.2	7.7
Aldehydes	3.5	2.2	7.4	14.5	40.5	2.3	tr	3.4	8.4	ND	15.2	2.4	ND
Alcohol	6.0	5.7	12.1	19.5	32.9	4.2	tr	3.3	8.5	ND	2.9	ND	ND

[a] Fatty acid composition data compiled from Emmanuel (1974); Katz and Keeney (1967); Williams and Dinusson (1973a) and adapted from Williams (1986). Individual fatty acid values are expressed as a percentage of the total fatty acids of the various lipid classes. ND, not detected; tr, trace amount present, insufficient to quantify.

[b] Number of carbon atoms in fatty acid : number of unsaturated bonds; br, branched-chain fatty acid.

al. 1962; Williams et al, 1963). Mixed *Isotricha* spp. incorporated linoleic acid at a rate of almost 3 pmol cell per hour. Linoleic acid concentrations of 0.014 mM are toxic to isolated holotrichs (Girard and Hawke, 1978).

The long-chain fatty acids taken up by the holotrichs are usually assimilated into the free fatty acid pool and lipid components of the cell, although oleic acid stimulated gas (carbon dioxide and hydrogen) but not short-chain volatile fatty acid formation by either *Isotricha prostoma* (Gutierrez et al, 1962) or *I. intestinalis* (Williams et al. 1963). Similarly, some short-chain (C2–C6) fatty acids, and the methyl derivatives of lauric acid (C12) and myristic acid (C14) were also shown to stimulate gas production by the individual *Isotricha* species. The triglyceride tributyrin, but not tripalmitin, also stimulated the fermentative acitvities and with *I. intestinalis* volatile fatty acids were recovered by steam distillation at the end of the incubation. There would thus appear to be some limited catabolic activity with exogenously supplied free fatty acids. However, most of the dietary long-chain fatty acids are present as constituents of phospholipids and the *Isotricha* spp. appear to have only limited ability to hydrolyze complex lipids (Girard and Hawke, 1978). After incubation with ^{14}C-labelled phosphatidylcholine only 10% of the substrate was recovered as free fatty acids; 89% of the incorporated label was present as phospholipid. However, in the presence of bacteria the proportion of the substrate taken up (42%) had increased threefold; two-thirds was present as intact phospholipid and 30% as free fatty acid. The implication is that *Isotricha* spp. are not primary degraders but are able to take up products arising from the activities of the lipolytic population and plant lipases. The holotrichs do, however, ingest and degrade chloroplasts (Stern et al., 1977). The ciliates form galactoside-hydrolyzing enzymes (Williams et al., 1984) and may be able to breakdown and utilize the galactosylglyceride (galactolipid) component of chloroplast lipids.

The involvement of the holotrich ciliates in biohydrogenation (i.e., the saturation of polyunsaturated fatty acids) has not been conclusively established. Wright (1959) demonstrated that a washed protozoal suspension, comprising mainly holotrichs and *Epidinium* spp. was able to hydrogenate linoleic acid, linseed oil, and chloroplast lipids. *Isotricha prostoma* (Gutierrez et al., 1962) and *I. intestinalis* (Williams et al., 1963) have been shown to hydrogenate oleic acid to stearic acid, while mixed suspensions of *Isotricha* spp. hydrogenated exogenously supplied linoleic acid before incorporation into cellular lipids occurred (Demeyer et al., 1978). However, in three other separate studies (Abaza et al., 1975; Chalupa and Kutches, 1968; Girard and Hawke, 1978) it was concluded that in the absence of bacteria the holotrich ciliates from the family Isotrichidae possessed little or no ability to hydrogenate added unsaturated fatty acid substrates. However, in co-incubations with bacteria, biohydrogenation has been observed (Chalupa and Kutches, 1968; Girard and Hawke, 1978) and in the latter report synergy was apparent in the conversion of oleic acid to stearic acid. The nature of the metabolic interaction necessary to effect biohydrogenation has not been established, although the holotrichs may function as hydrogenogens for the process.

The ability of the holotrichs to desaturate fatty acids is also equivocal. Emmanuel (1974) concluded that mixed rumen protozoa formed octadecenoic acids by direct desaturation of saturated acids. In addition, Abaza et al. (1975) observed a 90–100% increase in the iodine value of oleic acid during control incubations with washed suspensions of ciliates to which fermentable substrates had not been added. The endogenous desaturation activity over the 8 hour incubation period was enhanced by methionine. However, despite this endogenous desaturation activity in mixed protozoal preparations Abaza et al. (1975) were apparently not able to observe fatty acid desaturation with separated suspensions of either holotrich or entodiniomorphid protozoa.

Lipid Synthesis The holtrich ciliates are able to utilize preformed dietary lipids as well as having the capability to synthesize some cellular lipids de novo from various precursors, although it is likely that the lipid requirements of the protozoa are principally fulfilled by the assimilation of dietary and bacterial lipids. Short-chain volatile fatty acids are incorporated into structurally related long-chain fatty acids by holotrich-containing suspensions of rumen ciliate protozoa. Acetate and butyrate were assimilated principally into palmitate and other fatty acids containing an even number of carbon atoms, whereas propionate was incorporated into straight-chain acids possessing an odd number of carbon atoms, principally pentadecenoic acid and heptadecenoic acid (Emmanuel, 1974; Saini et al., 1977). Fatty acids were elongated by 2-C atoms and shortened by α-oxidation (1C-atom) or a restricted β-oxidation (2-C atoms). Monounsaturated octadecenoic acids were produced by direct desaturation of the saturated fatty acid whereas polyunsaturated fatty acids and branched long-chain fatty acids are incorporated into the protozoal lipids directly from dietary sources or following the digestion of ingested bacteria (Emmanuel, 1974).

The holotrich protozoa are able to sythesize fatty acids de novo using either acetate or monosaccharides as precursors. However, long-chain fatty acids are also taken up from the surrounding medium and assimilated into complex cellular lipids (phospholipids, sterol esters) and the free fatty acid pool (Demeyer et al., 1978; Girard and Hawke, 1978; Gutierrez et al., 1962; Williams et al., 1963). Biochemical studies have confirmed that *D. ruminantium* and *Isotricha* spp. (Demeyer et al., 1978; Harmeyer and Hekimoglu, 1968; Williams and Harfoot, 1976) incorporate ^{14}C-labelled acetate into the cellular lipids. Some incorporation of acetate into the C18 straight-chain polyunsaturated fatty acids linoleic acid and linolenic acid has been demonstrated by Viviani and Borgatti (quoted in Viviani, 1970). Approximately 20–30% of the [^{14}C] derived from [I-^{14}C]acetate present in the cellular components of *D. ruminantium* was located in the lipid fraction (Williams and

Harfoot, 1976). Direct incorporation from monosaccharides into polar lipids was much lower, however, and only 1.0% and $< 0.5\%$ of the label derived from [U-^{14}C]galactose and [U-^{14}C]glucose, respectively, was present in the lipid components (Williams, 1979a; Williams and Harfoot, 1976). The labelled glucose carbon was recovered principally in the glycerol component of the lipid fraction (Demeyer et al., 1978).

The synthesis of more complex structural lipids has only been examined with mixed holotrich suspensions that contained mainly mixed *Isotricha* spp. (Demeyer et al., 1978; Girard and Hawke, 1978). Linoleic acid taken up by the *Isotricha* spp. was utilized for synthesis of phospholipids and sterol esters. The highest proportion of the linoleic acid used was incorporated into phosphatidylcholine with smaller amounts being assimilated into the phosphatidylethanolanine and sterol ester moieties respectively. Some hydrogenation of unsaturated long-chain fatty acids has been reported to take place after uptake and prior to incorporation (Demeyer et al., 1978). In *I. prostoma* and *D. ruminantium* glycerol enters the principal diacylphosphoglycerides (diacylphosphatidylcholine, diacylphosphatidylethanolamine), and plasmalogens (ethanolamine plasmalogen) via the common intermediate *sn*-glycerol-3-phosphate rather than dihydroxyacetone phosphate (Prins and Van Golde, 1976). The holotrich ciliates, the anaerobic rumen bacterium *Selenomonas ruminantium*, and another strictly anaerobic microorganism, *Vibrio parvula*, thus utilize the same pathway for the incorporation of glycerol into plasmalogens.

The major sterol in the holotrichs, cholestanol, was presumed to be derived directly from, or by modification of, dietary sterols (Katz and Keeney, 1967; Keeney, 1970). However, some of the unesterified linoleic acid taken up by *Isotricha* spp. is incorporated into sterol esters (Demeyer et al., 1978), and the assimilation of acetate and mevalonate into cholestanol by mixed rumen ciliate preparations may indicate that this particular sterol is formed by de novo synthesis (Hino and Kametaka, 1978). Others (e.g., campestanol, stigmastanol) were apparently formed from the hydrogenation of exogenous unsaturated C28 and C29 dietary sterol precursors by mixed ciliate preparations.

8.5 Metabolism of Nucleic Acids

Although the holotrich ciliates ingest plant and bacterial cells, there is no information on their ability to degrade nucleic acids. The entodinomorphid ciliates do not synthesize purines or pyrimidnes de novo (Coleman, 1979b, 1980; see also Chapter 6). It is likely that the holotrichs are similar and obtain the nucleotide precursors for nucleic acid synthesis by assimilation and from the digestion of ingested bacterial nucleic acid material.

9

Factors Affecting Protozoal Populations In vivo

Young ruminants or ciliate-free adult animals can only become infected with a given rumen ciliate if this species is already present in a normal adult animal with which the young animal or ciliate-free adult has very close contact. There is no resistant phase in the life cycle of rumen ciliate protozoa (Strelkow et al., 1933). In the authors' experience some of the less common large ciliates can disappear from individual sheep, and presumably cattle, for no obvious reason. If this occurs, and if, during the normal management of a flock or herd, animals containing such species are culled, then this protozoal species is lost until a new animal containing this species is reintroduced. The absence of any protozoal species from a group of animals may mean that conditions, such as the nature of the host species or its food, are unsuitable for the growth of the protozoon or that the protozoon is unavailable for infection. The absence of *Polyplastron multivesiculatum* from New Zealand sheep (Clarke, 1964) may well be the result of its lack of introduction into that country. When sheep were first introduced, they would have had to endure the trauma of a long sea voyage with far-from-ideal feeding conditions and large protozoa could have been lost due to, for example, an acidosis. Those working with experimental animals that have been used for a variety of nutritional experiments should also be careful that, in a previous experiment, the large ciliates have not all been killed by some treatment.

9.1 Infection of Young Ruminants

Ciliate protozoa can normally be seen in the rumen of young lambs and calves in the United Kingdom two weeks after birth, provided that the ration favors a high rumen pH (>6.0 and preferably >6.5) (Eadie, 1962a). Entodinia become established first, followed by, in lambs, the larger entodiniomorphs and the holotrichs; in calves the holotrichs tend to become established before the large entodiniomorphs. Early inoculation with rumen contents confers no advantage over natural infection from the dam. In contrast, Naga et al. (1969) in studies in Egypt, where cattle and buffalo calves are fed on milk for 30 days followed by a starter containing corn grains, found that the protozoa did not appear for 30 days in cow calves. However, if the calves were drenched with buffalo, but not cow, rumen contents early on, then protozoa were visible after 4 days. These authors also found that buffalo calves became infected before cow calves.

Although it is always assumed that the young obtain their rumen ciliates from the dam, Coleman and Kemp, quoted by Coleman (1980), obtained evidence based on protozoal population type (see next section) that the larger protozoa at least were obtained from the flock at large rather than the dam. In the authors' experience of lambs kept under normal farm conditions at Babraham, England, it sometimes takes 3–6 months for the animals to obtain their full adult protozoal population. In three successive years, 1973–1975, 10 out of 20 lambs at 2–4 months, 17 out of 61 at 3–5 months, and 17 out of 22 at 2–4 months contained no large entodiniomorphid protozoa. Two five-year-old breeding ewes out of a flock of 28 contained no large protozoa.

9.2 Interrelationships Between Protozoal Species

Although *Entodinium* spp. and the holotrich protozoa are found in most sheep, cattle, and goats under normal farm conditions, some of the larger entodiniomorphid protozoa are mutually exclusive and population types have been described (Eadie, 1962b, 1967; Abou Akkada et al., 1969; Imai et al., 1978, 1979a). The A-type population is characterized by the presence of *Polyplastron multivesiculatum* which is usually, but not invariably, accompanied by *Diploplastron affine*. The B-type population is characterized by *Epidinium* spp. and/or *Eudiplodinium maggii*. The K-type is characterized by the presence of *Elytroplastron bubali* and O-type by none of these. There is, however, disagreement between different authors as to which other protozoa are normally found with the basic A-, B-, or K-type species, although these differences may reflect host differences and variations due to geographical distribution (Table 10.1). It is likely that the existence of the A and B types depends upon the predatory activity of *Polyplastron multivesiculatum* which engulfs and can completely eliminate from a rumen *Epidinium* spp.,

Table 9.1 A classification scheme for protozoal population types found in cattle, sheep, and goats

Type A	Type B	Type O	Type K[e] (cattle only)
Entodinium	*Entodinium*	*Entodinium*	*Entodinium*
Holotrichs	Holotrichs	Holotrichs	Holotrichs
Polyplastron multivesiculatum	*Epidinium* (except *E. tricaudatum*[a])		*Elytroplastron bubali*
Diploplastron affine	*Eudioplodinium maggii*		*Eremoplastron rostratum*
Ophryoscolex (cattle and sheep[a,b,c,d])	*Eremoplastron*[c,f]		*Diplodinium*
Ostracodinium (cattle and sheep[e])	*Ostracodinium* (sheep[c])		
Eremoplastron (rarely cattle[e])	*Ophryoscolex* (goat[e])		
Diplodinium	*Diplodinium*		
Enoploplastron[c]	*Enoploplastron* (rarely cattle[b])		
Eodinium[e]	*Metadinium*[a]		
	Eodinium[e]		

[a] From Abou Akkada et al. (1969).
[b] From Bonhomme-Florentin et al. (1978).
[c] From Eadie (1962b, 1967).
[d] From G.S. Coleman (unpublished observations).
[e] From Imai et al. (1979a).
[f] From Ogimoto and Imai (1981).

Eudiplodinium maggii, *Eremoplastron bovis*, and *Ostracodinium* spp. However, it is of interest that Abou Akkada et al. (1969) found that *Epidinium tricaudatum*, unlike *Epidinium caudatum*, could coexist with *Polyplastron multivesiculatum*.

The accidental introduction of *Polyplastron multivesiculatum*-containing rumen fluid through a rumen cannula into B-type rumen fluid results in the complete elimination of the B-type protozoa, so it is surprising that any of these protozoa exist. However, as flocks of sheep exist with approximately equal numbers of animals containing either A- or B-type protozoal populations, there must be some mechanism which allows *Epidinium* spp. but not *Polyplastron multivesiculatum* to reach the rumen of certain lambs. We have no idea what this mechanism is but it is tempting to speculate there must be some factor in the saliva of some sheep that kills or inactivates *Polyplastron multivesiculatum*. If there is, then this factor could be genetically determined.

9.3 Host Specificity

Specificity Within One Ruminant Species In any group of ruminants of one species kept in one place, the principal differences in the protozoal populations are those of the A-, B-, K-, or O-types described above. However, within these groups individual protozoal species, especially *Diplodinium* and *Ophryoscolex* spp., can appear and disappear from a group of animals kept under constant conditions for no apparent reason.

Individual sheep, and possibly other ruminants, can develop an apparent "immunity" to certain protozoal species. If a sheep is defaunated and refaunated with a single protozoal species and then defaunated again, it is very difficult, if not impossible to refaunate with the same protozoon by inoculating with cultured organisms through the rumen cannula (Coleman, 1987b). Protozoa of other genera will normally, but not invariably, grow in such animals.

Specificity Between Different Host Species It is not proposed to give detailed information on the many studies (Table 9.2) giving protozoal populations in different host species, but rather to concentrate on those investigations where different hosts in the same area have been shown to contain different species and on the results of transfaunation experiments.

There is good evidence that *Polyplastron multivesiculatum, Diploplastron affine, Epidinium* spp., and *Eudiplodinium maggii* are commonly, but not universally, present in domestic sheep and cows (Eadie, 1962b: Abou Akkada et al., 1969; Imai et al., 1979a). Imai et al. (1978, 1979a) found that, in Japanese animals, *Entodinium rostratum, Eodinium posterovesiculatum, Eremoplastron rostratum, Charonina equi,* and three *Ostracodinium* spp. were present in cattle but not sheep, with the reverse being true for *Ophryoscolex caudatus* and *Enoploplastron triloricatum* in animals on the same pasture. On introduction of cattle ciliates into a defaunated sheep rumen, the *Ostracodinium* spp. died, as did *Isotricha prostoma* and *Diplodinium dentatum*. If the same experiment was made with a normally faunated sheep, *Ostracodinium gracile,* but not *O. clipeolum,* survived. In contrast, *Isotricha prostoma, Dasytricha ruminantium, Charonina ventriculi,* and *Entodinium dilobum* all died on introduction as did *Polyplastron multivesiculatum* which was present in the sheep originally. All these protozoa were commonly found in Japanese sheep and cows. These results show how difficult it is to obtain meaningful results in this kind of study.

The evidence available suggests that *Eodinium, Ostracodinium, Eremoplastron,* and *Metadinium* spp. are probably associated with cattle rather than sheep (Clarke, 1964; Naga et al., 1969; Abou Akkada et al., 1969; Imai et al., 1978, 1979a) with the reserve true for *Ophryoscolex* spp. Goats in Japan (Imai et al., 1978, 1979a) contain *Polyplastron multivesiculatum, Ophryoscolex* spp., *Epidinium* spp., and *Eudiplodinium maggii* but not *Eremoplastron* spp. Eadie (1962b) believed that goats and kids had a strong similarity to sheep.

Dehority (1978) investigated this problem by inoculating protozoa-containing rumen contents from an alfalfa-fed steer into the rumens of six sheep reared protozoa-free from birth. Three of the sheep were fed on alfalfa (same ration as the steer) and three on concentrates. It was found that,

Table 9.2 Locations and references to the occurrence of entodiniomorphid protozoa in ruminants and some nonruminant species

Host animal	Location	Reference
Antelopes		
General	Africa	Buisson, 1923b, 1924; Dogiel, 1925a, 1932
Bleesbok (*Damaliscus dorcas phillipsi*)	Africa	Van Hoven, 1978
Bohor reedbuck (*Redunca* spp.)	Africa	Van Hoven, 1983
Bushbok (*Tragelaphus scriptus*)	Africa	Buisson, 1924; Dogiel, 1932
Coke's Hartebeest (*Bubalus cokei*)	Africa	Dogiel, 1927
Dik-dik (*Madoqua* spp.)	Africa	Dogiel, 1927
Forest duiker (*Cephalophus grimmia*)	Africa	Buisson, 1924
Harvey's duiker (*Cephalophus harveyi*)	Africa	Hoppe et al., 1981
Impala (*Aepyceros melampus*)	Africa	Dogiel, 1927
Lichtenstein's hartebeest (*Bubalus lichtensteini*)	Africa	Buisson, 1923b
Sable antelope (*Hippotragus niger*)	Africa	Van Hoven et al., 1979
Springbok (*Antidorcas marsupialis*)	Africa	Wilkinson and Van Hoven, 1976a
Steinbok (*Rhaphiceros* sp.)	Africa	Dogiel, 1927
Tsessebe (*Damaliscus lunatus lunatus*)	Africa	Van Hoven, 1975
Bison (*Bison bison*)	USA	Pearson, 1967; Towne et al., 1988b; Varel and Dehority, 1989
Dromedary (*Camelus dromedarius*)	Turkestan	Dogiel, 1926b, 1927, 1928
Capybara (*Hydrochoerus hydrochaeris*)	USA	Dehority, 1987
Cattle		
Bali (*Bos javanicus domesticus*)	Indonesia	Imai, 1985
Chinese	China	Hsiung, 1932
European (*Bos taurus*)	Japan	Imai et al., 1978, 1979a, 1982; Hukui, 1940
	New Zealand	Clarke, 1964
	Russia	Dogiel, 1927
	USA	Becker and Talbott, 1927; Abou Akkada et al., 1969; Dehority, 1978; Towne et al., 1988a
	Yugoslavia	Wertheim, 1935b
Gaur (*Bos gaurus*)	India	Kofoid and Christenson, 1933
Korean	Korea	Han, 1984
Zebu (*Bos indicus*)	Brazil	Dehority, 1986b
	India	Kofoid and MacLennan, 1930, 1932, 1933
		Banerjee, 1955
		Bhatia, 1936
	Philippines	Shimizu et al., 1983
	Sri Lanka	Kofoid and MacLennan, 1930, 1932, 1933
		Imai, 1986
	Thailand	Imai and Ogimoto, 1984
	West Africa	Latteur and Sulten, 1972; Bonhomme-Florentin et al., 1978
Deer		
Fallow (*Dama dama*)	Czechoslovakia	Sladecek, 1946
Hog (*Cervus porcinus*)	USA	Krascheninnokow, 1955
Mule (*Odocoileus hemionus*)	USA	MacLennan, 1935
Muntjak (*Muntiacus muntjak*)	India	Banerjee, 1955
Red (*Cervus elaphus*)	Czechoslovakia	Kubikova, 1935
		Wertheim, 1935c
		Sladecek, 1946
	Scotland	Hobson et al., 1975/6
Roe (*Capreolus capreolus*)	Czechoslovakia	Wertheim, 1933b; Sladecek, 1946
	New Zealand	Clarke, 1968
Sambar (*Rusa unicolor*)	India	Banerjee, 1955

Table 9.2 *Continued*

Host animal	Location	Reference
Wapiti (*Cervus canadensis ludorfi*)	Russia	Dogiel, 1927
White-tailed (*Odocoileus virginianus*)	USA	Pearson, 1965; Zielyk, 1961
African elephant (*Loxodonta africana*)	Africa	Eloff and Van Hoven, 1980
Giraffe (*Giraffa camelopardalis*)	West Africa	Noirot-Timothée, 1963
	South Africa	Kleynhans and Van Hoven, 1976
Goats		
Chamois (*Rupicapra rupicapra*)	Central Europe	Wertheim, 1934
		Christl, 1955
Domestic (*Capra* spp.)	Iran	Dogiel, 1927
	India	Bhatia, 1936
	Japan	Imai et al., 1978, 1979a
Ibex (*Capra ibex*)	Central Europe	Wertheim, 1933a; Noirot-Timothée and Couturier, 1957
		Christl, 1958a, 1958b
Indian (*Capra hircus*)	India	Das Gupta, 1935; Lubinsky, 1957d
Serow (*Capricornis crispus*)	Japan	Imai et al., 1981a
Hippopotamus (*Hippopotamus amphibius*)	Africa	Thurston and Noirot-Timothée, 1973; Van Hoven, 1974
Kaffir buffalo (*Syncerus caffer*)	Africa	Dogiel, 1932
Guanaco (*Lama glama*)	South America	Lubinsky, 1964
Alaskan moose (*Alces americana*)	Alaska	Dehority, 1974
	Finland	Westerling, 1969
	Russia	Dogiel, 1934
Musk-ox (*Ovibos moschatus*)	Alaska	Dehority, 1974
	Canada	Tener, 1965; Dehority, 1985
Reinder (*Rangifer tarandus tarandus*)	Alaska	Dehority, 1975a
	Canada	Lubinsky, 1958a, 1958b
	Finland	Westerling, 1970; Syrjala et al., 1973
	Russia	Dogiel, 1925d, 1935
	Scotland	Hobson et al., 1975/76
Caribou (*R. tarandus granti*)	Alaska	Dehority, 1975b, 1986c
Rhinoceros		
Black (*Diceros bicornis*)	Africa	Buisson, 1923c; Van Hoven et al., 1988
White (*Ceratotherium simum*)	Africa	Van Hoven et al., 1987, 1988
Sheep		
Bighorn (*Ovis canadensis sierrae*)	USA	Bush and Kofoid, 1948
Dall Mountain (*Ovis dalli*)	Alaska	Dehority, 1974
Domestic	Belgium	Devuyst et al., 1973
	China	Hsiung, 1931
	India	Banerjee, 1955
	Japan	Imai et al., 1978, 1979a
	Russia	Dogiel, 1927, 1934
	Scotland	Eadie; 1957, 1962a, 1962b; Hobson et al., 1975/76
	USA	Dehority, 1978
Ovis lervis	USA	Bush and Kofoid, 1948
Ovis nivicola nivicola	USA	Bush and Kofoid, 1948
Ovis orientalis cycloceros	USA	Bush and Kofoid, 1948
Water buffalo (*Bubalus bubalis*)	Brazil	Dehority, 1979
	Egypt	Naga et al., 1968
	Indonesia	Imai, 1985
	Philippines	Shimizu et al., 1983
	Russia	Dogiel, 1927
	Taiwan	Imai, 1981; Imai et al., 1981; Fujita et al., 1979
	Thailand	Imai and Ogimoto, 1984
	Turkestan	Dogiel, 1928

after 15 days, the percentage species composition in the rumen of the alfalfa-fed sheep was similar to that in the donor steer, but that only 12 species (compared with 26) established in concentrate-fed sheep. The difference was almost entirely due to the failure of *Diplodinium* spp. to grow. It was of interest that *Entodinium laterospinum,* which was present in the steer in numbers too small to be observed, established in concentrate- but not alfalfa-fed sheep. Dehority's results also confirmed that the present authors' observations that ciliate species can appear in, and disappear from, a rumen for no apparent reason.

Buffalo in Egypt and the Far East contain *Eudiplodinium maggii, Diplodinium* spp., and *Ostracodinium* spp., but tend to lack *Isotricha* spp., *Elytroplastron bubali, Ophryoscolex* spp., and *Polyplastron multivesiculatum* (Naga et al., 1969; Shimizu et al., 1983; Imai and Ogimoto, 1984; Imai, 1985). It is of interest that young buffalo calves only became infected with *Ostracodinium* spp. from adult buffalo, but not adult cow, rumen contents although the same protozoa were present in both.

Giesecke (1970) suggested, in a review of the literature, that grazing animals such as cattle contained more protozoal species than browsing ruminants such as deer, although Giesecke and Van Gylswyck (1975) found the reverse. Water buffalo contain fewer protozoal species than Zebu cattle in the same area (Shimizu et al., 1983; Imai and Ogimoto, 1984).

There is some evidence for host specificity for certain protozoal genera. *Caloscolex* spp. only occur in the camel, which is not a true ruminant (Dogiel, 1926b), and *Epiplastron* spp. and *Opisthotricum* spp. are only found in African antelope (Dogiel, 1932, Kofoid and MacLennan, 1933; Van Hoven, 1975). Dogiel (1927) believed that 10 protozoal species were specific to their reindeer, but Dehority (1986c) pointed out that nine of these have now been found in other hosts, some not specifically associated with the Arctic. However, a certain combination of species is characteristic of all wild Artic ruminants, and some species such as *Enoploplastron confluens, Eremoplastron spectabile, Metadinium magnum,* and *Epidinium gigas* have only been found in such animals. However, domestic reindeer have a population resembling that of domestic sheep and goats (Dehority, 1978, 1986c). This suggests possibly that the characteristic reindeer population reflects an early separation of these ruminants and their protozoa from others living in temperate areas rather than a specific protozoal population for the survival of Arctic ruminants.

9.4 Geographical Distribution

There have been very many reports on the protozoal populations found in the rumens of various host ruminants in different parts of the world. Although the results are of interest, they, of necessity, often just give in-

Table 9.3 Geographical distribution of three entodiniomorphid protozoal species in cattle and sheep

	Polyplastron multivesiculatum		*Diploplastron affine*		*Ophryoscolex* spp.	
Location	Cattle	Sheep	Cattle	Sheep	Cattle	Sheep
Africa—west	+		−		+	
Africa—south					+	+
Alaska		−		−		+
Brazil	−		−		−+	
China					+	+
England	+	51%	+	51%	+	10%[a]
India (Zebu)	−		−		−	
India (Gaur)	−		−		−	
Indonesia	−		37%		−	
Japan	42%	60%	25%	60%	−	30%
Korea	10%		−		−	
New Zealand	−		−		−	
Philippines	−		−		−	
Russia	+	+	+	+	+	+
Scotland	+	+	+	+	+	+
Scottish highlands		+		+		−
Sri Lanka	5%		5%		−	
Thailand	−		11%		−	
USA	+	+	+	+	+	+

The values indicate the percentage of animals studied which have the particular protozoon.
[a] Approximate value.

formation on animals in one flock or herd at one place and may not be typical of the country in general. In the authors' experience with sheep in England, only 3 animals out of 50 contained *Polyplastron multivesiculatum* in one flock whereas in two others within a 15 mile radius the proportions were 54 out of 69 and 16 out of 24. In a flock in Wales the proportion was 15 out of 53. The reasons for these differences are not clear, but they could depend on chance, determined by which animals are culled out of a flock or by the protozoal populations in animals introduced into the flock from outside. These points could apply equally between countries as between areas in countries. Therefore statements that a given species occurs or does not occur in any country only have validity when many animals have been examined from several different locations. However, notwithstanding these provisos, Table 9.3 has been prepared to show the distribution of two species and one genus (*Polyplastron multivesiculatum*, *Diploplastron affine*, and *Ophryoscolex* spp.) in cattle and sheep. Again with these provisos, comparison of the number, genera, and species present in the Far East and Japan shows that half the species were common to Japan and Thailand; six were specific to Thailand and three to Japan. Cattle in Thailand and the Philippines contained the same protozoal species as did those in India and

Sri Lanka and Indonesia, although they tended to differ from those found in Chinese cattle (Hsiung, 1932; Imai et al., 1978, 1979a; Shimizu et al., 1983; Imai and Ogimoto, 1984; Imai, 1985, 1986).

9.5 Effect of Feed Composition

Problems in Determining Protozoa Population Densities The first problem is that rumen contents are not homogenous but have a crust of solid material overlying a liquid phase. However, in animals fed once a day even this is not constant as just before feeding the whole becomes homogeneous. To obtain a uniform sample some researchers such as Warner (1962b) have removed a comparatively large volume (800 g from a sheep) mixed it, taken a sample, and returned the remainder to the rumen. Unfortunately this procedure takes time and destroys the structure of the contents so as to make a series of samples of dubious worth as a measure of what happens in a normal rumen. There is some advantage in just sampling the liquid, especially as Purser and Moir (1959) showed that there was little difference in the protozoal population density in samples taken from different parts of the rumen. There is, however, evidence that however the samples are taken, the protozoal population density will be lower than when the animal is killed and the rumen contents mixed before the sample is taken (Minor et al., 1977).

The second problem is the mass of plant debris to which the protozoa may be attached or which may overly the protozoa when a count is being made. This can be overcome by squeezing the rumen contents through one or two layers of cheesecloth to remove the solid material. Unfortunately this also removes all protozoal species to some extent, with the exact result depending on the time after feeding (Dehority, 1984). Some workers always shake the rumen sample with diluent while others blend the rumen contents in an attempt to dislodge attached protozoa, although the latter treatment reduces the number of protozoa (Dehority, 1984).

The third problem is that once the rumen contents have been diluted with fixative, for example, the protozoa settle out rapidly during the procedures necessary to produce a countable number on a slide. It is absolutely essential to use a pipette with a wide mouth, such as a 1 ml graduated pipette from which the tip has been removed, and to do all operations as rapidly as possible after mixing the tubes from which a sample is to be taken. Some workers (Dehority, 1984) have added 30% glycerol presumably to decrease the rapid sedimentation of the protozoa.

Two techniques have been adopted for the final counting. In the first, a known volume of liquid is pipetted into a counting chamber of some kind (e.g., a Manners or Fuchs-Rosenthal chamber) and the number of protozoa are counted in grids on the slide or in the eyepiece (Boyne et al., 1957; Clarke et al., 1982; Dehority, 1984). The protozoa are preserved in formaldehyde and sometimes stained with brilliant green. Another method, which is favoured by the authors, is to fix the protozoa in 0.1 M iodine (in KI), which also stains well-fed organisms dark blue, and to count all the protozoa in a known volume of material blown on to a slide. This method has the advantage over the first that the protozoa do not have to be distributed uniformly over the slide. In the authors' experience iodine is a more reliable preservative than formaldehyde (Coleman, 1987a).

Animal to Animal and Within Animal Variations Boyne et al. (1957) first showed that there was up to a threefold variation in protozoal population density in one animal kept on a constant ration under constant conditions. More recently Clarke et al. (1982) have measured protozoal population densities in each of a series of four animals fed and sampled differently. The results showed that while the animals in some groups behaved similarly, there could be large differences such as 9 g to 58 g protozoa per rumen, and smaller and larger differences when individual protozoal genera were considered. Results in which the effects of different feeding regimes have been determined using very few different animals must be treated with care.

Effect of Starch in the Ration Almost all rumen ciliates feed on starch grains and, within limits, any increase (provided that it is done slowly) in the amount of starch fed increases the protozoal population density (Abe et al., 1973), mainly due to an increase in the numbers of entodinia (Lyle et al., 1981; Dennis et al., 1983). However, if the animals are fed to appetite on starch-rich rations, then the rumen acidity increases and all the ciliate protozoa disappear (Eadie et al., 1967; Lyle et al., 1981); the entodinia are the least sensitive to low pH and are the last to disappear. Purser and Moir (1959) found a linear regression of mean daily protozoal population density and minimum daily pH with, under their conditions, 670×10^3/ml at rumen pH 5.9 and 290×10^3/ml at rumen pH 5.2. In contrast, feeding of a restricted high-grain ration (80% of appetite) produced very high ($> 2 \times 10^6$/ml) protozoal population densities (Christiansen et al., 1964; Eadie et al., 1970) and over 95% of these could be *Entodinium* spp. (Van der Wath and Myburgh, 1941). A wider range of protozoal genera and species are found in animals fed on hay rather than starch (Bonhomme and Durand, 1974).

Senaud et al. (1986) and Groliere et al. (1980) investigated the effect of diet on the establishment of three protozoal species in sheep and found important competition between *Polyplastron multivesiculatum* and *Isotricha prostoma* for starch but lesser competition between *Entodinium* spp. and *Isotricha prostoma*. On rations containing equal amounts of roughage and starch, *Polyplastron multivesiculatum* eliminates the entodinia, possibly by predation, but with rations containing a smaller proportion of starch these protozoa stimulate each other. For example, the introduction of *Polyplastron multivesiculatum* into a rumen containing 150,000 entodinia/ml results, after a week, in populations of 12,000 *P. multivesiculatum* and 130,000 entodinia/ml; introduction of entodinia into a rumen containing 8,500 *P. multivesiculatum* /ml results, after a week, in populations of 10,000 *P. multivesiculatum* and 90,000 entodinia/ml.

Although the proportion of starch in the ration is obviously important, the form in which it is fed also affects protozoal numbers. On a basal ration of alfalfa and maize, the protozoal population density was 9.1×10^4/ml if the ration was just mixed together, but 2.3×10^4/ml if it was pelleted. If the proportion of starch was increased, the corresponding values were 4.2×10^4/ml and 1×10^4/ml (Christiansen et al., 1964). These differences were associated with a rapid passage of material through the rumen in animals fed pelleted rations. The degree of grinding of the mixed ration was also important with the population density falling from 8.5×10^4/ml when coarsely ground to 4.1×10^4/ml when finely ground ration was fed.

Effect of Addition of Protein As all rations, however poor, contain some protein, the effect of added protein will depend to some extent on the amount present in the basal ration. However, the effect of adding additional protein is more complicated than this, and there is no agreement as to what happens to the protozoa when this is done. Purser and Moir (1966) showed that in sheep fed a probably nitrogen-deficient ration, the addition of wheat gluten so that the total nitrogen was 18 g/day increased the protozoal population density from 100,00 to 150,000/ml. Devuyst et al. (1973) added 2 or 5% torula yeast to the ration given to sheep and found small (up to 50% but usually smaller) increases in the population density of individual protozoal species. In contrast, Reichl (1961) showed that, in sheep fed lucerne with and without ground grain, there was a positive correlation between protozoal counts and the content of nitrogen-free compounds in the feed; there was a negative correlation with protein content. In these experiments there was a negative correlation between the relative number of entodinia and the relative number of other species. Again in contrast to all the above results, Bird et al. (1979) found that in lambs given a basal ration of oaten chaff and sugar, the addition of fish meal had no effect on the protozoal population (96% small entodinia) density.

There is also some indirect and inconclusive evidence that in wild ruminants such as mule deer (Klein, 1962, 1965) and in "browsing" animals such as roe deer, white-tailed deer, impala, springbok, and kudu which selectively eat material rich in protein, the animals have protozoal population densities up to five times higher than in "grazing" species which eat material much poorer in protein (Van der Wath and Myburgh, 1941; Giesecke and Van Gylswyk, 1975). However, Giesecke (1970) in a review of the literature came to the reverse conclusion, quoting results obtained with browsers from Europe and the USA where many animals contained no protozoa. It is of interest that holotrich protozoa are rarely found in browsers. Entodinial populations are 2–3 times higher than those of the large entodiniomorphs in browsers whereas the reverse is true in grazers which feed less selectively (Van der Wath and Myburgh, 1941).

Effect of Diets Rich in Free Sugars The population density of the holotrich protozoa (*Isotricha* spp. and *Dasytricha ruminantium*) tends to increase as the proportion of soluble sugars in the ration increases because these protozoa rapidly assimilate many such sugars (Oxford, 1955a; Purser and Moir, 1966). However not all workers have found high densities in animals fed sugars or molasses, although this could be due to lack of inoculation with the protozoa. Valdez et al. (1977) found that in cattle in Mexico fed on sugar cane, molasses, and urea, 95% of the protozoal biomass was holotrich protozoa, and there were less than 10^5 entodinia/ml rumen contents. Leng et al. (1981) found densities of 0.8×10^4 holotrichs and 1.8×10^4 entodinia/ml in cattle in the Dominican Republic fed on sugar cane/urea whereas Bird et al. (1979) found that small entodinia made up 96% of the population of 5×10^5/ml in Australian lambs fed on oaten chaff-sugar-urea. The reader should remember than *Isotricha* spp. can be one hundred times the volume of small entodinia.

Effect of the Addition of Urea As with most other variation in the ration fed, the effect of adding urea or replacing other feed constituents by urea, is complex and may depend on a variety of uncontrolled factors.

Purser and Moir (1966) showed that adding urea to a nitrogen-deficient ration increased the protozoal population density from 100 to 300×10^3/ml and that the urea encouraged the growth of diplodinia and holotrichs at the expense of entodinia. In contrast, Nour et al. (1979), who fed four ruminant species on straw plus cotton seed cake and progressively replaced the latter by rice starch and urea, found protozoal population densities ($\times 10^5$/ml) of 1.71, 1.07, 1.47, and 1.53 when 0, 36, 55, and 100% of the cake was replaced. As the proportion of starch-urea was increased, the proportion of entodinia in the population increased, while that of the other species declined to almost zero. The average protozoal generation time decreased from 5.1 h to 21.8 h as the urea proportion increased. Dennis et al. (1983), using semi-purified diets, showed that when soybean meal was replaced by urea the proportion of entodinia increased from 88 to 92%, that of charonina from 0.6 to 1.2% while that of the holotrichs decreased from 6.8 to 1.7%. Similar effects on holotrich numbers were observed by Punia et al. (1987) in cattle fed on hay supplemented with urea.

Effect of Feeding Purified and Semi-purified Rations Oltjen et al. (1966) first determined protozoal population densities in steers transferred from an alfalfa hay to a wood pulp-starch ration and found little change with low densities of 10^4/ml present on both. Slyter et al. (1971) in a similar study with diets rich in wood pulp found densities of $2–5 \times 10^3$/ml with urea, biuret or soy in the absence of starch and $20–40 \times 10^3$/ml in its presence. Tackechi et al. (1978) who began with sheep (fed on hay and concentrates) that had 10^6 protozoa/ml found that on transfer to rations of wood pulp alone, wood pulp + protein or wood pulp + starch the counts were decreased to 6×10^3, 25×10^3 and 120×10^3/ml respectively. The authors also made differential protozoal counts and concluded that protein was effective in stimulating the growth of entodinia.

Effect of Other Additives that Increase Protozoal Densities or Growth Rate

Diethylstilboestrol As implantation of diethylstilboestrol (DES) was used at one time to improve ruminant performance (Clegg and Cole, 1954; Ibrahim et al., 1968), it was suggested that this could be due to changes in rumen metabolism. However there is not agreement among those who have investigated this. Slyter et al. (1970) using steers implanted with DES and Dearth et al. (1974) who included DES in the ration fed to lambs found no affect of DES on protozoal population densities. In contrast Ibrahim et al. (1970) and Ibrahim and Ingalls (1972) found that the inclusion of DES in the ration increased the amount of protozoal biomass/ml 2–6 times and doubled the proportion of rumen protein represented by the protozoa.

Aureomycin The feeding of aureomycin (20–30 mg daily) to weathers increased the protozoal population by up to 180% (e.g., 650×10^3 to 1830×10^3/ml in animals fed a hay/concentrate ration (Klopfenstein et al., 1964; Purser et al., 1965).

Avoparcin When sheep are transferred from pasture to stalls and fed on a pelleted ration containing roughage and gelatinized starch, the protozoal population density drops due to a fall in rumen pH. The addition of avoparcin (45 mg/kg ration) prevents this decline in protozoal numbers probably due to its action in inhibiting the growth of lactic acid producing bacteria (Watson and Mottram, 1989).

Methionine hydroxyanalogue (DL α Hydroxy γ methyl mercaptobutyrate) When this is added to the feed it increases methionine concentration in the rumen and specifically increases the population density of entodinia up to 10 times while tending to decrease the number of other protozoa (De Vuyst et al., 1975).

Monensin Although the addition of monensin (50 mg/kg feed) either has no effect (Dinius et al., 1976; Van Nevel and Demeyer, 1977; Leng et al., 1984) or a decreased protozoal population densities (Oddy et al., 1978; Lyle et al., 1981; Hino 1981), there is evidence that monensin, either directly or indirectly, can increase protozoal growth rate under some conditions.

β-Sitosterol Hino and Kametaka (1974) have shown that when goats were fed a hay/amylopectin ration without β-sitosterol, the protozoal population density decreased from 2.9×10^5 to 4.0×10^4/ml in 12 days and returned to the original level when the ration was supplemented with β-sitosterol.

Tylosin The feeding of tylosin (15 mg daily) to wethers fed on a hay/concentrate ration increased the protozoal population by up to 70% without affecting the total bacterial viable count.

Effect of Other Additives that Decrease Protozoal Population Densities Very many inorganic and organic compounds including antibiotics have been tested for possible activity against rumen ciliates in vivo. Information on those which have been used to defaunate ruminants is given in Chapter 4 as is information on those which are inactive against protozoa and can be used to obtain axenic cultures. Details on other compounds are given by Willard and Kodras (1967), Clarke and Reid (1969), O'Connor et al. (1970, 1971), and Williams and Feil (1971).

Mycotoxins which may be a health hazard to domestic animals are not toxic to rumen ciliates and are degraded by them (Kiessling et al., 1984).

The feeding of moulded hay to ruminants has a generally harmful effect and depresses the population density of *Isotricha*, *Dasytricha*, and *Entodinium* spp. while increasing the proportion of diplodinia in the population (Mohanty et al., 1969).

Effect of Changes in Osmotic Pressure in the Rumen Restriction (from 1.5 to 1.0 litres/day) in the amount of water drunk by sheep can produce dramatic increases in the numbers of holotrich protozoa, e.g., from 2000 to 14000/ml. However, this effect is short lived (2 weeks) and is followed by a decline in the population density (Purser and Moir, 1966; Dehority and Purser, 1970). In contrast, Dehority and Males (1974) found that manipulation of feed and water intake, while altering the osmolality of the rumen fluid, had no effect on holotrich population densities.

The protozoa (species unknown) in black Bedouin goats living in desert areas are very tolerant to changes in osmotic pressure in that the population density dropped by only 30% when animals deprived of water for 4 days rapidly drank 7–8 litres of water and decreased the rumen osmotic pressure by 75% (Brosch et al., 1983). In contrast, exposure of entodiniomorphid protozoa, isolated from normal sheep in England, to hypotonic salt solution (30% of standard) for 2 min resulted in death of over 50% of the protozoa (Coleman, 1985b).

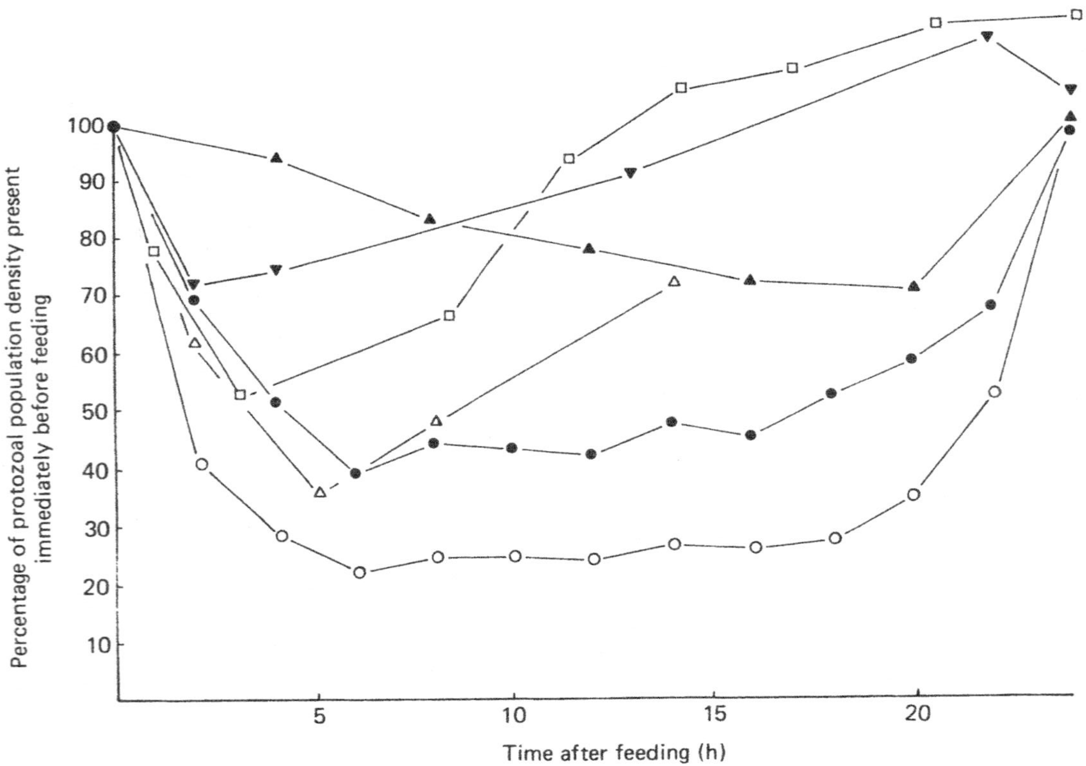

Fig. 9.1 Diurnal variation in rumen volume and protozoal population density found by different authors in animals fed once a day. The results are expressed as a percentage of the values obtained immediately before feeding. Three lines are for data obtained from the same sheep fed on hay and concentrates (Michalowski and Muszynski, 1978): ▲, rumen volume; ○, population density of *Ophryoscolex* sp.; and ●, population density of *Diplodinium* spp. Other lines: □, population density of *Entodinium* spp. in sheep fed wheaten chaff and lucerne chaff (Warner, 1966a); △, total protozoa in cattle fed fresh clover (Clarke, 1965a). ▼, total protozoa in sheep fed on alfalfa meal, corn cobs, and oats (Potter and Dehority, 1973).

9.6 Diurnal Variation in Densities of Entodiniomorphid Protozoa

Unlike the holotrich protozoa there is no evidence that the entodiniomorphid protozoa sequester in any part of the reticulorumen. The number obtained by sampling different parts of the rumen should therefore give a satisfactory measure of the protozoal population density given the inherent variability of the population density (see above).

In animals fed once a day the entodiniomorphid protozoal population density declines by over 50% in the 6 h after feeding and at some time during the remainder of the day increases slowly or rapidly (preceded or followed by a stable period) until the original prefeeding level is reached (Fig. 9.1; Warner, 1962b, 1966a; Michalowski, 1977; Michalowski and Muszynski, 1978). The classical explanation of this behavior is that, immediately after feeding, the rumen fluid is diluted by water, saliva, etc., and this causes the protozoal population density to fall. A short time after feeding, the protozoa, which are full of nutrients, begin to divide with the division rate reaching a maximum after 8 h, after which time it falls off to almost zero. It is of interest that in New Zealand cattle Clarke (1965a) found almost no change in the total number of entodiniomorphid protozoa present in the rumen over a 24 h period (Fig. 9.1).

The pattern found for *Ophryoscolex* sp., for example, by Michalowski and Muszynski (1978) with a rapid decline in protozoal density after feeding and a rapid increase just before feeding, cannot be explained just by changes in rumen fluid volume. The initial decline could be the result of an input of fluid into the rumen with a concomitant passage of fluid plus protozoa into the omasum. The final increase in density just before feeding in the absence of a fluid change is presumably due to division of the protozoa although it could be the result of a redistribution of protozoa within the rumen. The conditioned contractions of the reticulum associated with the routine of feeding (Abe et al., 1983) could cause a forced migration of some entodiniomorphid protozoa as has been postulated to occur with the holotrich protozoa.

If animals, normally fed once a day, are not fed for 48 h, the protozoal population density declines by 80–85% from the maximum, for example, from 1200×10^3 to 160×10^3/ml (Michalowski and Muszynski, 1978) or 750×10^3 to 130×10^3/ml (Potter and Dehority, 1973).

Animals fed twice a day or every 3 h showed a similar pattern to those fed once a day except that the cycles of peaks and troughs were shorter to fit the feeding pattern. The maximum population density was similar under all conditions (Michalowski and Muszynski, 1978; Warner, 1966b).

If sheep are fed roughage rations to appetite in pens or allowed to graze pasture freely, they tend to consume most of their daily intake in one period (Warner, 1966c). The pattern of changes in protozoal population density is therefore similar to that found in animals fed once a day (Warner,

1966c). However, Moir and Somers (1956) showed that feeding a hay ration four times a day rather than once increased the prefeeding protozoal population density from 1.15×10^6 to 3.14×10^6/ml.

9.7 Seasonal Variation

There is little information on the seasonal variation in protozoal numbers because most experiments are carried out under controlled conditions on animals kept in pens on constant rations. However, Van der Wath and Myburgh (1941) studied the protozoal population density of sheep grazing the veld in South Africa and found (average 20 sheep) that in July (winter), October (spring), January (summer), and April (autumn), the means counts were, respectively, 100, 277, 455, and 278×10^3 protozoa/ml. In winter 20.6% of the protozoa were large entodiniomorphs which can utilize lignified forage whereas in summer this figure was only 15.7% with a flourishing population of entodinia being present. Similar results were found by Westerling (1970) for ciliates in Finnish reindeer where the populations were 3500×10^3/ml in summer and 2000×10^3/ml in winter. This trend was shown by most protozoal species except *Entodinium anteronucleatum* which behaved in the opposite way, *Diplodinium* spp. which were most numerous in autumn and *Enopoplastron triloricatum* and *Epidinium gigas* which were least numerous in autumn.

10

Role of Protozoa in the Rumen

Measurement of various ruminal parameters in faunated, defaunated, and ciliate-free animals indicate that the ruminal environment can be markedly altered by the presence of protozoa (Table 10.1; Williams and Coleman, 1988). These changes in the ruminal ecosystem will have an effect on the bacterial activities, and hence the supply of metabolites to the host will be affected. The presence of protozoa in the rumen has been shown to influence the volume of the rumen and the retention time of the digesta, the concentration and proportion of the volatile fatty acids, the levels of other acidic metabolites and ammonia, the environmental pH, and the numbers and type of rumen bacteria present. Changes in any of these parameters will influence ruminal function with resultant beneficial or detrimental consequences for the host.

10.1 Metabolism of Cellulose and Hemicellulose

One of the first studies on the role of protozoa was that of Becker et al. (1929) who defaunated adult animals and found no difference in the rate of cellulose digestion. At around this time there were also a number of investigations on cultures of rumen ciliates and Margolin (1930) and Westphal (1934) concluded that the protozoa did not digest cellulose whereas Trier (1926), and Hungate (1942, 1943) had evidence that the ciliates produced cellulase.

In more recent years Yoder et al. (1966) incubated washed bacterial and protozoal fractions from steers with cellulose and concluded that the pro-

Table 10.1 The change in ruminal metabolism when protozoa are absent

	Change in activity when protozoa absent[a]		
	Increase	Decrease	No significant effect
Breakdown of:			
Organic matter	34, 49	2, 4, 14, 15, 16, 17, 20, 23, 27, 29, 30, 31, 32, 34, 39, 46, 47, 48, 54, 58, 59, 62, 63, 66, 67, 68	40, 43, 50, 72
Fiber[b]		15, 24, 27, 29, 31, 42, 46, 48, 62, 66, 67	35, 47
Cellulose	49	2, 6, 8, 10, 22, 23, 25, 28, 33, 37, 45, 61	3, 35, 40
Hemicellulose		45, 61	47
Starch		9, 22, 23, 28, 33, 34, 36, 66, 67	43
Protein degradation		19, 20, 26, 27, 39, 46, 56, 58, 59, 63	1, 2, 7, 16, 19, 26, 35, 40, 41, 43, 47, 48
Proteolytic activity		52, 56, 57, 71	26
Utilization of:			
Urea	55		
Nitrate/nitrite		73	
Selenomethionine			18
Fat digestion		20	

Biohydrogenation		33	11
Methanogenesis		12, 16, 17, 19, 27, 34, 60, 71	
Phospholipid formation		11, 44	
Absorption of Ca, Mg, P		5	
Availability of sulfur		17	
Rate of flow to lower tract of:			
Trace elements			21
Nitrogen	14, 31, 38, 39, 43, 48, 50, 51, 53, 58, 59, 63, 65, 66, 68, 69		
Efficiency of microbial protein synthesis	12, 13, 15, 29, 31, 38, 43, 53, 58, 59, 63, 64, 69		48

[a]References are as follows: 1, Abou Akkada (1965); 2, Abou Akkada and El Shazly (1965); 3, Becker et al. (1929); 4, Bird and Leng (1983); 5, Boila and Phillips (1988); 6, Broudiscou et al. (1988); 7, Chalmers et al. (1968); 8, Coleman (1986a); 9, Coleman (1986b); 10, Conrad et al. (1958); 11, Dawson and Kemp (1969); 12, Demeyer and Van Nevel (1979); 13, Demeyer and Van Nevel (1986); 14, Demeyer et al. (1982b); 15, Demeyer et al. (1986); 16, Eadie and Gill (1971); 17, Hegarty et al. (1989); 18, Hidiroglu and Jenkins (1974); 19, Itabashi et al. (1984); 20, Itabashi and Matsukawa (1979); 21, Ivan and Veira (1982); 22, Jouany and Senaud (1979); 23, Jouany and Senaud (1982); 24, Jouany and Senaud (1983); 25, Jouany et al. (1981a); 26, Jouany and Thivend (1983); 27, Jouany et al. (1981b); 28, Kaneko et al. (1989); 29, Kayouli et al. (1982); 30, Kayouli et al. (1984); 31, Kayouli et al. (1986); 32, Klopfenstein et al. (1966); 33, Kreuzer and Kirchgessner (1986a); 34, Kreuzer et al. (1986); 35, Kumar and Raghavan (1978); 36, Kurihara et al. (1968); 37, Kurihara et al. (1978); 38, Leng and Nolan (1984); 39, Lindsay and Hogan (1972); 40, Luther et al. (1966); 41, Males and Purser (1970); 42, McAllan et al. (1983); 43, Meyer et al. (1986); 44, Neil et al. (1979); 45, Onodera et al. (1988); 46, Perkins and Luther (1967); 47, Punia et al. (1986); 48, Punia et al. (1987); 49, Romulo et al. (1986); 50, Rowe et al. (1981); 51, Rowe et al. (1983); 52, Shinchi et al. (1986); 53, Sutton et al. (1983); 54, Szuecs et al. (1982); 55, Takahashi and Kametaka (1976); 56, Ushida and Jouany (1985); 57, Ushida and Jouany (1986); 58, Ushida et al. (1984); 59, Ushida et al. (1986a); 60, Ushida et al. (1986b); 61, Ushida et al. (1987a); 62, Ushida et al. (1989a); 63, Ushida et al. (1989b); 64, Van Nevel and Demeyer (1981); 65, Van Nevel et al. (1985); 66, Veira et al. (1981); 67, Veira et al. (1983); 68, Veira et al. (1984); 69, Verite et al. (1986); 70, Wallace et al. (1987); 71, Whitelaw et al. (1984); 72, Xie and Lin (1987); 73, Yoshida et al. (1982).
[b]Neutral detergent fiber (NDF) and acid detergent fiber (ADF).

Table 10.2 Effect of faunation on the digestion of acid detergent fiber (ADF) and starch in sheep

	Digestibility (%) in animal[a]		Digestibility (%) in rumen[b]	
	Ration		Ration	
Protozoal population	Hay ADF	Hay/barley ADF	Corn silage ADF	Ground corn Starch
Ciliate-free	68.8	56.5	33.6	84.0
Entodinium spp. only	70.1	60.3	40.4	87.1
Polyplastron multivesiculatum only	70.6	61.4		
Normal fauna	71.1	63.0	41.7	89.2

[a]Data based on Jouany and Senaud (1979b).
[b]Data based on Veira et al. (1983).

tozoa stimulated cellulolysis by the bacteria. At the same time Klopfenstein et al. (1966) showed that apparent dry matter digestion in lambs was higher in faunated than defaunated animals (73.7% compared with 64.2%). However, Luther et al. (1966) disagreed and found no differences in the digestion of rations that could be ascribed to protozoa. Subsequently studies in vitro indicated that approximately a quarter to one-third of fiber breakdown in the rumen was protozoal (Amos and Akin, 1978; Demeyer, 1981; Lubberding et al., 1987).

Many experiments of the following general types have been carried out to determine the effect of the presence or absence in the rumen of ciliate protozoa, on the breakdown of organic matter, acid detergent fiber (ADF), or cellulose (Table 10.2). In the first group, whole animals have been used and the overall digestibility and/or digestibility in the rumen have been measured. In these experiments the amount of breakdown that occurs in the hind gut is important and in some experiments this has been measured by difference between overall and rumen breakdown. It must also be remembered that when protozoa are removed from the rumen they leave an ecological niche which is probably filled in part by bacteria. The overall result therefore depends on at least two other factors apart from the protozoa. The second method is the estimation of the digestion of food materials contained in a nylon bag suspended in a faunated or protozoa-free rumen. This method also suffers from the problem of bacteria "replacing" protozoa. The third method is the preparation of homogenates or extracts of bacteria and/or protozoa and measuring the amount of a specific enzyme in these materials. The method is only reliable if all the enzyme under test can be liberated from all the microorganisms under all conditions and there is no evidence that this has been successfully done. However, if it is assumed to have been done, then the method has the big advantage that the activity in the bacteria in hosts containing a variety of protozoal populations can be measured.

The results given in Table 10.3 show that in whole animals the effect of defaunation is small and often barely significant. In contrast, in experiments on digestion in just the rumen, the effect of defaunation is often highly significant with a 20% decrease in the rate of breakdown. The inoculation of defaunated animals with only a single species of protozoa has been shown to be effective in restoring the extent of ruminal digestion (Jouany and Senaud, 1979b, 1982, 1983; Jouany et al., 1981a, 1981b; Ushida et al., 1987a). Although there have been relatively few experiments (Lindsay and Hogan, 1972; Ushida et al., 1986; Punia et al., 1987) the evidence available suggests that in animals lacking rumen ciliates the decreased rate of cellulose digestion in the rumen is compensated for by an increase in the amount digested in the hind gut. In young animals reared ciliate-free there is enlargement of the hind gut to provide space for the increased flow of material from the rumen and such animals develop a typical pot-bellied appearance (Becker, 1932; Pounden and Hibbs, 1950). However this appearance could also be due in part to increased rumen volume (Orpin, 1983/84). Ciliate-free young animals (calves) also have a rougher coat suggesting a possible nutritional deficiency. It should be noted that the decreased digestion of cellulose in the rumen in the absence of protozoa may be due to increased rate of passage of food materials (Kayouli et al., 1983/84; Ushida et al., 1986), although Orpin (1983/84) found the reverse effect. The activities of polysaccharolytic enzymes are increased after refaunation (Williams et al., 1988, 1990). However, there has only been one attempt to determine the effect of individual protozoal species on the overall concentration of cellulase in the rumen and on the distribution of this between the various rumen fractions (Table 10.4) (Coleman, 1985a, 1986a). The results show that there is appreciably less total cellulase in ciliate-free sheep than in animals containing the non-cellulolytic *Entodinium caudatum*, and that this is due to markedly increased (235%) bacterial cellulase under these conditions. This supports a similar finding made on sheep containing mixed entodinia by Kurihara et al. (1978). In sheep containing cellulolytic protozoa (*Epidinium caudatum*, *Eudiplodinium maggii*, or *Ostracodinium bilobum*) the overall concentration of cellulase is 60–100% higher than in the animal containing *Entodinium caudatum* and 56–81% of the rumen cellulase is associated with the protozoal cytoplasmic fraction. This suggests a very important role for these protozoa in cellulose digestion in the rumen. Results are also given in Table 10.4 for a sheep with a natural A-type population. The presence or absence of protozoa may also influence the effects of digestion manipulants as shown by the studies of Onodera et al. (1986, 1987) with salinomycin. Cellulose and hemicellulose digestibilities were lowered in faunated goats by salinomycin, whereas the additive improved breakdown in defaunated animals. Protozoa also affect the action of the gastric hormones pentagastrin and secretin which, when infused intravenously, in-

Table 10.3 Effect of defaunation on digestion in ruminants

Host	Ration	Experimental system	Substrate	Parameter measured
Cattle	Hay/corn	Rumen fractions	Cellulose	Cellulose
Lambs	Low conc.	Whole animal		Cellulose
	High conc.	Whole animal		Cellulose
Sheep	Corn cobs/ corn	Whole animal		Dry matter
	Alfalfa/corn	Whole animal		Dry matter
Sheep	Lucerne	Whole animal		Organic matter
	Red clover			Organic matter
Cattle	Pasture	In vitro	Orchard grass	Dry matter
			Bermuda grass	Dry matter
Sheep	Purified diet	Nylon bag	Wood pulp	Cellulose
Cattle	Hay/conc.		Hay	Crude fibre
			Rice straw	Crude fibre
Sheep	Red clover hay	Rumen extracts	Clover hay	Hemicellulase
Sheep	Hay/conc.	Whole animal		ADF
				ADF
				ADF
Sheep	Straw	Whole animal		Organic matter
	Alkali-treated straw	Whole animal		Organic matter
Sheep	Hay/conc.	Whole animal		Organic matter
Sheep	Lucerne/ barley	Whole animal		ADF
				ADF
				ADF
				ADF
Sheep	Hay/conc.	Nylon bag		ADF
				NDF
Sheep	Corn/silage	Whole animal		ADF
Cattle	Hay	Rumen fractions	CMC	Sugar released
			Cell wall	Sugar released
			Hemi- cellulose A	Sugar released
Goats	Hay/conc.	Whole animal		Energy
	Hay	Whole animal		Energy
Sheep	Hay/conc.	Whole animal		Organic matter
				ADF
Sheep	Hay/oats	Rumen fractions	CMC	Sugar released
				Sugar released
				Sugar released
				Sugar released
				Sugar released
Sheep	Hay/barley	Whole animal		Organic matter
	Straw/sugar	Whole animal		Organic matter
Sheep	Hay/conc.	Whole animal		Organic matter
				ADF

Protozoal population	Digestion in rumen (%)		Digestion in hind gut (%)		Reference
	Faunated	Defaunated	Faunated	Defaunated	
Mixed	65.2	38.1			Yoder et al., 1966
Mixed	52.7	53.7			Luther et al., 1966
Mixed	41.2	51.8			
Mixed	73.7	64.2			Klopfenstein et al., 1966
Mixed	78.0	73.7			
Mixed	41.0	35.0	24.0	27.0	Lindsay and Hogan, 1972
Mixed	45.0	41.0	31.0	34.0	
Epidinia	57.2	30.0			Amos and Akin, 1978
Epidinia	37.0	12.4			
Entodinia	97.0	49.0			Kurihara et al., 1978
Mixed	48.5	36.9			Itabashi and Matsukawa, 1979
Mixed	64.1	53.9			
Mixed	(0.90)	(0.52)			Bailey and Macrae, 1979
Entodinia	70.1	68.8			Jouany and Senaud, 1979a
Polyplastron	71.6	68.8			
Mixed	71.1	68.8			
Mixed	32.3	23.6	26.9	18.7	Collomber, 1981 (quoted by Demeyer et al., 1986)
Mixed	36.3	37.0	24.9	21.9	
Mixed	34.0	32.0	42.0	35.0	Rowe et al., 1981
Polyplastron	47.0	43.0			Jouany et al., 1981b
Isotricha	35.0	43.0			
Entodinia	41.0	43.0			
Mixed	48.5	43.0			
Mixed	40.0	30.0			Kayouli et al., 1982
Mixed	30.0	24.0			
Mixed	41.7	33.6			Veira et al., 1983
Mixed	(26.6)	(12.9)			Williams and Strachan, 1984
Mixed	(15.6)	(40.3)			
Mixed	(752)	(735)			
Entodinia	66.6	64.8			Itabashi et al. 1984
Entodinia	65.1	63.7			
Mixed	41.7	34.4	30.5	36.9	Van Nevel et al., 1984 (quoted by Demeyer et al., 1986)
	50.1	36.1	4.7	19.9	
Epidinia	(712)	(134)			Coleman, 1986a
Eu. maggii	(822)	(134)			
Ost. bilobum	(627)	(134)			
Ent. caudatum	(389)	(134)			
Mixed A type	(409)	(134)			
Mixed	43.0	29.0	18.0	30.0	Ushida et al., 1986b
Mixed	39.0	31.0	22.0	28.0	
	41.7	34.7	30.5	36.9	Kayouli et al., 1986
	59.2	45.9	26.0	25.3	
	50.1	36.1	4.7	19.9	

Table 10.3 *Continued*

Host	*Ration*	*Experimental system*	*Substrate*	*Parameter measured*
Sheep		Whole animal		ADF
Sheep		Whole animal		Organic matter
Cattle	Paspalum hay	Whole animal		Organic matter ADF NDF
Sheep	Wheat straw NH_3-wheat straw	Whole animal		Dry matter
Sheep	NH_3-straw NH_3-straw starch	Whole animal		ADF
Sheep	NH_3-straw NH_3-straw/ corn	Whole animal		ADF

ADF, acid detergent fiber; NDF, neutral detergent fiber

hibit feeding; the inhibitory response is more marked in defaunated calves than conventionally faunated animals (Kobayashi and Itabashi, 1986).

10.2 Metabolism of Starch

The effect of starch in increasing protozoal population densities was considered in Chapter 9.

There have been relatively few studies on the effect of faunation of the rumen on the breakdown of starch probably because, as shown by Veira et al. (1983), 84% of the starch in a ration containing 47% ground shelled maize was digested in the rumen in the absence of protozoa. This was increased to 89% in the presence of *Entodinium* and *Diplodinium* spp. Lindsay (1970), in a review of the earlier literature on the fermentation of starch by normally faunated animals, pointed out that only 20–30 g starch/day/sheep reached the small intestine; this was approximately 10% of the intake. The largest value recorded was 100 g/day for a sheep fed 80% ground maize where the total daily intake was 576 g.

Coleman (1986a) has studied the distribution of α-amylase between rumen fluid fractions of sheep (fed 800 g hay and 100 g oats daily) containing no ciliate protozoa or one of several difference species (Table 10.5).

The results show that, although the total amylase varied at most 2.3 under the various conditions, the proportion in the protozoal cytoplasm varied 16-fold, with the proportion in *Entodinium caudatum* the lowest. Little amylase was found free in the fluid or attached to plant debris. Most of the

Protozoal population	Digestion in rumen (%)		Digestion in hind gut (%)		Reference
	Faunated	Defaunated	Faunated	Defaunated	
Mixed	60.1	49.2	3.2	10.9	Kayouli et al., 1986
Mixed	54.2	47.6	23.9	28.7	Veira and Ivan (quoted in Veira, 1986)
Mixed	45.0	42.0	9.8	10.5	Punia et al. 1987
Mixed	54.0	49.0	0.9	2.2	
Mixed	55.0	50.0	3.5	5.1	
Mixed	32.0	37.0			Soetanto, 1986
Mixed	50.0	55.0			(quoted by Demeyer, 1989)
Mixed	59.0	58.0			Ushida et al., 1988
	60.0	45.0			(quoted by Demeyer, 1989)
B type	60.0	58.0	6.0	4.0	Ushida et al., 1989a
	60.0	45.0	7.2	13.2	

activity was associated with the bacteria and protozoa, with the concentration in the two inversely related. Among the single protozoal species *Epidinium caudatum* and *Eudiplodinium maggii* contained the highest concentrations of amylase.

It is of interest that the ratio of carboxymethylcellulase/amylase (Table 10.4 in arbitrary units) is of the order of 1 in the bacterial fraction but varies from 2.7–31.1 in the single protozoa, with *Epidinium caudatum* and *Ostracodinium bilobum* at the highest ratios. This suggests that these protozoa digest cellulose relatively more rapidly than starch and should be regarded as cellulolytic.

The rates of digestion of intracellular starch by intact protozoa are 64, 127 and 117 pg/ng protozoal protein/h for *Eudiplodinium maggii, Entodinium simplex*, and natural mixed A-type protozoa, respectively (Coleman, 1986d).

10.3 Metabolism of Bacteria

Bacterial Ingestion Eadie and Hobson (1962) first showed that when lambs (fed on hay and concentrates) that had been kept ciliate-free from birth were inoculated with rumen ciliates, the number of small bacteria declined from 36×10^9 to 14×10^9/ml which was the population density in normal faunated animals. Although there are differences in the absolute values found, all other workers in the field have found the same result (Fig. 10.1). However, differences in the behavior of individual bacterial types have been found.

Table 10.4 Distribution of carboxymethylcellulase among rumen fluid fractions taken immediately before feeding from sheep containing either no protozoa or single or mixed protozoal species

	Protozoon present					
Fraction[a]	None	*Epidinium caudatum*	*Eudiplodinium maggii*	*Ostracodinium bilobum*	*Entodinium caudatum*	Natural mixed A-type population
Clarified fluid	7	0	8	2	0	0
Plant debris	30	14	56	60	87	45
Bacteria	76	18	88	110	255	36
Protozoal cytoplasm	21	540	670	350	18	200
Protozoal particles	0	140	0	105	29	78
Total	134	712	822	627	389	409
Amount in protozoal cytoplasm (%)		76	81	56	5	61

Expressed per ml rumen contents relative to that of a standard preparation of cytoplasmic fraction of *Epidinium caudatum* $\times 10^4$; see Coleman (1986a).

Table 10.5 Distribution of amylase among rumen fluid fractions taken immediately before feeding from sheep containing either no protozoa or single or mixed protozoal species

Fraction	None	Protozoon present						
		Epidinium caudatum	*Eudiplodinium maggii*	*Ostracodinium bilobum*	*Entodinium caudatum*	*Entodinium simplex*	Natural mixed A-type population	Natural mixed B-type population
Clarified fluid	0.7[a]	2.0	1.1	1.2	0.4	0.5	2.1	0.4
Plant debris	1.1	1.4	2.2	2.9	2.9	0.4	8.6	3.6
Bacteria	26.9	3.9	24.8	32.4	39.3	16.9	10.3	3.6
Protozoal cytoplasm	0.9	13.7	12.6	3.4	2.0	5.7	16.3	21.7
Protozoal particles	2.6	5.1	3.1	1.2	1.7	6.3	5.1	32.1
Total	32.2	26.1	43.8	41.1	46.3	29.8	42.4	61.4
Amount in protozoal cytoplasm (%)	2.7	52.5	28.8	8.0	4.3	19.1	38.4	35.3
Carboxymethylcellulase/ amylase in bacteria[b]	1.1	2.7	1.7	0.7	1.4	2.1	0.8	0.8
Carboxymethylcellulase/ amylase in protozoal cytoplasm[b]		31.1	11.6	27.4	2.0	2.7	2.5	12.7

[a]Expressed as μmol maltose produced from amylose per ml rumen contents per h; see Coleman (1986b).
[b]Carboxymethylcellulase activity is expressed in arbitrary units.

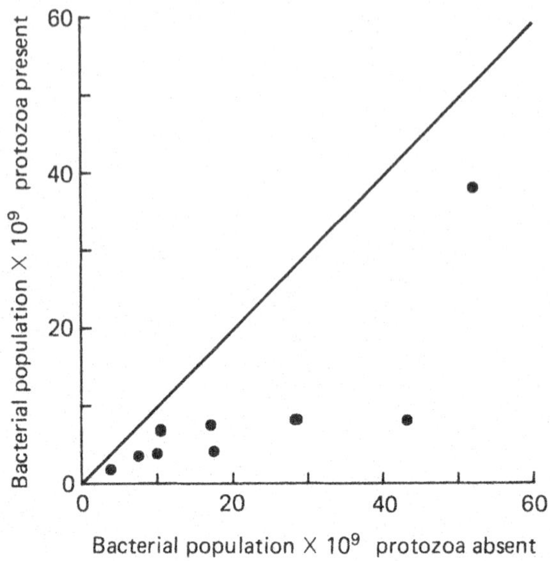

Fig. 10.1 The effect of the presence and absence of rumen ciliate protozoa on the size of the bacterial population in the ovine rumen. The line of equal values in faunated and ciliate-free animals is indicated. The sources of the data used are given in Table 10.7.

Eadie and Hobson (1962) found that on faunation the population density of *Oscillospira* declined from 150 × 10^5/ml to less than 10^5/ml and that there was a similar decline in the number of flagellates. In contrast, the numbers of ovals, large selenomonads, and certain large cocci were apparently unaffected. This result with *Oscillospira* has been confirmed by Kurihara et al. (1968) and Eadie and Gill (1971), but the situation with regard to ovals is less clear. Kurihara et al. (1968) and Eadie and Gill (1971) both found that Quin's oval was more common in faunated than defaunated animals and that the numbers tended to increase when more protozoal species were present. Large selenomonads behaved similarly. In contrast, the population density of Eadie's oval (Eadie, 1962a), now known as *Magnoovum eadii* (Orpin, 1976), is decreased in the presence of especially more than one protozoal species (Eadie and Gill, 1971; Kurihara et al., 1968). It should be noted that these two morphologically similar organisms are taxonomically unrelated. The rumen fungal population can also increase as the result of defaunation (Romulo et al., 1986, 1989).

Information on the effect of defaunation on the population densities of small culturable bacteria is sparse, although some patterns do emerge. Kurihara et al. (1968) faunated ciliate-free lambs first with entodinia, then with *Polyplastron multivesiculatum*, and finally with a mixed ciliate population. These authors found a decrease in the number of starch-digesting bacteria, especially in the presence of entodinia, and suggested that this result could be due not only to the uptake of free bacteria but also to the uptake of starch grains to which these bacteria had become attached. However, the results are not easy to interpret because *Streptococcus bovis*-like bacteria were more common in sheep containing entodinia than in ciliate-free animals but were less common in sheep containing *Polyplastron multivesiculatum* or mixed ciliates. The main result found was that the protozoa non-selectively engulfed small bacteria and that the decrease in bacterial volume on faunation corresponded closely with the increase in protozoal volume. It is of interest that Orpin and Letcher (1983/84) found that on defaunation the number of free, but not particle-associated bacteria, increased, suggesting that the protozoa only engulf free bacteria.

In animals on a restricted high-grain ration, where the protozoal numbers in faunated animals are high ($5–30 \times 10^6$/ml) the effect of defaunation is more marked with bacterial population densities increasing from 4×10^9 to 34×10^9/ml. However, the proportion of those that were culturable increased from 2.5 to 9.2%. The predominant bacteria were *Bacteroides* spp. and slender curved Gram-negative rods under both conditions (Whitelaw et al., 1972).

The most extensive study of the effect of faunation on bacterial species was made by Kurihara et al. (1978) who used sheep fed on purified diets. They found that the number of cellulolytic bacteria increased from 2.4×10^9/ml to 3.1×10^9/ml on faunation while that of amylolytic bacteria decreased from 6.50×10^9/ml to 0.59×10^9/ml under the same conditions. Furthermore the cellulolytic bacteria in faunated animals were entirely *Ruminococcus albus* which has greater activity than *R. flavefaciens* and which had made up 33% of the cellulolytic bacteria in the defaunated animal. It is possible that entodinia which were the principal protozoa present could ingest amylolytic bacteria attached to starch grains but were unable to take up the generally larger plant particles and their associated bacteria. There is also evidence from cultural studies that the density of viable bacteria on digesta particles is highest in faunated sheep and lowest in animals containing only *Epidinium* spp. and that the density of cellulolytic bacteria is highest on these particles in faunated sheep and lowest in ciliate-free animals (Argyle and Forster, 1989; Newbold et al, 1989a). These pronounced effects of the protozoal population on the rumen bacteria will be manifested in the bacterial activities in the rumen and metabolite supply to the host.

Teather et al. (1984) investigated protozoal and bacterial protein levels in 22 lactating cows fed on corn silage and concentrates. Individual animals maintained constant levels of bacterial and protozoal protein (standard deviation 28%) but there was massive variation between animals with protozoal protein varying from < 0.01 to 5.73 mg/ml and bacterial protein ranging from 0.58 to 5.76 mg/ml. However because of the negative correlation between bacterial and protozoal numbers the total microbial protein varied from only 1.33 to 6.42 mg/ml. It was shown that $y = 4.752 \log_{10} x$ where x = mg bacterial protein/ml and y = mg protozoal protein/ml.

Unfortunately all these calculations are complicated by the observation first made by Faichney and Griffiths (1978) and confirmed by Orpin and Letcher (1983/84) that on defaunation the rumen volume increases (mean 30%) and that the rumen flow decreases (mean 36%) with a resultant increase in the mass of digesta. The mean residence time of the suspended particulate matter was 0.75 days before defaunation and 1.52 days afterwards. However Veira et al. (1983) found no effect of defaunation on rumen volume or dilution rate; Kayouli et al. (1983/84) also found no effect on rumen volume but observed that the rate of flow of particles was doubled on defaunation.

Effect of Protozoa on Turnover of Bacterial Carbon and Nitrogen Nolan and Leng (1972) first showed that, in sheep fed lucerne hay, 4.3 g of the 14.2 g of nitrogen entering the rumen ammonia pool was recycled via microbial protein. This could have been caused by autolysis of the bacteria, lysis of the bacteria by bacteriophage, or engulfment and digestion of the bacteria by ciliate protozoa. Little is known about the first two, but Coleman (1975) pointed out that if entodinia grown in vivo took up, killed, and digested bacteria at the same rate as *Entodinium caudatum* grown in vitro, then in the presence of high protozoal populations (2×10^6/ml), 1% of the bacteria could be engulfed and digested each minute. After digestion, the bacterial amino acids (45 g/day in sheep) would be released into the rumen and be available as sources of energy, carbon, and nitrogen for the growth of the remaining bacteria. Subsequently, on the basis of studies on the engulfment of mixed rumen bacteria by protozoal species grown in vivo (Coleman and Sandford, 1979a), this estimate was revised downward to 0.27% of the bacteria (at 10^9 bacteria/ml) or 0.10% (at 10^{10} bacterial/ml) of the rumen bacteria digested each minute. These figures for bacterial engulfment and digestion would, of course, be considerably lower in animals fed on hay, where the population density would be approximately 2×10^5/ml, than in animals fed a restricted high barley ration.

Cottle et al. (1978) injected $^{15}NH_3$ into the rumen of a sheep fed on oaten chaff and sucrose and then measured the flow of nitrogen between the free ammonia and bacterial and protozoal protein fractions. Under these conditions the protozoa accounted for 38% of the microbial biomass, 1.3 g protozoal nitrogen, and 2.2 g bacterial nitrogen left in the rumen each day. The turnover of nitrogen directly between bacteria and protozoa amounted to 6 g (= 36 g protein) each day. Subsequently Leng and Nolan (1984) quoting Cottle (1980) showed that when 4% fish meal was added to the ration, 1.2 g protozoal nitrogen and 3.0 g bacterial nitrogen left the rumen each day and that the direct turnover of nitrogen between bacteria and protozoa was only 3.7 g/day.

Another approach to the measurement of turnover was made by measuring gross microbial nitrogen synthesis from the uptake of ^{32}P and the N/P ratio (Demeyer and Van Nevel, 1979; Van Nevel and Demeyer,

1977). The net synthesis was determined from the rate of disappearance of soluble nitrogenous compounds. These studies demonstrated that total nitrogen synthesis was unaffected by defaunation but that the net synthesis increased showing that the rate of degradation of total and net syntheses increased from 35 and 13 to 47 and 30 g nitrogen per kg organic matter fermented, respectively. When the results were expressed per mg microbial nitrogen in rumen fluid, the total synthesis increased from 106 to 164 μg nitrogen and the net synthesis from 44 to 101 μg nitrogen on defaunation which means the rate of degradation was almost unaffected by defaunation. The authors interpreted this as showing that the presence of protozoa did not alter the proportion of nitrogen recycling but inserted an energetically inefficient step in the overall synthesis of microbial matter. The rate of recycling found by Demeyer and Van Nevel (1979) was much higher (> 100%) than that found initially by Coleman (1975a).

Rates of engulfment and digestion (as percentage of the total bacterial biomass engulfed per minute) found by various authors are: 1.0 (Coleman, 1975); 0.05 (Cottle et al., 1978); 0.1–0.27 (Coleman and Sandford, 1979); 0.055% (Nolan and Stachiw, 1979); 2.5% (Demeyer and Van Nevel, 1979); 0.03% (Nolan and Leng, 1972).

Although it is assumed that the digestion of bacteria by rumen ciliates always involves engulfment by living protozoa and active release of digestion products, there are two other ways by which bacterial protein can be released into the medium.

Specific Lysis by Epidinial Enzyme The first involves the lysis of certain bacteria by extracellular enzymes released by *Epidinium* spp. (Coleman and Laurie, 1974b). Among these is a lysozyme-like enzyme which rapidly attacks the cell walls of bacteria such as *Bacillus megaterium*. Although sensitive bacteria (which include *Micrococcus lysodeikticus* and *Bacillus subtilis*) are not common rumen inhabitants, the enzymes released may have important actions on rumen bacteria.

Turnover as the Result of Protozoal Death The second is concerned with the low rate of passage from the rumen of protozoa compared with free liquid or bacteria. This can be as low as 10% (Leng et al., 1981) although it is more usually 20–40% (e.g., Harrison et al., 1979; Pilgrim et al., 1970). Leng et al. (1981) and Leng (1982) have postulated that the protozoa that remain in the rumen die there and lyse. It has been shown (Coleman 1985b) that this may not be due to chance but could be the result of exposure of the protozoa to air and/or hypotonic salt solutions for short times during drinking or ruminantion. If this occurred, the protozoa would die while full of bacteria, many of which would be damaged by protozoal enzymes and which would break up when the protozoa disintegrated. This mechanism would also release bacterial digestion products from those protozoal species which do not normally release bacterial digestion products into the medium. If there are 2×10^6 entodinia/ml and half of them die when each is full of 4000 bacteria (Coleman, 1964b), then $10^6 \times$ rumen volume (= 4500 ml) $\times$ protein in each bacterium (= 0.18 pg) $\times 4000 \div 10^{12} = 3.24$ g bacterial protein could be released from dying protozoa each day. In addition, if each entodinium contains 2 ng protein, $10^6 \times 2 \times 4500 \div 10^9 = 9$ g protozoal protein would be released at the same time. This combined release of about 12 g per day is similar to that of 10–45 g produced by the engulfment and digestion of bacteria by living protozoa. Both figures will increase or decrease as the numbers of protozoa rise or fall.

10.4 Metabolism of Protein

Effect of Protozoa on Protein Degradation Protein eaten by the host ruminant passes into the rumen where it is degraded by the microorganisms. The bacteria, protozoa, and fungi grow on these degradation products and then pass on down the alimentary tract of the host where they are killed and digested. These digestion products are then utilized by the host. The relative importance of the protozoa to the host therefore depends on the relative numbers of protozoa in the rumen, their growth rate, and hence the rate at which they synthesize protein, and finally on the rate at which the protozoa leave the rumen.

Factors, which are mostly dietary, that affect protozoal population density in the rumen are considered in Chapter 9. There is now good evidence that the protozoa leave the rumen more slowly than the bacteria and values of 20–40% of the rate have been commonly found although wide variations on either side have been reported. For detailed discussion of the results and the methodology see Chapter 9.

McDonald (1954, 1957) and Weller et al. (1958) first showed that feed protein was broken down and converted into microbial protein in the rumen. Blackburn and Hobson (1960) extended this work and showed that the proteolytic activity for the breakdown on casein had an optimum pH of 6–7 and that there was appreciable activity in all fractions including those containing all the protozoa. It was noted that after disruption of the protozoal fraction and removal of the debris by centrifugation, some soluble activity remained. Although the experiments to be described below are concerned with the digestion of feed proteins, it must be remembered that as far as the protozoa are concerned, bacteria are the principal source of amino acids for growth.

Since 1960 progress on investigations into the role of protozoa in rumen proteolysis has been slow. Many of the experiments undertaken have been carried out with washed suspensions of protozoa and these are considered in Chapters 6 and 8. These have shown that there is considerable confusion over the existence of proteolytic activity in protozoa and the optimum pH. There is evidence that entodiniomorphid protozoa take up and digest insoluble, but not soluble, casein whereas *Isotricha* spp. take up both soluble and insoluble casein (Onodera and Kandatsu, 1970b).

These studies have been followed up recently in vivo by Ushida and Jouany (1985, 1986) who studied the rate of degradation of four proteins placed in nylon bags in the rumens of faunated and ciliate-free sheep. The results showed that protozoa are unimportant in the degradation of soluble proteins but became progressively more important the more insoluble the proteins became. The results also suggested that the larger protozoa were more important than the small in the digestion of insoluble proteins. This is in contrast to the results obtained with individual species in vitro by Coleman (1983) for the breakdown of leaf fraction 1 protein and for the breakdown of the protein of intact bacteria by Wallace and McPherson (1987) where the small entodiniomorphs were found to have the highest activity.

In a continuation of this work, Ushida et al. (1986) compared the rates of degradation of feed protein in faunated and ciliate-free sheep given two isonitrogenous diets. The first (designated L) consisted of lucerne hay, the protein of which was of low rumen degradability but which supported a high population of entodinia and *Polyplastron multivesiculatum*. The other ration (S) was NaOH-treated straw, the protein of which was readily degradable in the rumen but which supported only 40–50% of the protozoal population of diet L. The authors measured non-ammonia nitrogen entering the duodenum and subtracted from this endogenous nitrogen flow and microbial nitrogen flow estimated by three different methods. Although there were differences between the methods, faunation increased feed nitrogen digestibility by 95% in animals given diet L and by 32% with diet S. However, in defaunated animals nitrogen digestion was 76% higher with diet S than L whereas in faunated animals it was only 18% higher. It is tempting to ascribe these differences to the action of the protozoa on feed proteins, but the changes in bacterial populations, the increased retention time of rumen material in the presence of protozoa amounting to 10–27% for liquid, 14–22% for solid material, and 47–99% for bacterial proteins, and the more extensive digestion of organic matter may all be factors in the above results.

Effect of Protozoa on Microbial Protein Synthesis In vivo It has been shown above ("Bacterial Ingestion" section) that on defaunation the bacterial population density increases; it might therefore be expected that the rate of bacterial protein synthesis would also increase, and this has been found in practice. Lindsay and Hogan (1972) found that in sheep fed on lucerne or red clover, the bacterial nitrogen leaving the rumen increased from 12.0 to 14.0 g/day and from 18.0 to 19.3 g/day respectively on defaunation. The efficiency (g nitrogen incorporated per kg organic matter fermented) also increased from 37 to 50 and from 53 to 60 respectively. Similar results were found by Demeyer and Van Nevel (1979) using sheep fed on hay and concentrates and by Ushida et al. (1986) using sheep fed on lucerne or alkali-treated straw.

Protozoal Protein Synthesis In vivo The amount of protein synthesis by the protozoa in the rumen depends on a variety of interrelated factors, and we are still lacking all the information needed. In the earlier work carried out before 1979, it was assumed that the population density of the ciliates gave an accurate measure of their importance to the nutrition of the host (e.g., Ibrahim and Ingalls, 1972). We now know that this is not true because the protozoa pass out of the rumen more slowly than the free liquid or bacteria. However, there have been a number of determinations of the proportion of the microbial protein present in protozoa and this usually varies from 20–40% depending on the protozoal population density (Table 10.6). Obviously, if the density is low because of the nature of the feed, the proportion of protein in the protozoa will be low, although, as shown by Punia and Leibholz (1984) with cattle fed rations with and without urea, the highest protozoal mass (g dry matter/liter) was associated with the lowest proportion of protozoal nitrogen in microbial nitrogen or in total nitrogen and vice versa.

In the more recent studies shown in Table 10.6, measurements have been made of the relative rates of protein synthesis in bacteria and protozoa in the rumen. Leng et al. (1981, 1986), in studies on cattle fed on sugar cane or freshly cut pasture, found that 4% and 9% respectively of the net microbial protein synthesis in the rumen was protozoal. However, in both these studies the population density of large protozoa (largely *Isotricha* spp.) was unusually high (50,000/ml) and even in the first, where 400,000 entodinia/ml were present, the authors estimated that 90% of the protozoal biomass was represented by large protozoa. In contrast, when most of the protozoa were entodinia (e.g., $1\text{–}3 \times 10^6$/ml with $1\text{–}10 \times 10^3$ *Polyplastron multivesiculatum*/ml—Leng, 1982) a much higher proportion (32%) of the microbial protein synthesis was protozoal, indicating a much higher growth rate of entodinia than holotrichs (Foulkes and Leng, quoted by Leng and Nolan, 1984).

10.5 Effect of Protozoa on Rumen Metabolites

The physical characteristics of the rumen environment are changed on defaunation, although the nature of the observed change is not always consistent. The rumen volume and retention time have been reported to decrease following defaunation (Kayouli et al., 1984; Ushida et al., 1986a). However, in other studies (Argyle and Forster, 1989; Demeyer, 1988; Faichney and Griffiths, 1978; Orpin and Letcher, 1984) both rumen volume and retention time increased in the absence of ciliate protozoa; the dry matter content of

Table 10.6 Microbial and protozoal protein synthesis in the rumen

Animal	Ration	Protozoal population	Microbial N present in protozoa (%)	Contribution of protozoa to microbial protein synthesis (%)	Protozoal N derived from bacteria (%)	Reference
Sheep		Mixed	33			Weller et al. (1958)
Sheep	Hay	Mixed	20			Pilgrim et al. (1970)
Sheep	Alfalfa pellets	Mixed	58			Hungate et al. (1971)
Cows	Semipurified	Mixed	44.4			Ibrahim and Ingalls (1972)
Sheep	Hay	Mixed	50.5			Weller and Pilgrim (1974)
Zebu bulls	Sugar cane	Large protozoa		4		Leng et al. (1981)
Sheep	Molasses-chaff	Entodinia		32		Leng (1982)
Cattle	NaOH straw	Mixed	37			Punia and Leibholz (1984)
	NaOH straw + urea		25			
Cattle	Grass	Large protozoa		9		Leng et al. (1986)
Sheep	Sunflower meal	Mixed	27.2		35–45	Faichney et al. (1989)

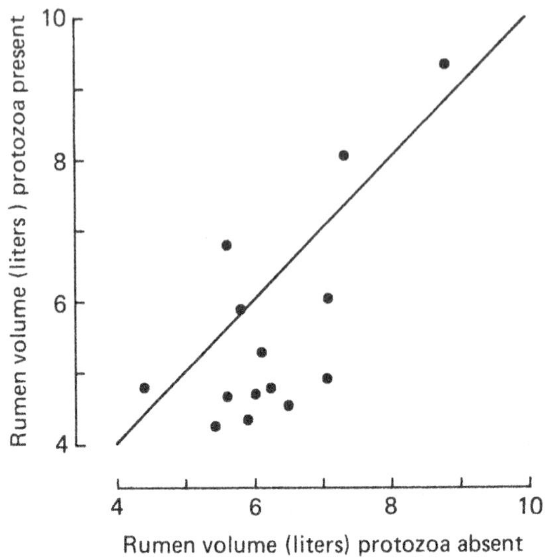

Fig. 10.2 The effect of the presence and absence of rumen ciliate protozoa on the ovine ruminal volume. The line of equal values in faunated and ciliate-free animals is shown. The data used were obtained from the appropriate references given in Table 10.7.

rumen contents was also higher (Faichney and Griffiths, 1978; Klopfenstein et al., 1966; Males and Purser, 1970). Such changes in rumen volume (Fig. 10.2) and retention time are likely to influence feed intake and the rate of digestion with eventual effects on animal productivity (Demeyer, 1989).

The total concentration and proportion of the acidic fermentation products are different in faunated and defaunated animals (Table 10.7). The rate of bacterial and protozoal volatile fatty acid production is similar (Michalowski, 1987), and, in the majority of studies, defaunation of the rumen is associated with a decrease in the total volatile fatty acid concentration (Fig. 10.3) although in some experiments VFA levels were higher in the defaunated state (Grummer et al. 1983; Stern and Hinkson, 1974) or alternatively were not affected by defaunation (Hinkson et al., 1976; Itabashi et al., 1982). The inoculation of a single protozoon species or genus (*Polyplastron multivesiculatum* or *Entodinium* spp.) to meroxenic lambs resulted in an increase in the ruminal VFA concentration and changes in the molar proportions of the principal acids acetate, propionate, and butyrate (Fonty et al., 1983). In defaunated animals there is also evidence to indicate that the higher population of rumen bacteria may result in increased fermentation of the VFA thereby reducing the availability of VFA, and hence energy, to the host (Rowe et al., 1985).

Changes in the molar proportions of the volatile fatty acids after defaunation are not consistent (Figure 10.3). Examination of the published data in the literature will show that the molar proportions of the volatile

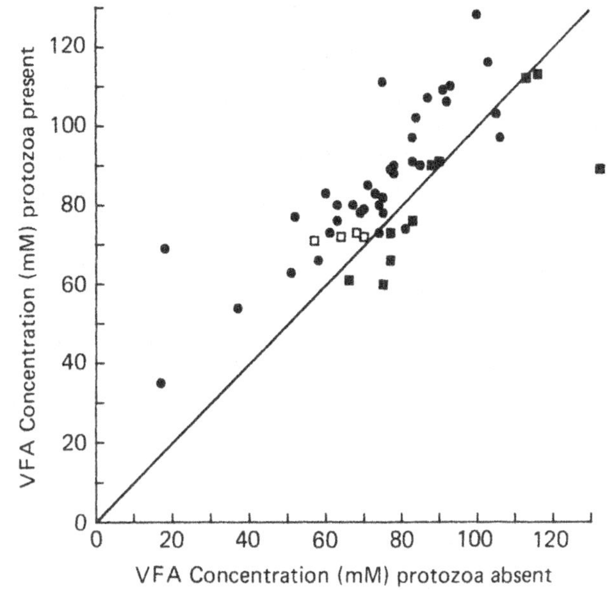

Fig. 10.3 The effect of the presence and absence of rumen ciliate protozoa on the concentration of volatile fatty acids (VFA) in the rumen of sheep (●), zebu and buffalo (□), and cattle (■). The line of equal values in ciliate-free and faunated animals is shown. The data for this figure were collated from studies reported in Table 10.7

fatty acids, in animals receiving similar diets, have been reported both to increase and decrease in the absence of protozoa (Figs. 10.4, 10.5, 10.6). It is, therefore, apparent that the observed changes in the fermentation profiles are not directly attributable to the protozoa but arise as a consequence of the concomitant changes in the rumen bacterial population that occur on defaunation. However, the volatile fatty acids produced by the rumen microbial population are absorbed and utilized by the host and hence alterations in the proportions of the acids formed will have consequences for the metabolic capabilities of the host. Defaunation of lactating cows resulted in lower ruminal concentrations of acetate and butyrate with a consequential depression of milk fat synthesis (Chalupa et al., 1967).

Defaunation also influences the ruminal concentrations of other microbial metabolites. A decrease in rumen ammonia levels as a result of defaunation is probably the most consistent of the observed effects of protozoal elimination (Fig. 10.7). The establishment of *P. multivesiculatum* or *Entodinium* spp. in the rumen of previously ciliate-free meroxenic lambs resulted in a marked increase in ruminal ammonia-nitrogen concentration (Fonty et al., 1983). Inoculation of ciliate-free lambs with *Diplodinium* spp., *Isotricha* spp., or *Ophryoscolex* spp. likewise resulted in increased ammonia and VFA concentrations (Christiansen et al., 1965).

Table 10.7 The change in environmental characteristics of the rumen when protozoa are absent

Environmental characteristic	Change in characteristic when protozoa absent		
	Increase	Decrease	No significant effect
Rumen volume	3, 18, 48[a]	34, 51, 60	5, 41, 44, 60, 65, 66
Retention time	3, 18, 48	34, 51, 60	41, 65, 66
Bacterial population	7, 14, 16, 17, 24, 34, 36, 39, 40, 46, 48, 53, 54, 55, 57, 61	3	
Fungal population	52		
ATP concentration		47, 68	
Rumen pH	12, 28, 29, 72	13, 16, 20, 28, 33, 43, 64, 65, 67, 69	10, 11, 15, 25, 32, 34, 35, 39, 40, 42, 44, 46, 48
VFA concentration	20, 21, 23, 35, 39, 43, 51, 58, 65, 66, 69	1, 2, 6, 10, 12, 16, 19, 22, 27, 28, 29, 32, 33, 34, 35, 39, 40, 41, 42, 46, 49, 55, 56, 59, 63, 66, 72	15, 25, 37, 38, 71
Molar proportions (mol%) of:			
Acetic acid	2, 12, 15, 21, 24, 29, 40, 64, 71	4, 9, 10, 14, 20, 23, 28, 34, 35, 40, 43, 58, 63, 69, 70, 72	25, 26, 32, 35, 46, 55, 60, 66

[a]References are as follows: 1, Abou Akkada (1965); 2, Abou Akkada and El Shazly (1964); 3, Argyle and Forster (1989); 4, Bird and Leng (1978); 5, Bird and Leng (1983); 6, Borhami et al. (1967); 7, Bryant and Small (1960); 8, Chalmers et al. (1968); 9, Chamberlain et al. (1981); 9, Chamberlain et al. (1983); 11, Chamberlain et al. (1985); 12, Christiansen et al. (1965); 13, Conrad et al. (1958); 14, Demeyer and Van Nevel (1979); 15, Demeyer et al. (1982b); 16, Eadie and Gill (1971); 17, Eadie and Hobson (1962); 18, Faichney and Griffiths (1978); 19, Fonty et al. (1983); 20, Grummer et al. (1976); 24, Itabashi and Katada (1976a); 25, Itabashi et al. (1982); 26, Itabashi et al. (1984); 27, Itabashi and Matsukawa (1979); 28, Jouany and Senaud (1982); 29, Jouany and Senaud (1983); 30, Jouany et al. (1981a); 31, Jouany and Thivend (1983); 32, Jouany et al. (1981b); 33, Kahlon et al. (1970); 34, Kayouli et al. (1984); 35, Kayouli et al. (1986); 36, Klopfenstein et al. (1966); 37, Kreuzer and Kirchgessner (1986b); 38, Kumar and Raghavan (1978); 39, Kurihara et al. (1968); 40,

Lactic acid levels are usually higher in defaunated animals (Fig. 10.8) and the rate of metabolism of both endogenously and exogenously supplied lactate in the rumen is slower (Chamberlain et al., 1981, 1983; Newbold et al., 1986a, 1986b). Lactic acid depresses fibre degradation (Fay and Ovejero, 1986) and elevated ruminal levels of lactic acid can lead to the onset of the potentially harmful nutritional disorder lactic acid acidosis. The protozoa consume lactate more rapidly than the bacteria (Newbold et al., 1987) and appear to be important in minimizing the potentially adverse effects of lactic acid.

Methanogenesis is also decreased as a consequence of defaunation (Demeyer and Van Nevel, 1979; Itabashi et al., 1984; Kreuzer et al., 1986; Ushida et al., 1986b). The ciliates alter the proportions of rumen volatile fatty acids formed and this metabolic shift has been suggested as the cause

Table 10.7 *Continued*

Environmental characteristic	Change in characteristic when protozoa absent: Increase	Decrease	No significant effect
Propionic acid	13, 14, 20, 23, 25, 26, 28, 32, 34, 35, 40, 43, 51, 58, 60, 63, 69, 70, 72	2, 9, 10, 11, 12, 15, 16, 21, 24, 29, 40, 42, 64, 71	4, 35, 46, 55, 66
Butyric acid	2, 4, 9, 10, 11, 15, 25, 40, 41, 55, 64, 69	11, 12, 13, 14, 19, 20, 21, 23, 25, 26, 27, 28, 29, 32, 34, 35, 36, 40, 41, 46, 51, 60, 63, 70, 71, 72	66
Concentration of:			
Formic acid		2	
Lactic acid	9, 10, 15, 32, 34, 45, 46, 59	23, 28, 29	35, 64
Bicarbonate	55		
Ammonia		1, 2, 6, 8, 11, 12, 14, 15, 16, 19, 21, 24, 25, 26, 27, 28, 29, 31, 33, 34, 35, 36, 39, 40, 41, 42, 43, 46, 48, 50, 51, 59, 60, 62, 64, 65, 66, 67	38, 55

Kurihara et al. (1978); 41, Lindsay and Hogan (1972); 42, Luther et al. (1966); 43, Males and Purser (1970); 44, Meyer et al. (1986); 45, Nagaraja et al. (1986); 46, Newbold et al. (1986b); 47, Nuzback et al. (1983); 48, Orpin and Letcher (1984); 49, Osman et al. (1970); 50, Perkins and Luther (1967); 51, Punia et al. (1987); 52, Romulo et al. (1986); 53, Rowe et al. (1981); 54, Rowe et al. (1983); 55, Rowe et al. (1985); 56, Sedloev (1970); 57, Singh and Makkar (1976); 58, Stern and Hinkson (1974); 59, Takahashi and Kametaka (1976); 60, Ushida et al. (1986a); 61, Ushida et al. (1989a); 62, Ushida et al. (1989b); 63, Ushida et al. (1986b); 64, Van Nevel et al. (1985); 65, Veira et al. (1981); 66, Veira et al. (1983); 67, Veira et al. (1984); 68, Wallace and West (1982); 69, Whitelaw et al. (1972); 70, Whitelaw et al. (1984); 71, Williams and Dinusson (1973b); 72, Youssef and Allen (1968).

of the differing rates of methane formation in faunated and ciliate-free animals (Whitelaw et al., 1984). However, the fate of the methanogenic population after defaunation has not been determined, although the protozoa are known to be hydrogenogens and interact with the methanogens in the rumen (Krumholz et al., 1983; Stumm et al., 1982). In addition, the protozoa consume oxygen and may have a significant role in ruminal oxygen scavenging and help to control oxygen fluxes that are detrimental to more susceptible microorganisms (Ellis et al., 1989a, 1989b). Oxygen levels increase transiently after feeding in defaunated sheep (Hillman et al., 1985b); these higher oxygen levels would also inhibit hydrogen formation and methanogenesis (Scott et al., 1983). Methanogenesis is an energetically wasteful process for the ruminant and its suppression may, therefore, be beneficial.

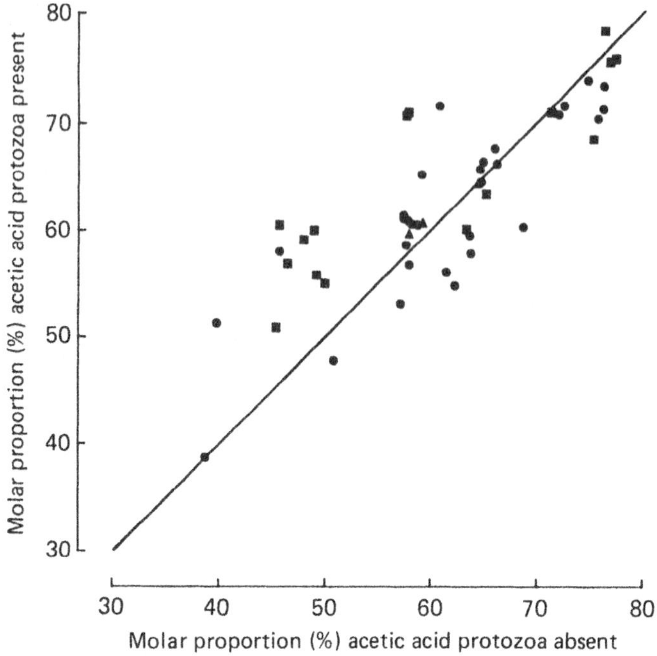

Fig. 10.4 The effect of the presence and absence of rumen ciliate protozoa on the molar proportion of acetic acid in the ruminal VFA of sheep (●), goats (▲), and cattle (■). The line of equal concentrations in ciliate-free and faunated animals is shown. The sources of the data for this figure are given in Table 10.7

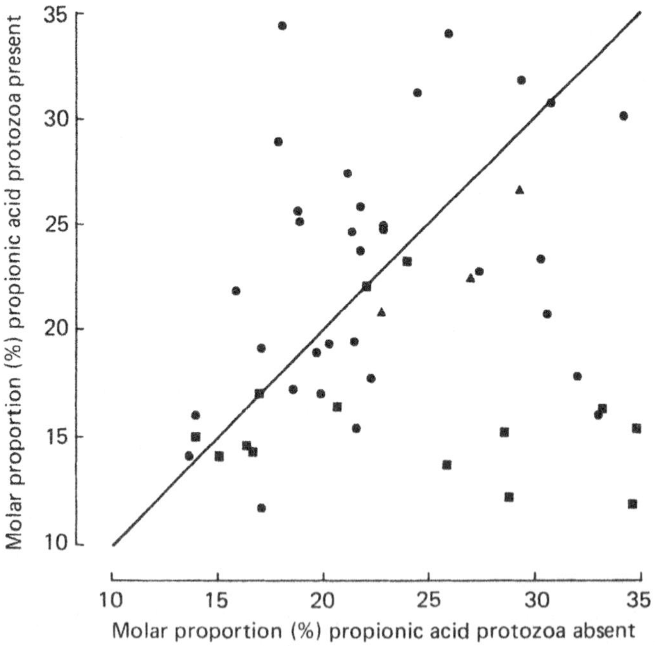

Fig. 10.5 The effect of the presence and absence of rumen ciliate protozoa on the molar proportion of propionic acid in the ruminal VFA of sheep (●), goats (▲), and cattle (■). The line of equal concentration in faunated and ciliate-free hosts is shown. All data used were collated from studies referred to in Table 10.7.

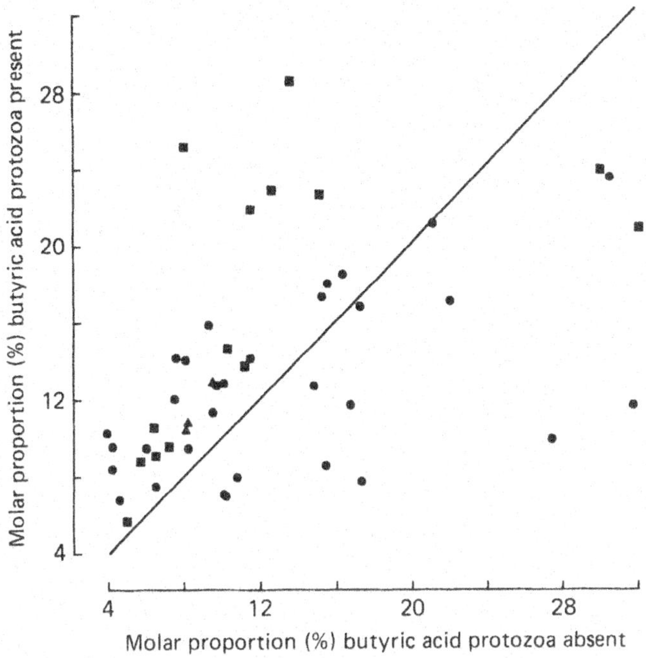

Fig. 10.6 The effect of the presence and absence of rumen ciliate protozoa on the molar proportion of butyric acid in the ruminal VFA of sheep (●), goats (▲), and cattle (■). The line of equal concentration in faunated and ciliate-free ruminants is indicated. All data in the figure were obtained from studies referred to in Table 10.7.

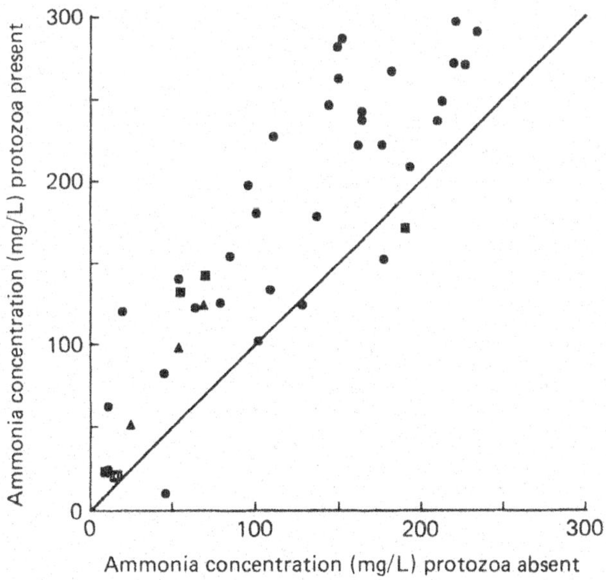

Fig. 10.7 The effect of the presence and absence of rumen ciliate protozoa on the concentration of ammonia in the rumen contents of sheep (●), zebu and buffalo (□), cattle (■), and goats (▲) receiving a range of diets. The line shows equal concentrations in faunated and ciliate-free hosts. The data sources for this figure are included in Table 10.7

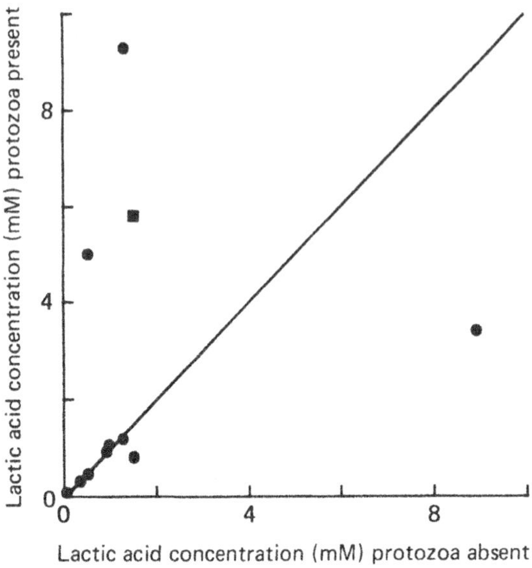

Fig. 10.8 The effect of the presence or absence of rumen ciliate protozoa on the concentration of lactic acid in sheep (●) and cattle (■) receiving a variety of rations. The line shows equal concentrations in ciliate-free and faunated animals. The data used were taken from the appropriate studies referred to in Table 10.7

The rumen pH, and the extent of the post-prandial fall in pH, differ in faunated and defaunated animals (Fig. 10.9). In general the pH of the ruminal contents of defaunated animals is lower irrespective of the ration consumed (Table 10.7). For example, in animals receiving a corn silage:corn (1:1) diet the post-feed variation in pH of the faunated animals was in the range pH 6.3–pH 6.6 whereas in the defaunated animals the range was 5.95–6.85 (Veira et al., 1983). However, about 8 hours after feeding, the pH of the rumen contents of the defaunated sheep was 0.2 pH units higher than the value measured in faunated animals (i.e., pH 6.8 and pH 6.6 respectively). The opposite effect has been observed, and in other studies (Christiansen et al., 1965; Eadie and Gill, 1971; Youssef and Allen, 1968) the pH of faunated rumen contents was 0.4–0.5 pH units lower. In the study of Christiansen et al. (1965) refaunation with a single protozoal genus was sufficient to lower the ruminal pH as compared to the defaunated control, although the extent of the pH changes during the post-feed period were not indicated. Protozoal stabilization of the environmental pH during the period of active ruminal fermentative activity may be significant as bacterial growth (Russell and Dombrowski, 1980; Russell et al., 1979) and enzymic activities (e.g., cellulolysis; Stewart, 1977) are subject to inhibition at lower pH values.

The effects of protozoa on ruminal lipid metabolism are less well-established. The protozoa contribute to the total microbial lipolytic activity

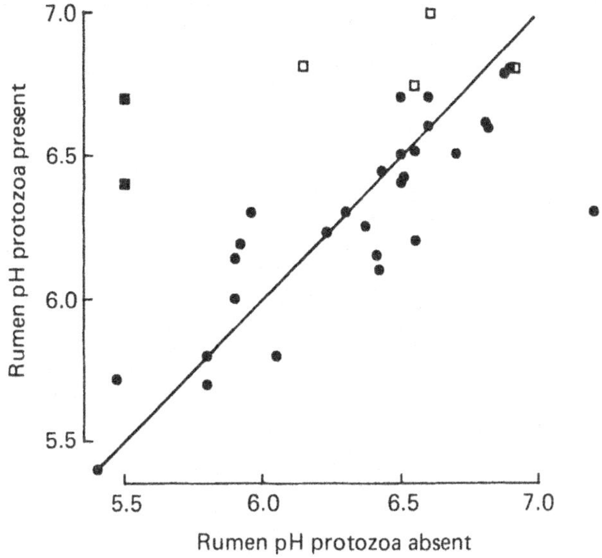

Fig. 10.9 Effect of the presence and absence of rumen ciliate protozoa on the ruminal pH of sheep (●), zebu and buffalo (□), and cattle (■). The line represents equal values in faunated and ciliate-free host ruminants. The data used were collated from references quoted in Table 10.7.

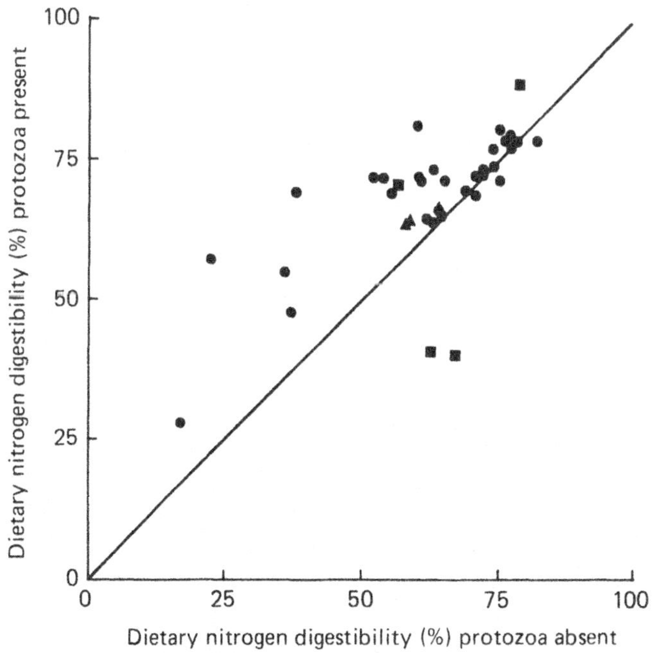

Fig. 10.10 Effect of the presence and absence of rumen ciliate protozoa on the digestion of dietary nitrogen in the rumen of sheep (●), goats (▲), and cattle (■). The line shows the position at which digestibility was equal in both the faunated and ciliate-free condition. The references from which the data were collated are included in Table 10.1

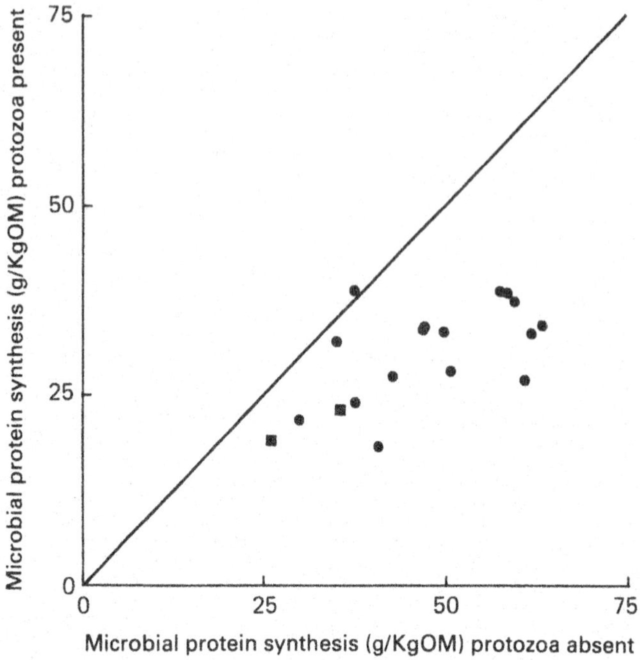

Fig. 10.11 The effect of the presence and absence of rumen ciliate protozoa on microbial protein synthesis in the rumen of sheep (●) and cattle (■). The line of equal values in faunated and ciliate-free ruminants is indicated. The data were compiled from relevant studies used in Table 10.1.

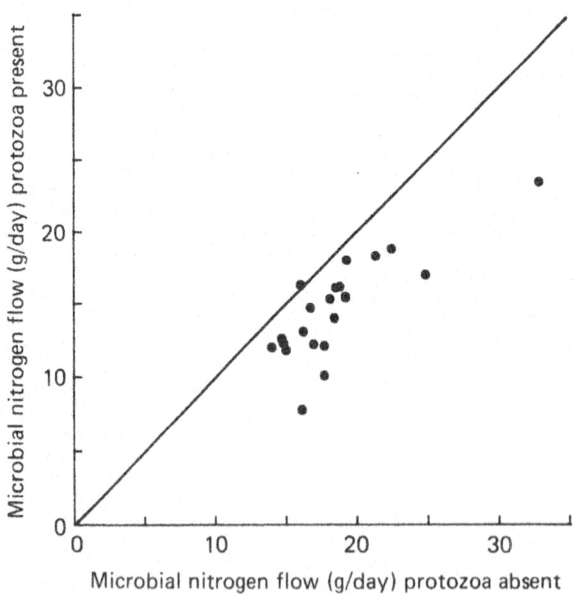

Fig. 10.12 The effect of the presence and absence of rumen ciliate protozoa on the flow of microbial nitrogen to the lower intestinal tract (duodenum) from the rumen of sheep. The line indicates the values at which microbial nitrogen flow was equal in the faunated and ciliate-free state. The data was obtained from the references included in Table 10.1

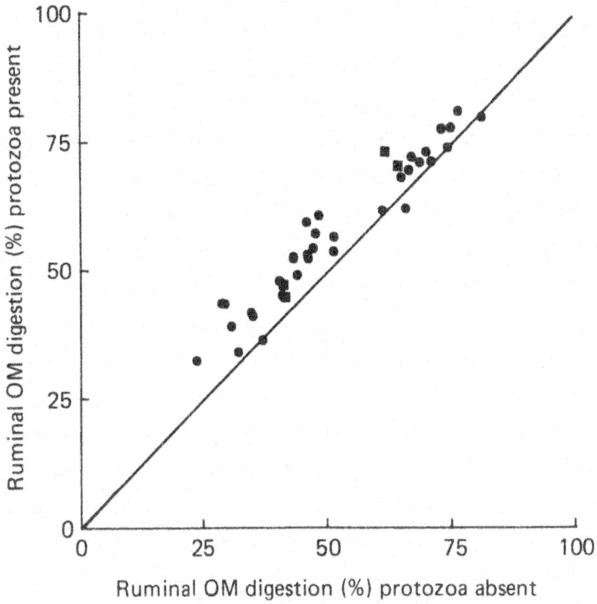

Fig. 10.13 Effect of the presence and absence of rumen ciliate protozoa on the digestion of organic matter in the rumen of sheep (●) and cattle (■). The line represents the points of equal value in the faunated and ciliate-free condition. The data used in this figure were obtained from studies referred to in Table 10.1.

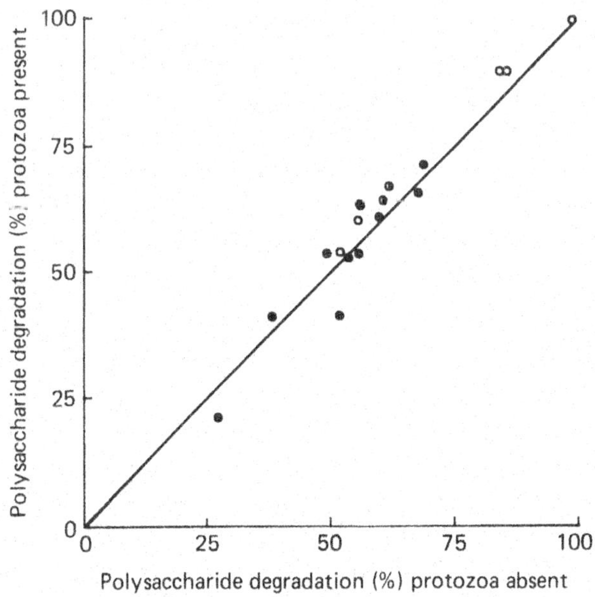

Fig. 10.14 Effect of the presence and absence of rumen ciliate protozoa on the digestion starch (○), cellulose (●), and hemicellulose (◑) in the ovine rumen. The line shows where digestion was equal in faunated and ciliate-free hosts. The data was obtained from references listed in Table 10.1.

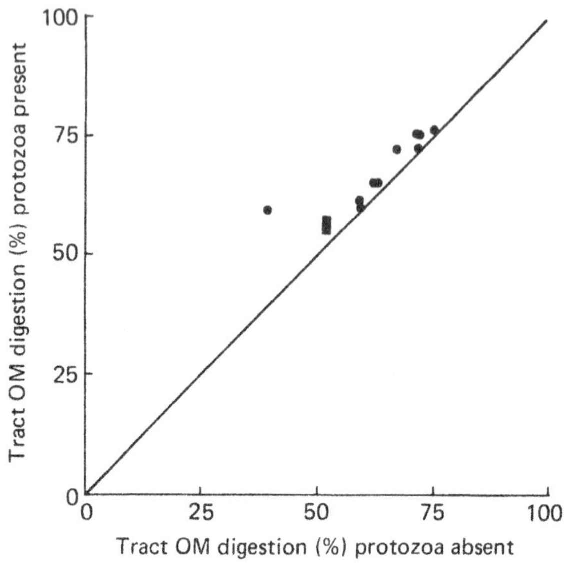

Fig. 10.15 The effect of the presence or absence of ciliate protozoa in the rumen on the digestibility of organic matter during passage along the complete gastrointestinal tract of sheep (●) or cattle (■). The line of equal digestibility values in the faunated and ciliate-free animals is shown. The data were obtained from studies referred to Tablet 10.1

(Latham et al., 1972), but their role in biohydrogenation is less well-defined. Although the protozoa have been shown to hydrogenate lipids in in vitro incubations (Wright, 1959), the ability of rumen contents to biohydrogenate was only slightly reduced following defaunation (Dawson and Kemp, 1969). However, Chalupa et al. (1967) believed that the increased levels of oleate and linoleate in the milk of animals with depressed milk fat tests were attributable to the virtual absence of protozoa in the rumens of these cattle. The effects of defaunation on the levels of polyunsaturated fatty acids in blood are variable (See Chapter 11).

The ingestion of bacteria and the proteolytic activities of the protozoa decrease the efficiency of protein utilization by the host (Demeyer and Van Nevel, 1986; Kreuzer and Kirchgessner, 1989; Leng and Nolan, 1984; Nolan, 1989). The protozoa increase the proteolytic activity of rumen fluid (Shinchi et al., 1986; Wallace et al., 1987) and have a positive effect, under certain dietary conditions, on protein degradability (Fig. 10.10; Matsumoto et al., 1989; Ushida and Jouany, 1986; Ushida et al., 1984, 1989b). The protozoa are more important in the degradation of insoluble proteins (Hino and Russell, 1987; Ushida and Jouany, 1985) and in the turnover of bacterial proteins (Wallace and McPherson, 1987). In defaunated animals, the levels of the protein breakdown product, ammonia, are lower and the quantity of non-ammonia nitrogen available in the lower intestinal tract is increased (Table 10.7), although amino acid disappearance from defaunated rumen

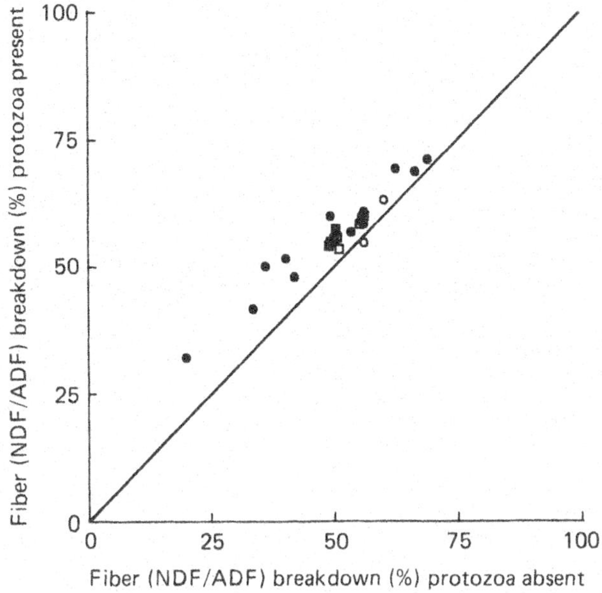

Fig. 10.16 The effects of the presence or absence of ciliate protozoa in the rumen on the breakdown of neutral detergent fiber (NDF) and acid detergent fiber (ADF) in rumen (closed symbols) or after passage along the complete gastrointestinal tract (open symbols) of sheep (●, ○) and cattle (■, □). The position of the line is that of equal digestion in the faunated and ciliate-free animal. The information used in this figure was obtained from studies referred to in Table 10.1.

contents is greater (Stuart et al., 1976). The metabolism of urea is also affected (Males and Purser, 1970; Takahashi and Kametaka, 1976); selenomethionine metabolism, however, is similar in faunated and defaunated animals (Hidiroglu and Jenkins, 1974). The outflow of protozoal protein from the rumen does not compensate the ruminant for the losses in dietary and bacterial nitrogen caused by the protozoal activities (Demeyer and van Nevel, 1979; Leng and Nolan, 1984). In consequence, defaunation decreases the degradation of dietary protein and increases microbial cell (i.e., protein) synthesis, so that an increased amount of protein is available for intestinal digestion by the host (Figs. 10.11, 10.12). The protozoa directly influence the proportions of protein and energy in the ruminal effluent that are available for metabolism and absorption in the intestinal tract.

A further effect of defaunation is to increase the extent of the post-ruminal fermentation. Although the ruminal digestion of starch and cell wall polysaccharides (organic matter) is reduced in protozoa-free animals (Figs. 10.3, 10.14), the differences in the total losses of material are less pronounced after passage along the whole intestinal tract (Figs. 10.15, 10.16). The fermentation at other locations in the small intestine and hind gut will not, however, completely negate the adverse effects of the decreased ruminal fermentation (Table 10.3).

11

Effect of Ciliate Protozoa on the Productivity of the Host Ruminant

Ciliate protozoa occur naturally in all wild and domesticated ruminants. A protozoal population becomes established in the rumen of young animals in the first few weeks of life through maternal contact and by transfer from other members of the flock or herd (Coleman, 1980; Chapter 9). However, the protozoa are not essential for the development and survival of ruminants because animals can be reared without a ciliate population developing and maintained ciliate-free providing all contact with faunated animals is avoided (Becker and Hsiung, 1929; Eadie, 1962a). Alternatively, protozoa may be eliminated from the rumen of mature animals by specific chemical or physical treatments (Section 4.2). Comparisons of ciliate-free (or defaunated) animals with (re)faunated animals have yielded much information on the effects of the protozoa (e.g., Jouany et al., 1988; Nolan et al., 1989; Veira, 1986; Williams and Coleman, 1988). The initial studies were undertaken over fifty years ago (Becker et al., 1929; Becker and Everett, 1930) and despite the many studies since then there is still no common concensus on the value of the protozoa to the host animal.

The protozoa contribute to the nutrition of the host through their metabolic activities; the metabolites and cells formed represent a potential source of nutrients for the host. The protozoal population influences the environmental characteristics of the rumen and modulates the size, composition, and activities of the bacterial population. The protozoa have also been implicated in the etiology of certain rumen disorders. The role of the protozoa in the rumen metabolic processes was discussed in the previous chapter.

11.1 Post-ruminal Digestion of Protozoa

The nutritional value of the cellular protein, carbohydrate, and lipid fractions of the protozoa to the host is dependent upon the rate of flow of protozoal biomass out of the rumen and the digestibility of the cellular constituents in the lower digestive tract. Most diets support a large protozoal population in the rumen which can represent over half of the microbial biomass present. Analysis of the protein and lipid composition of the ciliates confirmed the potential nutritional value of protozoal cells and it was, therefore, assumed that the protozoa were an important source of nutrients for the host. However, the protozoa are selectively retained within the rumen and the full nutritional value of the ciliates cannot be gained by the host ruminant.

Although the full protozoal biomass is not available for post-ruminal digestion, the protozoa do contribute in the nutrition of the host and approximately one quarter of the microbial protein available to the host is protozoal in origin (Coleman, 1979a). The biological value of bacterial and protozoal protein is similar, although the ciliate protein is more susceptible to digestion (Bergen et al., 1968; Johnson et al, 1944; McNaught et al., 1954). Fungal proteins may also have a high biological value (Connell et al., 1987; Gulati and Ashes, 1987; Kemp et al., 1985). The amino acid compositions of individual protozoal genera (Harmeyer, 1966; Holler and Harmeyer, 1964; Williams and Dinusson, 1973a) and mixed protozoal preparations from different hosts including the camel (Ghosal and Lahiri, 1986), goat (Holler and Harmeyer, 1964), and cattle (Williams and Dinusson, 1973a) have been determined and, although the same amino acids are present, the proportions differ. Rumen protozoal, fungal, and bacterial proteins are likewise similar (Table 11.1). However, lysine availability to the host is improved in faunated animals as the amino acid has been shown to be limiting in defaunated lambs (Klopfenstein et al., 1966). The amount of lysine in the rumen contents of faunated animals is higher than the levels measured in samples from defaunated animals (Onodera, 1986; Onodera and Koga, 1987; Onodera and Suzuki, 1987; Onodera et al., 1988). It has been speculated that the protozoa may accumulate and conserve amino acids that are deficient in plants (Coleman, 1975a).

The reserve polysaccharides are the principal carbohydrates present in the protozoa and the quantities available for digestion by the host are unlikely to be of nutritional significance (Heald, 1951). The deposition and subsequent utilization of the reserve material by the ciliates does, however, stabilize and prolong the period of volatile fatty acid production in the rumen.

The protozoal cellular lipids contain proportionally more unsaturated fatty acids than the bacterial lipid fraction and are a potential lipid source for the ruminant. Approximately 10–20% of the lipid present in rumen digesta is microbial and about three-quarters of this is protozoal (Katz and Keeney,

Table 11.1 Amino acid composition of the rumen microorganism[a]

Amino acid	Rumen contents (goat)[b]	Ciliate-free contents (goat)[b]	Bacteria	Mixed protozoa (camel)	*Isotricha intestinalis* + *I. prostoma*	*Isotricha* spp.
Alanine	6.2	6.5	7.8	0.32	5.9	4.4
Arginine	5.1	5.6	4.9	0.08	4.4	9.5
Aspartic acid	11.3	11.1	10.8	0.11	17.4	11.7
Cystine	ND	ND	0.9	0.79	0.9	2.0
Glutamic acid	12.7	13.0	13.7	0.13	12.5	10.6
Glycine	5.5	5.8	5.2	0.13	4.5	5.3
Histidine	2.3	2.5	1.7	0.24	1.6	8.1
Isoleucine	5.7	5.2	6.6	0.16	6.9	4.2
Leucine	8.2	9.2	7.5	0.12	8.2	5.4
Lysine	7.4	6.4	8.2	0.01	11.4	12.2
Methionine	2.2	1.6	2.1	0.04	2.2	1.3
Ornithine	ND	ND	ND	0.01	ND	ND
Phenylalanine	7.8	6.2	5.3	ND	2.6	2.2
Proline	4.8	5.7	4.6	0.002	3.6	3.8
Serine	4.8	5.3	4.7	0.16	4.1	4.9
Threonine	5.4	5.5	5.3	0.11	5.3	4.3
Tryptophan	ND	ND	ND	0.01	ND	ND
Tyrosine	4.4	3.9	3.9	0.08	3.4	2.2
Valine	6.1	6.5	6.8	0.63	5.1	3.3
Aminoethyl phosphonic acid	ND	ND	ND	ND	0.17	0.17

[a] Relative proportions expressed as g per 100 g or (a) mol per 100 mol. ND, not detected. Table compiled using data from: Connell et al. (1987); Ghosal and Lahiri (1986); Holler and Harmeyer (1964); Kemp et al.

1967; Keeney, 1970). Unsaturated dietary fatty acids may be partially protected from biohydrogenation by incorporation into the structural phospholipids of the protozoa. Girard and Hawke (1978) estimated that 20–25% of the linoleic acid in the rumen of a cow, receiving an all-hay diet, was present in the lipids of the relatively small *Isotricha* population (10^3–10^4 cells per ml). A proportion of the polyunsaturated fatty acids that reach the lower digestive tract may thus originate from protozoal lipids. The ciliate phospholipids are an important source of choline. Dietary choline is degraded rapidly in the rumen and following defaunation the choline-containing phospholipids, phosphatidylcholine, and sphingomyelin, could not be detected in the rumen microbial population (Dawson and Kemp, 1969; Neil et al., 1979). The rumen contents of defaunated sheep also do not contain phosphatidylhydroxyethylalanine- or ciliatine-containing phospholipids (diglyceride aminoethylphosphonate, ceramide aminoethylphosphonate). The role of protozoa in the synthesis and utilization of vitamins in the rumen has not been established, although thiamin (vitamin B_1) is present in, and broken down by, protozoal suspensions during incubations in vitro (Michalowski et al., 1981).

The protozoal contribution to the nutrition of the host is ultimately

Dasytricha spp.	*Ophryo-scolex* spp.	*Ento-dinium* spp.	*Ent. caudatum* + *Ent. simplex*	*Neocallimastix* spp.	*N. frontalis*	*Piromonas communis*
3.1	3.3	3.0	4.7	7.7	18.8	9.0
10.0	11.8	10.1	4.5	3.9	3.8	3.8
9.0	7.7	5.4	14.6	10.8	11.4	11.0
0.66	0.34	0.79	2.3	2.1	1.3	0.18
7.5	7.5	5.6	14.1	11.6	9.4	9.6
4.9	3.4	3.4	4.2	8.0	8.2	8.9
9.8	4.5	3.6	1.5	1.7	1.7	1.7
3.2	3.3	2.6	6.3	6.2	5.4	5.2
4.8	4.8	3.6	7.7	7.6	7.7	7.8
12.1	15.1	10.4	11.2	8.2	8.3	7.9
1.4	0.65	0.56	2.2	2.4	2.0	2.2
ND	ND	ND	ND	ND	ND	ND
2.5	2.5	2.1	5.3	4.1	3.9	3.9
4.4	2.1	1.1	3.7	4.5	3.1	5.4
4.0	4.4	3.4	4.3	6.0	8.9	7.1
3.0	3.2	2.8	4.8	6.0	7.2	6.7
ND	ND	ND	ND	ND	ND	ND
1.8	2.6	1.6	4.2	2.9	3.5	3.2
3.3	2.9	2.3	4.4	6.9	6.8	6.6
ND	0.13	0.24	ND	ND	ND	ND

(1985); Onodera and Koga (1987); Whitelaw et al. (1983); Williams and Dinusson (1973a).
[b]Two hours after feeding hay cube concentrate 3:1 ratio.

dependent upon the protozoal types and the biomass (i.e., numbers and size, Clarke et al., 1982) of the ciliates reaching the lower tract. It is now realized, however, that the protozoa sequester in the rumen. This retention in the rumen results in the proportion of the protozoal biomass being considerably greater in the rumen digesta than that in the effluent material reaching the lower gut (Table 11.2). Two different experimental approaches have confirmed selective protozoal retention in the rumen. Studies on the dynamics of ^{14}C-labelled protozoa in the rumen indicated that the ciliate protozoa were extensively retained in the organ. Only a small proportion of the ciliate population was found in the rumen effluent; approximately 65% of the small protozoa and practically all of the larger ciliates were retained and degraded in the rumen (Coleman et al., 1980a; Ffoulkes and Leng, 1988; Leng, 1982, 1989; Leng et al., 1981, 1986). Comparative measurements of the proportions of bacterial and protozoal nitrogen in rumen digesta and in the ruminal outflow reaching the lower tract have also confirmed differential passage of bacteria and protozoa from the rumen (Cockburn and Williams, 1984; Harrison et al., 1979; John and Ulyatt, 1984; Punia and Leibholz, 1984; Steinhour et al., 1982; Ushida et al., 1987b; Weller and Pilgrim, 1974). Data from such studies are summarized in Table 11.2 and

Table 11.2 Percentage of protozoal nitrogen in the total nitrogen or the microbial nitrogen reaching the lower gut

	% Protozoal N in		
Host	Total N	Microbial N	Reference
Cattle		16–30	Ling and Buttery (1978)
Cattle		14	Allam et al. (1982)
Cattle	22–41	41–48	Steinhour et al. (1982)
Cattle	9–56 (ave. 26)	27–56	Cockburn and Williams (1984)
Cattle	13–19 (ave. 17)		Punia and Leibholz (1984)
Cattle		24, 27	Punia et al. (1987)
Cattle		25	Theurer (1980) (quoted in Czerkawski, 1986)
Cattle	27		Firkins et al. (1987)
Sheep	1.2		Weller and Pilgrim (1974)
Sheep		<12, 50	Smith et al. (1978)
Sheep		21, 27	Harrison et al. (1979)
Sheep		15	Jouany and Thivend (1983)
Sheep		9–19	John and Ulyatt (1984)
Sheep		20	Ushida et al. (1984)
Sheep		43, 36, 36	Meyer et al. (1986)
Sheep		4, 20	Ushida et al. (1986a)
Sheep	37	61	Ushida et al. (1987b)

were obtained using various chemical markers (e.g., diaminopimelic acid, aminoethylphosphonic acid, RNA, phosphatidylcholine) or radioactive labels (e.g., ^{35}S, ^{15}N) to monitor the bacterial and protozoal populations at the respective intestinal sites. Direct attempts to measure the passage of protozoa from the rumen by counting the numbers of ciliates in omasal fluid and omasal effluent passing to the abomasum (Bird et al., 1978; Imai et al., 1981d; Michalowski and Harmeyer, 1983; Michalowski et al., 1986; Towne and Nagaraja, 1990; Weller and Gray, 1954; Weller and Pilgrim, 1974) confirmed that protozoal concentrations were only 20–30% of the rumen values. The relative proportions of the protozoal genera present in the rumen and omasum were also different. In sheep receiving a hay-concentrate diet at maintenace level the ratios of protozoal concentrations in the rumen and the omasal effluent for the following genera were: *Diplodinium* spp., 4.6; *Ophryoscolex*, 4.3; *Dasytricha ruminantium*, 4.0; *Isotricha* spp., 3.8; *Entodinium* spp., 3.6; and *Polyplastron multivesiculatum*, 2.6 (Michalowski et al., 1986). The apparent residence time in the rumen, under normal feeding conditions, was approximately 2.5 days, which was 4–6 times greater than the mean residence time of the inert chromium-EDTA flow rate marker.

The common conclusion from all of these studies, irrespective of the experimental methodology adopted, is that, whereas bacterial flows from the rumen are similar to predicted specific outflow rates, the actual flows of protozoa were generally 0.2–0.4 of the calculated liquid outflow rate from the rumen. The sequestration of the protozoa in the digesta solids and on the rumen wall will account for some of the difference and more studies are required to explain the apparent loss of protozoa between the rumen and abomasum in the omasum. It is apparent that some protozoal biomass is available for digestion in the lower tract, although the protozoal proportion of the total biomass is lower than in the rumen. The contribution of protozoal cellular components in the nutrition of the host will, therefore, be less significant than ruminal concentrations would indicate. However, there are still uncertainties concerning protozoal flow to the lower intestinal tract. Czerkawski (1987) reassessed the contribution of the protozoa to the microbial protein supply of the host ruminant, and suggested that the contribution may be greater than is now currently accepted.

11.2 Effect of Protozoa on Blood Metabolite Concentrations

The metabolite levels in the blood of faunated and defaunated animals under the same feeding regime have been observed to differ. Blood metabolite levels have, however, been monitored far less frequently than the corresponding rumen parameters and, although the findings are not always consistent, some trends do emerge (Table 11.3). In the majority of studies undertaken, plasma glucose and amino acid concentrations increase as a consequence of defaunation whereas the changes in the levels of urea are more variable. Post-feed insulin levels tend to be higher in defaunated calves (Itabashi et al., 1982). Whitelaw et al. (1972) observed that the increase in plasma glucose levels of defaunated animals was associated with higher ruminal propionic acid concentrations, while the increase in plasma urea of faunated animals was attributed to the elevated ammonia levels in the rumen which arise as a result of the increased protein turnover in the presence of an active ciliate population. It has, however, been argued (Abou Akkada, 1965; Abou Akkada and El Shazly, 1965) that the elevated levels of plasma urea and ammonia in defaunted animals are due to a less efficient conversion of ammonia into microbial protein.

Differences also exist in the plasma fatty acid levels before and after defaunation. The short-chain volatile fatty acid content of peripheral blood decreases following defaunation (Sedloev, 1970), and the rate of irreversible loss of bicarbonate + carbon dioxide from the blood is also reduced (Rowe et al., 1985). It is believed that in defaunated animals an increased bacterial fermentation of rumen volatile fatty acids and digestible carbohydrate occurs and this reduces the quantity of VFA available to the host for absorption.

The long-chain fatty acid composition of plasma lipids can also be affected by defaunation. Comparison of faunated and defaunated ovine plasma lipids indicated that the C16:0 (palmitic acid), C16:1 (palmitoleic

Table 11.3 Changes in blood components in the absence of ciliate protozoa

Blood or plasma component	Change in level when protozoa absent		
	Increase	Decrease	No significant effect
Hemoglobin	12, 23[a]	2, 3, 21	5
β-Globulin	18		
γ-Globulin		19	
Albumin		19	11
Amino acids	8, 11, 14, 17, 23	2, 3, 21	
Nonprotein nitrogen	2, 3		
Protein		2, 3	11
Ammonia	2, 3		
Urea	2, 3, 10, 21	11, 16, 22, 23	4, 9
Glucose	2, 3, 10, 20, 21, 22	6, 9	4, 5, 11, 14
Reducing sugar	2, 3		5
Bicarbonate			18
VFA		19	
Lactic acid		20	
Saturated fatty acid		1	
Fatty acids			
Palmitic C16:0			14, 15
Palmitoleic C16:1			14, 15
Stearic C18:0			14, 15
Oleic C18:1		14, 15	
Linoleic C18:2	14, 15		
Linolenic C18:3	14, 15		14
Ascorbic acid		7	
Carotenoid			7
Vitamin A			7
Insulin	10		
Magnesium, calcium			11
Copper	12, 13		
Ceruloplasmin (Ferroxidase)	13		

[a] References are as follows: 1, Abaza et al. (1975); 2, Abou Akkada (1965); 3, Abou Akkada and El Shazly (1965); 4, Borhami et al. (1967); 5, Eadie and Gill (1971); 6, Habib et al. (1989); 7, Hibbs and Pounden (1948); 8, Itabashi and Katada (1976b); 9, Itabashi et al. (1984); 10, Itabashi et al. (1982); 11, Itabashi and Matsukawa (1979); 12, Ivan (1988); 13, Ivan et al. (1986); 14, Klopfenstein et al. (1966); 15, Lough (1968); 16, Perkins and Luther (1967); 17, Purser et al. (1966); 18, Rowe et al. (1985); 19, Sedloev (1970); 20, Singh and Makkar (1976); 21, Thamman et al. (1971); 22, Veira et al. (1983); 23, Whitelaw et al. (1972).

acid), and C18:0 (stearic acid) levels were unchanged. However, the plasma concentrations of linolenic acid (C18:3) and linoleic acid (C18:2) were increased in defaunated animals whereas the concentration of oleic acid (C18:1) fell (Klopfenstein et al., 1966; Lough, 1968). The increase in the plasma levels of polyunsaturated fatty acids has also been reported in other defaunation studies (Abaza et al., 1975; Klopfenstein et al., 1966). These observations may indicate that dietary fatty acids are hydrogenated more effectively when ciliate protozoa are present in the rumen. Cud transfer to young calves has also resulted in higher plasma ascorbic acid levels with some diets. Carotenoid and vitamin A levels were, however, similar in the plasma of uninoculated and fully faunated animals (Hibbs and Pounden, 1948). Plasma copper levels are higher in defaunated animals (Ivan et al., 1986).

The presence of a ciliate population not only affects metabolite levels in the plasma but also causes changes in the proportions of the blood components. Hemoglobin levels were consistently higher in the blood of defaunated Friesian steers (Whitelaw et al., 1972) while in the three animals studied by Sedloev (1970) the albumin and γ-globulin fractions were reduced and the β-globulin fraction increased after defaunation. Over half of the studies on the blood components and metabolite levels have been conducted using sheep (Table 11.3), although cattle (Itabashi and Katada, 1976a, 1976b; Itabashi et al., 1982; Whitelaw et al., 1972), goats (Itabashi et al., 1984), and zebu cattle and buffalo (Singh and Makkar, 1976; Thamman et al., 1971) have also been used. Measurements of metabolite concentrations in the blood are of value in comparing the effects of the protozoal activities on the availability of precursors for metabolism in the host. The information available to date, however, is limited and further studies are warranted to explain the apparently contradictory results that have been obtained so far.

11.3 Effect of Protozoa on Animal Growth

The various ways in which the protozoa can influence ruminal digestion through their effects on the ecosystem and as a direct consequence of their own metabolic activities have been described in the preceding sections. Although the protozoa are not essential for the growth and development of the ruminant, their presence may have marked effects, and improvements in animal productivity have been obtained in comparative trials with both faunated and defaunated animals (Table 11.4). The protozoa adversely affect net protein synthesis in the rumen and this is reflected in the protein:energy ratio in the nutrients available to the host ruminant. However, an increased protein supply to the lower tract will only improve animal productivity when protein deficiency is the primary limitation. Improved growth rates in young rapidly growing animals with a high amino acid demand should be achieved after defaunation with diets that fail to fulfill the nitrogen demands while meeting the energy requirements of the animals. In older animals the benefits afforded by the protozoal population may be more significant, although wool growth and lactation place a high amino acid demand on the mature animal.

In view of the potential financial benefits to the producer, defaunation has been evaluated as a means of improving animal performance (Bird, 1989). However, the manipulation of population size or species composition by selected re-inoculation of defaunated animals to establish a single species or limited range of protozoal types with beneficial characteristics

Table 11.4 Changes in ruminal productivity in the absence of protozoa

	Effect of absence of protozoa on performance characteristics		
	Increase	Decrease	No significant effect
Appearance/physical condition		2, 11, 19, 24, 31[a]	4
Susceptibility to:			
Bloat		14, 28	
Scours	30		
Copper toxicity	25, 27		
Lactic acid acidosis	29		
Animal performance:			
Liveweight gain	3, 5, 6, 7, 8, 9, 17, 18, 21, 22, 23, 27, 33	1, 2, 10, 13, 24, 25, 26, 30, 32	7, 11, 12, 19, 20, 31, 34
Wool growth	6, 8, 9, 15, 16, 21, 22, 23		
Carcass fat		33	32
Food conversion efficiency	6, 7, 8, 9, 17, 23, 32	10, 25, 30, 32	21, 22

[a]References are as follows: 1, Abou Akkada (1965); 2, Abou Akkada and El Shazly (1964); 3, Becker and Everett (1930); 4, Becker et al. (1929); 5, Bird et al. (1978); 6, Bird et al. (1979); 7, Bird and Leng (1978); 8, Bird and Leng (1983); 9, Bird and Leng (1984); 10, Borhami et al. (1967); 11, Bryant and Small (1960); 12, Chalmers et al. (1968); 13, Christiansen et al. (1965); 14, Clarke et al. (1969); 15, Cottle (1988a); 16, Cottle (1988b); 17, Demeyer et al. (1982b); 18, Demeyer et al. (1986); 19, Eadie (1962a); 20, Eadie and Gill (1971); 21, Forster and Leng (1989a); 22, Forster and Leng (1989b); 23, Habib et al. (1989); 24, Itabashi and Matsukawa (1979); 25, Ivan (1988); 26, Ivan et al. (1986); 27, Kreuzer and Kirchgessner (1987b); 28, Moate (1989); 29, Newbold et al. (1986a); 30, Osman et al. (1970); 31, Pounden and Hibbs (1950); 32, Ramprasad and Raghavan (1981); 33, Van Nevel et al. (1985); 34, Williams and Dinusson (1973b).

(e.g., lactate consumers, cellulolytic, etc.) has not been examined in depth and this approach may warrant further research effort. There have been many studies since the initial work of Becker some sixty years ago (Becker et al., 1929; Becker and Everett, 1930) to examine the effects of the ciliates on the growth of young ruminants. Many of the earlier studies concluded that inoculation with protozoa either had no significant effect on animal growth (Chalmers et al., 1968; Eadie, 1962a; Eadie and Gill, 1971; Pounden and Hibbs, 1950; Williams and Dinusson, 1973b) or resulted in increased growth (Abou Akkada, 1965; Abou Akkada and El Shazly, 1964; Borhami et al., 1967; Bryant and Small, 1960; Christiansen et al., 1965; Osman et al., 1970; Ramaprasad and Raghavan, 1981). Several of the more recent studies from Europe and Australia (Bird and Leng, 1978, 1983, 1984; Bird et al., 1979; Demeyer et al., 1982b; Habib et al., 1989; Kayouli et al., 1982; Van Nevel et al., 1985) support the initial findings of Becker and Everett (1930) that defaunated animals have higher growth rates than conventionally faunated animals (Fig. 11.1). The apparent disparity of these results was discussed in detail in the reviews of Veira (1986) and Coleman (1988) who

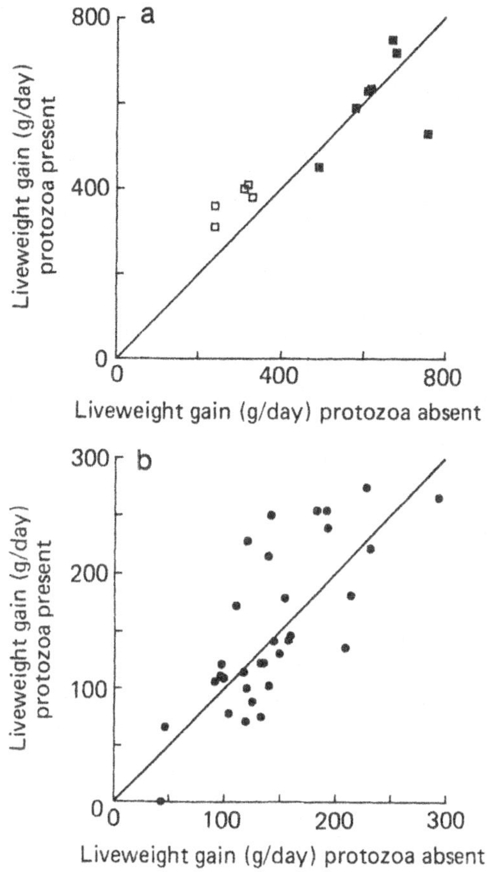

Fig. 11.1 The effect of the presence and absence of rumen ciliate protozoa on the liveweight gain in (a) zebu and buffalo (□) and cattle (■); and (b) sheep (●). The line of equal liveweight gain in faunated and ciliate-free ruminants is shown. The data used were collated from studies referred to in Table 11.4

were able to explain the findings in terms of the known nutritional effects of the protozoa. It was concluded that defaunation was beneficial to young animals receiving medium to high energy diets that had adequate ruminal degradable nitrogen but were deficient in nondegradable nitrogen. Animals that failed to respond to defaunation were initially too mature to benefit from the increased nitrogen availability after defaunation and were subsequently disadvantaged by the associated decrease in ruminal digestibility. These animals also had a depressed feed intake unlike the defaunates in the other group which exhibited increased food conversion efficiency (Table 11.4, Fig. 11.2; Bird and Leng, 1978, 1983, 1984; Bird et al, 1979; Demeyer et al, 1982b). In the experiments of Abou Akkada and El Shazly (1964) and Borhami et al. (1967), young faunated lambs and buffalo calves grew more rapidly than their defaunated counterparts as the diet was energy- rather

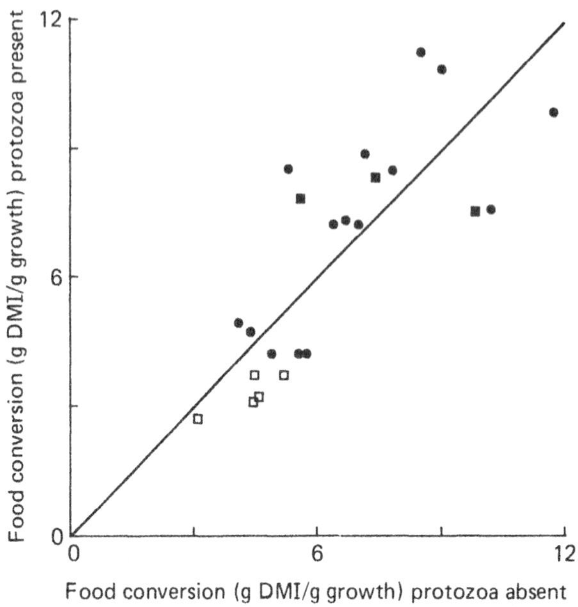

Fig. 11.2 The effect of the presence and absence of ciliate protozoa on the efficiency of food utilization in cattle (■), zebu and buffalo (□), and sheep (•). The position of equal values of feed conversion efficiency in faunated and ciliate-free ruminants is indicated. The data for this figure were obtained from the studies referred to in Table 11.4.

than protein-limited. The inoculated animals grew faster and had a higher feed conversion efficiency (Christiansen et al., 1965; Osman et al., 1970).

Comparative trials have established that the protozoa, in addition to affecting the tendency to gain body weight, are likely to exert some influence in other areas of ruminant productivity, and hence the financial gain to the producer. There is thus a need to establish the consequences and cost benefits of protozoal activities in more mature animals. Abou Akkada (1965) indicated that faunated male and female sheep had tended to reach puberty before defaunated animals and were able to breed earlier. However, a later study with grazing pregnant ewes suggested that defaunation could be beneficial (Bird and Leng, 1983). Although the defaunation was only completely effective for the first 10 weeks of a 23-week trial period, comparison with re-inoculated animals confirmed that defaunation was associated with a 12% higher birth weight of single lambs. However, the average weight of twin lambs and the body-weight changes of the ewes were similar for both groups.

Two studies in Australia with lambs that were either given low quality diets or were allowed to graze in a green oats pasture demonstrated that the elimination of protozoa from the rumen stimulated wool growth (Bird et al., 1979; Bird and Leng, 1983, 1984). The extra wool production was not gained at the expense of extra food intake, suggesting that defaunation may have improved nutrient availability to the host and the subsequent efficiency of utilization of the absorbed nutrients. Wool growth is dependent on the amount of protein reaching the duodenum and, in particular, it requires an adequate supply of sulfur-containing amino acids. The improved wool growth in defaunated animals was attributed to the resultant improvements in protein flow to the lower tract that ensue in the absence of protozoa (Bird et al., 1979; Bird and Leng, 1983, 1984; Fenn and Leng, 1989; Leng and Nolan, 1984). The improved wool growth after defaunation occurred with both low- and high-protein diets and the treatment offers potentially major benefits to wool producers if the improvements can be achieved widely. More recent studies (Cottle, 1988a, 1988b; Forster and Leng, 1989a, 1989b; Habib et al., 1989) also achieved increased wool production by defaunation. It was estimated by Bird and Leng (1983) that the increased wool production in Australia would be worth an extra 10–100 million Australian dollars annually. In addition to sheep, other more exotic ruminants are now being evaluated because of the potential value of their coats. The South American camelids (guanaco, alpaca, llama) produce a high quality fleece while goats can also be reared in traditional upland sheep areas for the fine fiber content (cashmere) of their coats. The potential value of defaunation to improve fleece yields in these animals may, therefore, be considered in view of the success that has been obtained with sheep.

The observed effects of protozoa on live-weight gain have obvious implications in intensive husbandary systems for meat production. When the young animal is removed from the mother at birth or soon after, and is denied direct contact with mature animals, it is unlikely that a protozoal population will establish. Borhami et al. (1967) concluded that early inoculation of early weaned calves was advantageous if the young were separated from the dam. However, it is now apparent that the decision to inoculate the young ruminant will be determined by the maturity of the animal, the nature of the diet, and the husbandary objectives. Animals reared intensively for meat production should require less protein supplementation if they are given high energy diets after defaunation, although the ciliate-free animals may be more prone to acidosis (see Section 12.4). The carcass quality and dressing percentages of faunated and defaunated stock were similar (Bird and Leng, 1984; Van Nevel et al., 1985, confirming that the increased live-weight gains of the defaunated animals resulted from tissue deposition and were not artifacts that could be attributed to any increased weight of rumen contents due to increased rumen capacity or ruminal fill. The carcasses of defaunated sheep tended to have more meat and less fat, although the proportions of meat and bone were the same in carcasses of faunated and defaunated animals (Demeyer et al., 1982b; Ramaprasad and Raghavan, 1981; Van Nevel et al., 1985).

Defaunation may thus offer some advantage in the intensive husbandry of young animals for meat production and in the less intensive growth regimes for wool formation. Advantage may also be gained in tropical climates where the diets used are characteristically high-energy and low-protein formulations. Outside these particular geographical locations and husbandry requirements, the merits of faunation versus defaunation have to be evaluated. Veira (1986) was of the opinion that stock reared non-intensively to produce replacement animals would be unlikely to benefit from defaunation. However, the effects of protozoa on the productivity of animals reared and maintained for milk production have not been quantified other than in animals receiving a daily treatment with a chemical defaunating agent (Yang and Varga, 1989). The high milk yields now required of dairy cattle necessitate the use of costly dietary protein supplements. Yield can be increased by improving amino acid availability in the lower intestinal tract (Oldham, 1981) and it would thus appear that the potential exists to reduce the cost of protein supplementation in intensive dairy systems. In less-developed agricultural economies where dietary supplementation is not an option, it may become possible to improve milk production by the alternative expedient of eliminating the rumen ciliate protozoa.

There are thus certain production areas in which the use of ciliate-free animals may offer some advantage. However, the defaunating agents currently available are not always completely effective, may be injurious to the health of the animal or even cause death (Bird and Leng, 1983; Van Nevel et al., 1985). The most commonly used agents are not specific and can have a short-term adverse effect on the bacterial population (Eadie and Shand, 1981; Orpin, 1977c). The voluntary feed intake of treated animals is reduced, milk production and milk lactose concentration are also reduced, and the animal may not take food for several days. There is a variable period of time in which the weight losses caused by the defaunation treatment may counteract the improvements in live-weight gain that accrue (Bird et al., 1979; Bird and Leng, 1984; Demeyer et al., 1982b). There is thus a need to identify a safer and more specific antiprotozoal agent, although a treatment protocol that significantly reduced the protozoal biomass without rendering the animal ciliate-free may prove to be equally satisfactory. In this latter situation repeated dosing or the use of drug entrapment in slow release capsules would have to be contemplated to prevent the protozoal population from proliferating and re-establishing. Re-inoculation by contact with wild ruminants or untreated domestic animals could easily occur and specific husbandary practices will be necessary for maintaining large numbers of animals free from protozoa.

12

Protozoa and Health of the Host Animal

In addition to their nutritional and digestive effects, the metabolic activities of the protozoa can influence the aetiology of some ruminant disorders, with important consequences for the health and general well being of the host animal.

12.1 Appearance of Animal

In early studies observations were made on the comparative development of young animals that were either inoculated with rumen contents from mature animals or were raised in isolation. Although the uninoculated protozoa-free animals did not suffer any obvious ill-effects, the general appearance of the inoculated animals was observed to be superior (e.g., Abou Akkada and El Shazly, 1964). Pounden and Hibbs (1950) reported that the abdomens of uninoculated calves were slightly enlarged or distended and the coat hair of the animal was much coarser. Eadie (1962a) also commented on the rougher coat of ciliate-free calves. The characteristic "pot-bellied" appearance of lambs and calves raised without ciliate protozoa is a well-established observation (Becker, 1932; Bryant and Small, 1960; Eadie, 1962a; Itabashi and Matsukawa, 1979). The pot-bellied appearance may be due to a greater feed intake by the animals (Bryant and Small, 1960) or as a consequence of the increased digestion that occurs in the lower tract of defaunated animals (Demeyer et al., 1982b).

12.2 Detoxification

The protozoal population is able to detoxify certain dietary toxins ingested by the host. The protozoal fraction has been shown to be more active than the bacterial fraction in reducing nitrates and detoxifying certain mycotoxins. The protozoa, therefore, may confer some benefit to animals fed contaminated material. Mycotoxins are potential health hazards for domestic ruminants, and some of the trichothecene mycotoxins formed by *Aspergillus* and *Fusarium* spp. are partially degraded in the rumen. Zearalenone, T-2 toxin, diacetoxyscirpenol, and ochratoxin A were degraded in in vitro incubations in which the protozoa were more active than the bacteria (Kiessling et al., 1984; Westlake et al., 1989). The peptide bond in ochratoxin A was cleaved to yield nontoxic ochratoxin α and phenylalanine (Hult et al., 1976) whilst zearalenone was reduced to α-zearalenol and β-zearalenol. Diacetoxyscirpenol and T-2 toxin were deacetylated to monoacetoxyscirpenol and HT-2 toxin respectively; the deacetylated products are less active inhibitors of protein synthesis than the original trichothecene mycotoxins.

The use of the tropical legume *Leucanea leucocephala* has been limited by the presence of a toxic non-protein amino acid β-[N-(3-hydroxy-4-pyridone)]-α-aminopropionic acid, mimosine, which is readily hydrolyzed in the rumen to 3-hydroxy-4-(1H)-pyridone (DHP), a potent goitrogen (Hegarty et al., 1964, 1976). Mimosine is degraded completely with some DHP production by both protozoa-enriched and bacterial fractions of rumen fluid (Shiroma and Akashi, 1976). In animals receiving concentrate supplementations most of the mimosine-degrading activity was found in the bacterial fractions (Kumar et al., 1987; Tangendjaja et al., 1983). However, with sheep fed lucerne, hay, or *Digitaria pentzii*, the degradative activity, although lower, was located principally in the protozoa-rich fraction (Tangendjaja et al., 1983). The rumen protozoa are also more effective than the bacteria in reducing dietary nitrates and nitrites (Yoshida et al., 1982). The results from this study indicated that lactic acid was used as a hydrogen donor during the potentially important protozoal reduction of nitrate and nitrite in the rumen.

12.3 Lactic Acid Acidosis

Lactic acid is a normal intermediate in ruminal metabolism but it may also enter the rumen from exogenous dietary sources such as silage. Lactate does not usually accumulate in the rumen. However, under certain dietary conditions an unbalanced fermentation can occur which results in increased ruminal lactate concentrations with consequential adverse effects on rumen function, animal health, and productivity (Slyter, 1976). Lactic acid acidosis typically occurs when readily fermentable carbohydrates are included in the diet without an intervening period of adaptation. In these circumstances the increased rate of fermentation leads to an enhanced production of volatile fatty acids and a sudden reduction in rumen pH, which precedes peak levels of lactic acid. The low ruminal pH favors lactate absorption with a concomitant increase in the amount of undissociated lactic acid in the blood. Other consequences of high ruminal lactate levels include anorexia, rumen stasis, diarrhea, and dehydration (Huber, 1976). Other ailments common in acidoic animals include rumenitis, laminitis, and liver abcesses. Feedlot animals receiving high energy diets or increased grain feeding without adaptation are particularly prone to acidosis. Within the United Kingdom the most common management situation leading to acidosis is the addition of starch-based concentrates to dairy cattle rations in order to meet the increased energy demands of early lactation. A condition similar to acidosis has also been reported when cattle are introduced to very acidic silage (Tutt, 1972).

The protozoa have an important stabilizing effect on the ruminal fermentation and have an important role in regulating ruminal lactate metabolism and in preventing the occurrence of lactic acid acidiosis. Protozoal ingestion of starch and conversion of soluble sugars to reserve polymers prevents the alternative rapid acidogenic bacterial fermentation. The protozoal fermentation of the ingested starch and storage polysaccharide is less rapid, and a detrimental VFA build-up does not occur. Conditions within the rumen remain more stable when an established protozoal population is present (Whitelaw et al., 1972). The protozoa would appear to exert some control over the rate of volatile fatty acid formation and lactic acid removal. The rapid protozoal uptake of starch and soluble sugars also has a stabilizing effect as the alternative bacterial fermentation to lactic acid is prevented (Kariya et al., 1989; Mackie et al., 1978). The subsequent protozoal fermentation of the ingested starch and storage polysaccharide stabilizes the rate of VFA formation and allows their synthesis over a longer period of time. When the transition to high grain or concentrate rations is achieved with a period of adaptation, the protozoal population can increase in size and may ingest much of the dietary starch so that the magnitude of the potentially harmful amylolytic bacterial fermentation is reduced (Hungate, 1978). This stabilizing effect of the protozoa in the rumen in animals changing from forage to grain rations is important in preventing the onset of acidosis. The protozoa are, however, sensitive to falling pH (Newbold et al., 1986b) and may be eliminated from animals when the dietary transition is uncontrolled.

The retention of an active protozoal population is necessary as the protozoa not only reduce the rate of lactate formation in the rumen but they also have an important role in the ruminal metabolism of lactic acid (Table 12.1). Uptake studies in vitro demonstrated that the rate of disappearance of both the D(−) and L(+) isomers of lactic acid was 15-fold higher in incubations with the protozoal fraction as compared to the rate of disappearance in bacterial incubations (Newbold et al., 1987). The entodiniomorphid

Table 12.1 Postprandial ruminal lactic acid levels[a] in faunated and defaunated sheep fed molassed sugar beet pulp

	Time after feed intake					
Animal status	0.5	1	1.5	2	3	5
Faunated	3.1	3.9	3.2	3.3	3.2	3.1
Defaunated	3.9	6.6	11.1	13.9	7.8	6.1

[a] Lactic acid concentration is given in mM. Data from Newbold et al. (1986a).

ciliates were responsible for the lactate uptake. The activity appeared to be diet-dependent and was higher in populations recovered from animals fed silage (i.e., lactic acid present in feed) or when the diet contained readily fermentable carbohydrate (e.g., molassed sugar beet pulp), which resulted in increased lactate formation in the rumen. Ruminal lactate is cleared less effectively from defaunated animals (Chamberlain et al., 1983; Newbold et al., 1986a). The likelihood of digestive disturbances due to an accumulation of lactic acid in the rumen is diminished by the maintenance of an active protozoal population.

The occurrence of other intestinal disorders can also be affected by the rumen ciliate population. Pounden and Hibbs (1950) noted a lower incidence of scours in faunated animals as uninoculated calves were apparently more prone to suffer recurrent mild diarrhoea.

12.4 Copper Toxicity

Although there is little information available on the role of protozoa in mineral metabolism, defaunated sheep are apparently more susceptible to chronic copper toxicity (Ivan, 1989a). The absorption of copper and accumulation in the liver is higher in ciliate-free animals, as are the plasma copper levels and ferroxidase activity (Tables 11.3, 12.2; Ivan, 1989b; Ivan et al., 1986). Hepatic copper levels are usually below 0.5 mg per g dry matter and toxicity problems occur at levels of 1–3 mg per g dry matter. In their comparative trial of faunated and ciliate-free sheep, Ivan et al. (1986) reported that the hepatic copper levels in all the protozoa-free animals exceeded 1 mg per g dry matter, rendering the animals more susceptible to stress-induced haemolytic crisis. An outbreak of chronlc toxicity, with a high mortality rate, had previously occurred in a 10–12 month old flock that had been raised protozoa-free from birth. It would therefore appear that under certain dietary situations the rumen ciliate protozoa can have an important role in the prevention of chronic copper toxicity in sheep by reducing the total hepatic accumulation of dietary copper (Ivan, 1988, 1989b). However, Kreuzer and Kirchgessner (1990) did not observe any pronounced body copper retention in wethers fed semi-purified diets.

Table 12.2 Effect of defaunation on ovine hepatic copper concentrations in animals receiving supplemented maize silage rations

Ration	Duration of experiment (days)	Hepatic copper concentration ($\mu g/g$ DM)		
			Final	
		Initial	Defaunated	Faunated
Maize silage	92	621	796	599
+ soybean meal	90	744	931	719
+ soybean meal	184	745	1684	962
+ soybean meal	197[a]	431	583	460
+ soybean meal	197[b]	431	900	732
+ casein	90	744	676	720
+ casein	165	604	503	501
+ fishmeal	165	604	633	588

[a] Dietary copper at 7 $\mu g/g$ DM.
[b] Dietary copper at 15 $\mu g/$ DM.
Data from Ivan (1988, 1989) and Ivan et al. (1986).

The mechanism by which the protozoa decrease copper availability has not been determined. Ivan et al. (1986), however, proposed that as the protozoa increased the breakdown of dietary protein, sulfide formation would also increase as a result of the degradation of sulfur-containing amino acids in the rumen. The sulfide and copper react to form insoluble copper sulfide which is unavailable for absorption and utilization. The protozoal effect on copper metabolism is apparently indirect and dependent on the type of protein present in the diet (Ivan, 1989b). The flow of copper to the small intestine is, however, similar in faunated and defaunated sheep (Ivan and Veira, 1982).

The availability of other mineral cations could also be affected by analogous complex formation. The presence of a rumen protozoal population, however, does not affect the utilization of selenomethionine in sheep (Hidiroglou and Jenkins, 1974) and is not essential for the process of digestion and release of zinc, manganese, iron, or copper from the feed (Ivan and Veira, 1982). In this latter experiment, the soluble proportions of trace metals in rumen and duodenal digesta and the total flow of trace elements to the small intestine were not significantly different in faunated and defaunated animals. The effects of defaunation on trace element excretion were also small (Kreuzer and Kirchgessner, 1990) and the authors, therefore, concluded that the protozoa did not have a critical role in trace metal metabolism in the rumen of sheep. However, another study (Ivan et al., 1986) did suggest that the protozoa could influence the bioavailability of trace elements (Fig. 12.1) and this role warrants further examination in view of its importance in animal production. Protozoa increase the availability of sulfur in the rumen (Hegarty et al., 1989) and influence the secretion and absorption of calcium, magnesium, and phosphorus in the digestive tract of sheep (Boila and Phillips, 1988). Excessive dietary supplementation with

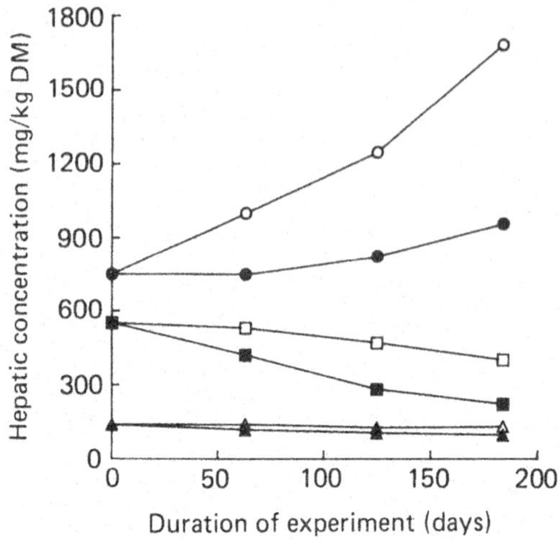

Fig. 12.1 Effect of defaunation on the changes in ovine hepatic concentrations of copper (○), iron (□), and zinc (△). Values for defaunated animals are shown by open symbols and faunated as closed symbols. Date obtained from Ivan et al. (1986).

either copper (Szuecs et al., 1982) or zinc (Durand and Kawashima, 1979) can reduce the protozoal population with a resultant and potentially harmful increase in the absorption and hepatic retention of the elements. Alternatively the problems arising with copper-deficient diets could be exacerbated as a consequence of the protozoal activities. The trace mineral requirements of faunated and ciliate-free animals may not be identical and should be determined so that toxic problems are not encountered when defaunation is used by the farming industry to manipulate animal productivity.

12.5 Bloat

Bloat is a disorder of ruminants caused by excessive gas retention in the rumen and can afflict both domesticated and wild ruminants, although it is most prevalent in cattle. The complaint results in reduced feed intake, lower weight gain, and lower milk yields, and can result in animal death. Nonfoamy or free-gas bloat is less common than either feedlot or legume (pasture) bloat. Pasture bloat occurs mainly on legume forages (e.g., alfalfa, clover) although various other crops are bloat-provoking. Feedlot bloat occurs in beef cattle receiving rations that are high in concentrates and low in roughages. The incidence of bloat is influenced by the genetically predetermined susceptibility of the animal, and by the nature of the ration and the microbial population in the rumen (Clarke and Reid, 1974).

The rumen protozoa appear to have an important role in the occurrence of legume bloat. This form of bloat is caused by the formation of a stable proteinaceous foam which entraps the fermentation gases preventing eructation. Plant, salivary, and microbial proteins contribute to foam formation. The protozoal involvement in the etiology of bloat is principally related to their gas production and enhancement of foam stability (El Fouly and Leng, 1981). The cell contents of the holotrich ciliates can contribute to foaming in the rumen. Ruminal lysis of holotrich ciliates has been reported in animals consuming diets rich in readily fermentable soluble carbohydrates (Clarke, 1965a, 1965b, 1965c). The holotrich proteins produce rigid foams that are stable and persistent within the normal rumen pH range (i.e., pH 5.5–6.5) (Jones and Lyttleton, 1972). The released amylopectin storage grains are also effective stabilizers of rumen foams (Clarke and Reid, 1974). In addition, the ciliates ingest plant chloroplasts which contain glycolipids that may have antifoaming properties (Mangan, 1959; Oxford, 1958, 1959), and Moate (1989) concluded that the incidence of bloat in cows on clover-dominated pasture was associated with high numbers of *Epidinium* spp. in the rumen. Attempts have been made, therefore, to control legume bloat by eliminating the protozoa, and bloat severity was reduced by treatment with anti-protozoal agents. Limited success was obtained following treatment with dimetridazole (1,2-dimethyl-5-nitroimidazole), copper sulphate, surfactant compounds, and antibiotics (Bartley, 1965; Clarke, 1966; Clarke and Reid, 1974; Clarke et al., 1969; Davis and Essig, 1972; Katz et al., 1986). Even though defaunation was achieved, bloat still occurred. The protozoa, therefore, are not necessary for the onset of bloat to occur, although both the incidence and severity are greater when protozoa are present in the rumen. Although antiprotozoal agents have not been completely successful in controlling bloat, the disorder can be treated by the administration of antifoaming agents that also possess antiprotozoal properties (e.g., poloxalene, phenol ethoxylates, and alcohol ethoxylates—Clarke and Reid, 1974).

Feedlot bloat, like pasture bloat, is caused by excessive foaming of the rumen contents and characteristically develops in feedlot cattle fed high-grain ($>50\%$ concentrate) rations. The symptoms develop slowly and can become chronic. Stable foam formation again occurs and the foaming agents appear to be microbial in origin (Bartley et al., 1975; Gutierrez et al., 1961), although no specific protozoal effect has been suggested. The pH of the rumen contents of bloated animals is not generally lower except where the onset of bloat was associated with an acid indigestion (Hironaka et al., 1973). The protozoal activity would be reduced at lower pH values (Newbold et al., 1986b) and Mishra (1969) reported that protozoal populations were reduced in bloating animals. The protozoal population density and the degree of feedlot bloat severity appeared to be inversely related (Mishra, 1964, 1969).

13

The Flagellate Protozoa in the Rumen

The flagellate protozoa have been largely neglected by those describing, culturing, and studying the metabolism of rumen protozoa with the result that very little information is available.

13.1 Description and Classification

From the time that rumen protozoa were first studied up to about 1970, there were thought to be many species of rumen flagellates (Becker and Talbott, 1927; Das Gupta, 1935), but then Orpin (1977a, 1977b) showed that *Neocallimastix frontalis, Piromonas communis,* and *Sphaeromonas communis* were zoospores of phycomycete fungi rather than protozoa. A summary of the morphology of the true flagellates in given in Table 13.1. All belong to the Class Zoomastigophorea.

Chilomastix caprae (Order Retortamonadida; Family Retortamonadidae) was first described from the goat rumen by Fonseca (1915) and was also reported from the Indian goat by Das Gupta (1935). *C. caprae* has three anteriorly directed flagella and a fourth recurrent one which undulates in the cytostomal cleft.

Monocercomonas ruminantium (Braune, 1913) (Order Trichomonadida; Family Monocercomonadidae) is believed by Levine (1961) and Hungate (1966) to be the same as *Trichomonas ruminantium*. *M. ruminantium* has three anterior flagella and one recurrent flagellum but no undulating membrane, although *T. ruminantium* is reported by Becker and Talbott (1927) and Das

Table 13.1 Characteristics of the species of flagellate protozoa found in the rumen

Species	Shape	Size	Number of flagella	Axostyle	Undulating membrane	Cytostomal groove	Pelta	Size of nucleus (μm)	References[a]
Chilomastix caprae	Piriform	8.3 × 4.4	4	Absent	Absent	Present, large	–		2, 3, 4
Monocercomonas ruminantium	Piriform	4.8 × 4.1	4	Present	Absent	Absent	+	1.8 × 1.6	1, 3, 4
Monocercomonoides bovis	Elliptical	5.4 × 2.8	4	Slender	Absent	Absent	+	1.6 × 1.4	3
Monocercomonoides caprae	Elliptical	9 × 6	4	Present	Absent	Absent	+		2, 3, 4
Pentatrichomonas hominis	Elliptical	7.5 × 5.6	5	Present, protruding	Present	Absent	+	2.5 × 2.0	3, 4
Tetratrichomonas buttreyi	Elliptical	5.3 × 4.8	4	Present, protruding	Present	Absent	+	2.0 × 1.7	3
Trichomonas ruminantium	Elliptical	12 × 10	3	Present, protruding slightly	Present	Absent	?		1, 2

[a]References are as follows: 1, Becker and Talbott, 1927; 2, Das Gupta (1935); 3, Jensen and Hammond (1964); 4, Ogimoto and Imai (1981).

Gupta (1935) to have one. *M. ruminatium* is described by Jensen and Hammond (1964) and Ogimoto and Imai (1981).

Monocercomonoides bovis (Order Oxymonadida; Family Polymastigidae) was first described from the bovine caecum by Jensen and Hammond (1964) but can also occur in the rumen. It has four flagella arising in pairs from separate blepharoplasts. One pair is directed anteriorly while one of the second pair is anterior and the other is a recurrent flagellum. It has an axostyle which protrudes very slightly posteriorly.

Monocercomonoides caprae was first described inadequately from the Indian goat as *Monocercomonas caprae* by Das Gupta (1935). This organism is larger (6–12 μm long compared with 3.7–7.0 μm) than the preceding organism and has no recurrent flagellum.

Pentatrichomonas hominis (Order Trichomonadida; Family Trichomonadidae) is one of the most common rumen flagellates having been described by Jensen and Hammond (1964) and Ogimoto and Imai (1981). It has five anterior flagella and an undulating membrane along which a sixth flagellum runs that exists posteriorly as a free trailing flagellum. *P. hominis* has a thick axostyle that protrudes posteriorly one-third to one-half the length of the body.

Tetratrichomonas buttreyi (Order Trichomonadida; Family Trichomonadidae) was originally described from swine by Hibler et al. (1960). It was also found in the bovine rumen by Jensen and Hammond (1964) who described the rumen organism in detail. *T. buttreyi* has four anterior flagella, an undulating membrane, and a fine axostyle with a long protruding tip.

Trichomonas ruminantium (Braune, 1913), which may be the same as *Monocercomonas ruminantium* (Levine, 1961; Hungate, 1966), has been described from cattle and sheep rumen by Becker and Talbott (1927) and Eadie (1962a) and from the Indian goat by Das Gupta (1935). *T. ruminantium* has three long anterior flagella, an undulating membrane with a bordering flagellum, and an axostyle. Jensen and Hammond (1964) believe this organism could be a pentatrichomonad.

13.2 Infection of the Host and Population Density In vivo

Direct contact between infected and non-infected animals is not necessary for the transfer of flagellate protozoa which develop in the rumens of ciliate-free sheep and cattle. However, the pH must be greater than 6.5 as the flagellates, like the ciliates are intolerant of low pH. *Trichomonas ruminantium* with smaller numbers of *Chilomastix* spp. can be observed in calves as early as six days after birth (Lengemann and Allen, 1959; Bryant and Small, 1960; Eadie, 1962a). The numbers decrease when ciliate protozoa become established. Warner (1962b; 1966a) showed that the population density of "polymastigates" increased markedly after feeding and then deceased again to reach a minimum after 6–12 h The ratio of maximum to minimum could exceed 25. Unfortunately, many of the organisms counted by Warner could have been the zoospores of phycomycete fungi which are released from sporangia after feeding and which develop into a vegetative phase an hour later (Orpin, 1975). There is therefore little firm information on flagellate numbers in the rumen but there are probably less than 10^5/ml.

13.3 Cultivation

Jensen and Hammond (1964) succeeded in culturing pentatrichomonads from the bovine rumen in several media including cecal extract, Diamond's media, and thioglycollate media (Diamond, 1957). *Tetratrichomonas buttreyi* could only be grown in caecal extract medium. *Monocercomonoides bovis* grew only in caecal extract medium containing serum and *Monocercomonas ruminantium* could not be cultivated for more than a few days in any medium. However attempts by the author and others to grow flagellates from the ovine rumen have been unsuccessful.

References

Aafjes, J.H. and Nijhof, J.K. 1967. A simple artificial rumen giving production of volatile fatty acids. *British Veterinary Journal* 123: 436–446.

Abaza, M.A., Abou Akkada, A.R. and El Shazly, K. 1975. Effect of rumen protozoa on dietary lipid in sheep. *Journal of Agricultural Science* 85: 135–143.

Abe, M. and Iriki, T. 1978. Effects of diet on the protozoa population in permeable continuous cultures of rumen contents. *British Journal of Nutrition* 39: 255–264.

Abe, M. and Iriki, T. 1989. Mechanism whereby holotrich ciliates are retained in the reticulo-rumen of cattle. *British Journal of Nutrition* 62: 579–587.

Abe, M. Iriki, T, Tobe, N. and Shibui, H. 1981. Sequestration of holotrich protozoa in the reticulo-rumen of cattle. *Applied and Environmental Microbiology* 41: 758–765.

Abe, M. and Kumeno, F. 1973. In vitro simulation of rumen fermentation: apparatus and effects of dilution rate and continuous dialysis on fermentation and protozoal population. *Journal of Animal Science* 36: 941–948.

Abe, M. and Kurihara, Y. 1984. Long-term cultivation of certain rumen protozoa in a continuous fermentation system supplemented with sponge materials. *Journal of Applied Bacteriology* 56: 201–214.

Abe, M., Shibui, H., Iriki, T. and Kumeno, F. 1973. Relation between diet and protozoal population in the rumen. *British Journal of Nutrition* 29: 197–202.

Abe, M., Suzuki, Y., Okano, H. and Iriki, T. 1983. Specific difference in fluctuation pattern of holotrich protozoa in the rumen of cattle, goat and sheep. *Japanese Journal of Zootechnical Science* 54: 457–462.

Abou Akkada, A.R. 1965. The metabolism of ciliate protozoa in relation to rumen function, pp. 335–345 in Dougherty, R.W., Allen, R.S., Burroughs, R.W., Jacobson, N.L. and McGilliard, A.D. (editors), *Physiology of Digestion in the Ruminant*, Butterworths, Washington.

Abou Akkada, A.R., Bartley, E.E., Berube, R., Fina, L.R., Meyer, R.M., Henricks, D. and Julius, F. 1968a. Simple method to remove completely ciliate protozoa of adult ruminants. *Applied Microbiology* 16: 1475–1477.

Abou Akkada, A.R., Bartley, E.E. and Fina, L.R. 1969. Ciliate protozoa in the rumen of the lactating cow. *Journal of Dairy Science* 52: 1088–1091.

Abou Akkada, A.,R., Eadie, J.M. and Howard, B.H. 1963. The biochemistry of rumen protozoa. 7. The carbohydrases of *Polyplastron multivesiculatum* (Dogiel and Fedorowa). *Biochemical Journal* 89: 268–272.

Abou Akkada, A.R. and El Shazly, K. 1964. Effect of absence of ciliate protozoa from the rumen on microbial activity and growth of lambs. *Applied Microbiology* 12: 384–390.

Abou Akkada, A.R. and El Shazly, K. 1965. Effect of presence or absence of rumen ciliate protozoa on some blood components, nitrogen retention and digestibility of food constituents in lambs. *Journal of Agricultural Science* 64: 251–255.

Abou Akkada, A.R. and Howard, B.H. 1960. The biochemistry of rumen protozoa. 3. The carbohydrate metabolism of *Entodinium*. *Biochemical Journal* 76: 445–451.

Abou Akkada, A.R. and Howard, B.H. 1961. The biochemistry of rumen protozoa . 4 . Decomposition of pectic substances. *Biochemical Journal* 78: 512–517.

Abou Akkada, A.R. and Howard, B.H. 1962. The biochemistry of rumen protozoa. 5. The nitrogen metabolism of *Entodinium*. *Biochemical Journal* 82: 313–320.

Abou Akkada, A.R., Messmer, D.A., Fina, L.R. and Bartley, E.E. 1968b. Distribution of 2-aminoethylphosphonic acid in some rumen microorganisms. *Journal of Dairy Science* 51: 78–81.

Ahuja, S.P. and Sarmah, T.C. 1979. Studies on the activity of rumen protozoa. 2. Utilisation of U-^{14}C-L-leucine by rumen protozoa. *Zentralblatt für Veterinärmedizin Reihe A* 26: 551–557.

Akin, D.E. and Amos, H.E. 1979. Mode of attack on Orchardgrass leaf blades by rumen protozoa. *Applied and Environmental Microbiology* 37: 332–338.

Allam, S.M., Rohr, K., Brandt, M. and Lebzien, P. 1982. Beiträge zur quantifizierung der N-umsetzungen in den vormagen von milthkühen. *Zeitschrift für Tierphysiologie, Tierernährung und Futtermittelkunde* 48: 58–64.

Amos, H.E. and Akin, D.E. 1978. Rumen protozoal degradation of structurally intact forage tissues. *Applied and Environmental Microbiology* 36: 513–522.

Ankrah, P., Loerch, S.C. and Dehority, B.A. 1989. Occurrence of 2-aminoethylphosphonic acid in feeds, ruminal bacteria and duodenal digesta from defaunated sheep. *Journal of Animal Science* 67: 1061–1069.

Ankrah, P., Loerch, S.C. and Dehority, B.A. 1990. Sequestration migration and lysis of protozoa in the rumen. *Journal of General Microbiology* 136: 1869–1875.

Appleby, J.C., Eadie, J.M. and Oxford, A.E. 1956. Inter-relationships between ciliate protozoa and bacteria in the sheep's rumen. *Journal of Applied Bacteriology* 19: 166–172.

Argyle, J.L. and Forster, R.J. 1989. Effects of ciliate protozoa on rumen bacterial populations, pp. 317–319 in Nolan, J.V., Leng, R.A. and Demeyer, D.I. (editors), *The Roles of Protozoa and Fungi in Ruminant Digestion*, Penambul Books, Armidale, Australia.

Asada, T., Konno, T. and Katsuki, T. 1980. Studies on the contents of the rumen in Shiba goats. 1. Comparison of the contents of the rumen between Shiba goats and Saanen goats under the same feeding conditions. *Bulletin of the Nippon Veterinary and Zootechnical College* 29: 69–73.

Aspinall, G.O. and Hirst, E.L. 1965. Inulin: preparation of inulin from Dahlia tubers. *Methods in Carbohydrate Chemistry* 5: 157–158.

Awerinzew, S. and Mutafowa, R. 1914. Material zur Kenntnis der Infusorien aus dem Magen der Wiederkaüer. *Archiv für Protistenkunde* 38: 109–118.

Bailey, R.W. 1958. Bloat in cattle. X. The carbohydrases of the cattle rumen ciliate *Epidinium ecaudatum* Crawley isolated from cows fed on red clover (*Trifolium pratense* L.). *New Zealand Journal of Agricultural Research* 1: 825–833.

Bailey, R.W. and Clarke, R.T.J. 1963a. Carbohydrase activity of *Entodinium* species from sheep on a starch-free diet. *Nature (London)* 198: 787.

Bailey, R.W. and Clarke, R.T.J. 1963b. Carbohydrases of the rumen oligotrich *Eremoplastron bovis*. *Nature (London)* 99: 1291–1292.

Bailey, R.W., Clarke, R.T.J. and Wright, D.E. 1962. Carbohydrases of the rumen ciliate *Epidinium ecaudatum* (Crawley). Action on plant hemicellulose. *Biochemical Journal* 83: 517–523.

Bailey, R.W. and Gaillard, B.D.E. 1965. Carbohydrases of the rumen ciliate *Epidinium ecaudatum* (Crawley). Hydrolysis of plant hemicellulose fractions and β-linked glucose polymers. *Biochemical Journal* 95: 758–766.

Bailey, R.W. and Gaillard, B.D.E. 1969. Hydrolysis of plant mannans by protozoal enzymes. *Australian Journal of Biological Sciences* 22: 267–269.

Bailey, R.W. and Howard, B.H. 1962. Preparation of enzymes from rumen protozoa by indole disintegration. *Archives of Biochemistry and Biophysics* 99: 299–303.

Bailey, R.W. and Howard, B.H. 1963. The biochemistry of rumen protozoa. 6. The maltases of *Dasytricha ruminantium, Epidinium ecaudatum* (Crawley) and *Entodinium caudatum*. *Biochemical Journal* 86: 446–452.

Bailey, R.W. and MacRae, J.C. 1970. The hydrolysis by rumen and caecal microbial enzymes of hemicellulose in plant and digesta particles. *Journal of Agricultural Science* 75: 321–326.

Banerjee, A.K. 1955. Studies on parasitic ciliates from Indian ruminants. *Proceeding of the Zoological Society, Calcutta* 8: 87–101.

Bartley, E.E. 1965. Bloat in cattle. VI. Prevention of legume bloat with a nonionic surfacant. *Journal of Dairy Science* 48: 102–104.

Bartley, E.E., Lippke, H., Pfost, H.B., Nijweide, R.J., Jacobson, N.L. and Meyer, R.M. 1965. Bloat in cattle. X. Efficacy of poloxalene in controlling alfalfa bloat in diary steers and lactating cows in commercial dairy herds. *Journal of Dairy Science* 48: 1657–1662.

Bartley, E.E., Meyer, R.M. and Fina, L.R. 1975. Feedlot or grain bloat, pp. 551–562 in McDonald, I.W. and Warner, A.C. (editors), *Digestion and Metabolism in the Ruminant*, University of New England Publishing Unit, Armidale Australia.

Bauchop, T. 1971. Mechanism of hydrogen formation in *Trichomonas foetus*. *Journal of General Microbiology* 68: 27–33.

Bauchop, T. 1979. The rumen ciliate *Epidinium* in primary degradation of plant tissues. *Applied and Environmental Microbiology* 37: 1217–1223.

Bauchop, T. 1980. Scanning electron microscopy in the study of microbial digestion of plant fragments in the gut, pp. 305–326 in Ellwood, D.C., Hedger, J.N., Latham, M.J., Lynch, J.M. and Slater, J.H. (editors), *Contemporary Microbial Ecology*, Academic Press, London.

Bauchop, T. 1989. Colonization of plant fragments by protozoa and fungi, pp. 83–96 in Nolan, J.V., Leng, R.A. and Demeyer, D.I. (editors). *Roles of Protozoa and Fungi in Ruminant Digestion*, Penambul Books, Armidale, Australia.

Bauchop, T. and Clarke, R.T.J. 1976. Attachment of the ciliate *Epidinium* Crawley to plant fragments in the sheep rumen. *Applied and Environmental Microbiology* 32: 417–422.

Bauchop, T. and Mountfort, D.O. 1981. Cellulose fermentation by a rumen anaerobic fungus in both the absence and the presence of rumen methanogens. *Applied and Environmental Microbiology* 42: 1103–1110.

Becker, E.R. 1929. Methods of rendering the rumen and reticulum of ruminants free from their normal infusorian fauna. *Proceedings of the National Academy of Sciences, U.S.A.* 15: 435–438.

Becker, E.R. 1932. The present status of problems relating to the ciliates of ruminants and equidae. *Quarterly Review of Biology* 7: 282–297.

Becker, E.R. and Everett, E.C. 1930. Comparative growths of normal and infusoria-free lambs. *American Journal of Hygiene* 11: 362–370.

Becker, E.R. and Hsiung, T.S. 1929. The method by which ruminants acquire their fauna of infusoria, and remarks concerning experiments on the host-specificity of these protozoa. *Proceedings of the National Academy of Sciences, U.S.A.* 15: 684–690.

Becker, E.R., Schulz, J.A. and Emmerson, M.A. 1929. Experiments on the physiological relationships between the stomach infusoria of ruminants and their hosts, with a bibliography. *Iowa State College Journal of Science* 4: 215–251.

Becker, E.R. and Talbott, M. 1927. The protozoan fauna of the rumen and reticulum of American cattle. *Iowa State College Journal of Science* 1: 345–371.

Bergen, W.G., Purser, D.B. and Cline, J.H. 1968. Determination of limiting amino acids of rumen isolated microbial protein fed to rat. *Journal of Dairy Science* 51: 1698–1700.

Bhatia, B.L. 1936. Protozoa: ciliophora, pp. 156–357 in Sewell, R.B.S. (editor), *The Fauna of British India including Ceylon and Burma,* Taylor and Francis, London.

Bird, S. H. 1989. Production from ciliate-free ruminants, pp. 233–246 in Nolan, J.V., Leng, P.A. and Demeyer, D.I. (editors), *The Roles of Protozoa and Fungi in Ruminant Digestion*, Penambul Books, Armidale, Australia.

Bird, S., Baigent, D.R., Dixon, R. and Leng, R.A. 1978. Ruminal protozoa and growth of lambs. *Proceedings of the Australian Society of Animal Production* 12: 137.

Bird, S.H., Hill, M.K. and Leng, R.A. 1979. The effect of defaunation of the rumen on the growth of lambs on low-protein high-energy diets. *British Journal of Nutrition* 42: 81–87.

Bird, S.H. and Leng, R.A., 1978. The effects of defaunation of the rumen on the growth of cattle on low-protein high-energy diets. *British Journal of Nutrition* 40: 163–167.

Bird, S.H. and Leng, R.A. 1983. The influence of the absence of rumen protozoa on ruminant production, pp. 110–118 in Farrell, D.J. and Vohra, P. (editors), *Recent Advances in Animal Production in Australia,* University of New England Publishing Unit, Armidale Australia.

Bird, S.H. and Leng, R.A., 1984. Further studies on the effects of the presence or absence of protozoa in the rumen on live-weight gain and wool growth of sheep. *British Journal of Nutrition* 52: 607–611.

Blackburn, T.H. and Hobson, P.N. 1960. Proteolysis by whole and fractionated rumen contents. *Journal of General Microbiology* 22: 272–281.

Boila, R.J. and Phillips, G.D. 1988. Effects of faunation, protein source and surgical modification of the intestinal tract upon flow of calcium, magnesium and phosphorus in the digestive tract of sheep. *Canadian Journal of Microbiology* 68: 853–863.

Bonhomme, A. 1968. Mise en évidence de l'action cellulolytique des ciliés oligotriches du genre *Entodinium*. *Protistologica* 4: 61–66.

Bonhomme, A. and Durand, M. 1974. Variations qualitatives et quantiatives de la faune du rumen du mouton en fonction de la nature du régime. *Annales Biologie Animales Biochemica Biophysica* 14: 679–687.

Bonhomme, A., Durand, M., Quintana, C. and Halpern, S. 1982a. Influence du cobalt et de la vitamine B_{12} sur la croissance et la survie des ciliés du rumen in vitro en fonction de la population bacterienne. *Reproduction, Nutrition Developpement* 22: 107–122.

Bonhomme, A., Fonty, G., Foglietti, M.J. and Weber, M. 1986. Endo-1,4-glucanase and β-glucosidase of the ciliate *Polyplastron multivesiculatum* free of cellulolytic bacteria. *Canadian Journal of Microbiology* 32: 219–225.

Bonhomme, A., Fonty, G. and Senaud, J. 1982a. Obtention de *Polyplastron multivesiculatum* (cilié entodiniomorphe du rumen) en condition axeniqué. *Journal of Protozoology* 29: 231–233.

Bonhomme, A., Fonty, G. and Senaud, J. 1982b. Essai d'obtention et de survie de ciliés entodiniomorphes du rumen en cultures axénique. *Annales de Microbiologie, Institute Pasteur* 133B: 335–341.

Bonhomme-Florentin, A. 1967. Les méthodes d'isolement du genre *Entodinium* cilié oligotriche. Action de ce cilié dans la digestion de la cellulose. *Protistologica* 3: 113–120.

Bonhomme-Florentin, A. 1975a. Activité cellulolytique des ciliés entodiniomorphes. *Journal of Protozoology* 22: 447–451.

Bonhomme-Florentin, A. 1975b. Ureolytic activity of endocommensal Cilata Entodiniomorpha of ruminants. *Protistologica* 11: 325–329.

Bonhomme-Florentin, A., Blancou, J. and Latteur, B. 1978. Étude des variations saisonnières de la microfaune du rumen de zebus. *Protistologica* 14: 283–289.

Bonhomme-Florentin, A. and Hempel-Zawitkowska, J. 1977. Comportment des populations ciliaires du rumen en culture dans le milieux saccharose et saccharose-urée. *Annales de la Nutrition et de l'Alimentation* 31: 331–348.

Borhami, B.E.A., El Shazly, K., Abou Akkada, A.R. and Ahmed, I.A. 1967. Effect of early establishment of ciliate protozoa in the rumen on microbial activity and growth of early weaned buffalo calves. *Journal of Dairy Science* 50: 1654–1660.

Boyne, A.W., Eadie, J.M. and Raitt, K. 1957. The development and testing of a method of counting rumen ciliate protozoa. *Journal of General Microbiology* 17: 414–423.

Braune, R. 1913. Untersuchungen über die im Wiederkäuermagen vorkommenden Protozoen. *Archiv für Protistenkunde* 32: 111–170.

Bressler, R. and Wakil, S.J. 1961. Studies on the mechanism of fatty acid synthesis. IX. The conversion of malonyl coenzyme A to long chain fatty acids. *Journal of Biological Chemistry* 236: 1643–1651.

Bretschneider, L.H. 1934. Beitrage zur Struckturiehre der Ophryoscolesciden. II. *Archiv für Protistenkunde* 82: 298–330.

Bretschneider, L.H. 1960a. Elektronenmikroskopische Untersuchung des Peristomapparates einiger Ophryoscoleciden. I. *Proceedings Koniglich Nederlandse Akademie van Wetenschappen* 63C: 291–305.

Bretschneider, L.H. 1960b. Elektronenmikroskopische Untersuchung des Peristomapparates einiger Ophryoscoleciden. II. *Proceedings Koniglich Nederlandse Akademie van Wetenschappen* 63C: 306–317.

Bretschneider, L.H. 1962. Das Paralabialorgan der Ophryoscoleciden. *Proceedings Koniglich Nederlandse Akademie van Wetenschappen* 65C: 423–452.

Bretschneider, L.N. and Van Vorstenbosch, C.J.A. 1964. Das Vorkommen von intraplasmatischen und intranuklearen Mikroorganismen in einigen Pansenciliaten. *Proceedings Koninklijke Nederlandse Akademie van Wetenschappen*, Series C, 67: 313–319.

Broad, T.E. and Dawson, R.M.C. 1973. Formation of ceremide phosphorylethanolamine from phosphatidylethanolamine in the rumen protozoon *Entodinium caudatum*. *Biochemical Journal* 134: 695–662.

Broad, T.E. and Dawson, R.M.C. 1975. Phospholipid biosynthesis in the anaerobic protozoon *Entodinium caudatum*. *Biochemical Journal* 146: 317–328.

Broad, T.E. and Dawson, R.M.C. 1976. Role of choline in the nutrition of the rumen protozoon *Entodinium caudatum*. *Journal of General Microbiology* 92: 391–397.

Brock, F.M., Forsberg, C.W. and Buchanan-Smith, J.G. 1982. Proteolytic activity of rumen microorganisms and effects of proteinase inhibitors. *Applied and Environmental Microbiology* 44: 561–569.

Brosch, A., Sneh, B. and Shkolnok, A. 1983. Effect of severe dehydration on the activity of the rumen microbial population of black Bedouin goats. *Journal of Agricultural Science* 100: 413–421.

Broudiscou, L., Van Nevel, C.J., Demeyer, D.I. and Jouany, J.P. 1988. Addition d'hydrolysat d'huile de soja dans la ration de mouton. Effect sur la dègradation in sacco de la paille et de la cellulose. *Reproduction Nutrition Developpement* 28: 159–160.

Brüggeman, J. von, Giesecke, D. and Walser-Kärst, K. 1967. Beiträge zur Wildbiologie und vergleichenden Tierphysiologie. II. Mikroorganismen in Pansen von Rothirsch (*Cervus elaphus*) und Reh (Capreolus capreolus). *Zeitschrift für Tierphysiologie, Tierernährung und Futtermittelkunde* 23: 143–151.

Bryant, M.P. and Small, N. 1960. Observations on the ruminal microorganisms of isolated and inoculated calves. *Journal of Dairy Science* 43: 654–667.

Buckelew, T.P. and Kontir, D.M. 1977. The ultrastructure of two cow rumen ciliates, the holotrich *Isotricha* spp., and the spirotrich *Epidinium ecaudatum*. *Proceedings of the Pennsylvania Acadamy of Science* 51: 95.

Buisson, J. 1923a. Les infusoires ciliés du tube digestif de l'homme et des mammifères. *Thèse Paris* (Paris Le Gall).

Buisson, J. 1923b. Infusoires nouveaux parasites d'antilopes africaines. *Comptes Rendus des Seances de la Societe de Biologie. Paris* 89: 1217–1219.

Buisson, J. 1923c. Sur quelques infusoires nouveaux ou peu connus parasites des mammifères. *Annales de Parasitologie Humaine et Comparee* 1: 209–246.

Buisson, J. 1924. Quelques infusoires parasites d'antilopes africaines. *Annales de Parasitologie Humaine et Comparee* 2: 155–160.

Burggraaf, W. and Leng, R.A. 1980. Antiprotozoal effects of surfactant detergents in the rumen of sheep. *New Zealand Journal of Agricultural Research* 23: 287–291.

Bush, M. and Kofoid, C.A. 1948. Ciliates from the Sierra Nevada bighorn (*Ovis canadensis sierrae* Grinnell). *University of California Publication in Zoology* 53: 237–261.

Bütschli, O. 1889. Protozoa Abt. III. Infusoria und System der Radiolaria, pp. 1098–2035 in Bronn, H.G. (editor), *Klassen und Ordnung des Thiers-Reichs* volume 1, C.F. Winter, Leipzig.

Bygrave, F.L. and Dawson, R.M.C. 1976. Phosphatidylcholine biosynthesis and choline transport in the anaerobic protozoon *Entodinium caudatum*. *Biochemical Journal* 160: 481–490.

Campbell, A.J., Cumming, G.J., Graham, C.A. and Leng, R.A. 1982. An in vitro assay for compounds toxic to rumen protozoa. *New Zealand Journal of Agricultural Research* 25: 535–540.

Campbell, A.S. 1929. The structure of *Isotricha prostoma*. *Archiv für Protistenkunde* 66: 331–339.

Campbell, A.S. 1930. Fission in *Isotricha prostoma*. *Archiv für Protistenkunde* 72: 141–146.

Carnie, J.A. and Porteous, J.W. 1959. The kinetic properties of invertases obtained from different genera of sheep rumen holotrich protozoa. *Biochemical Journal* 73: 47p.

Chalmers, M.I., Davidson, J., Eadie, J.M. and Gill, J.C. 1968. Some comparisons of performance of lambs with and without rumen ciliate protozoa. *Proceedings of the Nutrition Society* 27: 29A.

Chalupa, W. and Kutches, A.J. 1968. Biohydrogenation of linoleic-1-^{14}C-acid by rumen protozoa. *Journal of Animal Science* 27: 1502–1508.

Chalupa, W., O'Dell, G.D., Kutches, A.J. and Lavker, R. 1967. Changes in rumen chemical characteristics and protozoa populations of animals with depressed milk fat tests. *Journal of Dairy Science* 50: 1002.

Chamberlain, D.G., Thomas, P.C. and Anderson, F.J. 1981. Lactic acid metabolism in the rumen of animals given silage diets, pp. 25–26 in Harkess, R.D. and Castle, M.E. (editors), *Proceedings of the VI Silage Conference*, University of Edinburgh.

Chamberlain, D.G., Thomas, P.C. and Anderson, F.J. 1983. Volatile fatty acid proportions and lactic acid metabolism in the rumen of sheep and cattle receiving silage diets. *Journal of Agricultural Science* 101: 47–58.

Chamberlain, D.G., Thomas, P.C., Wilson, W., Newbold, C.J. and MacDonald, J.C. 1985. The effects of carbohydrate supplements on ruminal concentrations of ammonia in animals given diets of grass silage. *Journal of Agricultural Science* 104: 331–340.

Chavarria, M.C. 1933. Estuios protistologicos. I. Fauna del tubo digestivo die tor (*Bos taurus* Linn.). *Anales del Instituto de Biologia, Universidad Nacionale de Mexico* 4: 109–142.

Chen, M. and Wolin, M.J. 1977. Influence of methane production by *Methanobacterium ruminantium* on the fermentation of glucose and lactate by *Selenomonas ruminantium. Applied and Environmental Microbiology* 34: 756–759.

Cheng, K.-J., McCowan, R.P. and Costerton, J.W. 1979. Adherent epithelial bacteria in ruminants and their roles in digestive tract function. *American Journal of Clinical Nutrition* 32: 139–148.

Chesters, J.K. 1968. Cell-free protein synthesis by rumen protozoa. *Journal of Protozoology* 15: 509–512.

Christiansen, W.C., Kawashima, R. and Burroughs, W. 1965. Influence of protozoa upon rumen acid production and liveweight gains in lambs. *Journal of Animal Science* 24: 730–734.

Christiansen, W.C., Wood, W. and Burroughs, W. 1964. Ration characteristics influencing rumen protozoal populations. *Journal of Animal Science* 23: 984–988.

Christie, A.O. and Porteous, J.W. 1957. An invertase from the holotrich protozoa of sheep rumen liquor *Biochemical Journal* 67: 19p.

Christl, H. 1955. Beitrage zur Kenntis der Panseninfusorien einheimischer Weiderkauer. I Gemse (*Rupicapra rupicapra* L.). *Zeitschrift für Parasitenkunde* 17: 185–192.

Christl, H. 1958a. *Entodinium alpinum* n.sp. (*Caliata*) aus dem pansen eines Alpensteinbockes (*Capra ibex ibex*). *Zeitschrift für Parasitenkunde* 18: 297–307.

Christl, H. 1958b. Neue Entodinien (*Ciliata, Entodiniomorpha*) aus dem Alpensteinbock (*Capra ibex ibex*). *Zeitschrift für Parasitenkunde* 18: 403–418.

Clarke, R.T.J. 1963. The cultivation of some oligotrich protozoa. *Journal of General Microbiology* 33: 401–408.

Clarke, R.T.J. 1964. Ciliates of the rumen of domestic cattle (*Bos taurus* L.). *New Zealand Journal of Agricultural Research* 7: 248–257.

Clarke, R.T.J. 1965a. Diurnal variations in the number of ciliate protozoa in cattle. *NewZealand Journal of Agricultural Research* 8: 1–9.

Clarke, R.T.J. 1965b. Quantitative studies of digestion in the reticulo-rumen. 3. Fluctuations in the number of protozoa and their possible role in bloat. *Proceedings of the New Zealand Society of Animal Production* 25: 96–103.

Clarke, R.T.J. 1965c. Role of rumen ciliates in bloat in cattle. *Nature, London* 205: 95–96.

Clarke, R.T.J. 1966. Bloat in cattle. XXVII. Preliminary observations with the anti-protozoal agent 1,2-dimethyl-5-nitroimidazole for prevention of bloat. *New Zealand Journal of Agricultural Research* 9: 437–443.

Clarke, R.T.J. 1968. The ophryoscolecid ciliates of Red Deer (*Cervus elaphus* L.) in New Zealand. *New Zealand Journal of Science* 11: 686–692.

Clarke, R.T.J. 1977. Protozoa in the rumen ecosystem, pp. 251–275 in Clarke, R.T.J. and Bauchop, T. (editors), *Microbial Ecology of the Gut*, Academic Press, London.

Clarke, R.T.J. and Hungate, R.E. 1966. Culture of the rumen holotrich ciliate *Dasytricha ruminantium* Schuberg. *Applied Microbiology* 14: 340–345.

Clarke, R.T.J. and Reid, C.S.W. 1969. Bloat in cattle. XXXI. The effect of dimetridazole on the rumen ciliate protozoa of dry and lactating cows. *New Zealand Journal of Agricultural Research* 12: 437–445.

Clarke, R.T.J. and Reid, C.S.W. 1974. Foamy bloat of cattle. A review. *Journal of Dairy Science* 57: 753–785.

Clarke, R.T.J. Reid, C.S.W. and Young, P.W. 1969. Bloat in cattle. XXXII. Attempts to prevent bloat in dry and lactating cows by partial or complete elimination of the holotrich protozoa with dimetridazole. *New Zealand Journal of Agricultural Research* 12: 446–466.

Clarke, R.T.J., Ulyatt, M.J. and John, A. 1982. Variation in numbers and mass of ciliate protozoa in the rumens of sheep fed chaffed alfalfa (*Medicago sativa*). *Applied and Environmental Microbiology* 43: 1201–1204.

Clegg, M.T. and Cole, H.H. 1954. The action of stilboestrol on growth response in ruminants. *Journal of Animal Science* 13: 108–130.

Cockburn, J.E. and Williams, A.P. 1984. The simultaneous estimation of protozoal, bacterial and dietary nitrogen entering the duodenum of steers *British Journal of Nutrition* 51: 111–132.

Coleman, G.S. 1958. Maintenance of oligotrich protozoa *in vitro*. *Nature* (*London*) 182: 1104–1105.

Coleman, G.S. 1960a. The cultivation of sheep rumen oligotrich protozoa *in vitro*. *Journal of General Microbiology* 22:555–563.

Coleman, G.S. 1960b. Effect of penicillin on the maintenance of rumen oligotrich protozoa. *Nature* (*London*) 187: 518.

Coleman, G.S. 1962. The preparation and survival of almost bacteria-free suspensions of *Entodinium caudatum*. *Journal of General Microbiology* 28: 271–281.

Coleman, G.S. 1963. The growth and metabolism of rumen ciliate protozoa, pp. 298–324, in *Symbiotic Associations* Nutman, P.S. and Mosse, B. (editors), Thirteenth Symposium of Society for General Microbiology, Cambridge University Press.

Coleman, G.S. 1964a. The metabolism of ^{14}C-glycine and ^{14}C-bicarbonate by washed suspensions of the rumen ciliate *Entodinium caudatum*. *Journal of General Microbiology* 35: 91–103.

Coleman, G.S. 1964b. The metabolism of *Escherichia coli* and other bacteria by *Entodinium caudatum*. *Journal of General Microbiology* 37: 209–223.

Coleman, G.S. 1967a. The metabolism of free amino acids by washed suspensions of the rumen ciliate *Entodinium caudatum*. *Journal of General Microbiology* 47: 433–447.

Coleman, G.S. 1967b. The metabolism of the amino acids of *Escherichia coli* and other bacteria by the rumen ciliate *Entodinium caudatum*. *Journal of General Microbiology* 47: 449–464.

Coleman, G.S. 1968. The metabolism of bacterial nucleic acid and of free components of nucleic acid by the rumen ciliate *Entodinium caudatum*. *Journal of General Microbiology* 54: 83–96.

Coleman, G.S. 1969a. The cultivation of the rumen ciliate *Entodinium simplex*.

Journal of General Microbiology 57: 81–90.
Coleman, G.S. 1969b. The metabolism of starch, maltose, glucose and some other sugars by the rumen ciliate *Entodinium caudatum*. *Journal of General Microbiology* 57: 303–332.
Coleman, G.S. 1972. The metabolism of starch, glucose, amino acids, purines, pyrimidines and bacteria by the rumen ciliate *Entodinium simplex*. *Journal of General Microbiology* 71:117–131.
Coleman, G.S. 1975a. The inter-relationships between rumen ciliate protozoa and bacteria, pp. 149–164 in McDonald, I.W. and Warner, A.C. (editors), *Digestion and Metabolism in the Ruminant*, University of New England Publishing Unit, Armidale Australia.
Coleman, G.S. 1975b. The role of bacteria in the metabolism of rumen entodiniomorphid protozoa, pp. 553–558, in *Symbiosis*, Jennings, D.H. and Lee, D.L. (editors), Symposium of the Society for Experimental Biology, Volume 29, Cambridge University Press.
Coleman, G.S. 1978a. Rumen entodiniomorphid protozoa, pp. 39–54 in Taylor, A.E.R. and Baker, J.R. (editors), *Methods of Cultivating Parasites in vitro*, Academic Press, London.
Coleman, G.S. 1978b. The metabolism of cellulose, glucose and starch by the rumen ciliate *Eudiplodinium maggii*. *Journal of General Microbiology* 107: 359–366.
Coleman, G.S. 1978c. Methods for the study of the metabolism of rumen ciliate protozoa and their closely associated bacteria, pp. 143–163, in *Techniques for the study of mixed populations*, Lovelock, D.W. and Davies, R. (editors), Society for Applied Bacteriology Technical Series, Volume 11, Academic Press, London.
Coleman, G.S. 1979a. The role of rumen protozoa in the metabolism of ruminants given tropical feeds. *Tropical Animal Production* 4: 199–213.
Coleman, G.S. 1979b. Rumen ciliate protozoa, pp. 381–408 in Levandowsky, M., Hutner, S.H. and Provasoli, L. (editors), *Biochemistry and Physiology of Protozoa*, Volume 2, Academic Press, London.
Coleman, G.S. 1980. Rumen ciliate protozoa. *Advances in Parasitology* 18: 121–173.
Coleman, G.S. 1981. In Alternate metabolic pathways in protozoal energy metabolism (Workshop 1, EMPO 3). *Parasitology* 82: 5–6, 24–25.
Coleman, G.S. 1983a. Hydrolysis of fraction 1 leaf protein and casein by rumen entodiniomorphid protozoa. *Journal of Applied Bacteriology* 55: 111–118.
Coleman, G.S. 1983b. Preliminary observations on the separation of the cellulases of the rumen ciliates *Eremoplastron bovis* and *Eudiplodinium maggii*. *Journal of Protozoology* 30: 36A
Coleman, G.S. 1985a. The cellulase content of 15 species of entodiniomorphid protozoa, mixed bacteria and plant debris isolated from the ovine rumen. *Journal of Agricultural Science* 104: 349–360.
Coleman, G.S. 1985b. Possible causes of the high death rate of ciliate protozoa in the rumen. *Journal of Agricultural Science* 105: 39–43.
Coleman, G.S. 1986a. The distribution of carboxymethylcellulase between fractions taken from the rumen of sheep containing no protozoa or one of five different protozoal populations. *Journal of Agricultural Science* 106: 121–127.
Coleman, G.S. 1986b. The amylase activity of 14 species of entodiniomorphid protozoa and the distribution of amylase in digesta fractions of sheep containing no protozoa or one of seven different protozoal populations. *Journal of Agricultural Science* 107: 709–721.
Coleman, G.S. 1986c. The metabolism of rumen ciliate protozoa. *FEMS Microbiology Reviews* 39: 321–344.

Coleman, G.S. 1986d. The rate of synthesis of amylopectin from cellulose and of the degradation of intracellular 'starch' by rumen ciliate protozoa. *Journal of Protozoology* 33: 101.

Coleman, G.S. 1987a. Rumen entodiniomorphid protozoa, pp. 29–51, in Taylor, A.E.R. and Baker, J.R. (editors), *In vitro Methods for Parasite Cultivation*, Academic Press, London.

Coleman, G.S. 1987b. Rumen ciliate protozoa, pp. 30–47, in U. B. Singh (editor) *Advanced Animal Nutrition for Developing Countries*, Indo-Vision Private Ltd, Ghaziabad, India.

Coleman, G.S. 1988. The importance of rumen ciliate protozoa in the growth and metabolism of the host ruminant. *International Journal of Animal Science* 3: 75–95.

Coleman, G.S. 1989. Protozoal-bacterial interaction in the rumen, pp. 13–27 in Nolan, J.V., Leng, R.A. and Demeyer, D.I. (editors), *The Role of Protozoa and Fungi in Ruminant Digestion*, Penambul Books, Armidale, Australia.

Coleman, G.S., Davies, J.I. and Cash, M.A. 1972. The cultivation of the rumen ciliates *Epidinium ecaudatum caudatum* and *Polyplastron multivesiculatum in vitro*. *Journal of General Microbiology* 75: 509–521.

Coleman, G.S., Dawson, R.M.C. and Grime, D.W. 1980a. The rate of passage of ciliate protozoa from the ovine rumen. *Proceedings of the Nutrition Society* 39: 6A.

Coleman, G.S. and Hall, F.J. 1969. Electron microscopy of the rumen ciliate *Entodinium caudatum* with special reference to the engulfment of bacteria and other particulate matter. *Tissue and Cell* 1: 607–618.

Coleman, G.S. and Hall, F.J. 1971a. A study in the light and electron microscope of the extruded peristome and related structures of the rumen ciliate *Entodinium caudatum*. *Tissue and Cell* 3: 371–380.

Coleman, G.S. and Hall, F.J. 1971b. Studies in the electron microscope of the adoral zone of membranelles of the rumen ciliate *Entodinium caudatum* using the technique of negative staining. *Tissue and Cell* 3: 381–388.

Coleman, G.S. and Hall, F.J. 1972. Fine structural studies on the digestion of bacterial species in the rumen ciliate *Entodinium caudatum*. *Tissue and Cell* 4: 37–48.

Coleman, G.S. and Hall, F.J. 1974. The metabolism of *Epidinium ecaudatum caudatum* and *Entodinium caudatum* as shown by autoradiography in the electron microscope. *Journal of General Microbiology* 85: 265–273.

Coleman, G.S. and Hall, F.J. 1978. Digestion of *Epidinium ecaudatum caudatum* by the rumen ciliate *Polyplastron multivesiculatum* as shown by studies in the electron microscope. *Society for General Microbiology Quarterly* 6: 29–30.

Coleman, G.S. and Hall, F.J. 1979. Digestion of *Entodinium caudatum* by the rumen ciliate *Entodinium bursa* as shown by studies in the electron microscope. *Society for General Microbiology Quarterly* 6: 81–82.

Coleman, G.S. and Hall, F.J. 1984. The uptake and utilization of *Entodinium caudatum*, bacteria, free amino acids and glucose by the rumen ciliate *Entodinium bursa*. *Journal of Applied Bacteriology* 56: 283–294.

Coleman, G.S., Kemp, P. and Dawson, R.M.C. 1971. The catabolism of phosphatidylethanolamine by the rumen protozoon *Entodinium caudatum* and conversion into the N-(1-carboxyethyl) derivative. *Biochemical Journal* 123: 97–104.

Coleman, G.S. and Laurie, J.I. 1974a. The metabolism of starch, glucose, amino acids, purines, pyrimidines and bacteria by three *Epidinium* spp. isolated from the rumen. *Journal of General Microbiology* 85: 244–256.

Coleman, G.S. and Laurie, J.I. 1974b. The utilization of *Bacillus megaterium* and the release of a lytic enzyme by three *Epidinium* spp. isolated from the rumen. *Journal of General Microbiology* 85: 257–264.

Coleman, G.S. and Laurie, J.I. 1976. The uptake and metabolism of glucose, maltose and starch by the rumen ciliate *Epidinium ecaudatum caudatum*. *Journal of General Microbiology* 95: 364–374.

Coleman, G.S. and Laurie, J.I. 1977. The metabolism of starch, glucose, amino acids, purines, pyrimidines and bacteria by the rumen ciliate *Polyplastron multivesiculatum*. *Journal of General Microbiology* 98: 29–37.

Coleman, G.S., Laurie, J.I. and Bailey, J.E. 1977. The cultivation of the rumen ciliate *Entodinium bursa* in the presence of *Entodinium caudatum*. *Journal of General Microbiology* 101: 253–258.

Coleman, G.S., Laurie, J.I., Bailey, J.E. and Holdgate, S.A. 1976. The cultivation of cellulolytic protozoa isolated from the rumen. *Journal of General Microbiology* 95: 144–150.

Coleman, G.S. and Reynolds, D.J. 1982a. The effect of sterols and haemin on the growth of *Ophryoscolex caudatus* and some other Entodiniomorphid protozoa. *Journal of Applied Bacteriology* 52: 129–134.

Coleman, G.S. and Reynolds, D.J. 1982b. The uptake of bacteria and amino acids by *Ophryoscolex caudatus, Diploplastron affine* and some other rumen Entodiniomorphid protozoa. *Journal of Applied Bacteriology* 52: 135–144.

Coleman, G.S. and Sandford, D.C. 1979a. The engulfment and digestion of mixed rumen bacteria and individual bacterial species by single and mixed species of rumen ciliate protozoa grown *in vivo*. *Journal of Agricultural Science* 92: 729–742.

Coleman, G.S. and Sandford, D.C. 1979b. The uptake and utilization of bacteria, amino acids and nucleic acid components by the rumen ciliate *Eudiplodinium maggii*. *Journal of Applied Bacteriology* 47: 409–419.

Coleman, G.S. and Sandford, D.C. 1980. The uptake and metabolism of bacteria, amino acids, glucose and starch by the spined and spineless forms of the rumen ciliate *Entodinium caudatum*. *Journal of General Microbiology* 117: 411–418.

Coleman, G.S., Sandford, D.C. and Beahon, S. 1980b. The degradation of polygalacturonic acid by rumen ciliate protozoa. *Journal of General Microbiology* 120: 295–300.

Coleman, G.S. and White, R.W. 1970. Re-establishment of *Entodinium caudatum*, cultured *in vitro*, in the rumen of a defaunated sheep. *Journal of General Microbiology* 62: 265–266.

Connell, P.J., Gulati, S.K. and Ashes, J.R. 1987. Improved techniques in the measurement of amino acids and amino sugars in rumen anaerobic fungi *Proceedings of the Nutrition Society of Australia* 12: 92–95.

Conrad, H.R., Hibbs, J.W. and Frank, N. 1958. High roughage systems for raising calves based on early development of rumen function. IX. Effect of rumen inoculations and chlortetracycline on rumen function of calves fed high roughage pellets. *Journal of Dairy Science* 41: 1248–1261.

Conrad, H.R., Hibbs, J.W., Pounden, W.D. and Sutton, T.S. 1950. The effect of rumen inoculations on the digestibility of roughages in young dairy calves. *Journal of Dairy Science* 33: 585–592.

Constantinescu, E. and Dragos, N. 1984. Unele aspecte de structura electronmicroscopica a infuzorilor ruminali de la ovine. *Buletinul Institutului Agronomic Cluj-Napoca, Seria Zootehnie si Medicina Veterinara* 38: 57–60.

Corliss, J.O. 1956. On the evolution and systematics of ciliated protozoa. *Systematic Zoology* 5: 68–91, 121–140.

Corliss, J.O. 1979. *The Ciliated Protozoa: Characterization, Classification and Guide to the Literature*, Pergamon Press, Oxford. 455 p.

Coste, M. 1864. Développement des Infusoires ciliés dans une macération de foin. *Comptes Rendus Hebdomadaire des Seances de l'Academie des Sciences, Paris* 59: 149–155.

Cottle, D. 1980. The synthesis, turnover and outflow of ruminal microorganisms. PhD Thesis, University of New England, Armidale, Australia quoted by Leng and Nolan (1984).

Cottle, D.J. 1988a. Effects of defaunation of the rumen and supplementation with amino acids on the wool production of housed Saxon Merinos. I. Lupins and extruded lupins. *Australian Journal of Experimental Agriculture* 28: 173–178.

Cottle, D.J. 1988b. Effects of defaunation of the rumen and supplementation with amino acids on the wool production of housed Saxon Merinos. 2. Methionine and protected methionine. *Australian Journal of Experimental Agriculture* 28: 179–185.

Cottle, D.J., Nolan, J.V. and Leng, R.A. 1978. Turnover of protozoa and bacteria in the rumen of sheep. *Proceedings of the Australian Society of Animal Production* 12: 138.

Counotte, G.H.M. 1979. Kinetic parameters of lactate dehydrogenases of some rumen bacterial species, the anaerobic rumen ciliate *Isotricha prostoma* and mixed rumen microorganisms. *Antonie van Leeuwenhoek Journal of Microbiology and Serology* 45: 614.

Counotte, G.H.M., De Groot, M. and Prins, R.A. 1980. Kinetic parameters of lactate dehydrogenases of some rumen bacterial species, the anaerobic ciliate *Isotricha prostoma* and mixed rumen microorganisms, *Antonie van Leeuwenhoek Journal of Microbiology and Serology* 46: 363–381.

Crawford, R.J., Hoover, W.H. and Knowlton, P.H. 1980. Effects of solids and liquid flows on fermentation in continuous culture. 1. Dry matter and fiber digestion, VFA production and protozoa numbers. *Journal of Animal Science* 51: 975–985.

Crawley, H. 1923. Evolution in the family *Ophryoscolecidae*. *Proceedings of the National Academy of Natural Science. Philadelphia* 75: 393–412.

Cunha, A.M. da 1914a. Über die Ziliaten, Weiche in Brazilien im Magen von Rindern und Schafen vorkommen. *Memorias do Instituto Oswaldo Cruz* 6: 58–67.

Cunha, A.M. da 1914b. Über die Darmziliaten der Säugetiere. *Memorias do Instituto Oswaldo Cruz* 6: 212–216.

Cunha, A.M. da 1914c. Contribuicao par o conhecimento dos ciliados parasitos dos mammiferos brazileiros. Thesis Rio de Janeiro, Rodrigues.

Cunha, A.M. da 1917. Sobre os ciliados do tubo digestivo dos mammiferos pp. 383–390 in *Proceedings of the Conference of the Society of South American Hygiene (Buenos Aires)*.

Cunha, A. da and Muniz, J. 1928. Nouveau cilié parasite du caecum du *Tapirus americanus*. Description d'un nouveau genere. *Comptes Rendus de la Societe de Biologie* 98: 631–632.

Curds, C.R., Roberts, D. McL. and Chih-Hua, W. 1978. The use of continuous cultures and electronic sizing devices to study the growth of two species of ciliated protozoa, pp. 165–178 in Lovelock, D.W. and Davies, R. (editors), *Techniques for the study of mixed populations*, Academic Press, London.

Czerkawski, J.W. 1986. *An Introduction to Rumen Studies*, Pergamon Press, Oxford. 236 p.

Czerkawski, J.W. 1987. Reassessment of the contribution of protozoa to the microbial protein supply of the host ruminant animal. *Journal of Theoretical Biology* 126: 335–341.

Czerkawski, J.W. and Breckenridge, G. 1977. Design and development of a long-term rumen simulation technique (Rusitec). *British Journal of Nutrition* 38: 371–384.

Czerkawski, J.W. and Breckenridge, G. 1979 Experiments with the long-term rumen simulation technique (Rusitec); use of soluble food and an inert solid matrix. *British Journal of Nutrition* 42: 229–245.

Das Gupta, M. 1935. Preliminary observations on the protozoan fauna of the rumen of the Indian goat, *Capra hircus Linn. Archiv für Protistenkunde* 85: 153–172.

Dauvrin, T. and Thines-Sempoux, D. 1989a. Invertosomes in the rumen holotrich ciliate, *Isotricha prostoma*, pp. 56–65 in Lloyd, D., Coombs, G.H. and Paget, T.A. (editors), *The Biochemistry and Molecular Biology of Anaerobic Protozoa*, Harwood Academic Publishers, Chur, Switzerland.

Dauvrin, T. and Thines-Sempoux, D. 1989b. Multimeric form and microheterogeneity of the invertase from the rumen holotrich *Isotricha prostoma. Journal of Protozoology* 36: 16A.

Dauvrin, T. and Thines-Sempoux, D. 1989c. Purification and characterization of a heterogeneous glycosylated invertase from the rumen holotrich ciliate *Isotricha prostoma. Biochemical Journal* 264: 721–727.

Davis, J.D. and Essig, H.W. 1972. Comparison of three bloat preventing compounds for cattle grazing clover. *Canadian Journal of Animal Science* 52: 329–335.

Dawson, R.M.C. and Kemp, P. 1967. The aminoethylphosphonate-containing lipids of rumen protozoa. *Biochemical Journal* 105: 837–842.

Dawson, R.M.C. and Kemp, P. 1969. The effect of defaunation on the phospholipids and on the hydrogenation of unsaturated fatty acids in the rumen. *Biochemical Journal* 115: 351–352.

Dearth, R.N., Dehority, B.A. and Potter, E.L. 1974. Rumen microbial numbers in lambs as affected by level of feed intake and dietary diethylstilboestrol. *Journal of Animal Science* 38: 991–996.

Dehority, B.A. 1970. Occurrence of the ciliate protozoa *Butschlia parva* Schuberg in the rumen of the ovine. *Applied Microbiology* 19: 179–181.

Dehority, B.A. 1974. Rumen ciliate fauna of Alaskan Moose (*Alces americana*), Musk-Ox (*Ovibos moschatus*) and Dall Mountain sheep (*Ovis dalli*). *Journal of Protozoology* 21: 26–32.

Dehority, B.A. 1975a. Rumen ciliate protozoa of Alaskan Reindeer and Caribou (*Rangifer tarandus* L.), pp. 241–250 in *Proceedings of the 1st International Reindeer and Caribou Symposium*, Special Report Number 1, Biological Papers of the University of Alaska, Fairbanks.

Dehority, B.A. 1975b. *Diplodinium (Ostracodinium) minorum* sp.n., ciliate from the rumen of domestic sheep. *Journal of Protozoology* 22: 328–330.

Dehority, B.A. 1978. Specificity of rumen ciliate protozoa in cattle and sheep. *Journal of Protozoology* 25: 509–513.

Dehority, B.A. 1979. Ciliate protozoa in the rumen of Brazilian water buffalo *Bubalus bubalis* Linnaeus. *Journal of Protozoology* 26: 536–544.

Dehority, B.A. 1984. Evaluation of subsampling and fixation procedures used for counting rumen protozoa. *Applied and Environmental Microbiology* 48: 182–185.

Dehority, B.A. 1985. Rumen ciliates of Musk-oxen (*Ovibos moschatus* Z.) from the Canadian arctic. *Journal of Protozoology* 32: 246–250.

Dehority, B.A. 1986a. Protozoa of the digestive tract of herbivorous mammals. *Insect Science and its Application* 7: 279–296.

Dehority, B.A. 1986b. Rumen ciliate fauna of some Brazilian cattle: occurrence of several ciliates new to the rumen including the cycloposthid *Parentodinium africanum. Journal of Protozoology* 33: 416–421.

Dehority, B.A. 1986c. Microbes in the foregut of arctic ruminants, pp. 307–325 in Milligan, L.P., Grovum, W.L. and Dobson, A. (editors), *Control of Digestion and Metabolism in Ruminants*, Prentice-Hall, Englewood Cliffs N.J.

Dehority, B.A. 1987. Rumen Ophryoscolecid protozoa in the hindgut of the Capybara (*Hydrochoerus hydrochaeris*). *Journal of Protozoology* 34: 143–145.

Dehority, B.A., Damron, W.S. and McLaren, J.B. 1983. Occurrence of the rumen ciliate *Oligoisotricha bubali* in domestic cattle (*Bos taurus*). *Applied and Environmental Microbiology* 45: 1394–1397.

Dehority, B.A. and Males, J.R. 1974. Rumen fluid osmalality: evaluation of its influence upon the occurrence and numbers of holotrich protozoa in sheep. *Journal of Animal Science* 38: 865–870.

Dehority, B.A. and Mattos, W.R.S. 1978. Diurnal changes and effect of ration on concentrations of the rumen ciliate *Charon ventriculi*. *Applied and Environmental Microbiology* 36: 953–958.

Dehority, B.A. and Potter, E.L. 1974. *Diplodinium flabellum*: Occurrence and numbers in the rumen of sheep with a description of two new subspecies. *Journal of Protozoology* 21: 686–693.

Dehority, B.A. and Purser, D.B. 1970. Factors affecting the establishment and numbers of holotrich protozoa in the ovine rumen. *Journal of Animal Science* 30: 445–449.

Dehority, B.A. and Tirabasso, P.A. 1989. Factors affecting the migration and sequestration of rumen protozoa in the family Isotrichidae. *Journal of General Microbiology* 135: 539–548.

Delfosse, J. 1977. Ultrastructure de l'endoplasme des Ophryoscolescides. *Cellule* 72: 77–90.

Delfosse-Debusscher, J., Thines-Sempoux, D., Vanbelle, M. and Latteur, B. 1979a. Contribution of protozoa to rumen cellulolytic activity. *Annales de Recherches Veterinaires* 10: 255–257.

Delfosse-Debusscher, J., van Hoof, F., Hellings, P. and Thines-Sempoux, D. 1979b. Hydrolytic activities of rumen ciliates. *Annales de Recherches Veterinaires* 10: 258–260.

Dellow, D.W., Hume, J.D., Clarke, R.T.J. and Bauchop, T. 1988. Microbial activity in the forestomach of free-living macropodid marsupials: comparisons with laboratory studies. *Australian Journal of Zoology* 36: 383–395.

Demeyer, D.I. 1981. Rumen microbes and digestion of plant cell walls. *Agriculture and Environment* 6: 295–337.

Demeyer, D.I. 1982. Influence of calcium peroxide on fermentation pattern and protozoa in the rumen. *Archiv für Tierernährung* 32: 579–593.

Demeyer, D.I. 1988. Interdépendance ces effets de la défaunation sur l'activité muralytique, le volume et la vitesse de renouvellement du contenu du rumen. Resultats préliminaires et hypothèse. *Reproduction Nutrition Developpement* 28: 161–162.

Demeyer, D.I. 1989. Effect of defaunation on rumen fibre digestion and digesta kinetics, pp. 171–179 in Nolan, J.V., Leng, R.A. and Demeyer, D.I. (editors). *The Roles of Protozoa and Fungi in Ruminant Digestion*, Penambul Books, Armidale, Australia.

Demeyer, D.I., Henderson, C. and Prins, R.A. 1978. Relative significance of exogenous and *de novo* synthesized fatty acids in the formation of rumen microbial lipids *in vitro*. *Applied and Environmental Microbiology* 35: 24–31.

Demeyer, D.I., Todorov, N., Van Nevel, C. and Vets, J. 1982a. α-ϵ-Diaminopimelic acid (DAPA) in a sheep rumen infused with a synthetic diet of sugars and urea: evidence for degradation of bacteria. *Zeitschrift für Tierphysiologie, Tierernährung und Futtermittelkunde* 48: 21–32.

Demeyer, D.I. and Van Nevel, C.J. 1979. Effect of defaunation on the metabolism of rumen microorganisms. *British Journal of Nutrition* 42: 515–524.

Demeyer, D.I. and Van Nevel, C.J. 1986. Influence of substrate and microbial interaction on efficiency of rumen microbial growth. *Reproduction Nutrition Developement* 26: 161–179.

Demeyer, D.I., Van Nevel, C.J., Teller, E. and Godeau, J.M. 1986. Manipulation of rumen digestion in relation to the level of production in ruminants. *Archives of Animal Nutrition, Berlin* 36: 132–143.

Demeyer, D.I., Van Nevel, C.J. and Van de Voorde, G. 1982b. The effect of defaunation on the growth of lambs fed three urea-containing diets. *Archiv für Tierernährung* 32: 595–604.

Denholm, A.M. and Ling, J.R. 1984. *In vitro* metabolism of bacterial cell wall by rumen ciliate protozoa. *Canadian Journal of Animal Science* 64 (suppl.): 18–19.

Dennis, S.M., Arambel, M.J., Bartley, E.E. and Dayton, A.E. 1983. Effect of energy concentration and source of nitrogen on numbers and types of rumen protozoa. *Journal of Dairy Science* 66: 1248–1254.

Dennis, S.M., Arambel, M.J., Bartley, E.E., Riddell, D.O. and Dayton, A.D. 1982. Effect of heated or unheated soybean meal with or without niacin on rumen protozoa. *Journal of Dairy Science* 65: 1643–1646.

Dennis, S.M., Nagaraja, T.G. and Dayton, A.D. 1986. Effect of lasolocid, monensin and thiopeptin on rumen protozoa. *Research in Veterinary Science* 41: 251–256.

Devuyst, A., Mordenti, A., Vanbelle, M. and Crevecoeur, E. 1968. Variation du pH et de la teneur en acides gras volatils dans le liquide du rumen du mouton induites par l'addition de glucides a la ration de base. 1. Action du glucose et de l'amidon. *Rivista di Zootecnia Veterinaria* 6: 339–352.

Devuyst, A., Jaramillo, D., Vanbelle, M., Arnould, R., Ska, P. and Moreels, A. 1973. Comparison des effets de diverses rations sur le developpement des ciliates du rumen. *Zeitschrift für Tierphysiologie, Tierernährung und Futtermittelkunde* 31: 136–149.

Devuyst, A., Vanbelle, M., Jossart, J.N. and Baguette, A. 1975. The effect of methionine hydroxyanalogue supplementation of the diet on the concentration of ciliate protozoa in the rumen of sheep. *Zeitschrift für Tierphysiologie, Tierernährung und Futtermittelkunde* 35: 316–321.

Diamond, L.S. 1957. The establishment of various trichomonads of animals and man in axenic cultures. *Journal of Parasitology* 43: 488–490.

Dinius, D.A., Simpson, M.E. and Marsh, P.B. 1976. Effect of monensin fed with forage on digestion and the ruminal ecosystem of steers. *Journal of Animal Science* 42: 229–234.

Doflein, F. 1901. *Die Protozoen als Parasiten und Krankheitserreger nach biologischen Gesichtspunkten dargestellt.* G. Fisher, Jena. pp. 274.

Doflein, F. and Reichenow, E. 1929. *Lehrbuch der Protozoenkunde. Eine Darstellung der Naturgeschichte der Protozoen mit besonderer Berücksichtigung der parasitischen und pathogenen Formen.* 5th edition, G. Fisher, Jenna. pp. 1264.

Dogiel, V.A. 1923a. Die Artbildung in der Infusioren-familie *Ophryoscolecidae*. *Archives Russes de Protistologie* 2: 89–104.

Dogiel, V.A. 1923b. Cellulose ais Bestandteil des Skelletts bei einigen Infusioren. *Biolisches Zentrablatt* 43: 289–291.

Dogiel, V.A. 1925a. Nouveaux infusiores de la famille de *Ophryoscolecidae* parasites d'antilopes africaines. *Annales de Parasitologie Humaine et Comparee* 3: 116–142.

Dogiel, V.A. 1925b. On the influence of the macronucleus on the formation of new morphological characters in infusoria. *Quarterly Journal of Microscopical Science* 69: 611–617.

Dogiel, V.A. 1925c. Die Gwschlechtprozesse bei Infusioren (speziall bei dem Ophryoscoleciden) neue theoretische Erwagungen. *Archiv für Protistenkunde* 50: 282–442.

Dogiel, V.A. 1925d. Neue parasistische Infusorien aus dem Magen des Renntieres (*Rangifer tarandus*). *Archives Russes de Protistologie* 4: 43–65.

Dogiel, V.A. 1925e. Über die Art der Nahrung und der Nahrungsaufnahme bei den Darm der Huftiere parasitierende Infusorien. *Transactions of the Leningrad Natural History Society* 54: 69–93.

Dogiel, V. 1926a. Une nouvelle espece du genere *Blepharocorys*, B. bovis N.sp. habitant l'estomac du boeuf. *Annales de Parasitologie* 4: 61–64.

Dogiel, V. 1926b. Sur quelques infusoires nouveaux habitant l'estomac du dromadaire (*Camelus dromedarius*). *Annales de Parasitologie* 4: 241–271.

Dogiel, V.A. 1927. Monographie der Familie *Ophryoscolecidae*. *Archiv für Protistenkunde* 59: 1–288.

Dogiel, V. 1928. La faune d'infusoires inhabitant l'estomac du buffle et du dromadaire. *Annales de Parasitologie* 6: 323–338.

Dogiel, V. 1932. Beschreibung einiger neuer vertreter der Familie Orphryoscolecidae aus afrikanishen Antilopen nebst Revision der infusioren Fauna afrikanischer Wiederkäuer. *Archiv für Protistenkunde* 77: 92–107.

Dogiel, V. 1934. Angaben ber die Ophryoscolecidae des Wildschafs aus kamtschatka des Elches und des Yaks, nebst deren zoogeographischen Verwertung. *Archiv für Protistenkunde* 82: 290–297.

Dogiel, V. 1935. Eine Notiz ber die Infusiorien des Renntiermagens. *Transactions of the Arctic Institute, Leningrad* 24: 144–148.

Dogiel, V. 1947. The phylogeny of the stomach infusorians of ruminants in the light of palaeontological and parasitological data. *Quarterly Journal of Microscopical Science* 88: 337–343.

Dogiel, V. and Fedorowa, T. 1925a. A note on the reproduction of *Isotricha (Dasytricha) ruminantium*. *Ruskii Arkhiv Protistologii (Archives Russes de Protistologie)* 4: 75–82.

Dogiel, V.A. and Fedorowa, T. 1925b. Über den Bau und die Funktion des inneren Skeletts der Ophryoscolesciden. *Zoologischer Anzeiger* 62: 97–107.

Dogiel, V. and Winogradowa-Fedorawa, T. 1930. Experimentelle Untersuchungen zur Biologie der Infusioren des Wiederkäuermagens. *Wissenschaftliche Archiv für Landwirtschaftlichen Abt. B. Archiv für Tierernährung und Tierzuecht* 3: 172–188.

Dufva, G.S., Bartley, E.E., Arambel, M.J., Galitzer, S.J. and Dayton, A.D. 1982. Content of 2-aminoethylphosphonic acid in feeds, bacteria and protozoa and its role as a rumen protozoal marker. *Journal of Animal Science* 54: 837–840.

Durand, M. and Kawashima, R. 1980. Influence of minerals in the rumen microbial digestion, pp. 375–408 in Ruckebusch, Y. and Thivend, P. (editors), *Digestive Physiology and Metabolism in Ruminants*, MTP Press, Lancaster.

Eadie, J.M. 1957. The mid-winter rumen microfauna of the seaweed-eating sheep of North Ronaldshay. *Proceedings of the Royal Society of Edinburgh, Series B* 66: 276–287.

Eadie, J.M. 1962a. The development of rumen microbial populations in lambs and calves under various conditions of management. *Journal of General Microbiology* 29: 563–578.

Eadie, J.M. 1962b. Inter-relationships between certain rumen ciliate protozoa. *Journal of General Microbiology* 29: 579–588.

Eadie, J.M. 1967. Studies on the ecology of certain rumen ciliate protozoa. *Journal of General Microbiology* 49: 175–194.

Eadie, J.M. and Gill, J.C. 1971. The effect of the absence of rumen ciliate protozoa on growing lambs fed on a roughage-concentrate diet. *British Journal of Nutrition* 26: 155–167.

Eadie, J.M. and Hobson, P.N. 1962. Effect of the presence or absence of rumen ciliate protozoa on the total rumen bacterial count in lambs. *Nature, London* 193: 503–505.

Eadie, J.M., Hobson, P.N. and Mann, S.O. 1967. A note on some comparisons between the rumen contents of barley-fed steers and that of young calves also fed on a high concentrate ration. *Animal Production* 9: 247–250.

Eadie, J.M., Hyldegaard-Jensen, J., Mann, S.O., Reid, R.S. and Whitelaw, F.G. 1970. Observations on the microbiology and biochemistry of the rumen in cattle given different quantities of a pelleted barley ration. *British Journal of Nutrition* 24: 157–177.

Eadie, J.M., Mann, S.O. and Oxford, A.E. 1956. A survey of physically active organic infusoricidal compounds and their soluble derivatives with special reference to their action on the rumen microbial system. *Journal of General Microbiology* 14: 122–133.

Eadie, J.M., Manners, D.J. and Stark, J.R. 1963. The molecular structure of a reserve polysaccharide from *Entodinium caudatum*. *Biochemical Journal* 89: 91p–92p.

Eadie, J.M. and Oxford, A.E. 1954. A remarkable disintegrative effect of skatole upon certain rumen ciliate protozoa. *Nature, London* 174: 973.

Eadie, J.M. and Oxford, A.E. 1955. Factors involved in the production of a novel kind of derangement of storage mechanism in living holotrich ciliate protozoa from sheep rumen. *Journal of General Microbiology* 12: 298–310.

Eadie, J.M. and Oxford, A.E. 1957. A simple and safe procedure for the removal of holotrich ciliates from the rumen of an adult fistulated sheep. *Nature, London* 179: 485.

Eadie, J.M. and Shand, W.J. 1981. The effect of synperonic NP9 upon ciliate-free and faunated sheep. *Proceedings of the Nutrition Society* 40: 113A.

Eberlein, R. 1895. Über die im Wiederkäuermagen vorkommenden Ciliaten Infusorien. *Zeitschrift für Wissenschaftliche Zoologie* 59: 233–304.

Edwards, M. 1864. Remarques de M. Milne Edwards à l'occasion de la précédente communication: Développement des infusoires ciliés dans une macération de foin. *Comptes Rendus Hebdomadaire de Seances de l'Acadamie des Sciences*, Paris 59: 155–160.

Einszporn, T. 1961. Effect of proteins on the population size of *Entodinium longincleatum* Dogiel in culture. *Acta Parasitologica Polonica* 9: 193–210.

Ellis, J.E., Lloyd, D. and Williams, A.G. 1989a. Protozoal contribution to ruminal oxygen utilization, pp. 32–41 in Lloyd, D., Coombs, G.H. and Paget, T.A. (editors), *Biochemistry and Molecular Biology of Anaerobic Protozoa*, Harwood Academic Press, Chur, Switzerland.

Ellis, J.E., McIntyre, P.S., Saleh, M., Williams, A.G., and Lloyd, D. 1991. *Journal of General Microbiology* 137. (in press)

Ellis, J.E., Williams, A.G. and Lloyd, D. 1989b. Oxygen consumption by ruminal microorganisms: protozoal and bacterial contributions. *Applied and Environmental Microbiology* 55: 2583–2587.

Eloff, A.K. and Van Hoven, W. 1980. Intestinal protozoa of the African Elephant *Loxodonta africana* (Blumenbach). *South Aftrican Journal of Zoology* 15: 83–90.

Emmanuel, B. 1974. On the origin of rumen protozoan fatty acids. *Biochimica et Biophysica Acta* 337: 404–413.

Endo, M. and Kandatsu, M. 1981. Degradation of tryptophan by rumen ciliate protozoa. *Journal of the Agricultural Chemical Society of Japan* 55: 817–819.

Faichney, G.J. and Griffiths, D.A. 1978. Behaviour of solute and particle markers in the stomach of sheep given a concentrate diet. *British Journal of Nutrition* 40: 71–82.

Faichney, G.J., Tagari, H., Teleki, E. and Boston, R.C. 1989. Contribution of bacterial-N to rumen protozoal-N in sheep supplemented with untreated or HCHO-treated sunflower seed meal, pp. 301–304 in Nolan, J.V., Leng, R.A. and Demeyer, D.I. (editors), *The Roles of Protozoa and Fungi in Ruminant Nutrition*, Penambul Books, Armidale, Australia.

Fantham, H.B. 1926. Some parasitic protozoa found in South Africa. IX. *South African Journal of Science* 23: 560–570.

Fay, J.P. and Ovejero, F.M.A. 1986. Effect of lactate on the *in vitro* digestion of *Agropyron elongatum* by rumen microorganisms. *Animal Feed Science and Technology* 16: 161–167.

Fenn, P. and Leng, R.A. 1989. Effects of bentonite on wool growth of faunated and fauna-free sheep, pp. 327–329 in Nolan, J.V., Leng, R.A. and Demeyer, D.I. (editors), *The Roles of Protozoa and Fungi in Ruminant Digestion*, Penambul Books, Armidale, Australia.

Fiorentini, A. 1889. *Intorno ai Protisti Stomaco dei Bovini.* frat Fusi, Pavia, Italy.

Fiorentini, A. 1890. Sur les protistes de l'estomac de bovidés. Journal de Micrographie 14: 23–28, 79–83, 178–183.

Firkins, J.L., Berger, L.L., Merchen, N.R., Fahey, G.C. and Mulvaney, R.L. 1987. Ruminal nitrogen metabolism in steers as affected by feed intake and dietary urea concentration. *Journal of Dairy Science* 70: 2302–2311.

Ffoulkes, D. and Leng, R.A. 1988. Dynamics of protozoa in the rumen of cattle. *British Journal of Nutrition* 59: 429–436.

Fonseca, O.O.R. da 1915. Sobre os flagellados dos mammiferos de Brazil. um novo parasito do homen. *Brazil-Medico* 29: 281.

Fonty, G., Jouany, J.-P., Thivend, P, Gouet, Ph. and Senaud, J. 1983. A descriptive study of rumen digestion in meroxenic lambs according to the nature and complexity of the microflora. *Reproduction Nutrition Developpement* 23: 857–873.

Forsberg, C.W., Lovelock, L.K.A., Krumholz, L. and Buchanan-Smith, J.G. 1984. Protease activities of rumen protozoa. *Applied and Environmental Microbiology* 47: 101–110.

Forster, R.J. and Leng, R.A. 1989a. The effect of varying protozoal population and diet supplementation on wool growth, pp. 321–322 in Nolan, J.V., Leng, R.A. and Demeyer, D.I. (editors), *The Role of Protozoa and Fungi in Ruminant Digestion*,Penambul Books, Armidale, Australia.

Forster, R.J. and Leng, R.A. 1989b. The effect on sheep of varying rumen protozoal population and bentonite supplementation of the drinking water, pp. 331–332 in Nolan, J.V., Leng, R.A. and Demeyer, D.I. (editors), *The Roles of Protozoa and Fungi in Ruminant Digestion*, Penambul Books, Armidale, Australia.

Forsyth, G. and Hirst, E.L. 1953. Protozoal polysaccharides. Structure of the polysaccharides produced by the holotrich ciliates present in sheep's rumen. *Journal of the Chemical Society* pp. 2132–2135.

Fouly, H.A. El and Leng, R.A. 1981. The metabolism of ^{35}S-labelled protozoa in the rumen of bloated sheep. Isotopes and Radiation Research 13: 45–50.

Fuchigami, M., Senshu, T. and Horiguchi, M. 1989. A simple continuous culture system for rumen microbial digestion study and effects of defaunation and dilution rates. *Journal of Dairy Science* 72: 3070–3078.

Fujita, J., Imai, S. and Ogimoto, K. 1979. Bacterial flora, protozoal fauna and volatile fatty acids in the rumen of the water buffalo in Taiwan. *Japanese Journal of Zootechnical Science* 50: 850–854.

Furness, D.N. and Butler, R.D. 1983. The cytology of sheep rumen ciliates. I. Ultrastructure of *Epidinium caudatum* Crawley. *Journal of Protozoology* 30: 676–687.

Furness, R.D. and Butler, R.D. 1985a. The cytology of sheep rumen ciliates. I. Ultrastructure of *Eudiplodinium maggii*. *Journal of Protozoology* 32: 205–214.

Furness, D.N. and Butler, R.D. 1985b. The cytology of sheep rumen ciliates. III. Ultrastructure of *Entodinium* (Stein). *Journal of Protozoology* 32: 699–707.

Furness, D.N. and Butler, R.D. 1986. Morphogenesis of infraciliature in the entodiniomorphid ciliate *Eudiplodinium maggii*. *Journal of Ultrastructure and Molecular Structure Research* 94: 16–29.

Furness, D.N. and Butler, R.D. 1988. The functional and evolutionary significance of the ultrastructure of the Ophryoscolecidae (Order Entodiniomorphida). *Journal of Protozoology* 35: 34–38.

Gaumont, R. and Grain, J. 1967. L'anaerobiose et les mitchondries chez les protozoaires du tube digestif. *Annales de l'universite et de l'ARERS (Association Regional pour l'Etude et de la Recherche Scientifique)* 5: 174–176.

Gelei, J. and Sebestyen, O. 1932. Einige Bemerkungen zum Bau und Funktion der Syncilien bei den Darmciliaten, besonders der Entodiniomorpha. *Acta Biologica* 2: 141–161.

Genskow, R.D., Lynch, D.L. and Hess, E.A. 1969. Evaluation of the contribution of forage material to the *in vitro* culture of *Isotricha*. *Transactions of the Illinois State Academy of Science* 62: 80–84.

Gerassimova, Z.P. 1981. The ultrastructure and systematic position of the rumen ciliate *Isotricha intestinalis* Stein 1859. *Tsitologiya* 23: 861–866.

Ghoshal, A.K. and Lahiri, S.B. 1986. Some observations on protozoal protein and amino acids from rumen of camel. *Indian Veterinary Journal* 63: 729–731.

Giesecke, D. 1970. Comparative microbiology of the alimentary tract, pp. 306–318 in Philipson, A.T., Annison, E.F., Armstrong, D.G., Balch, C.C., Comline, R.S., Hardy, R.N., Hobson, P.N. and Keynes, R.D. (editors), *Physiology of Digestion and Metabolism in the Ruminant*, Oriel Press, Newcastle-upon-Tyne.

Giesecke, D. and Van Gylswyck, N.O. 1975. A study of feeding types and certain rumen functions in six species of South African wild ruminants. *Journal of Agricultural Science* 85: 75–83.

Gijzen, H.J. 1987. Anaerobic digestion of cellulosic waste by a rumen-derived process. *Ph.D. Thesis University of Nijmegen*, The Netherlands. 173 pp.

Gijzen, H.J., Zwart, K.B., van Gelder, P.T. and Vogels, G.D. 1986. Continuous cultivation of rumen microorganisms, a system with possible application to the anaerobic degradation of lignocellulosic waste materials. *Applied Microbiology and Biotechnology* 25: 155–162.

Gijzen, H.J., Zwart, K.B., Teunissen, M.J. and Vogels, G.D. 1988. Anaerobic digestion of cellulose fraction of domestic refuse by means of rumen microorganisms. *Biotechnology and Bioengineering* 32: 749–755.

Girard, V. and Hawke, J.C. 1978. The role of holotrichs in the metabolism of dietary linoleic acid in the rumen. *Biochimica et Biophysica Acta* 528: 17–27.

Grain, J. 1966a. Étude cytologique de quelques ciliés holotriches endocommensaux des ruminants et des équidés (Première partie). Protistologica 2 (Part 1): 59–141.

Grain, J. 1966b. Étude cytologique de quelques ciliés holotriches endocommensaux des ruminants et des équidés. Microscopie electronique et considérations sur la phylogénie des holotriches gymnostomes et trichostomes endocommensaux. *Protistologica* 2 (Part 2): 5–51.

Grain, J. and Senaud, J. 1984. New data on the degradation of fresh lucerne fragments by the rumen ciliate *Epidinium ecaudatum*, attachment, ingestion and digestion. *Canadian Journal of Animal Science* 64 (Suppl.): 26.

Grain, J. and Senaud, J. 1985. Degradation de fragments de lucerne fraiche par le cilié du rumen *Epidinium ecaudatum*: attachement, ingestion et digestion. *Protistologica* 21: 447–466.

Groleau, D. and Forsberg, C.W. 1981. Cellulolytic activity of the rumen bacterium *Bacteroides succinogenes*. *Canadian Journal of Microbiology* 27: 517–530.

Groliere, C.A., Senaud, J., Jouany, J.P., Grain, J. and de Puytorac, P. 1980. Implantation et développement des populations de protozoaires ciliés *(Polyplastron multivesiculatum, Entodinium* sp., *Isotricha prostoma)* dans le rumen de moutons recevant différents régimes alimentaires. II. Régimes a base de foin, enrichis en céréales. *Protistologica* 16: 385–394.

Gredzinski, F.M. 1975. The effect of urea on population density of the ciliates *Entodinium* cultivated *in vitro*. *Bulletin de l'Academie Polonaise des Sciences, Serie des Sciences Biologiques* 23: 631–636.

Gruby, D. and Delafond, H.M.O. 1843. Recherches sur des animalcules se devéllopant en grand nombre dans l'estomac et dans les intestins pendant la digestion des animaux herbivores et carnivores. *Comptes Rendus Hebdomadaire des Seances de l'Academie des Sciences. Paris* 17: 1304–1308.

Grummer, R.R., Staples, C.R. and Davis, C.L. 1983. Effect of defaunation on ruminal volatile fatty acids and pH of steers fed a diet high in dried whole whey. *Journal of Dairy Science* 66: 1738–1741.

Gulati, S.K. and Ashes, J.R. 1987. Digestibility of methionine from rumen fungal proteins. *Proceedings of the Nutrition Society of Australia* 12: 91.

Gunther, A. 1899. Untersuchungen über die im Magen unserer Hauswiederkäuer vorkommenden Wimperinfusioren. *Zeitschrift für Wissenschaftliche Zoologie* 65: 529–572.

Gutierrez, J. 1955. Experiments on the culture and physiology of holotrichs from the bovine rumen. *Biochemical Journal* 60: 517–522.

Gutierrez, J. 1958. Observations on bacterial feeding by the rumen ciliate *Isotricha prostoma*. *Journal of Protozoology* 5: 122–126.

Gutierrez, J. and Davis, R.E. 1962. Culture and metabolism of the rumen ciliate *Epidinium ecaudatum* Crawley. *Applied Microbiology* 10: 305–308.

Gutierrez, J., Davis, R.E. and Lindahl, J.L. 1961. Some chemical and physical properties of a slime from the rumen of cattle. *Applied Microbiology* 9: 209–214.

Gutierrez, J. and Hungate, R.E. 1957. Inter-relationships between certain bacteria and the rumen ciliate *Dasytricha ruminantium*. *Science* 126: 511.

Gutierrez, J. Williams, P.P., Davis, R.E. and Warwick, E.J. 1962. Lipid metabolism of rumen ciliates and bacteria. 1. Uptake of fatty acids by *Isotricha prostoma* and *Entodinium simplex*. *Applied Microbiology* 10: 548–551.

Gyulai, F. 1983. Isolation of the ciliate *Ophryoscolex caudatus* from ovine rumen contents by filtration and cultivation of the rumen ciliate *Entodinium simplex*. *Journal of Protozoology* 30: 106A.

Gyulai, F. and Baran, M. 1988. Effect of monensin on rumen ciliate protozoa in sheep. *Archives of Animal Nutrition, Berlin* 38: 153–157.

Gyulai, F. and Hudouskova, G. 1987. Effects of metronidazole and monensin on the rumen ciliate *Entodinium simplex*, pp. 365–371 in Boda, K. (editor), *Physiology of Ruminant Nutrition*, Slovak Academy of Sciences, Kosice.

Habib, G., Nolan, J.V. and Leng, R.A. 1989. Fermentative digestion and metabolism in faunated or fauna-free lambs fed roughage-based diets, pp. 323–326 in Nolan, J.V., Leng, R.A. and Demeyer, D.I. (editors), *The Roles of Protozoa and Fungi in Ruminant Digestion*, Penambul Books, Armidale, Australia.

Hall, F.J., West, J. and Coleman, G.S. 1974. Fine structural studies on the digestion of chloroplasts in the rumen ciliate *Entodinium caudatum*. *Tissue and Cell* 6: 243–253.

Han, S.-S. 1984. Rumen ciliate protozoal fauna of the native cattle in Korea. *Japanese Journal of Zootechnical Science* 55: 279–286.

Harmeyer, J. 1965. Fixation of carbon dioxide in amino acids by isolated rumen protozoa (*Isotricha prostoma* and *I. intestinalis*). *Zentralblatt für Veterinärmedizin Reihe A* 12: 10–17.

Harmeyer, J. 1966. Free amino acids in rumen protozoa. *Zeitschrift für Tierphysiologie, Tierernährung und Futtermittelkunde* 21: 211–218.

Harmeyer, J. 1971a. Amino acid metabolism of isolated rumen protozoa (*Isotricha prostoma* and *I. intestinalis*). 1. Catabolism of amino acids. *Zeitschrift für Tierphysiologie, Tierernährung und Futtermittelkunde* 28: 65–75.

Harmeyer, J. 1971b. Amino acid metabolism of isolated rumen protozoa (*Iso-*

tricha prostoma and *I. intestinalis*). 2. Excretion of amino acids. *Zeitschrift für Tierphysiologie, Tierernährung und Futtermittelkunde* 28:75–85.

Harmeyer, J. and Hekimoglou, H. 1968. Acetate incorporation by isolated rumen protozoa. *Zentralblatt für Veterinärmedizin Reihe A* 15: 242–254.

Harrison, D.G., Beever, D.E. and Osbourn, D.F. 1979. The contribution of protozoa to the protein entering the duodenum of sheep. *British Journal of Nutrition* 41: 521–527.

Hayer, G.S., Punj, M.L. and Bhatia, I.S. 1976. Studies on hemicellulase of rumen protozoa from buffalo. *Indian Journal of Animal Science* 46: 587–592.

Heald, P.J. 1951. The assessment of glucose containing substances in rumen microorganisms during a digestion cycle in sheep. *British Journal of Nutrition* 5: 84–93.

Heald, P.J. and Oxford, A.E. 1953. Fermentation of soluble sugars by anaerobic holotrich ciliate protozoa of the genera *Isotricha* and *Dasytricha*. *Biochemical Journal* 53: 506–512.

Heald, P.J., Oxford, A.E. and Sugden, B. 1952. A convenient method for preparing massive suspensions of virtually bacteria-free ciliate protozoa of the genera *Isotricha* and *Dasytricha* for manometric studies. *Nature, London* 169: 1055–1056.

Hegarty, M.P., Court, R.D., Christie, M.D. and Lee, C.P. 1976. Mimosine in *Leucaena leucocephala* is metabolised to a goitrogen in ruminants. *Australian Veterinary Journal* 52: 490.

Hegarty, M.P., Schinkel, P.G. and Court, R.D. 1964. Reaction of sheep to the consumption of *Leucaena glauca* (Benth) and its toxic principle mimosine. *Australian Journal of Agricultural Research* 15: 153–167.

Hegarty, R.S., Nolan, J.V. and Leng, R.A. 1989. Evidence for protozoa influencing rumen sulphur availability, pp. 305–306 in Nolan, J.V., Leng, R.A. and Demeyer, D.I. (editors), *The Roles of Protozoa and Fungi in Ruminant Digestion*, Penambul Books, Armidale, Australia.

Hellings, P. Debois, M. Daubresse, P. and Thines-Sempoux, D. 1981. Subcellular distribution of hydrolytic activities in the the rumen ciliates *Ophryoscolecidae*. *Archives Internationales de Physiologie et Biochimie* 89: B170–B172.

Hellings, P. Debois, M. Daubresse, P. and Thines-Sempoux, D. 1982. Cellulolytic system of rumen ciliates *Orphryoscolecidae*. *Archives Internationales de Physiologie et Biochimie* 90: B33–B34.

Hibbs, J.W. and Pounden, W.D. 1948. The influence of the ration and early rumen development on the changes in the plasma carotenoids, vitamin A and ascorbic acid of young dairy calves. *Journal of Dairy Science* 31: 1055–1061.

Hibler, C.P., Hammond, D.M., Caskey, F.H., Johnson, A.E. and Fitzgerald, P.R. 1960. The morphology and incidence of the trichomonads of swine, *Tritrichomonas suis* (Gruby and Delfond), *Tritrichomonas rotunda*, n.sp. and *Trichomonas buttreyi*, n.sp *Journal of Protozoology* 10: 20–63.

Hidiroglu, M. and Jenkins, K.J. 1974. Influence de la défaunation sur l'utilisation de la sélénométhionine chez le mouton. *Annales de Biologie Animale Biochimie Biophysique* 14: 157–165.

Hillman, K. 1987. Studies on metabolism in rumen protozoa. *Ph.D. Thesis University of Wales*. 188 pp.

Hillman, K., Lloyd, D., Scott, R.I. and Williams, A.G. 1985a. The effects of oxygen on hydrogen production by rumen holotrich protozoa, as determined by membrane-inlet mass spectrometry, pp. 271–277 in Poole, R.K. and Dow, C.S. (editors), *Microbial Gas Metabolism, Mechanistic and Biotechnological Aspects*. Academic Press, London.

Hillman, K. Lloyd, D. and Williams, A.G. 1985b. Use of a portable quadrupole mass spectrometer for the measurement of dissolved gas concentrations in

ovine rumen liquor *in situ*. *Current Microbiology* 12: 335–340.
Hillman, K. Lloyd, D. and Williams, A.G. 1985c. Continuous monitoring of fermentation gases in an artifical rumen system (RUSITEC) using a membrane inlet probe on a portable quadrupole mass spectrometer, pp. 201–206 in Degn, H., Cox, R.P. and Toftlund, H. (editors), *Gas Enzymology*, D. Reidel, Dordrecht, The Netherlands.
Hillman, K., Lloyd, D. and Williams, A.G. 1988. Interactions between the methanogen *Methanosarcina barkeri* and rumen holotrich ciliate protozoa. *Letters in Applied Microbiology* 7: 49–53.
Hillman, K., Williams, A.G. and Lloyd, D. 1990. Effects of various headspace gases on the production of volatile fatty acids by rumen ciliate protozoa, pp. 395–397, in Belaich, J.P., Bruchi, M. and Garcia, J.L. (editors), *Microbiology and Biochemistry of Strict Anaerobes involved in Interspecies Hydrogen Transfer*, Plenum Press.
Hinkson, R.S., Stern, M.D. and Gray, H.G. 1976. Effect of defaunation on acid concentrations in the rumen of the mature bovine. *Science of Biology Journal* 2: 116–123.
Hino, T. 1981. Action of monensin on rumen protozoa. *Japanese Journal of Zootechnical Science* 52: 171–179.
Hino, T., 1983. Influence of hydrogen on the fermentation of rumen protozoa. *Japanese Journal of Zootechnical Science* 54: 320–328.
Hino, T. and Kametaka, M. 1974. Effects of diets on the number of protozoa in the rumen, with special reference to the effect of purified diets and sterol in diets. *Japanese Journal of Zooechnical Science* 45: 223–232.
Hino, T. and Kametaka, M. 1975. Sterols in rumen oligotrich protozoa. *Japanese Journal of Zootechnical Science* 46: 693–705.
Hino, T. and Kametaka, M. 1977. Gnotobiotic and axenic cultures of a rumen protozoon, *Entodinium caudatum*. *Journal of General and Applied Microbiology* 23: 37–48.
Hino, T. and Kametaka, M. 1978. Metabolic origin of sterols in rumen protozoa. *Journal of General and Applied Microbiology* 24: 173–184.
Hino, T., Kametaka, M. and Kandatsu, M. 1973a. The cultivation of rumen oligotrich protozoa. 1. Factors affecting the life of *Entodinium*. *Journal of General and Applied Microbiology* 19: 305–315.
Hino, T., Kametaka, M. and Kandatsu, M. 1973b. The cultivation of rumen oligotrich protozoa. 2. Growth of entodinia *in vitro*. *Journal of General and Applied Microbiology* 19: 325–337.
Hino, T., Kametaka, M. and Kandatsu, M. 1973c. The cultivation of rumen oligotrich protozoa. 3. White clover factors which stimulate the growth of entodinia. *Journal of General and Applied Microbiology* 19: 397–413.
Hino, T. and Russell, J.B. 1987. Relative contributions of ruminal bacteria and protozoa to the degradation of protein *in vitro*. *Journal of Animal Science* 64: 261–270.
Hironaka, R., Miltimore, J.E., McArthur, J.M., McGregor, D.R. and Smith, E.S. 1973. Influence of particle size of concentrate on rumen conditions associated with feedlot bloat. *Canadian Journal of Animal Science* 53: 75–80.
Hobson, P.N. and Jouany, J.P. 1989. Models, mathematical and biological, of the rumen function, pp. 461–511, in Hobson, P.N. (editor), *The Rumen Microbial Ecosystem*, Elsevier Aplied Science, London.
Hobson, P.N., Mann, S.O. and Summers, 1975/76. Rumen micro-organisms in red deer, hill sheep and reindeer in the Scottish Highlands. *Proceedings of the Royal Society of Edinburgh Series B* 75: 171–180.
Hohl, H.R. 1965. Nature and development of membrane systems in food vacuoles of cellular slime moulds predatory upon bacteria. *Journal of Bacteriol-*

ogy 90: 755–765.
Holler, H. and Harmeyer, J. 1964. Nitrogen and amino aicd content of rumen protozoa. *Zentralblatt für Veterinärmedizin Reihe B* 11: 244–252.
Honigberg, B.M., Balamuth, W., Bovee, E.C., Corliss, J.O., Gojdics, M., Hall, R.P., Kudo, R.R., Levine, N.D., Loeblich, A.R., Weiser, J. and Wenrich, D.H. 1964. A revised classification of the phylum protozoa. *Journal of Protozoology* 11: 7–20.
Hoover, W.H., Crooker, B.A. and Sniffen, C.J. 1976. Effects of differential solid-liquid removal rates on protozoa numbers in continuous cultures of rumen contents. *Journal of Animal Science* 43: 528–534.
Hoppe, P.P., Gwynn, M.D. and Van Hoven, W. 1981. Nutrients, protozoa and volatile fatty acids in the rumen of Harvey's duiker *Cephalophus harveyi*. *South African Journal of Wildlife Research* 11: 110–111.
Horiguchi, M. and Kandatsu, M. 1959. Isolation of 2-aminoethane phosphonic acid from rumen protozoa. *Nature, London* 184: 901–902.
Hoshino, S. and Sugiyama, S.-I. 1974. Protein synthesis in a cell-free preparation from rumen protozoa. *Comparative Biochemistry and Physiology* 48B: 39–45.
Hoshino, S., Wakita, M. and Iida, A. 1982. Glycogen phorphorylase and synthase activities of rumen ciliates of the genus *Entodinium*. *Journal of Protozoology* 29: 543–546.
Hotten, P.M., Jones, K.L. and Grainger, J.M. 1983. The application of a mathematical model to an appraisal of the cellulose-azure method for determining cellulase activity. *European Journal of Applied Microbiology and Biotechnology* 18: 344–349.
Howard, B.H. 1957. Biochemical studies of individual genera of rumen holotrich protozoa. *Biochemical Journal* 67: 18P.
Howard, B.H. 1959a. The biochemistry of rumen protozoa. 1. Carbohydrate fermentation by *Dasytricha* and *Isotricha*. *Biochemical Journal* 71: 671–675.
Howard, B.H. 1959b. The biochemistry of rumen protozoa. 2. Some carbohydrases in cell-free extracts of *Dasytricha* and *Isotricha*. *Biochemical Journal* 71: 675–680.
Howard, B.H. 1963. Metabolism of carbohydrates by rumen protozoa. *Biochemical Journal* 89: 89P.
Hsiung, T.-S. 1929. A survey of the protozoan fauna of the large intestine of the horse. *Journal of Parasitology* 16: 99.
Hsiung, T.-S. 1930a. A monograph on the protozoa of the large intestine of the horse. *Iowa State College Journal of Science* 4: 359–423.
Hsiung, T.-S. 1930b. Some new ciliates from the large intestine of the horse. *Transactions of The American Microscopical Society* 49: 34–41.
Hsiung, T.-S. 1931. The protozoan fauna of the rumen of chinese sheep. *Bulletin of the Fan Memorial Institute of Biology* 2: 29–41.
Hsiung, T.-S. 1932. A general survey of the protozoan fauna of the rumen of chinese cattle. *Bulletin of the Fan Memorial Institute of Biology* 3: 87–103.
Huber, T.L. 1976. Physiological effects of acidosis on feedlot cattle. *Journal of Animal Science* 43: 902–909.
Hukui, T. 1940. Untersuchungen über die drei neuen und vierzehn bekannten Ciliaten von *Bos taurus* var *domesticus* Gmelin in West Japan. *Journal of Science of the Hiroshima University, Series B* 7: 169–181.
Hukui, T. and Nisida, K. 1954. On *Dasytricha hukuokaensis* n.sp. *Zoological Magazine (Tokyo)* 63: 367–369.
Hult, K., Teiling, A. and Gatenbeck, S. 1976. Degradation of Ochratoxin A by a ruminant. *Applied and Environmental Microbiology* 32: 443–444.
Hume, I.D. and Warner, A.C.I. 1980. Evolution of microbial digestion in mammals, pp. 665–684, in Ruckebusch, Y. and Thivend, P. (editors), *Digestive*

Physiology in Ruminants, MTP Press, Lancaster.

Hungate, R.E. 1942. The culture of *Eudiplodinium neglectum* with experiments on the digestion of cellulose. *Biological Bulletin* (Woods Hole) 83: 303–319.

Hungate, R.E. 1943. Further experiments on cellulose digestion by the protozoa in the rumen of cattle. *Biological Bulletin* (Woods Hole) 84: 157–163.

Hungate, R.E. 1966. The Rumen and its Microbes. Academic Press, London. pp. 533.

Hungate, R.E. 1978. The rumen protozoa, pp. 655–695 in Kreier, J.P. (editor), *Parasitic Protozoa*, Volume 2, Academic Press, London.

Ibrahim, E.A., El-Shazly, K., Karam, H.A. and Hamada, M.K.O. 1968. A note on the effect of diethylstilboestrol on the performance of lambs. *Animal Production* 10: 217–219.

Ibrahim, E.A. and Ingalls, J.R. 1972. Microbial protein synthesis in the rumen. *Journal of Dairy Science* 55: 971–978.

Ibrahim, E.A., Ingalls, J.R. and Stanger, N.E. 1970. Effect of dietary diethylstilboestrol on populations of ciliate protozoa in dairy cattle. *Canadian Journal of Animal Science* 50: 101–106.

Ikwuegbu, O.A. and Sutton, J.D. 1982. The effect of varying the amount of linseed oil supplementation on rumen metabolism in sheep. *British Journal of Nutrition* 48: 365–375.

Imai, S. 1981. Four new rumen ciliates, *Entodinium ogimotoi* sp.n., *E. bubalum* sp.n., and *E. tsunodai* sp.n., and *Oligoisotricha bubali* (Dogiel, 1928) n. comb. *Japanese Journal of Veterinary Science* 43: 201–209.

Imai, S. 1984. New rumen ciliates, *Polymorphella bovis* sp.n. and *Entodinium longinucleatum* forma *spinolobum* f.n. from zebu cattle in Thailand. *Japanese Journal of Veterinary Science* 46: 391–395.

Imai, S. 1985. Rumen ciliate protozoal faunae of Bali cattle (*Bos javanicus domesticus*) and water buffalo (*Bubalus bubalis*) in Indonesia, with the description of a new species *Entodinium javanicum* sp. nov. *Zoological Science, Tokyo* 2: 591–600.

Imai, S. 1986. Rumen ciliate protozoal fauna of zebu cattle (*Bos taurus indicus*) in Sri Lanka, with the description of a new species, *Diplodinium sinhalicum* sp.nov. *Zoological Science, Tokyo* 3: 699–706.

Imai, S. 1988. Ciliate protozoa in the rumen of Kenyan zebu cattle, *Bos taurus indicus*, with the description of four new species. *Journal of Protozoology* 35: 130–136.

Imai, S. Abe, M. and Ogimoto, K. 1981a. Ciliate protozoa from the rumen of the Japanese serow *Capricornis crispus* (Temminck). *Japanese Journal of Veterinary Science* 43: 359–367.

Imai, S., Chang, C.-H., Wang, J.S., Ogimoto, K. and Fujita, J. 1981b. Rumen ciliate protozoal fauna of the water buffalo (*Bubalus bubalis*) in Taiwan. *Bulletin of the Nippon Veterinary and Zootechnical College* 29: 77–81.

Imai, S., Katsuno, M. and Ogimoto, K. 1978. Distribution of rumen ciliate protozoa in cattle, sheep and goat and experimental transfaunation of them. *Japanese Journal of Zootechnical Science* 49: 494–505.

Imai, S., Katsuno, M. and Ogimoto, K. 1979a . Type of the pattern of the rumen ciliate composition of the domestic ruminants and the predator-prey interaction of the ciliates. *Japanese Journal of Zootechnical Science* 50: 79–87.

Imai, S. and Ogimoto, K. 1978. Scanning electron and fluorescent microscopic studies on the attachment of spherical bacteria to ciliate protozoa in the ovine rumen. *Japanese Journal of Veterinary Science* 40: 9–19.

Imai, S. and Ogimoto, K. 1983. *Parabundleia ruminantium* gen.n. sp.n., *Diplodinium mahidoli* sp.n. with two formae and *Entodinium parvum* forma *monospinosum* forma n. from the zebu cattle (*Bos indicus*., 1758) in Thailand. *Japanese Journal of Veterinary Science* 45: 585–591.

Imai, S. and Ogimoto, K. 1984. Rumen ciliate protozoal fauna and bacterial flora of the zebu cattle (*Bos indicus*) and the water buffalo (*Bubalus bubalis*) in Thailand. *Japanese Journal of Zootechnical Science* 55: 576–583.

Imai, S., Ogimoto, K. and Fujita, K. 1981c. Rumen ciliate protozoal fauna of water buffalo *Bubalis bubalis* (Linnaeus) in Okinawa Japan. *Bulletin of the Nippon Veterinary and Zootechnical College* 29: 82–85.

Imai, S., Ogimoto, K., Katsuno, M. 1979b. Studies on conditions for a simple culture method of rumen ciliate protozoa *in vitro*. 1. Examination on the composition of salt solution, injected gas and the number of ciliates inoculated. *Bulletin of the Nippon Veterinary and Zootechnical College* 28: 69–74.

Imai, S., Shimizu, M., Kinoshita, M., Toguchi, M., Ishi, T. and Fujita, J. 1982. Rumen ciliate protozoal fauna and composition of the cattle in Japan. *Bulletin of the Nippon Veterinary and Zootechnical College* 31: 70–74.

Imai, S., Tashiro, H. and Ishii, T. 1983. Scanning electron microscopy of the adoral ciliary zone of *Entodinium* Stein (*Ciliophora Entodiniomorphida*). *Journal of Protozoology* 30: 466–472.

Imai, S . and Tsunoda, K. 1972 . Scanning electron microscopic observations on the surface structures of ciliated protozoa in sheep rumen. *National Institute of Animal Health Quarterly (Yatabe)* 12: 74–88.

Imai, S., Yanagida, N., Katagiri, K. and Ogimoto, K. 1981d. Distribution of rumen ciliate protozoa, anaerobic bacteria and their products, volatile fatty acids and lactic acid in the gut of cattle. *Japanese Journal of Zootechnical Science* 52: 606–613.

Itabashi, H. and Katada, A. 1976a. Nutritional significance of rumen ciliate protozoa in cattle. 1. Effects of protozoa on microbial populations and rumen metabolism. *Bulletin of the Tohoku National Agricultural Experiment Station (Morioka)* 52: 161–168.

Itabashi, H. and Katada, A. 1976b. Nutritional significance of rumen ciliate protozoa in cattle. 2. Effects of protozoa on amino acid concentrations in some rumen fractions and blood plasma. *Bulletin of the Tohoku National Agricultural Experiment Station (Morioka)* 52: 169–176.

Itabashi, H., Kobayashi, T. and Matsumoto, M. 1984. The effects of rumen ciliate protozoa on some constituents in rumen fluid and blood plasma of goats. *Japanese Journal of Zootechnical Science* 55: 248–256.

Itabashi, H., Kobayashi, T., Morii, R. and Okamoto, S. 1982. Effects of ciliate protozoa on the concentrations of ruminal and duodenal volatile fatty acids and plasma glucose and insulin after feeding. *Bulletin of the National Institute of Animal Industry (Ibaraki)* 39: 21–32.

Itabashi, H. and Kurihara, Y. 1972. Maintenance of rumen microbial populations in an improved artificial rumen. *Japanese Journal of Ecology* 22: 262–265.

Itabashi, H. and Matsukawa, T. 1979. Studies on nutritional significance of rumen ciliate protozoa in cattle. 3. Influence of protozoa on growth rate, food intake, rumen fermentation and various plasma components in calves under different planes of nutrition. *Bulletin of the Tohoku National Agricultural Experiment Station (Morioka)* 59: 111–128.

Ivan, M. 1988. Effect of faunation on ruminal solubility and liver content of copper in sheep fed low or high copper diets. *Journal of Animal Science* 66: 1496–1501.

Ivan, M. 1989a. The involvement of rumen protozoa in copper metabolism and animal health, pp. 199–210 in Nolan, J.V., Leng, R.A. and Demeyer, D. (editors), *The Roles of Protozoa and Fungi in Ruminant Digestion*. Penambul Books, Armidale, Australia.

Ivan, M. 1989b. Effects of faunation and type of dietary protein on gastric solubility and liver content of copper in sheep. *Journal of Animal Science* 67: 3028–3035.

Ivan, M. and Veira, D. 1982. Duodenal flow and soluble proportions of zinc, manganese, copper and iron in the rumen fluid and duodenal digesta of faunated and defaunated sheep. *Canadian Journal of Animal Science* 62: 979–982.

Ivan, M., Veira, D. and Kelleher, C.A. 1986. The alleviation of chronic copper toxicity in sheep by ciliate protozoa. *British Journal of Nutrition* 55: 361–367.

Jameson, A.P. 1925a. A new ciliate, *Charon ventriculi* n.g., n.sp. from the stomach of ruminants. *Parasitology* 17: 403–405.

Jameson, A.P. 1925b. A note on the ciliates from the stomach of the mouse deer (*Tragulus meminna* Milne—Edwards) with the description of *Entodinium ovalis* n.sp. *Parasitology* 17: 406–409.

Janis, C. 1976. The evolutionary strategy of the Equidae and the origins of rumen and cecal digestion. *Evolution* 30: 757–774.

Jarvis, B.D.W. 1974. A method of obtaining axenic rumen ciliate protozoa. *Journal of General and Applied Microbiology* 20: 385–387.

Jensen, E.A. and Hammond, D.M. 1964. A morphological study of trichomonads and related flagellates from the bovine digestive tract. *Journal of Protozoology* 11: 386–394.

Jirovec, O. 1933. Beobachtungen über die Fauna des Rinderpansen. *Zeitschrift für Parasitenkunde* 5: 584–591.

John, A. and Ulyatt, M.J. 1984. Measurement of protozoa, using phosphatidyl choline, and of bacteria using nucleic acids, in the duodenal digesta of sheep fed chaffed lucerne hay (*Medicago sativa* L.) diets. *Journal of Agricultural Science* 102: 33–44.

Johnson, B.C., Hamilton, T.S., Robinson, W.B. and Garey, J.C. 1944. On the mechanism of non protein nitrogen utilisation by ruminants. *Journal of Animal Science* 3: 287–298.

Jones, W. T. and Lyttleton, J.W. 1972 . Bloat in cattle. XXXVII. The foaming properties of bovine salivary secretions and protozoal proteins. *New Zealand Journal of Agricultural Research* 15: 506–511.

Jouany, J.P. 1989. Effects of diet on populations of rumen protozoa in relation to fibre digestion, pp. 59–74 in Nolan, J.V., Leng, R.A. and Demeyer, D.I. (editors), *The Roles of Protozoa and Fungi in Ruminant Digestion*, Penambul Books, Armidale Australia.

Jouany, J.P., Demeyer, D.I. and Grain, J. 1988. Effect of defaunating the rumen. *Animal Feed Science and Technology* 21: 229–265.

Jouany, J.P. and Senaud, J. 1979a. Defaunation du rumen de mouton. *Annales de Biologie Animale Biochimie Biophysique* 19: 619–624.

Jouany, J.P. and Senaud, J. 1979b. Role of rumen protozoa in the digestion of food cellulosic materials. *Annales de Recherches Veterinaires* 10: 261–263.

Jouany, J.P. and Senaud, J. 1982. Influence des ciliés du rumen sur la digestion de différent glucides chez le mouton. 1. Utilisation des glucides pariétaux (cellulose et hémicelluloses) et de l'amidon. *Reproduction Nutrition Developpement* 22: 735–752.

Jouany, J.P. and Senaud, J. 1983. Influence des ciliés du rumen sur l'utilisation digestive de différents régimes riches en glucides solubles et sur les protduits terminaux formés dans le rumen. 2. Règimes contenant de l'inuline, du saccharose et du lactose. *Reproduction Nutrition Developpement* 23: 607–623.

Jouany, J.P., Senaud, J., Groliere, C.A., Thivend, P. and Grain, J. 1981a. Influence du traiment par la souche et de l'inoculation du cilié *Polyplastron multivesiculatum* dans le rumen sur la digestion d'un regime riche en glucides pari etaux. *Reproduction Nutrition Developpement* 21: 866.

Jouany, J.P. and Thivend, P. 1972. Evolution postprandiale de la composition glucidique des corps microbiens du rumen es fonction de la nature des glucides du régime. I. Les protozoaires. *Annales de Biologie Animale Biochemie et*

Biophysique 12: 673–677.

Jouany, J.P., and Thivend, P. 1983. Influence of protozoa on nitrogen digestion in the ruminant, pp. 287–290 in Pion, R., Arnal, M. and Bonin, D. (editors), *Protein Metabolism and Nutrition*, volume 2 Communications, INRA, Paris.

Jouany, J.P., Zainab, B., Senaud, J., Groliere, C.A., Grain, J. and Thivend, P. 1981b. Role of the rumen ciliate protozoa *Polyplastron multivesiculatum*, *Entodinium* sp. and *Isotricha prostoma* in the digestion of the mixed diet in sheep. *Reproduction Nutrition Development* 21: 871–884.

Kahlon, T.S., Ranhotra, G.S. and Langar, P.N. 1970. Partial defaunation of the rumen and its influence on the biochemical activity. 1. Effect of partial defaunation on the intraruminal environment of the buffalo and zebu. *Indian Journal of Animal Science* 40: 593–599.

Kandatsu, M. and Takahashi, N. 1955a. Studies on recticulo-rumen digestion. Part 2. On the artificial culture of some *Entodinia* I. *Journal of the Agricultural Chemistry Society, Japan* 29: 883–839.

Kandatsu, M. and Takahashi, N. 1955b. Studies on reticulo-rumen digestion. Part 3. On the artificial culture of some *Entodinia* II. *Journal of the Agricultural Chemistry Society, Japan* 29: 915–920.

Kandatsu, M. and Takahashi, N. 1956. Studies on reticulo-rumen digestion. Part 4. On the artificial culture of some *Entodinia* III. *Journal of the Agricultural Chemistry Society, Japan* 30: 96–101.

Kaneko, T., Ushida, K. and Kojima, Y. 1989. The effect of starch on cellulolysis by rumen microbial populations with or without protozoa, pp. 313–315 in Nolan, J.V., Leng, R.A. and Demeyer, D.I. (editors), *The Roles of Protozoa and Fungi in Ruminant Digestion*, Penambul Books, Armidale, Australia.

Kariya, R., Morita, Z., Oura, R. and Sekine, J. 1989. The *in vitro* study on the rates of starch consumption and volatile fatty acid production by rumen fluid with or without ciliates. *Japanese Journal of Zootechnical Science* 60: 609–613.

Katz, I. and Keeney, M. 1967. The lipids of some rumen holotrich protozoa. *Biochimica et Biophysica Acta* 144: 102–112.

Katz, M.P., Nagaraja, T.G. and Fina, L.R. 1986. Ruminal changes in monensin- and lasalocid-fed cattle grazing bloat-provocative alfalfa pasture. *Journal of Animal Science* 63: 1246–1257.

Kayouli, C., Demeyer, D. and Van Nevel, C. 1982. La défaunation du rumen: Technique pour ameliorer la production des ruminant avec aliment tropicaux. *Proceedings of the International Colloquium on Tropical Animal Production for the Benefit of Man*, pp. 302–308, Prince Leopold Institute of Tropical Medicine, Antwerp.

Kayouli, C., Demeyer, D.I., Van Nevel, C.J. and Dendooven, R. 1983/84. Effect of defaunation on straw digestion in sacco and on particle retention in the rumen. *Animal Feed Science and Technology* 10: 165–172.

Kayouli, C., Van Nevel, C.J., Dendooven, R. and Demeyer, D.I. 1986. Effect of defaunation and refaunation of the rumen on fermentation and N-flow in the duodenum of sheep. *Archives of Animal Nutrition, Berlin* 36: 827–837.

Keeney, M. 1970. Lipid metabolism in the rumen, pp. 489–503 in Phillipson, A.T., Annison, E.F., Armstrong, D.G., Balch, C.C. Comline, R.S., Hardy, R.N., Hobson, P.N. and Keynes, R.D. (editors), *Physiology of Digestion and Metabolism in the Ruminant*, Oriel Press, Newcastle-upon-Tyne.

Kemp, P. and Dawson, R.M.C. 1969a. Isolation of a new phospholipid, phosphatidyl-N-(2, hydroxethyl) alanine from rumen protozoa. *Biochemical Journal* 113: 555–558.

Kemp, P. and Dawson, R.M.C. 1969b. Characterization of N-(2, hydroxyethyl)-alanine as a component of a new phospholipid isolated from rumen protozoa. *Biochimica et Biophysica Acta* 176: 678–679.

Kemp, P., Jordan, D.J. and Orpin, C.G. 1985. The free- and protein-amino acids of the rumen phycomycete fungi *Neocallimastix frontalis* and *Piromonas communis*. *Journal of Agricultural Science* 105: 523–526.

Kiessling, K.-H., Pettersson, H., Sandholm, K. and Olsen, M. 1984. Metabolism of aflatoxin, ochratoxin, zearalenone and three trichothecenes by intact rumen fluid, rumen protozoa and rumen bacteria. *Applied and Environmental Microbiology* 47: 1070–1073.

Klein, D.R. 1962. Rumen contents analysis as an index to range quality. *Transactions of the North American Wildlife Conference* 27: 150–164.

Klein, D.R. 1965. Ecology of deer range in Alaska. *Ecological Monographs* 35: 259–284.

Kleynhans, C.J. and Van Hoven, W. 1976. Rumen protozoa of the giraffe with a description of two new species. *East African Wildlife Journal* 14: 203–214.

Klopfenstein, T.L., Purser, D.B. and Tyznik, W.T. 1964. Influence of aureomycin on rumen metabolism and blood metabolites. *Journal of Animal Science* 23: 490–495.

Klopfenstein, T.J., Purser, D.B. and Tyznik, W.J. 1966. Effect of defaunation on feed digestibility, rumen metabolism and blood metabolites. *Journal of Animal Science* 25: 765–773.

Knoth, M. 1928. Neue Versuche zur Züchtung der im Pansen von Wiederkäuern lebenden Ophryoscoleciden (Ciliata). *Zeitschrift für Wissenschaftliche Biologie,* Abt F 1: 262–270.

Kobayashi, T. and Itabashi, H. 1986. Control of feeding and gastric hormones in calves. *Kachiku Seikagaku Kenkyukaiho* 18: 11–22.

Kofoid, C.A. and Christenson, J.F. 1933. Ciliates from *Bos-gaurus* H. Smith. *University of California Publications in Zoology* 39: 341–391.

Kofoid, C.A. and MacLennan, R.F. 1930. Ciliates from *Bos indicus* Linn. I. The genus *Entodinium* Stein. *University of California Publications in Zoology* 33: 471–544.

Kofoid, C.A. and MacLennan, R.F. 1932. Ciliates from *Bos indicus* Linn. II. A revision of *Diplodinium* Schuberg. *University of California Publications in Zoology* 37: 53–152.

Kofoid, C.A. and MacLennan, R.F. 1933. Ciliates from Bos indicus Linn. 3. *Epidinium* Crawley, *Epiplastron* gen.nov., and *Ophryoscolex* Stein. *University of California Publications in Zoology* 39: 1–34.

Kopperi, A.J. 1926. Märehtijöitten rapamahan infusoreista (Zusammenfassung: Die mageninfusorien der Wiederkäuer). *Anales de la Sociedad Zoologica-Botanico, Vanamo* 4: 225–238.

Kopperi, A.J. 1929. Über eine Diplodinium-Art aus den Magen des Renntieres. *Annales Zoologici Societatis Zoologicae-Botanicae Fennicae, Vanamo* 8: 27–33.

Kopperi, A.J. 1935. Über die nicht-pathogene protozoen fauna des blinddarms einiger nagetiere. *Annales Zoologici Societatis Zoologicae-Botanicae Fennicae Vanamo* 3: 1–92.

Krascheninnikow, S. 1955. Observations on the morphology and division of *Eudiplodinium neglectum* Dogiel (Ciliata Entodioniomorpha) from the stomach of a moose (*Alces americana*). *Journal of Protozoology* 2: 124–134.

Kreuzer, M. and Kirchgessner, M. 1986a. The role of rumen ciliate protozoa in digestion of nutrients, N retention and partition of faecal and urinary nitrogen when feeding different amounts and sources of starch to wethers. *Journal of Animal Physiology and Animal Nutrition (Zeitschrift für Tierphysiologie, Tierernährung und Futtermittelkunde)* 56: 217–231.

Kreuzer, M. and Kirchgessner, M. 1986b. Role of protozoa in rumen fermentation, digestibility and nitrogen retention of lambs supplied with or without rumen-protected lysine (DHML-calcium). *Landwirtschaftliche Forschung* 39: 218–229.

Kreuzer, M. and Kirchgessner, M. 1987a. Investigations on the nutritive defaunation of the rumen of ruminants. *Archives of Animal Nutrition, Berlin* 37: 489–503.

Kreuzer, M. and Kirchgessner, M. 1987b. Growth response of lambs supplied with rumen protected lysine (DHML-Ca) under various conditions of ruminal protozoa status and different energy conditions of the ration. *Zeitschrift für Tierphysiologie, Tierernährung und Futtermittelkunde* 58: 188–200.

Kreuzer, M. and Kirchgessner, M. 1989. Effect of rumen protozoa on metabolism and retention in ruminants with special reference to diet characteristics, pp. 189–197 in Nolan, J.V., Leng, R.A. and Demeyer, D.I. (editors), *The Roles of Protozoa and Fungi in Ruminant Digestion*, Penambul Books, Armidale, Australia.

Kreuzer, M. and Kirchgessner, M. 1990. Faecal and urinary excretion of iron, copper, zinc and manganese in faunated and defaunated wethers fed semi-purified diets of different carbohydrate composition. *Animal Feed Science and Technology* 28: 135–143.

Kreuzer, M., Kirchgessner, M. and Müller, H.L. 1986. Effects of defaunation on the loss of energy in wethers fed different quantities of cellulose and normal or steamflaked maize starch. *Animal Feed Science and Technology* 16: 233–241.

Krishnamurti, C.R. and Kitts, W.D. 1969. Preparation and properties of cellulases from rumen microorganisms. *Canadian Journal of Microbiology* 15: 1373–1379.

Krogh, N. 1959. Studies on alterations in the rumen fluid of sheep, especially concerning the microbial composition, when readily available carbohydrates are added to the food. 1. Sucrose *Acta Veterinaria Scandinavica* 1: 74–97.

Krumholz, L.R., Forsberg, C.W. and Veira, D.M. 1983. Association of methanogenic bacteria with rumen protozoa. *Canadian Journal of Microbiology* 29: 676–680.

Kubikova, M. 1935. Infusiorien aus dem Pansen von *Cervus elaphus*. *Zoologischer Anzeiger, Leipzig* 111: 175–177.

Kubo, T. and Kandatsu, M. 1968. Carbohydrate metabolism of rumen ciliate protozoa. 1. Utilization of soluble sugars by mixed rumen protozoa. *Japanese Journal of Zootechnical Science* 39: 476–480.

Kubo, T. Kobayashi, H., Ishigami, H. Sekiya, A. and Mashiyama, F. 1980. The cultivation of the rumen ciliate *Dasytricha ruminantium*. *Bulletin of the College of Agriculture Utsunomiya University* 11: 9–16.

Kubo, T. Yamaguchi, K., Torigai, H. and Kono, K. 1981. Effects of variations in cultural conditions of rumen protozoa in the presence of oxyclozanide on their viability. *Bulletin of the College of Agriculture Utsunomiya University* 11 (2): 9–14.

Kumar, O.M. and Raghavan, G.V. 1978. Digestibility of nutrients and concentration of rumen metabolites in faunated and defaunated lambs. *Indian Veterinary Journal* 55: 775–780.

Kumar, S., Singh, A., Sareen, V.K., Mann, G.S. and Singh, S. 1987. Degradation of mimosine by rumen fluid and its fractions from buffalo. *Indian Journal of Animal Science* 57: 1128–1131.

Kurihara, Y., Eadie, J.M., Hobson, P.N. and Mann, S.O. 1968. Relationship between bacteria and ciliate protozoa in the sheep rumen. *Journal of General Microbiology* 51: 267–288.

Kurihara, Y. Takechi, T. and Shibata, F. 1978. Relationship between bacteria and ciliate protozoa in the rumen of sheep fed on a purified diet. *Journal of Agricultural Science* 90: 373–381.

Latham, M.J., Storry, J.E. and Sharp, M.E. 1972. Effect of low roughage diets on the microflora and lipid metabolism in the rumen. *Applied Microbiology* 24: 871–877.

Latteur, B. 1966a. *Epidinium dactylodonta* n.sp. Ciliate Ophryoscolecide du rumen de l'antelope *Tragelaphus scriptus* Pallas. *Bulletin de l'Institut Royal de Sciences Natutelles de Belgique* 42: 1–27.

Latteur, B . 1966b. Cytologie comparée de quelques Ciliates Ophryoscolecides. *La Cellule* 46: 7–37.

Latteur, B. 1966c. *Diplodinium archon* n.sp. Ciliate Ophryoscolecide du rumen de l'antelope *Tragelaphus scriptus* Pallas. *Annales de la Societe Belge de Medicine Tropicale* 46: 727–740.

Latteur, B. 1966d. Contribution à la systématique de la Famille des *Ophryoscolecidae* Stein. *Annales de la Societe Royal Zoologique de Belgique* 96: 117–144.

Latteur, B. 1967. *Helicozoster indicus* n.gen., n.sp., ciliate holotriche du caecum de l'elephant des indes. *Acta Zoologica et Pathologica Antverpiensia* 43: 93–106.

Latteur, B. 1968. Revision systematique de la Famille des *Ophryoscolecidae* Stein, 1858: sous-Famille des *Entodiniinae* Lubinsky, 1957, Genre *Entodinium* Stein 1858. *Annales de la Societe Royal Zoologique de Belgique* 98: 1–41.

Latteur, B. 1969. Revision systématique de la Famille des *Ophryoscolecidae* Stein, 1858: sous-Famille des *Entodiniinae* Lubinsky, 1957, Genre *Entodinium* Stein, 1958. *Annales de la Societe Royal Zoologique de Belgique* 99: 3–25.

Latteur, B. 1970. Revision systématique de la famille des *Ophryoscolecidae* Stein, 1858. Sous-famille des *Diplodiniinae* Lubinsky, 1957. Genre *Diplodinium* (Schuberg, 1888) *sensu novo*. *Annales de la Societe Royal Zoologique de Belgique* 100: 275–312.

Latteur, B. and Sulten, E. 1972. Contribution à l'étude de la spécificité écologique des Ciliés Ophryoscolecides. *Acta Zoologica et Pathologica Antverpiensia* 55: 41–54.

Lee, J.J., Hutner, S. and Bovee, E.C. 1985. *An illustrated Guide to the Protozoa,* Society of Protozoologists, Lawrence Kansas. pp. 629.

Leng, R.A. 1982. Dynamics of protozoa in the rumen of sheep. *British Journal of Nutrition* 48: 399–415.

Leng, R.A. 1989. The dynamics of protozoa in the rumen, pp. 51–58 in Nolan, J.V., Leng, R.A. and Demeyer, D.I. (editors), *The Roles of Protozoa and Fungi in Ruminant Digestion*, Penambul Books, Armidale, Australia.

Leng, R.A., Dellow, D. and Waghorn, G. 1986. Dynamics of large ciliate protozoa in the rumen of cattle fed on diets of freshly cut grass. *British Journal of Nutrition* 56: 455–462.

Leng, R.A., Gill, M., Kempton, T.J., Rowe, J.B., Nolan, J.V., Stachiw, S.J. and Preston, T.R. 1981. Kinetics of large ciliate protozoa in the rumen of cattle given sugar cane diets. *British Journal of Nutrition* 46: 371–384.

Leng, R.A. and Nolan, J.W. 1984. Nitrogen metabolism in the rumen. *Journal of Dairy Science* 67: 1072–1089.

Leng, R.A., Nolan, J.V., Cumming, G., Edwards, S.R. and Graham, C.A. 1984. The effects of monensin on the pool size and turnover rate of protozoa in the rumen of sheep. *Journal of Agricultural Science* 102: 609–613.

Lengemann, F.W. and Allen, N.N. 1959. Development of rumen function in the dairy calf. II. Effect of diet upon characteristics of the rumen flora and fauna of young calves. *Journal of Dairy Science* 42: 1171–1181.

Levine, N.D. 1961. *Protozoan parasites of domestic animals and of man*. Burgess Publishing Co: Minneapolis, Minn.

Levine, N.D., Corliss, J.O., Cox, F.E.G., Deroux, G., Grain, J., Honigberg, B.M., Leedale, G.F., Loeblich, A.R., Lom, J., Lynn, D., Merinfeld, E.G., Page, F.C., Poljansky, G., Sprague, V., Vavra, J. and Wallace, F.G. 1980. A newly revised classification of the protozoa. *Journal of Protozoology* 27: 37–58.

Liang, C.R. and Rosenberg, H. 1968. The biosynthesis of the carbon phosphorus bond in Tetrahymena. *Biochimica et Biophysica Acta* 156: 437–439.

Lindsay, D.B. 1970. Carbohydrate metabolism in ruminants, pp. 438–451, in Phillipson, A.T. (editor), *Physiology of Digestion and Metabolism in the Ruminant*, Oriel Press: Newcastle upon Tyne, England.

Lindsay, J.R. and Hogan, J.P. 1972. Digestion of two legumes and rumen bacterial growth in defaunated sheep. *Australian Journal of Agricultural Research* 23: 321–330.

Ling, J.R. and Buttery, P.J. 1978. The simultaneous use of ribonucleic acid, ^{35}S, 2,6-diaminopimelic acid and 2-aminoethylphosphonic acid as markers of microbial nitrogen entering the duodenum of sheep. *British Journal of Nutrition* 39: 165–179.

Lloyd, D., Williams, J., Yarlett, N. and Williams, A.G. 1982. Oxygen affinities of the hydrogenosome-containing protozoa *Tritrichomonas foetus* and *Dasytricha ruminantium*, and two aerobic protozoa determined by bacterial bioluminescence. *Journal of General Microbiology* 128: 1019–1022.

Lloyd, D., Yarlett, N., Hillman, K., Paget, T.A., Chapman, A., Ellis, J.E. and Williams, A.G. 1989. Oxygen and aerotolerant protozoa: studies using non-invasive techniques, pp. 1–21 in Lloyd, D., Coombs, G.H. and Paget, T.A. (editors), *Biochemistry and Molecular Biology of Anaerobic Protozoa*, Harwood Academic Press, Chur, Switzerland.

Lockwood, B., Coombs, G.H. and Williams, A.G. 1988. Proteinase activity in rumen ciliate protozoa. *Journal of General Microbiology* 134:2605–2614.

Lockwood, B., North, M.J., Williams, A.G. and Coombs, G.H. 1989. Proteinases of anaerobic protozoa: trichomonads and rumen ciliates, pp. 66–77 in Lloyd, D., Coombs, G.H. and Paget, T.A. (editors), *Biochemistry and Molecular Biology of Anaerobic Protozoa*, Harwood Academic Press, Chur, Switzerland.

Lough, A.K. 1968. Component of fatty acids of plasma lipids of lambs with and without rumen ciliate protozoa. *Proceedings of the Nutrition Society* 27: 30A–31A.

Lovelock, L.K.A., Buchanan-Smith, J.G. and Forsberg, C.W. 1982. Difficulties in defaunation of the ovine rumen. *Canadian Journal of Animal Science* 62: 299–303.

Lubberding, H.J., Gijzen, H.J., Gerhardus, M.J. and Vogels, G.D. 1987. Fibre degradation and activities in an artificial rumen system in the presence and the absence of rumen ciliates, pp. 301–315 in Boda, K. (editor), *Physiology of Ruminant Nutrition*, Slovak Acadamy of Sciences, Kosice.

Lubinsky, G. 1955a. On some parasites of parasitic protozoa. I. *Sphaerita hoari* sp.n. A chytrid parasitizing *Eremoplastron bovis*. *Canadian Journal of Microbiology* 1: 440–450.

Lubinsky, G. 1955b. On some parasites of parasitic protozoa. II. *Sagittospora cameroni* gen.n., sp.n. A phycomycete parasitizing *Ophryoscolecidae*. *Canadian Journal of Microbiology* 1: 675–684.

Lubinsky, G. 1957a. Studies on the evolution of *Ophryoscolecidae* (Ciliata: Oligotricha). I. A new species of *Entodinium* with *'caudatum'*, *'loboso-spinosum'* and *'dubardi'* forms and some evolutionary trends in the genus *Entodinium*. *Canadian Journal of Zoology* 35: 111–133.

Lubinsky, G. 1957b. Studies on the evolution of the *Ophryoscolecidae* (Ciliata: Oligotricha). II. On the origin of the higher *Ophryoscolecidae*. *Canadian Journal of Zoology* 35: 135–140.

Lubinsky, G. 1957c. Studies on the evolution of the *Ophryoscolecidae* (Ciliata: Oligotricha). III. Phylogeny of the *Ophryoscolecidae* based on their comparative morphology. *Canadian Journal of Zoology* 35: 141–159.

Lubinsky, G. 1957d. Note on the phylogenetic significance of predatory habits in the Ophryoscolecidae (Ciliata: Oligotricha). *Canadian Journal of Zoology* 35: 579–580.

Lubinsky, G. 1958a. Ophryoscolecidae (Ciliata: Entodiniomorphida) of reindeer (*Rangifer tarandus* L.) from the Canadian Arctic. I. *Entodiniinae*. *Canadian Journal of Zoology* 36: 819–835.

Lubinsky, G. 1958b. Ophryoscolecidae (Ciliata: Entodiniomorphida) of the reindeer (*Rangifer tarandus* L.) from the Canadian Arctic. II. *Diplodiniinae*. *Canadian Journal of Zoology* 36: 937–959.

Lubinsky, G. 1963. *Metadinium caudatum* sp.n.–A rumen ciliate of musk ox from Northern Canada. *Canadian Journal of Zoology* 41: 29–32.

Lubinsky, G. 1964. *Ophryoscolecidae* of a guanaco from the Winnipeg zoo. *Canadian Journal of Zoology* 42: 159.

Luther, R. Trenkle, A. and Burroughs, W. 1966. Influence of rumen protozoa on volatile acid production and ration digestibility in lambs. *Journal of Animal Science* 25: 1116–1122.

Lyle, R.R., Johnson, R.R., Wilhite, J.V. and Backus, W.R. 1981. Ruminal characteristics in steers as affected by adaptation from forage to all-concentrate diets. *Journal of Animal Science* 53: 1383–1390.

Mackie, R.I. 1987. Microbial digestion of forages in herbivores. pp. 233–265 in Hacker, J.B. and Ternouth, J.H. (editors), *The Nutrition of Herbivores*, (editors), *The Nutrition of Herbivores*, Academic Press, Sydney.

Mackie, R.I., Gilchrist, F.M.C., Robberts, A.M., Hannah, P.E. and Schwartz, H.M. 1978. Microbiological and chemical changes in the rumen during the stepwise adaptation of sheep to high concentrate diets. *Journal of Agricultural Science* 90: 241–254.

MacLennan, R.F. 1933. The pulsatory cycle of the contractile vacuoles in the Ophryoscolecidae, ciliates from the stomach of cattle. *University of California Publications in Zoology* 39: 205–249.

MacLennan, R.F. 1935. Ciliates from the stomach of mule-deer. *Transactions of the American Microscopical Society* 54: 181–188.

Mah, R.A. 1962. *Ophryoscolex purkynei*, its culture and physiology. PhD Thesis, University of California, Davis. Quoted by Hungate (1978).

Mah, R.A. 1964. Factors influencing the *in vitro* culture of the rumen ciliate *Ophryoscolex purkynei* (Stein). *Journal of Protozoology* 11: 546–552.

Mah, R.A. and Hungate, R.E. 1965. Physiological studies on the rumen ciliate, *Ophryoscolex purkynei* (Stein). *Journal of Protozoology* 12: 131–136.

Males, J.R. and Purser, D.B. 1970. Relationship between rumen ammonia levels and the microbial population and volatile fatty acid proportions in faunated and defaunated sheep. *Applied Microbiology* 19: 485–490.

Mangan, J.L. 1959. Bloat in cattle. XI. The foaming properties of proteins, saponins and rumen liquor. *New Zealand Journal of Agricultural Research* 2: 47–61.

Mangan, J.L. and West, J. 1977. Ruminal digestion of chloroplasts and the protection of protein by glutaraldehyde treatment. *Journal of Agricultural Science* 89: 3–15.

Mangold, E. and Schmitt-Krahmer, C. 1927. Die Stickstoffverteilung im Pansen die Wiederkäuer bei Fütterung und Hunger und ihre Beziehung zu den Pansen-infusorien. *Biochemische Zeitschrift* 191: 411–422.

Mangold, E. and Usuelli, F. 1930. Die schädliche Wirkung der Milch und der Veränderungen der H-Ionenkonzentration auf die Infusorien des Wiederkäuermagens. *Archiv für Tierernährung und Tierzücht* 3: 189–201.

Margolin, S. 1930. Methods for the cultivation of cattle ciliates. *Biological Bulletin* 59: 301–305.

Masson, H.A. and Ling, J.R. 1986. The *in vitro* metabolism of free and bacterially-bound 2,2′-diaminopimelic acid by rumen micro-organisms. *Journal of Applied Bacteriology* 60: 341–350.

Masson, F.M. and Oxford, A.E. 1951. The action of the ciliates of the sheep's rumen upon various water soluble carbohydrates, including polysaccharides. *Journal of General Microbiology* 5: 664–672.

Masson, M.J., Iggo, A., Reid, R.S. and Mann, G.F. 1952. Galvanotropic fractionation of rumen ciliates. *Journal of the Royal Microscopical Society* 72: 67–69.

Matsumoto, M., Kobayashi, T. and Itabashi, H. 1989. Effects of rumen ciliate protozoa on apparent digestibilities of dietary amino acids in goats. *Japanese Journal of Zootechnical Science* 60: 987–992.

McAllan, A.B., Knight, R. and Sutton, J.D. 1983. The effect of free and protected oils on the digestion of dietary carbohydrates between the mouth and the duodenum of sheep. *British Journal of Nutrition* 49: 433–440.

McDonald, I.W. 1954. The extent of conversion of food protein to microbial protein in the rumen of sheep. *Biochemical Journal* 56: 120–125.

McDonald, I.W. and Hall, R.J. 1957. The conversion of casein into microbial protein in the rumen.*Biochemical Journal* 67: 400–406.

McNaught, M.L., Owen, E.C., Henry, K.M. and Kon, S.K. 1954. The utilization of non protein nitrogen in the bovine rumen. 8. The nutritive value of the proteins of preparations of dried rumen bacteria, rumen protozoa and brewers yeast for rats. *Biochemical Journal* 56: 151–156.

Merricks, D.L. and Salsbury, R.L. 1973. The involvement of vitamin B_6 in dethiomethylation by rumen microorganisms *in vitro*. *Journal of Animal Science* 37: 368.

Merricks, D.L and Salsbury, R.L. 1974. Involvement of vitamin B_6 in the dethiomethylation of methionine by rumen microorganisms. *Applied Microbiology* 28: 106–111.

Merricks, D.L. and Salsbury, R.L. 1976. Dethiomethylation of methionine by an extract of rumen protozoa: general substrate specificity. *Journal of Animal Science* 42: 955–959.

Merry, R.J., Smith, R.H. and McAllan, A.B. 1987. Studies of rumen function in an *in vitro* continuous culture system. *Archives of Animal Nutrition*, Berlin 37: 475–488.

Mertens, E., van Schaftingen, E. and Müller, M. 1989. Presence of a fructose-2,6-bisphosphate-insensitive pyrophosphate:fructose-6-phosphate phosphotransferase in the anaerobic protozoa *Tritrichomonas foetus*, *Trichomonas vaginalis* and *Isotricha prostoma*. *Molecular and Biochemical Parasitology* 37: 183–190.

Meyer, J.H.F., van der Walt, S.I. and Schwartz, H.M. 1986. The influence of diet and protozoal numbers on the breakdown and synthesis of protein in the rumen of sheep. *Journal of Animal Science* 62: 509–520.

Michalowski, T. 1975a. Effect of different diets on the diurnal concentrations of ciliate protozoa in the rumen of water buffalo. *Journal of Agricultural Science* 85: 145–150.

Michalowski, 1975b. The effect of certain feedingstuffs on rumen ciliate protozoa *in vitro*. *Journal of Agricultural Science* 85: 151–158.

Michalowski, T. 1977. Diurnal changes in concentrations of rumen ciliates and in occurrence of dividing forms in water buffalo (*Bubalus bubalis*) fed once daily. *Applied and Environmental Microbiology* 33: 802–804.

Michalowski, T. 1979. A simple system for continuous culture of rumen ciliates. *Bulletin de l'Academie Polonaise des Sciences, Serie des Sciences Biologiques* 27: 581–583.

Michalowski, T. 1987. The volatile fatty acid production by ciliate protozoa in the rumen of sheep. *Acta Protozoologica* 26: 335–345.

Michalowski, T., Bochenek, J. and Muszynski, P. 1979. Survival of rumen ciliate protozoa in a continuous culture system. *Bulletin de l'Academie Polonaise des Sciences, Serie des Sciences Biologique* 27: 657–662.

Michalowski, T., Breves, G. and Höller, H. 1981. Thiamin (vitamin B1) in suspensions of protozoa isolated from the sheep rumen, Abstract 248, pp. 31–32, in *Proceedings of the VI International Congress of Protozoology, Warsaw.*

Michalowski, T. and Harmeyer, J. 1983. Selective outlfow of protozoa from the rumen of sheep, pp. 291–294 in Pion, R., Arnal, M. and Bonin, D. (editors), *Protein Metabolism and Nutrition,* (volume 2 Communications), INRA, Paris.

Michalowski, T., Harmeyer, J. and Breves, G. 1986. The passage of protozoa from the reticulo-rumen through the omasum of sheep. *British Journal of Nutrition* 56: 625–634.

Michalowski, T. and Muszynski, P. 1978. Diurnal variations in the number of ciliate protozoa in the rumen of sheep fed once and twice daily. *Journal of Agricultural Science* 90: 1–5.

Minor, S., Macleod, N.A. and Preston, T.R. 1977. Effect of sampling by fistula or at slaughter on estimation of rumen protozoa. *Tropical Animal Production* 2: 62–67.

Mishra, B.D. 1964. Ph.D. Thesis Kansas State University.

Mishra, B. 1969. Role of protozoa in rumen metabolism with special reference to bloat. *Indian Veterinary Journal* 46: 42–49.

Moate, P. 1989. Rumen protozoa and bloat in cows grazing white clover and perennial ryegrass, pp. 337–339 in Nolan, J.V., Leng, R.A. and Demeyer, D.I. (editors), *The Roles of Protozoa and Fungi in Ruminant Digestion,* Penambul Books, Armidale, Australia.

Mohanty, G.P., Jorgensen, N.A., Luther, R.M. and Voelker, H.H. 1969. Effect of molded alfalfa hay on rumen activity, performance and digestibility in dairy steers. *Journal of Dairy Science* 52: 79–83.

Moir, R.J. and Somers, M. 1956. A factor influencing the protozoal population in sheep. *Nature (London)* 178: 1472.

Moriggi, M. 1941. Ricerche sugli infusori dei ruminanti italiani. *Archivio Zoologico Italiano* 29: 369–412.

Moriggi, M. 1950. Protozoa ciliata parastica. *Rivista di Biologia* 10: 103–134.

Mould, F.L. and Orskov, E.R. 1983/84. Manipulation of rumen fluid pH and its influence on cellulololysis is sacco, dry matter degradation and the rumen microflora of sheep offered either hay or concentrate. *Animal Feed Science and Technology* 10: 1–14.

Mould, D.L. and Thomas, G.J. 1957. Amylases of holtrich protozoa from sheep rumen. *Biochemical Journal* 67: 18p.

Mould, D.L and Thomas, G.J. 1958. The enzymic degradation of starch by holotrich protozoa from sheep rumen. *Biochemical Journal* 69: 327–337.

Mountfort, D.O., Asher, R.A. and Bauchop, T. 1982. Fermentation of cellulose to methane and carbon dioxide by a rumen anaerobic fungus in triculture with *Methanobrevibacter* sp. strain RAI and *Methanosarcina barkeri. Applied and Environmental Microbiology* 44: 128–134.

Muller, M. 1980. The hydrogenosome, pp. 127–142 in Gooday, G.W., Lloyd, D. and Trinci, A.P.J. (editors), *The Eukaryotic Microbial Cell,* Cambridge University Press, Cambridge.

Muller, M. 1982. Energy metabolism of anaerobic protozoa, pp. 113–116 in Mettrick, D.F. and Desser, S.S. (editors), *Parasites–their world and ours.* Elsevier Biomedical Press, Amsterdam.

Muller, M. 1988. Energy metabolism of protozoa without mitochondria. *Annual Reviews of Microbiology* 42: 465–488.

Murphy, M.R., Drone, P.E. and Woodford, S.T. 1985. Factors stimulating migration of holotrich protozoa into the rumen. *Applied and Environmental Microbiology* 49: 1329–1331.

Muszynski, P. and Michalowski, T. 1979. The effect of protein and starch from cereal seeds upon the number of rumen protozoa *in vitro. Bulletin de l'Academie Polonaise des Sciences Seri des Sciences Biologique* 27: 585–590.

Naga, M.A., Abou Akkada, A.R. and El Shazly, K. 1969. Establishment of rumen ciliate protozoa in cow and water buffalo (*Bos bubalus* L.) calves under late and early weaning systems. *Journal of Dairy Science* 52: 110–112.

Naga, M.A. and El-Shazly, K. 1968. The metabolic characterization of the ciliate protozoon *Eudiplodinium medium* from the rumen of buffalo. *Journal of General Microbiology* 53: 305–315.

Nagaraja, T.G., Dennis, S.M., Galitzer, S.J. and Harmon, D.L. 1986. Effect of lasalocid, monensin, and thiopeptin on lactate production from *in vitro* rumen fermentation of starch. *Canadian Journal of Animal Science* 66: 129–139.

Nakamura, F. and Kurihara, Y. 1978a. Maintenance of a certain rumen protozoal population in a continuous *in vitro* fermentation system. *Applied and Environmental Microbiology* 35: 500–506.

Nakamura, F. and Kurihara, Y. 1978b. Two factors associated with the maintenance of ciliate protozoal concentrations in the rumen. *Journal of Fermentation Technology* 56: 708–711.

Neill, A.R., Grime, D.W., Snoswell, A.M., Northrop, A.J., Lindsay, D.B. and Dawson, R.M.C. 1979. The low availability of dietary choline for the nutrition of the sheep. *Biochemical Journal* 180: 559–565.

Newbold, C.J. and Chamberlain, D.G. 1988. Lipids as rumen defaunating agents. *Proceedings of the Nutrition Society* 47: 154A.

Newbold, C.J. and Hillman, K. 1990. The effect of ciliate protozoa on the turnover of bacterial and fungal protein in the rumen of sheep. *Letters in Applied Microbiology* 11: 100–102.

Newbold, C.J., Chamberlain, D.G. and Williams, A.G. 1986a. The effects of defaunation on the metabolism of lactic acid in the rumen. *Journal of the Science of Food and Agriculture* 37: 1083–1090.

Newbold, C.J., Chamberlain, D.G. and Williams, A.G. 1986b. The role of protozoa in lactic acidosis, pp. 154–157 in *Proceedings of the Sixth International Conference on Production Diseases in Farm Animals,* Queens University, Belfast.

Newbold, C.J., Griffin, P.W. and Wallace, R.J. 1989a. Interactions between rumen bacteria and ciliate protozoa in the attachment to barley straws. *Letters in Applied Microbiology* 8: 63–66.

Newbold, C.J., McKain, N. and Wallace, R.J. 1989b. The role of protozoa in ruminal peptide metabolism, pp. 42–55 in Lloyd, D., Coombs, G.H. and Paget, T.A. (editors), *Biochemistry and Molecular Biology of Anaerobic Protozoa,* Harwood Academic Publishers, Chur, Switzerland.

Newbold, C.J., Wallace, R.J., Watt, N.D. and Richardson, A.J. 1988. Effect of the novel ionophone Tetronasin (ICI139603) on ruminant microorganisms. *Applied and Environmental Microbiology* 54: 544–547.

Newbold, C.J., Williams, A.G. and Chamberlain, D.G. 1987. The *in vitro* metabolism of D,L-lactic acid by rumen microorganisms. *Journal of the Science of Food and Agriculture* 38: 9–18.

Noirot-Timothée, C. 1958. L'ultrastructure de la limite ectoplasme-endoplasme et des fibres formant le caryophore chez les ciliés du genre *Isotricha* Stein (Holotriches Trichostomes). *Comptes Rendus de l'Academie des Sciences* 247: 692–695.

Noirot-Timothée, C. 1959. *Diplodinium moucheti* n.sp. Remarques sur l'evolution des *Ophrysocolecidae* en Afrique. *Annales des Sciences Naturelles Zoologie et Biologie Animale* 1: 331–337.

Noirot-Timothée, C. 1960. Etude d'une famille de ciliés: les Ophryoscolecidae. Structures et ultrastructures. *Annales des Sciences Naturelles Zoologie et Biologie Animale* 2: 527–718.

Noirot-Timothée, C. 1963. Sur les ciliés du rumen de *Giraffa camelopardalis* L.

Comptes Rendus de l'Academie des Sciences 256: 5400–5401.

Noirot-Timothée, C. and Couturier, M. 1957. Sur deux nouvelles formes d'Infusoires *Ophryoscolecidae* trouvees chez le Bouquetin des Alpes (*Capra ibex ibex* L. 1758). *Comptes Rendus Hebdomadairedes Seances de l'Academie des Sciences, Paris* 244: 804–906.

Nolan, J.V. 1989. The implications of protozoa and fungi for the protein nutrition of ruminants, pp. 211–221 in Nolan, J.V., Leng, R.A. and Demeyer, D.I. (editors), *The Roles of Protozoa and Fungi in Ruminant Digestion*, Penambul Books, Armidale, Australia.

Nolan, J.V. and Leng, R.A. 1972. Dynamic aspects of ammonia and urea metabolism in sheep. *British Journal of Nutrition* 27: 177–194.

Nolan, J.V., Leng, P.A. and Demeyer, D.I. 1989. *The Roles of Protozoa and Fungi in Ruminant Digestion*. Penambul Books, Armidale, Australia.

Nolan, J.V. and Staciw, S. 1979. Fermentation and nitrogen dynamics in Merino sheep given a low-quality-roughage diet. *British Journal of Nutrition* 42: 63–80.

Nour, A.M., Abou Akkada, A.R., El-Shazly, K., Naga, M.A., Borhami, B.E. and Abaza, M.A. 1979. Effect of increased levels of urea in the diet on ruminal protozoal counts in four ruminant species. *Journal of Animal Science* 49: 1300–1305.

Nouzarede, M. 1978. Sur la présence d'une bactérie endocellulaire chez *Isotricha prostoma* Stein. *Protistologica* 14: 177–181.

Nummi, M., Fox, P.C., Niku-Paavola, M-L. and Enari, T-M. 1981. Nephalometric and turbimetric assays of cellulase activity. *Analytical Biochemistry* 116: 133–136.

Nuzback, D.E., Bartley, E.E., Dennis, S.M., Nagaraja, T.G., Galitzer, S.J. and Dayton, A.D. 1983. Relation of rumen ATP concentration to bacterial and protozoal numbers. *Applied and Environmental Microbiology* 46: 533–538.

Obendorf, D.L. 1984. The macropodid oesophagus. IV. Observations of the protozoan fauna of the macropodid stomach and oesophagus. *Australian Journal of Biological Science* 37: 117–122.

O'Connor, J.J., Myers, G.S., Maplesden, D.C. and Van der Noot, G.W. 1970. Chemical additives in rumen fermentations: *in vitro* effects of various drugs on rumen volatile fatty acids and protozoa. *Journal of Animal Science* 30: 812–818.

O'Connor, J.J, Myers, G.S. Maplesden, D.C. and Van der Noot, G.W. 1971. Chemical additives in rumen fermentations: *in vitro* effects of drugs, inoculum source and substrate on rumen volatile fatty acids and protozoa. *Journal of Animal Science* 32: 994–998.

Oddy, V.H., Cook, J.B., Jones, A.W. and Wells, B.A. 1978. The effect of monensin on energy metabolism of pregnant ewes. *Proceedings of the Australian Society of Animal Production* 12: 144.

Ogimoto, K. and Imai, S. 1981. *Atlas of Rumen Microbiology*, Japan Scientific Societies Press, Tokyo. pp. 231.

Ogimoto, K., Imai, S., Asada, T. and Fujita, J. 1983. Bacterial flora, protozoal fauna and volatile fatty acids in the rumen of water buffalo (*Bubalus bubalis*) in tropical Asia. *South African Journal of Animal Science* 13: 59–61.

Ohba, S., Komaki, H. and Ohta, K. 1981. Rumen protozoa. 1. *In vitro* utilization of urea-^{15}N added to the culture media for rumen protozoa. *Bulletin of the College of Agriculture and Veterinary Medicine Nihon University* 38: 228–237.

Ohnishi, T., Lloyd, D., Lindmark, D.G. and Muller, M. 1980. Respiration of *Tritrichomonas foetus* components detected in hydrogenosomes and intact cells by paramagnetic resonance spectrometry. *Molecular and Biochemical Parasitology* 2: 39–50.

Oldham, J.D. 1981. Amino acid requirements for lactation in high-yielding dairy cows, pp. 33–65 in Haresign, W. (editor), *Recent Advances in Animal Nutrition —1980*, Butterworths, London.

Oltjen, R.R., Gutierrez, J., Lehmann, R.P. and Davis, R.E. 1966. Rumen chemical and microbial characteristics of steers fed a purified and a natural diet. *Journal of Animal Science* 25: 521–525.

Olumeyan, D.B., Nagaraja, T.G., Miller, G.W., Frey, R.A. and Boyer, J.E. 1986. Rumen microbial changes in cattle fed diets with or without salinomycin. *Applied and Environmental Microbiology* 51: 340–345.

Onodera, R. 1986. Contribution of protozoa to lysine synthesis in the *in vitro* rumen microbial ecosystem. *Applied and Environmental Microbiology* 51: 1350–1351.

Onodera, R. and Henderson, C. 1980. Growth factors of bacterial origin for the culture of the rumen oligotrich protozoon *Entodinium caudatum*. *Journal of Applied Bacteriology* 48: 125–134.

Onodera, R. and Kandatsu, M. 1968. Amino acid and protein metabolism of rumen ciliate protozoa. 1. Consumption of amino acids. *Japanese Journal of Zootechnical Science* 39: 206–211.

Onodera, R. and Kandatsu, M. 1969a. Occurrence of L-(-)-pipecolic acid in the culture medium of rumen ciliate protozoa. *Agricultural and Biological Chemistry* 33: 113–115.

Onodera, R. and Kandatsu, M. 1969b. Amino acid and protein metabolism of rumen ciliate protozoa. 2. Incorporation of [^{14}C]-amino acids and [^{14}C]-labelled rumen bacteria in ciliates. *Japanese Journal of Zootechnical Science* 40: 205–211.

Onodera, R. and Kandatsu, M. 1970a. Amino acid and protein metabolism of rumen ciliate protozoa. III. Ingestion of particulate substances by ciliates. *Japanese Journal of Zootechnical Science* 40: 14–21.

Onodera, R. and Kandatsu, M. 1970b. Amino acids and protein metabolism of rumen ciliate protozoa. 4. Metabolism of casein. *Japanese Journal of Zootechnical Science* 41: 307–313.

Onodera, R. and Kandatsu, M. 1970c. Amino acid and protein metabolism of rumen ciliate protozoa. V. Investigation of artificial buffer solutions for cultivation of ciliates. *Japanese Journal of Zootechnical Science* 41: 343–348.

Onodera, R. and Kandatsu, M. 1970d. Amino acid and protein metabolism of rumen ciliate protozoa. VI. Endogenous nitrogen compounds of rumen ciliates. *Japanese Journal of Zootechnical Science* 41: 349–357.

Onodera, R. and Kandatsu, M. 1972. Conversion of lysine to pipecolic acid by rumen ciliate protozoa. *Agricultural and Biological Chemistry* 36: 1989–1995.

Onodera, R. and Kandatsu, M. 1974. Formation of lysine from α,ε-diaminopimelic acid and negligible synthesis of lysine from some other precursors by rumen ciliate protozoa. *Agricultural and Biological Chemistry* 38: 913–920.

Onodera, R. and Koga, K. 1987. Effect of inhabitation by rumen protozoa on the nutritive value of protein in rumen contents. *Agricultural and Biological Chemistry* 51: 1417–1424.

Onodera, R. and Migita, R. 1985. Metabolism of threonine, methionine and related compounds by mixed rumen ciliate protozoa. *Journal of Protozoology* 32: 326–329.

Onodera, R., Miura, K. and Fukuda, H. 1982. Occurrence of 2-aminobutanoic acid in the incubation medium of mixed rumen ciliate protozoa. *Journal of Protozoology* 29: 122–124.

Onodera, R., Nakagawa, Y. and Kandatsu, M. 1977a. Ureolytic activitiy of the washed cell suspension of rumen ciliate protozoa. *Agricultural and Biological Chemistry* 41: 2177–2182.

Onodera, R., Ohno, T., Nakayama, M., Tanaka, T. and Yamasaki, N. 1987. Feeding experiment of salinomycin with faunated or defaunated goat. III. Effect on the nutritive value of protein protein in the rumen. *Bulletin of the Faculty of Agriculture Miyazaki University* 34: 41–46.

Onodera, R., Shinjo, T. and Kandatsu, M. 1974. Formation of lysine from α, ε-diaminopimelic acid contained in rumen bacterial cell walls by rumen ciliate protozoa. *Agricultural and Biological Chemistry* 38: 921–926.

Onodera, R. and Suzuki, H. 1987. Quantity and quality of microbial protein produced by an agnotobiotic culture ecosystem of a rumen protozoon, *Entodinium caudatum*. *Agricultural and Biological Chemistry* 51: 1463–1465.

Onodera, R. Tanaka, T., Ono, T., Yamasaki, N. and Nakayama, M. 1986. Feeding experiment with salinomycin on faunated or defaunated goats. II. Effect on fiber digestibility in the rumen. *Bulletin of the Faculty of Agriculture Miyazaki University* 33: 19–22.

Onodera, R., Tsutsumi, W. and Kandatsu, M. 1977b. Formation of δ-aminovaleric acid from proline, ornithine and arginine by rumen ciliate protozoa. *Agricultural and Biological Chemistry* 41: 2169–2175.

Onodera, R. and Ushijima, T. 1982. Formation of 2-aminobutanoic acid from threonine and methionine by mixed rumen ciliate protozoa. *Journal of Protozoology* 29: 547–550.

Onodera, R., Yamaguchi, H., Eguchi, C. and Kandatsu, M. 1977c. Limits of the survival of the mingled rumen bacteria in the washed cell suspension of rumen ciliate protozoa. *Agricultural and Biological Chemistry* 41: 2465–2466.

Onodera, R., Yamaguchi, Y. and Morimoto, S. 1983. Metabolism of arginine, citrulline, ornithine and proline by starved rumen ciliate protozoa. *Agricultural and Biological Chemistry* 47: 821–828.

Onodera, R., Yamasaki, N. and Murakami, K. 1988. Effect of inhabitation by ciliate protozoa on the digestion of fibrous materials in vivo in the rumen of goats and in an in vitro rumen microbial digestion. *Agricultural and Biological Chemistry* 52: 2635–2637.

Opperdoes, F.R. 1987. Compartmentation of carbohydrate metabolism in trypansosomes. *Annual Review of Microbiology* 41: 127–151.

Orpin, C.G. 1975. Studies on the rumen flagellate *Neocallimastix frontalis*. *Journal of General Microbiology* 91: 249–262.

Orpin, C.G. 1976. The characterization of the rumen bacterium Eadie's oval, *Magnoovum* gen.nov *eadii* sp.nov. *Archiv für Mikrobiologie* 111: 155–159.

Orpin, C.G. 1977a. The rumen flagellate *Piromonas communis*: its life history and invasion of plant material in the rumen. *Journal of General Microbiology* 99: 107–117.

Orpin, C.G. 1977b. The occurrence of chitin in the cell walls of the rumen organisms *Neocallimastix frontalis, Piromonas communis* and *Sphaeromonas communis*. *Journal of General Microbiology* 92: 214–218.

Orpin, C.G. 1977c. Studies on the defaunation of the ovine rumen using dioctyl sodium sulphosuccinate. *Journal of Applied Bacteriology* 43: 309–318.

Orpin, C.G. 1983/84. The role of ciliate protozoa and fungi in the rumen digestion of plant cell walls. *Animal Feed Science and Technology* 10: 121–143.

Orpin, C.G. 1985. Association of rumen ciliate populations with plant particles *in vitro*. *Microbial Ecology* 11: 59–69.

Orpin, C.G., Greenwood, Y., Hall, F.J. and Patterson, I.W. 1985. The rumen microbiology of seaweed digestion in Orkney sheep. *Journal of Applied Bacteriology* 59: 585–596.

Orpin, C.G. and Hall, F.J. 1977. Attachment of the rumen holotrich protozoon *Isotricha intestinalis* to grass particles. *Proceedings of the Society for General Microbiology* 4: 82–83.

Orpin, C.G. and Hall, F.J. 1983. Surface structures of the rumen holotrich protozoon *Isotricha intestinalis* with particular reference to the attachment zone. *Current Microbiology* 8: 321–325.

Orpin, C.G. and Letcher, A.J. 1978. Some factors controlling the attachment of the rumen holotrich protozoa *Isotricha intestinalis* and *I. prostoma* to plant particles *in vitro*. *Journal of General Microbiology* 106: 33–40.

Orpin, C.G. and Letcher, A.J. 1983/84. Effect of absence of ciliate protozoa on rumen fluid volume, flow rate and bacterial populations in sheep. *Animal Feed Science and Technology* 10: 145–153.

Orpin, C.G. and Mathiesen, S.D. 1986. *Microcetus lappus* gen.nov., sp.nov.: new species of ciliated protozoon from the bovine rumen. *Applied and Environmental Microbiology* 52: 527–530.

Osman, H. El Sayed, Abou Akkada, A.R. and Agabawi, K.A. 1970. Influence of rumen ciliate protozoa on conversion of food and growth rate in early-weaned zebu calves. *Animal Production* 12: 267–271.

Owen, R.W. and Coleman, G.S. 1976. The cultivation of the rumen ciliate *Entodinium longinucleatum*. *Journal of Applied Bacteriology* 41: 341–344.

Owen, R.W. and Coleman, G.S. 1977. The uptake and utilisation of bacteria, amino acids and carbohydrates by the rumen ciliate *Entodinium longinucleatum* in relation to the sources of amino acids for protein synthesis. *Journal of Applied Bacteriology* 43: 67–74.

Oxford, A.E. 1951. The conversion of certain soluble sugars to a glucosan by holotrich ciliates in the rumen of sheep. *Journal of General Microbiology* 5: 83–90.

Oxford, A.E. 1955a. The bacteriology and protozoology of ruminant digestion. *Journal of the Science of Food and Agriculture* 6: 413–418.

Oxford, A.E. 1955b. The rumen ciliate protozoa. Their chemical composition, metabolism, requirements for maintenance and culture and physiological significance for the host. *Experimental Parasitology* 4: 569–605.

Oxford, A.E. 1958. Bloat in cattle. IX. Some observations on the culture of the cattle rumen ciliate *Epidinium ecaudatum* Crawley occurring in quantity in cows fed on red clover (*Trifolium pratense* L.). *New Zealand Journal of Agricultural Research* 1: 809–824.

Oxford, A.E. 1959. Bloat in cattle. XV. Further observations concerning the ciliate *Epidinium ecaudatum*, an inhabitant of the rumens of cows liable to legume bloat. *New Zealand Journal of Agricultural Research* 2: 365–374.

Panjarathinam, R., Rayudu, G.V.N. and Shukla, P.C. 1973. A note on the supplementation of oxytetracycline in the rations with urea on the rumen microorganisms of lactating cows. *Indian Journal of Animal Science* 43: 955–957.

Patterson, J.A. and Kung, L. 1988. Metabolism of DL-methionine and methionine analogs by rumen microorganisms. *Journal of Dairy Science* 71: 3292–3301.

Paul, R.G., Butler, R.D. and Williams, A.G. 1987. Isolation and ultrastructure of rumen holotrich protozoa. *Journal of Protozoology*: 34 (supplement): Abstract 100.

Paul, R.G., Butler, R.D. and Williams, A.G. 1989. Ultrastructure of the rumen holotrich ciliate *Dasytricha ruminantium* Schuberg. *European Journal of Protistology* 24: 204–215.

Paul, R.G., Williams, A.G. and Butler, R.D. 1990. Hydrogenosomes in the rumen entodiniomorphid ciliate *Polyplastron multivesiculatum*. *Journal of General Microbiology* 136: 1981–1989.

Pearson, H.A. 1965. Rumen organisms in white-tailed deer from south Texas. *Journal of Wildlife Management* 29: 493–496.

Pearson, H.A. 1967. Rumen microorganisms in buffalo from southern Utah.

Applied Microbiology 15: 1450–1451.

Perkins, J.L. and Luther, R.M. 1967. Effect of defaunation and energy level on ration utilization in sheep. *Journal of Animal Science* 26: 928.

Perty, M. 1852. *Zur Kenntniss kleinster Lebensformen nach Bau, Funktionen, Systematik mit Specialverzeichniss der in der Schweiz beobachteten*. Jent and Reinert, Bern. pp. 228.

Pilgrim, A.F., Gray, F.V., Weller, R.A. and Belling, C.B. 1970. Synthesis of microbial protein from ammonia in the sheep's rumen and the proportion of dietary nitrogen converted into microbial nitrogen. *British Journal of Nutrition* 24: 589–598.

Poche, F. 1913. Das System der Protozoa. *Archiv für Protistenkunde* 30: 125–131.

Poljansky, G. and Strelkow, A. 1938. Étude expérimentale sur la variabilité de quelques Ophryoscolécidés. *Archives de Zoologie Experimentale et Generale* 80: 1–123.

Poos, M.I., Hanson, T.L. and Klopfenstein, T.J. 1979. Monensin effects on diet digestibility, ruminal protein bypass and microbial protein synthesis. *Journal of Animal Science* 48: 1516–1524.

Potter, E.L. and Dehority, B.A. 1973. Effect of changes in feed level, starvation, and level of feed after starvation upon the concentration of rumen protozoa in the ovine. *Applied Microbiology* 26: 692–698.

Potter, V.R. and Elvehjem, C.A. 1936. A modified method of the study of tissue oxidations. *Journal of Biological Chemistry* 114: 495–502.

Pounden, W.D. and Hibbs, J.W. 1950. The development of calves raised without protozoa and certain other characteristic rumen microorganisms. *Journal of Dairy Science* 33: 639–644.

Prins, R.A. 1977. Biochemical activities of gut microorganisms, pp. 73–183 in Clarke, R.T.J. and Bauchop, T. (editors), *Microbial Ecology of the Gut*, Academic Press, London.

Prins, R.A. and Geelen, M.J.H. 1971. Rumen characteristics of red deer, fallow deer and roe deer. *Journal of Wildlife Management* 38: 673–680.

Prins, R.A. and Prast, E.R. 1973. Oxidation of NADH in a coupled oxidase-peroxidase reaction and its significance for the fermentation in the rumen protozoa of the genus *Isotricha*. *Journal of Protozoology* 20: 471–477.

Prins, R.A. and van Golde, L.M.G. 1976. Entrance of glycerol into plasmalogens of some strictly anaerobic bacteria and protozoa. *FEBS Letters* 63: 107–111.

Prins, R.A. and Van Hoven, W. 1977. Carbohydrate fermentation by the rumen ciliate *Isotricha prostoma*. *Protistologica* 13: 549–556.

Prins, R.A., van Rheenen, D.L. and van't Klooster, A. Th. 1983. Characterization of microbial proteolytic enzymes in the rumen. *Antonie van Leeuwenhoek Journal of Microbiology and Serology* 49: 585–595.

Punia, B.S. and Leibholz, J. 1984. Protozoal nitrogen in the stomach of cattle. *Canadian Journal of Animal Science* 64 (supplement): 24–25.

Punia, B.S., Leibholz, J. and Faichney, G.J. 1986. Effects of rumen protozoa on digestion in cattle fed Paspalum hay. *Proceedings of the Australian Society of Animal Production* 16: 427.

Punia, B.S., Leibholz, J. and Faichney, G.J. 1987. The role of rumen protozoa in the utilization of paspalum (*Paspalum dilatatum*) hay by cattle. *British Journal of Nutrition* 57: 395–406.

Punj, M.L. Kochar, A.S. and Bhatia, I.S. 1970. Degradation and utilisation of inulin by rumen microorganisms. *Indian Journal of Nutrition and Dietetics* 7: 247–251.

Purser, D.B. 1961. A diurnal cycle for holotrich protozoa of the rumen. *Nature, London* 190: 831–832.

Purser, D.B., Klopfenstein, T.J. and Cline, J.H. 1965. Influence of tylosine and aureomycin upon rumen metabolism and the microbial population. *Journal of Animal Science* 24: 1039–1044.

Purser, D.B., Klopfenstein, T.J. and Cline, J.H. 1966. Dietary and defaunation effects upon plasma amino acid concentrations in sheep. *Journal of Nutrition* 89: 226–234.

Purser, D.B. and Moir, R.J. 1959. Ruminal flora studies in the sheep. IX. The effect of pH on the ciliate population of the rumen *in vivo*. *Australian Journal of Agricultural Research* 10: 555–564.

Purser, D.B. and Moir, R.J. 1966. Dietary effects upon concentrations of protozoa in the rumen. *Journal of Animal Science* 25: 668–674.

Purser, D.B. and Tompkin, R.B. 1965. The influence of rumen fluid source upon viable bacterial counts and the cultivation of oligotrich and holotrich rumen protozoa. *Life Sciences* 4: 1493–1501.

Purser, D.B. and Weiser, H.H. 1963. Influence of time of addition of antibiotic on the *in vitro* life of rumen holotrich protozoa. *Nature, London* 200: 290.

Puytorac, P. de, Batisse, A., Bohatier, J., Corliss, J.O., Deroux, G., Didier, P., Dragesco, J., Fryd-Versavel, G., Grain, J., Grolière, C.-A., Hovasse, R., Iftode, F., Laval, M., Roque, M., Savoie, A. and Tuffrau, M. 1974. Proposition d'une classification du phylum Ciliophora Doflein 1901. *Comptes Rendus de l'Academie des Sciences* 278: 2799–2802.

Quin, L.Y. 1962. Continuous culture of ruminal microorganisms in chemically defined medium. 1. Design of continuous culture apparatus. *Applied Microbiology* 10: 580–582.

Quin, L.Y., Burroughs, W. and Christiansen, W.C. 1962. Continuous culture of ruminal microorganisms in chemically defined medium. 2. Culture medium studies. *Applied Microbiology* 10: 583–592.

Quin, L.Y., Farrell, F. and Oates, R.P. 1968. Culture of ruminal *Isotricha* in chemically defined medium. 1. Statistical evaluation of culture media. *U.S. Government Research and Development Reports* 68: 49.

Railliet, A. 1890. Protozoaires, pp. 281–321, in Bouley, H., Sansen, A., Trasbot, L. and Nocard, E. (editors), *Nouveau Dictionaire de Medicine de Chirugie et d'Hygiene Veterinaires,* Volume 18, Paris.

Rahman, S.A., Purser, D.B. and Tyznik, W.J. 1964. Cultivation of rumen protozoa. I. Factors influencing the cultivation of rumen oligotrich protozoa *in vitro*. *Journal of Protozoology* 11: 51–56.

Ramaprasad, J. and Raghavan, G.V. 1981. Note on the growth rate and body composition of faunated and defaunated lambs. *Indian Journal of Animal Science* 51: 570–572.

Rao, A.M., Punj, M.L. and Bhatia, I.S. 1977. Studies on epidinial hemicellulase from the rumen of goat. *Indian Journal of Animal Science* 47: 539–544.

Reichl, J. 1961. Relations of the occurrence of protozoa in the sheep rumen to feed and feed utilization, pp. 537–542, in *Proceedings of the First International Conference on Protozoology, Prague*.

Richardson, L.F., Potter, E.L. and Cooley, C.O. 1978. Effect of monensin on ruminal protozoa and volatile fatty acids. *Journal of Animal Science* 47 (supplement): 45.

Roberts, R.B., Abelson, P.H., Cowie, D.E., Bolton, E.T. and Britten, R.J. 1955. *Studies on Biosynthesis in Escherichia coli. Publications of the Carnegie Institution* No. 607.

Rock, R.C. 1971. Incorporation of ^{14}C-labelled non-lipid precursors into lipid of *Plasmodium knowlesi in vitro*. *Comparative Biochemistry and Physiology* 40B: 657–669.

Romulo, B., Bird, S.H. and Leng, R.A. 1986. The effects of defaunation on digestibility and rumen fungi counts in sheep fed high fibre diets. *Proceedings of the*

Australian Society on Animal Production 16: 327–330.

Romulo, B., Bird, S.H. and Leng, R.A. 1989. The effects of defaunation and protein supplementation on intake, digestibility, N retention and fungal numbers in sheep fed straw-based diets, pp. 285–288 in Nolan, J.V., Leng, R.A. and Demeyer, D.I. (editors), *The Roles of Protozoa and Fungi in Ruminant Digestion*, Penambul Books, Armidale, Australia.

Roth, L.E. and Shigenaka, Y. 1964. The structure and formation of cilia and filaments in rumen protozoa. *Journal of Cell Biology* 20: 208–213.

Rowe, J.B., Davies, A. and Broome, A.W.J. 1981. Quantitative effects of defaunation on rumen fermentation and digestion in the sheep. *Proceedings of the Nutrition Society* 40: 49A.

Rowe, J.B., Davies, A. and Broome, A.W.J. 1983. Changing rumen fermentation by chemical means, pp. 102–109 in Farrell, D.J. and Vohra, P. (editors), *Recent Advances in Animal Nutrition in Australia*, University of New England Publishing Unit, Armidale Australia.

Rowe, J.B., Davies, A. and Broome, A.W.J. 1985. Quantitative effects of defaunation on rumen fermentation and digestion in sheep. *British Journal of Nutrition* 54: 105–119.

Rufener, W.H., Nelson, W.O. and Wolin, M.J. 1963. Maintenance of rumen microbial populations in continuous culture. *Applied Microbiology* 11: 196–201.

Russell, J.B. and Dombrowski, D.B. 1980. Effect of pH on the efficiency of growth by pure cultures of rumen bacteria in continuous culture. *Applied and Environmental Microbiology* 39: 604–610.

Russell, J.B., Sharp, W.M. and Baldwin, R.L. 1979. The effect of pH on maximal bacterial growth rate and its possible role as a determinant of bacterial competition in the rumen. *Journal of Animal Science* 48: 251–255.

Saini, R., Ahuja, S.P. and Bhatia, I.S. 1977. *In vitro* incorporation of 2-[^{14}C]-acetate, 1-[^{14}C]-butyrate, U-[^{14}C]-glucose and [^{32}P]-disodium hydrogen phosphate into protozoal lipids. *Zentralblatt für Veterinärmedizin Reihe A* 24: 580–590.

Sarmah, T.C. and Ahuja, S.P. 1979. A note on species variation in the degradation of proteins in the rumen, the patterns of protozoal and their excretory amino acids. *Indian Journal of Animal Science* 49: 948–951.

Schedrina, Z.G. 1926. Sur les infusoires de l'ordre holotriche provenant de la panse du dromedaire et du renne. *Archives russes de Protistologie* (trans. *Ruskii Archiv Protistologii*) 5: 57–61.

Scheifinger, C.C., Linehan, B. and Wolin, M.J. 1975. Hydrogen production by *Selenomonas ruminantium* in the absence and presence of methanogenic bacteria. *Applied Microbiology* 29: 480–483.

Scherbaum, O. 1957. The division index and multiplication in a mass culture of *Tetrahymena* following inoculation. *Journal of Protozoology* 4: 257–259.

Schewiakoff, W. 1896. The organisation and systematics of the infusoria *Aspirotricha* (Holotricha auctorum). *Memoirs or the Imperial Academy of St. Petersburg* 4: 1–395.

Schrenk, H.-G. and Bardele, C.F. 1987. The fine structure of the paralabial organelle in the rumen ciliate *Ophryoscolex purkinjei* Stein, 1858. *Journal of Protozoology* 34: 97–103.

Schuberg, A. 1888. Die Protozoen des Wiederkäuermagens. 1. *Butschlia, Isotricha, Dasytricha, Entodinium. Zoologische Jahrbücher Abteilung für Systematik Oekologie und Geographie der Tiere* 3: 365–418.

Schuberg, A. 1981. Über einige organizations Verhältinisse der Infusorien des Wiederkäuermagens. *Sitzungsberichte der Physikalische-medizinischen Gesellschaft zu Wurzburg* pp. 122–137.

Schulze, P. 1924. Der Nachweis und die Verbreitung des Chitins mit einem

Anhan über das komplizierte Verdauungsystem der Ophryoscoleciden. *Zeitschift für Morphologie und Okologie der Trier* 2: 643–666.

Schumacher, I.C. 1915. On *Blepharocorys equi* sp.nov., a new ciliate from the caecum of the horse. *University of California Publications in Zoology* 16: 95–106.

Schwartz, H.M. and Gilchrist, F.M.C. 1975. Microbial interactions with the diet and the host animal, pp. 165–179, in McDonald, I.W. and Warner, A.C.I. (editors), *Digestion and Metabolism in the Ruminant*. University of New England Publishing Unit, Armidale, Australia.

Scott, R.I., Yarlett, N., Hillman, K., Williams, T.N., Williams, A.G. and Lloyd, D. 1983. The presence of oxygen in rumen liquor and its effects on methanogenesis. *Journal of Applied Bacteriology* 55: 143–149.

Sedloev, N. 1970. Effect of defaunation on volatile fatty acid levels, cellogel electrophoretic serum protein spectrum and rumen movement of sheep. *Nauchni trudove Vissh veterinarnomeditsinski institut Sofiya* 22: 301–306.

Sénaud, J., Bohatier, J. and Grain, J. 1986. Comparisons du comportement alimentaire de 2 protozoaires ciliés du rumen, *Epidinium ecaudatum* and *Polyplastron multivesiculatum*: méchanismes d'ingestion et de digestion. *Reproduction Nutrition Development* 26: 287–289.

Sénaud, J., Jouany, J-P., Grain, J. and de Puytorac, P. 1973. Dynamique d'une population de *Polylastron multivesiculatum* (Cilié, Oligotriche) en equilibre dans le rumen de mouton. *Comptes Rendus de l'Academie des Sciences de Paris* 227: 197–200.

Sharp, R.G. 1914. *Diplodinium ecaudatum* with an account of its neuromotor apparatus. *University of California Publications in Zoology* 13: 43–122.

Shimizu, M., Kinoshita, M., Fujita, J. and Imai, S. 1983. Rumen ciliate protozoal fauna and composition of the zebu cattle, *Bos indicus*, and water buffalo, *Bubalus bubalis*, in Phillippines. *Bulletin of the Nippon Veterinary and Zootechnical College* 32: 83–88.

Shinchi, S., Itoh, T., Abe, M. and Kandatsu, M. 1986. Effect of rumen ciliate protozoa on the proteolytic activity of cell free rumen liquid. *Japanese Journal of Zootechnical Science* 57: 89–96.

Shinchi, S. and Kandatsu, M. 1980. The extracellular proteolytic activity of rumen ciliate protozoa. 1. Proteolytic activitiy of cell free rumen fluid, incubation medium of rumen ciliate protozoa and their extracts. *Bulletin of Azabu University Veterinary Medicine* 1: 47–53.

Shinchi, S. and Kandatsu, M. 1981. On some properties of extracellular and intracellular proteolytic activity of rumen ciliate protozoa. *Japanese Journal of Zootechnical Science* 52: 861–868.

Shinchi, S. and Kandatsu, M. 1983. On some properties of extracellular proteolytic activity of rumen ciliate protozoa. *Japanese Journal of Zootechnical Science* 54: 290–296.

Shiroma, S. and Akashi, A. 1976. Degradation of mimosine in *Leucaena leucocephala* de Wit by goat rumen microorganisms. *Japanese Journal of Zootechnical Science* 47: 739–747.

Singh, S. and Makkar, G.S. 1976. Effect of partial defaunation on the utilization of carbohydrates in buffalo and zebu cattle. *Indian Journal of Animal Science* 46: 463–467.

Sladecek, F. 1946. Ophryoscolecidae z bachoru jelena (*Cervus elaphus* L.), dan'ka (*Dama dama* L.) a srnce (*Capreolus capreolus* L.). *Vestnik Ceskoslovenske Zoologicke Spolecnosti* 10: 201–231.

Slyter, L.L. 1976. Influence of acidosis on rumen function. *Journal of Animal Science* 43: 910–929.

Slyter, L.L., Kern, D.L., Weaver, J.M., Oltjen, R.R. and Wilson, R.L. 1971. Influence of starch and nitrogen sources on ruminal microorganisms of steers fed

high fibre purified diets. *Journal of Nutrition* 101: 847–853.
Slyter, L.L., Nelson, W.O. and Wolin, M.J. 1964. Modifications of a device for maintenance of the rumen microbial population in continuous culture. *Applied Microbiology* 12: 374–377.
Slyter, L.L., Oltjen, R.R., Kern, D.L. and Blank, F.C. 1970. Influence of type and level of grain and diethylstilbestrol on the rumen micobial populations of steers fed all-concentrate diets. *Journal of Animal Science* 31: 996–1002.
Small, E.B. 1976. A proposed subphyletic division of the phylum Ciliphora Doflein 1901. *Transactions of the American Microscopical Society* 95: 739–751.
Small, E.B. and Lynn, D.H. 1981. A new macrosystem for the phylum Ciliophora Doflein, 1901. *Biosystems* 14: 387–401.
Smith, J.D. and Law, J.H. 1970. Phosphatidylcholine biosynthesis in *Tetrahymena pyriformis*. *Biochimica et Biophysica Acta* 202: 141–152.
Smith, R.H., McAllan, A.B., Hewitt, D. and Lewis, P.E. 1978. Estimation of the amounts of microbial and dietary nitrogen compounds entering the duodenum of cattle. *Journal of Agricultural Science* 90: 557–568.
Snyers, L., Hellings, P. Bovy-Kesler, C. and Thines-Sempoux, D. 1982. Occurrence of hydrogenosomes in the rumen *Ophryoscolecidae*. *FEBS Letters* 137: 35–39.
Spinsi, D., Frateschi, T.L., Mariani, A.P., Martelli, F., Preziuso, F. and Sighieri, C. 1980. Improvements in artificial culture methods of rumen protozoa and determination of their nutritive values owing to high lysine contents. 1. Effects of organic substrates added to rumen protozoa cultures using an artificial rumen apparatus. *Annali della Facolta di Medicina Veterinari della Pisa Universita degli Studi di Pisa* 33: 49–60.
Stein, F. 1858. Über mehrere neue im Pansen der Widerkäuer lebende Infusionstiere. *Abhandlungen der k. Bohmischen Gesellschaft der Wissenschaften* 10: 69–70.
Stein, F. 1859. Characteristik neuer infusorien Gattungen. *Lotos* 9: 1–5 and 57–60.
Steinhour, W.D., Stokes, M.R., Clark, J.N., Rogers, J.A. and Davis, C.L. 1982. Estimation of the proportion of non-ammonia nitrogen reaching the lower gut of the ruminant derived from bacterial and protozoal nitrogen. *British Journal of Nutrition* 48: 417–431.
Stern, M.D. and Hinkson, R.S. 1974. Effect of defaunation and faunation on intraruminal factors. *Journal of Animal Science* 39: 253.
Stern, M.D., Hoover, W.N. and Leonard, J.B. 1977. Ultrastructure of rumen holotrichs by electron microscopy. *Journal of Dairy Science* 60: 911–918.
Stewart, C.S. 1977. Factors affecting the cellulolytic activity of rumen contents. *Applied and Environmental Microbiology* 33: 497–502.
Stewart, D.G., Warner, R.G. and Seeley, H.W. 1961. Continuous culture as a method for studying rumen fermentation. *Applied Microbiology* 9: 150–156.
Strand, E. 1928. Miscelanea nomenclatorica et paleontologica. I und II. *Archiv für Naturgeschichte* 92: 30–75.
Strelkov, A. 1939 . Parasitic infusioria from the intestine of *Ungulata* belonging to the family *Equidae*. *Uchenye Zapiski Leningradskogo Gosudarstvennogo Pedagogicheskogo Instituta Imeni, Gertsena* 17: 1–262.
Strelkov, A., Poljansky, G. and Issakowa-Keo, M. 1933. Über die Infektionswege der im Pansen und in der Haube der Wiederkäuer befindlichen Infusorien. *Archiv für Tierernährung and Tierzücht* 9: 679–697.
Stuart, R.L., Schelling, G.T., Mitchell, G.E. and Tucker, R.E. 1976. *In vitro* amino acid metabolism in defaunated ruminal fluid. *Journal of Animal Science* 42: 278.
Stumm, C.K., Gijzen, H.J. and Vogels, G.D. 1982. Association of methanogenic bacteria with ovine rumen ciliates. *British Journal of Nutrition* 47: 95–99.

Sugden, B. 1953. The cultivation and metabolism of oligotrich protozoa from the sheep's rumen. *Journal of General Microbiology* 9: 44–53.

Sugden, B. and Oxford, A.E. 1952. Some cultural studies with holotrich ciliate protozoa from the sheep's rumen. *Journal of General Microbiology* 7: 145–153.

Sutton, J.D., Knight, R., McAllan, A.B. and Smith, R.H. 1983. Digestion and synthesis in the rumen of sheep given diets supplemented with free and protected oils. *British Journal of Nutrition* 49: 419–432.

Svendsen, P.E. 1975. Experimental studies of gastro-intestinal anatomy in ruminants, pp. 563–575 in McDonald, I.W. and Warner, A.C.I. (editors), *Digestion and Metabolism in the Ruminant*, University of New England Publishing Unit, Armidale, Australia.

Syrjala, L., Kossila, V. and Sipila, H. 1973. A study of the nutritional status of Finnish Reindeer (*Rangifer tarandus* L.) in different months. 1. Composition and volume of rumen microbiota. *Journal of the Scientific Agricultural Society of Finland* 45: 534–541.

Syrjala, L., Saloniemi, H. and Laalahti, L. 1976. Composition and volume of the rumen of sheep fed on grass silage with different sucrose, starch and cellulose supplements. *Journal of the Scientific Agricultural Society of Finland* 48: 138–153.

Szuecs, E., Regius, A., Weber, A., and Szoelloesi, I. 1982. Effect of various nutrient supply and copper supplementation on the performance and carcass value of growing-finishing bulls. *Zeitschrift für Tierphysiologie, Tierernährung und Futtermittelkunde* 47: 276–288.

Takahashi, F. and Kametaka, M. 1976. Effect of absence of rumen ciliate protozoa on the utilization of urea by goats. *Japanese Journal of Zootechnical Science* 47: 192–196.

Takechi, T., Shibata, F. and Kurihata, Y. 1978. Microbiological and chemical characteristics of the rumen ingesta in sheep fed on purified diets. *Japanese Journal of Ecology* 28: 85–96.

Tangendjaja, B., Hogan, J.P. and Wills, R.B.H. 1983. Degradation of mimosine by rumen contents: Effects of feed composition and *Leucaena* substrates. *Australian Journal of Agricultural Research* 34: 289–293.

Teather, R.M., Mahadevan, S., Erfle, J.D. and Sauer, F.D. 1984. Negative correlation between protozoal and bacterial levels in rumen samples and its relation to the determination of dietary effects on the rumen microbial population. *Applied and Environmental Microbiology* 47: 566–570.

Teather, R.M. and Sauer, F.D. 1988. A naturally compartmented rumen simulation system for the continuous culture of rumen bacteria and protozoa. *Journal of Dairy Science* 71: 666–673.

Ten Kate, C.G.B. 1928. Das Fibrillensystem der Isotrichen. *Archiv für Protistenkunde* 62: 328–354.

Tener, J.S., 1965. *Musk Oxen in Canada.* Canadian Wildlife Service, Queen's Printer, Ottawa, Canada.

Thamman, O.P., Ranhotra, G.S. and Langar, P.N. 1971. Partial defaunation of the rumen and its influence on the biochemical activity. 2. Effect of partial defaunation on some blood components of buffalo and zebu. *Indian Journal of Animal Science* 41: 73–77.

Thines-Sempoux, D. and Dauvrin, T. 1989. Invertase secretion by the rumen ciliate *Isotricha prostoma*. *Journal of Protozoology* 36: 16A.

Thines-Sempoux, D., Delfosse-Debusscher, J. and Latteur, B. 1980. Mechanism of cellulose digestion by the rumen ciliates. *Archives International de Physiologie et Biochemie* 48: B102–B104.

Thomas, G.J. 1960. Metabolism of the soluble carbohydrates of grasses in the rumen of the sheep. *Journal of Agricultural Science* 54: 360–372.

Thurston, J.P. and Grain, J. 1971. Holotrich ciliates from the stomach of *Hippopo-*

tamus amphibius, with descriptions of two new genera and four new species. *Journal of Protozoology* 18: 131–141.

Thurston, J.P. and Noirot-Timothee, C. 1973. Entodiniomorph ciliates from the stomach of *Hippopotamus amphibius*, with descriptions of two new genera and three new species. *Journal of Protozoology* 20: 652–565.

Tipton, C.L. and Swords, M.D. 1966. Biosynthesis of phosphatidylcholine in *Euglena gracilis*. *Journal of Protozoology* 13: 469–472.

Tomasic, J., Jennings, H.J. and Glaudemans, C.P.J. 1978. Evidence for a single type of linkage in a fructofuranan from *Lolium perenne*. *Carbohydrate Research* 62: 127–133.

Tompkin, R.B., Purser, D.B. and Weiser, H.H. 1966. Influence of rumen fluid source upon establishment and cultivation *in vitro* of the rumen protozoon *Entodinium*. *Journal of Protozoology* 13: 55–58.

Towne, G. and Nagaraja, T.G. 1989. Occurrence and diurnal population fluctuations of the ruminal protozoan *Microcetus lappus*. *Applied and Environmental Microbiology* 55: 91–94.

Towne, G. and Nagaraja, T.G. 1990. Omasal ciliated protozoa in cattle, bison and sheep. *Applied and Environmental Microbiology* 56: 409–412.

Towne, G., Nagaraja, T.G., Cochran, R.C., Harmon, D.L., Owensby, C. E. and Kaufman, D.W. 1988a. Comparisons of ruminal fermentation characteristics and microbial populations in bison and cattle. *Applied and Environmental Microbiology* 54: 2510–2514.

Towne, G., Nagaraja, T.G. and Kemp, K.K. 1988b. Ruminal ciliated protozoa in bison. *Applied and Environemtal Microbiology* 54: 2733–2736.

Trier, H.J. 1926. Der Kohlehydrastoffwechsel der Panseninfusorien und die Bedeutung der grünen Pflanzenteile für diese Organismen. *Zeitschrift für vergleichende Physiologie* 4: 305–330.

Trebst, A. and Geike, F. 1967. Zur Biosynthese von Phosphonoaminoauren; die Verteilung der Radioaktivität in Aminoäthylphosphonaure nach Einbau von positions-markierter Glucose durch *Tetrahymena*. *Zeitschrift für Naturforschung* 22b: 989–991.

Tsutsumi, W., Onodera, R. and Kandatsu, M. 1975. Occurrence of δ-aminovaleric acid in the culture medium of rumen ciliate protozoa. *Agricultural and Biological Chemistry* 39: 711–714.

Tutt, J.B. 1972. Acid silage as a probable cause of enteritis in a dairy herd. *Veterinary Record* 90: 91–92.

Ushida, K. and Jouany, J.-P. 1985. Effect of protozoa on rumen protein degradation in sheep. *Reproduction Nutrition Developpement* 25: 1075–1081.

Ushida, K. and Jouany, J.-P. 1986. Influence des protozoaires sur la dégradation des protéines measurée *in vitro* et *in sacco*. *Reproduction Nutrition Developpement* 26: 293–294.

Ushida, K., Jouany, J.-P., Kayouli, C. and Demeyer, D.I. 1989a. Effect of defaunation on fibre digestion in sheep fed ammonia-treated straw based diets, pp. 307–308 in Nolan, J.V., Leng, R.A. and Demeyer, D.I. (editors), *The Roles of Protozoa and Fungi in Ruminant Digestion*, Penambul Books, Armidale, Australia.

Ushida, K., Jouany, J.-P., Lassalas, B. and Thivend, P. 1984. Protozoal contribution to nitrogen digestion in sheep. *Canadian Journal of Animal Science* 64 (supplement): 20–21.

Ushida, K., Jouany, J.-P. and Thivend, P. 1986a. Role of protozoa in nitrogen digestion in sheep given two isonitrogenous diets. *British Journal of Nutrition* 56: 407–419.

Ushida, K., Kaneko, T. and Kojina, Y. 1987a. Effect of presence of large entodiniomorphid protozoa on the rumen bacterial flora, fauna composition of small entodinia and *in vitro* cellulolysis and xylanolysis. *Japanese Journal of*

Zootechnical Science 58: 893–902.
Ushida, K., Miyazaki, A. and Kawashima, R. 1986b. Effect of defaunation on ruminal gas and VFA production *in vitro*. *Japanese Journal of Zootechnical Science* 57: 71–77.
Ushida, K., Ohta, K., Yano, F., Miyazaki, A. and Kawashima, R. 1987b. HPLC determination of phosphatidylcholine and its application for estimating protozoal protein at the duodenum. *Japanese Journal of Zootechnical Science* 58: 285–287.
Ushida, K., de Smet, S., Kayouli, C. and Jouany, J.-P. 1989b. Effect of defaunation on nitrogen digestion in sheep fed ammonia-treated straw with or without maize, pp. 309–311 in Nolan, J.V., Leng, R.A. and Demeyer, D.I. (editors), *The Roles of Protozoa and Fungi in Ruminant Digestion*, Penambul Books, Armidale, Australia.
Usuelli, F. 1930. Gli infusori ciliati che vivono nell' apparato digerente degli erbivori. Cenni storici sistematici e descrittivi, esperienze e considerazioni sur loro ruolo nella digestione e nella nutrizione degli erbivori. *La Clinica Veterinaria* 543–570, 625–645, 787–805.
Valdez, R.E., Alvarez, F.J., Ferreiro, H.M., Guerra, F., Lopez, J., Priego, A., Blackburn, T.H., Leng, R.A. and Preston, T.R. 1977. Rumen function in cattle given sugar cane. *Tropical Animal Production* 2: 260–272.
Van der Wath, J.G. and Myburgh, S.J. 1941. Studies on the alimentary tract of merino sheep in South Africa. VI. The role of infusoria in ruminal digestion with some remarks on ruminal bacteria. *Onderspoort Journal of Veterinary Science and Animal Industry* 17: 61–88.
Van Hoven, W. 1974. Ciliate protozoa and aspects of the nutrition of the hippopotamus in the Kruger National Park. *South African Journal of Science* 70: 107–109.
Van Hoven, W. 1975. Rumen ciliates of the Tsessebe (*Damaliscus lunatus lunatus*) in South Africa. *Journal of Protozoology* 22: 457–462.
Van Hoven, W. 1978. Development and seasonal changes in the rumen protozoan population in young blesbok (*Damaliscus dorcas phillipsi* Harper 1939). *South African Journal of Wildlife Research* 8: 127–130.
Van Hoven, W. 1983. Rumen ciliates with description of two new species from three African reedbuck species. *Journal of Protozoology* 30: 688–691.
Van Hoven, W., Gilchrist, F.M.C. and Hamilton-Attwell, V.L. 1987. Intestinal ciliated protozoa of African rhinoceros: two new genera and five new species from the white rhino (*Ceratotherium simum* Burchell, 1817). *Journal of Protozoology* 34: 338–342.
Van Hoven, W., Gilchrist, F.M.C. and Hamilton-Attwell, V.L. 1988. A new family, genus, and seven new species of Entodiniomorphida (protozoa) from the gut of African rhinoceros. *Journal of Protozoology* 35: 92–97.
Van Hoven, W., Hamilton-Attwell, V.L. and Grobler, J.H. 1979. Rumen ciliate protozoa of the sable antelope *Hippotragus niger*. *South African Journal of Zoology* 14: 37–42.
Van Hoven, W. and Prins, R.A. 1977. Carbohydrate fermentation by the rumen ciliate *Dasytricha ruminantium*. *Protistologica* 13: 599–606.
Van Nevel, C.J. and Demeyer, D.I. 1977. Determination of rumen microbial growth *in vitro* from ^{32}P-labelled phosphate incorporation. *British Journal of Nutrition* 38: 101–114.
Van Nevel, C.J. and Demeyer, D.I. 1981. Effect of methane inhibitors on the metabolism of rumen microbes *in vitro*. *Archiv für Tierernährung* 31: 141–151.
Van Nevel, C.J., Demeyer, D.I. and Van de Voorde, G. 1985. Effect of defaunating the rumen on growth and carcass composition of lambs. *Archives of Animal Nutrition, Berlin* 35: 331–337.

Van Nevel, C.J., Kayouli, C., Demeyer, D.I., Vermander, C., Faquaet, M. and Dendooven, R. 1984. Effect of defaunation and refaunation on rumen fermentation and N-flow in the duodenum of sheep. *Canadian Journal of Animal Science* 64 (supplement): 359.

Varel, V.H. and Dehority, B.A. 1989. Ruminal cellulolytic bacteria and protozoa from bison, cattle-bison hybrids and cattle fed three alfalfa-corn diets. *Applied and Environmental Microbiology* 55: 148–153.

Veira, D.M. 1986. The role of ciliate protozoa in nutrition of the ruminant. *Journal of Animal Science* 63: 1547–1560.

Veira, D.M., Ivan, M. and Jui, P.Y. 1981. The effects of partial defaunation on rumen metabolism in sheep. *Canadian Journal of Animal Science* 61: 1086–1087.

Veira, D.M., Ivan, M. and Jui, P.Y. 1983. Rumen ciliate protozoa: effects on digestion in the stomach of sheep. *Journal of Dairy Science* 66: 1015–1022.

Veira, D.M., Ivan, M. and Jui, P.Y. 1984. The effect of ciliate protozoa on the flow of amino acids from the stomach of sheep. *Canadian Journal of Animal Science* 64 (supplement): 22–23.

Vérité, R., Durand, M. and Jouany, J.-P. 1986. Influence des facteurs alimentaires sur la protéosynthèse microbienne dans le rumen. *Reproduction Nutrition Developpement* 26: 181–201.

Vigues, B., Grolière, C.-A. 1985. Evidence for Ca^{2+}-binding protein associated to non-actin microfilamentous systems in two ciliated protozoans. *Experimental Cell Research* 159: 366–376.

Vigues, B., Méténier, G. and Grolière, C.-A. 1984. Biochemical and immunological characterization of the microfibrillar ecto-endoplasmic boundary in the ciliate *Isotricha prostoma*. *Biology of the Cell* 51: 67–78.

Vigues, B., Méténier, G., Grolière, C.-A., Grain, J. and Sénaud, J. 1985. Biochemical study of proteins of cortical cytoskeleton in the ciliate *Isotricha prostoma*. *Journal of Protozoology* 32: 38–44.

Viviani, R. 1970. Metabolism of long-chain fatty acids in the rumen. *Advances in Lipid Research* 8: 267–346.

Vogels, G.D., Hoppe, W.F. and Stumm, C.K. 1980. Association of methanogenic bacteria with rumen ciliates. *Applied and Environmental Microbiology* 40: 608–612.

Wagener, S., Stumm, C.K. and Vogels, G.D. 1986. Electromigration, a tool for studies on anaerobic ciliates. *FEMS Microbiology Ecology* 38: 197–203.

Wakita, M. and Hoshino, S. 1979. Changes of reserve polysaccharide contents in sheep rumen ciliates of genus *Entodinium* after feeding different diets. *Japanese Journal of Zootechnical Science* 50: 1–7.

Wakita, M. and Hoshino, S. 1980. Physiochemical properties of a reserve polysaccharide from sheep rumen ciliate genus *Entodinum*. *Comparative Biochemistry and Physiology* 65B: 571–574.

Wallace, R.J., Broderick, G.A. and Brammall, M.L. 1987. Microbial protein and peptide metabolism in rumen fluid from faunated and ciliate-free sheep. *British Journal of Nutrition* 58: 87–93.

Wallace, R.J., McKain, N. and Newbold, C.J. 1990. Metabolism of small peptides in rumen fluid. Accumulation of intermediates during hydrolysis of alanine oligomers, and comparison of peptidolytic activities of bacteria and protozoa. *Journal of the Science of Food and Agriculture* 50: 191–199.

Wallace, R.J. and McPherson, C.A. 1987. Factors affecting the rate of breakdown of bacterial protein in rumen fluid. *British Journal of Nutrition* 58: 313–323.

Wallace, R.J. and West, A.A. 1982. Adenosine-5′-triphosphate and adenylate energy charge in sheep digesta. *Journal of Agricultural Science* 98: 523–528.

Wallis, O.C. and Coleman, G.S. 1967. Incorporation of ^{14}C-labelled components of *Escherichia coli* and of amino acids by *Isotricha intestinalis* and *Isotricha pro-*

stoma from the sheep rumen. *Journal of General Microbiology* 49: 315–323.

Warren, W.A. 1968. Biosynthesis of phosphonic acids in *Tetrahymena. Biochimica et Biophysica Acta* 156: 341–346.

Warner, A.C.I. 1962a. Enumeration of rumen microorganisms. *Journal of General Microbiology* 28: 119–128.

Warner, A.C.I. 1962b. Some factors influencing the rumen microbial population. *Journal of General Microbiology* 28: 129–146.

Warner, A.C.I. 1965. Factors influencing numbers and kinds of microorganisms in the rumen, pp. 346–359 in Dougherty, R.W., Allen, R.S., Burroughs, W., Jacobson, N.L. and McGilliard, A.D. (editors), *Physiology and Digestion in the Ruminant*, Butterworths, Washington, D.C.

Warner, A.C.I. 1966a. Diurnal changes in the concentrations of microorganisms in the rumens of sheep fed limited diets once daily. *Journal of General Microbiology* 45: 213–235.

Warner, A.C.I. 1966b. Periodic changes in the concentrations of microorganisms in the rumen of a sheep fed a limited ration every three hours. *Journal of General Microbiology* 45: 237–241.

Warner, A.C.I. 1966c. Diurnal changes in the concentrations of microorganisms in the rumens of sheep fed to appetite in pens or at pasture. *Journal of General Microbiology* 45: 243–251.

Watson, M.J. and Mottram, M.J. 1989. Influence of avoparcin on protozoal numbers in ruminal fluid of sheep given pelleted diets, pp. 341–343 in Nolan, J.V., Leng, R.A. and Demeyer, D.I. (editors), *The Roles of Protozoa and Fungi in Ruminant Digestion*, Penambul Books, Armidale, Australia.

Weill, R. 1929. La présence d'un infusoire du genre *Isotricha* (*I. caulleryi* n.sp.) chez un insecte (*Periplaneta americana*) et sa signification possible. *Archives de Zoologie Experimentale et Generale Notes et Revue* 69: 21–26.

Weineck, E. 1934. Die Celluloseverdauung bei den Ciliaten des Wiederkaüermagens. *Archiv für Protistenkunde* 82: 169–202.

Weller, R.A. and Gray, F.V. 1954. The passage of starch through the stomach of sheep. *Journal of Experimental Biology* 31: 40–48.

Weller, R.A., Gray, F.V. and Pilgrim, A.F. 1958. The conversion of plant nitrogen to microbial nitrogen in the rumen of sheep. *British Journal of Nutrition* 12: 421–429.

Weller, R.A. and Pilgrim, A.F. 1974. Passage of protozoa and volatile fatty acids from the rumen of the sheep and from a continuous *in vitro* fermentation system. *British Journal of Nutrition* 32: 341–351.

Wertheim, P. 1933a. Ein Beitrag zu den Untersuchungen über die Ophryoscolescidenfauna aus dem Magen von *Capra ibex* L. *Zoologischer Anzeiger* 104: 15–25.

Wertheim, P. 1933b. Über die Beschaffenheit der Infusorien magen Fauna von *Capreolus capreolus* L. *Zoologischer Anzeiger* 106: 67–70.

Wertheim, P. 1934. Les infusoires de l'estomac de *Rupicapra rupicapra* L. *Annales de Parasitologie* 12: 390–393.

Wertheim, P. 1935a. A new ciliate, *Entodinium bovis* n.sp., from the stomach of *Bos taurus* L. with the revision of *Entodinium exiguum*, *E. nanellum*, *E. simplex*, *E. dubardi dubardi* and *E. parvum*. *Parasitology* 27: 226–230.

Wertheim, P. 1935b. Infusorien aus dem Widerkäuermagen von Gebiete Jugoslawiens nebst einer Übersicht dieser Tierchen von Balkanhalbinsel-Bereich und ein kurzer Bericht über die Pferdedarminfusorien, zugleich Revision der Familie *Ophryoscolecidae*. *Veterinarski Archiv* 5: 386–536.

Wertheim, P. 1935c. Über die Infusorienfauna aus dem Magen von *Cervus elaphus* L. *Zeitschrift für Parasitenkunde* 7: 395–400.

Westerling, B. 1969. Vamciliatfaunan hos tamboskap i finska Lappland, med speciell hansyn till arter ansedda som specifika for ren. *Nordisk Veterinaermedicin* 21: 14–19.

Westerling, B. 1970. Rumen ciliate fauna of semi-domestic reindeer (*Rangifer tarandus* L.) in Finland: composition, volume and some seasonal variations. *Acta Zoologica Fennica*: 127: 1–76.

Westlake, K., Mackie, R.I. and Dutton, M.F. 1989. *In vitro* metabolism of mycotoxins by bacterial, protozoal and ovine ruminal fluid preparations. *Animal Feed Science and Technology* 25: 169–178.

Westphal, A. 1934. Studien über Ophryoscoleciden in der Kultur. *Zeitschrift für Parasitenkunde* 18: 71–117.

White, R.W. 1969. Viable bacteria inside the rumen ciliate *Entodinium caudatum*. *Journal of General Microbiology* 56: 403–408.

Whitelaw, F.G., Bruce, L.A., Eadie, J.M. and Shand, W.J. 1983. 2-Aminoethylphosphonic acid concentrations in some rumen ciliate protozoa. *Applied and Environmental Microbiology* 46: 951–953.

Whitelaw, F.G., Eadie, J.M., Bruce, L.A. and Shand, J.W. 1984. Methane formation in faunated and ciliate-free cattle and its relationship with rumen volatile fatty acid proportions. *British Journal of Nutrition* 52: 261–275.

Whitelaw, F.G., Eadie, J.M., Mann, S.O. and Reid, R.S. 1972. Some effects of rumen ciliate protozoa in cattle given restricted amounts of a barley diet. *British Journal of Nutrition* 27: 425–437.

Wilkinson, R.C. and Van Hoven, W. 1976a. Rumen ciliate fauna of the springbok (*Antidorcas marsupialis*) in Southern Africa. *Zoologica Africana*11: 1–22.

Wilkinson, R.C. and Van Hoven, W. 1976b. Geographical distribution and population structure of springbok (*Antidorcas marsupialis*) rumen protozoa in Southern Africa. *Koedoe* 19: 17–26.

Willard, F.L. and Kodras, R. 1967. Survey of chemical compounds tested *in vitro* against rumen protozoa for possible control of bloat. *Applied Microbiology* 15: 1014–1019.

Williams, A.G. 1979a. The selectivity of carbohydrate assimilation in the anaerobic rumen ciliate *Dasytricha ruminantium*. *Journal of Applied Bacteriology* 47: 511–520.

Williams, A.G. 1979b. Exocellular carbohydrase formation by rumen holotrich ciliates. *Journal of Protozoology* 26: 665–672.

Williams, A.G. 1982. The metabolism and significance of ciliate protozoa in the rumen ecosystem, pp. 93–110 in *Annual Report of the Hannah Research Institute*, Hannah Research Institute, Ayr.

Williams, A.G. 1986. Rumen holotrich ciliate protozoa. *Microbiological Reviews* 50: 25–49.

Williams, A.G. 1989. The metabolic activities of rumen protozoa, pp. 97–126 in Nolan, J.V., Leng, R.A. and Demeyer, D.I. (editors), *The Roles of Protozoa and Fungi in Ruminant Digestion*, Penambul Books, Armidale, Australia.

Williams, A.G. and Coleman, G.S. 1985. Hemicellulose degrading enzymes in rumen ciliate protozoa. *Current Microbiology* 12: 85–90.

Williams, A.G. and Coleman, G.S. 1988. Rumen ciliate protozoa, pp. 77–128 in Hobson, P.N. (editor), *The Rumen Microbial Ecosystem*, Elsevier Applied Science Publishers, Amsterdam.

Williams, A.G. and Ellis, A.B. 1985. Subcellular distribution of glycoside hydrolase and polysaccharide depolymerase enzymes in the rumen entodiniomorphid ciliate *Polyplastron multivesiculatum*. *Current Microbiology* 12: 175–182.

Williams, A.G., Ellis, A.B. and Coleman, G.S. 1986. Subcellular distribution of polysaccharide depolymerase and glycoside hydrolase enzymes in rumen ciliate protozoa. *Current Microbiology* 13: 139–147.

Williams, A.G. and Harfoot, C.G. 1976. Factors affecting the uptake and metabolism of soluble carbohydrates by the rumen ciliate *Dasytricha ruminantium* isolated from ovine rumen contents by filtration. *Journal of General Microbiology* 96: 125–136.

Williams, A.G. and Morrison, I.M. 1982. Studies on the production of saccharinic acids by the alkaline treatment of young grass and their effectiveness as substrates for mixed rumen microorganisms *in vitro*. *Journal of the Science of Food and Agriculture* 33: 21–29.

Williams, A.G. and Strachan, N.H. 1984. The distribution of polysaccharide-degrading enzymes in the bovine rumen digesta ecosystem. *Current Microbiology* 10: 215–220.

Williams, A.G., Strachan, N.H. and Ellis, A.B. 1988. Factors affecting the xylanolytic enzyme activity of microbial populations recovered from the liquid and solid phases of the rumen ecosystem, pp. 597–602 in Megusar, F. and Gantar, M. (editors), *Perspectives in Microbial Ecology*, Slovene Society for Microbiology, Ljubljana.

Williams, A.G. and Withers, S.E. 1990. Effect of ciliate protozoa on the activity of polysaccharide-degrading enzymes and fibre breakdown in the rumen ecosystem. *Journal of Applied Bacteriology* 70: 144–155.

Williams, A.G. Withers, S.E. and Coleman, G.S. 1984. Glycoside hydrolases of rumen bacteria and protozoa. *Current Microbiology* 10: 287–294.

Williams, A.G. and Yarlett, N. 1982. An improved technique for the isolation of holotrich protozoa from rumen contents by differential filtration with defined aperture textiles. *Journal of Applied Bacteriology* 52: 267–270.

Williams, P.P., Davis, R.E., Doetsch, R.N. and Gutierrez, J. 1961. Physiological studies on the rumen protozoan *Ophryoscolex caudatus* Eberlein. *Applied Microbiology* 9: 405–409.

Williams, P.P. and Dinusson, W.E. 1973a. Amino acid and fatty acid composition of bovine ruminal bacteria and protozoa. *Journal of Animal Science* 36: 151–155.

Williams, P.P. and Dinusson, W.E. 1973b. Ruminal volatile fatty acid concentrations and weight gains of calves reared with and without ruminal ciliate protozoa. *Journal of Animal Science* 36: 588–591.

Williams, P.P. and Feil, V.J. 1971. Identification of trifluralin metabolites from rumen microbial cultures. Effect of trifluralin on bacteria and protozoa. *Agricultural and Food Chemistry* 19: 1198–1204.

Williams, P.P., Gutierrez, J. and Davis, R.E. 1963. Lipid metabolism of rumen ciliates and bacteria. 2. Uptake of fatty acids and lipid analysis of *Isotricha intestinalis* and rumen bacteria with further information on *Entodinium simplex*. *Applied Microbiology* 11: 260–264.

Williams, V.J. and McKenzie, D.D.S. 1965. The absorption of lactic acid from the reticulo-rumen of the sheep. *Australian Journal of Biological Sciences* 18: 917–934.

Wolska, M. 1967a. Study on the family *Blepharocorythidae* Hsiung. II. *Charonina ventriculi* (Jameson). *Acta Protozoologica* 4: 279–283.

Wolska, M. 1967b. Study on the family *Blepharocorythidae* Hsiung. 3. *Raabena bella* gen.n., sp.n. from the intestine of Indian elephant. *Acta Protozoologica* 4: 285–290.

Wolska, M. 1968. Study on the family *Blepharocorythidae* Hsiung. 4. *Pararaabena dentata* gen.n., sp.n. from the intestine of Indian elephant. *Acta protozoologica* 5: 219–224.

Wolska, M. 1971. Studies on the family *Blepharocorythidae* Hsiung. V. A review of genera and species. *Acta Protozoologica* 9: 23–40.

Wright, D.E. 1959. Hydrogenation of lipids by rumen protozoa. *Nature, London* 184: 875–876.

Wright, D.E. 1960. Pectic enzymes in rumen protozoa. *Archives of Biochemistry and Biophysics* 86: 251–254.

Wright, D.E. 1961. Bloat in cattle. XX. Lipase activity of rumen microorganisms. *New Zealand Journal of Agricultural Research* 4: 216–223.

Wright, D.E. and Curtis, M.W. 1976. Bloat in cattle. XLII. The action of surface active chemicals on ciliated protozoa. *New Zealand Journal of Agricultural Research* 19: 19–23.

Wright, D.E. and Hungate, R.E. 1967. Amino acid concentrations in rumen fluid. *Applied Microbiology* 15: 148–151.

Xie, C.-W. and Lin, Q.-H. 1987. Studies on protozoa in the rumen of Ningxia sheep. *Proceedings of the Nutrition Society of Australia* 12: 90.

Yang, C.-M.J. and Varga, G.A. 1989. Effect of chemical defaunation on milk production and composition by Holstein cows. *Journal of Dairy Science* 72 (supplement): 414–415.

Yarlett, N., Coleman, G.S., Williams, A.G. and Lloyd, D. 1984. Hydrogenosomes in known species of rumen entodiniomorphid protozoa. *FEMS Microbiology Letters* 21: 15–19.

Yarlett, N., Hann, A.C., Lloyd, D. and Williams, A.G. 1981. Hydrogenosomes in the rumen protozoon *Dasytricha ruminantium* Schuberg. *Biochemical Journal* 200: 365–372.

Yarlett, N., Hann, A.C., Lloyd, D. and Williams, A.G. 1983a. Hydrogenosomes in a mixed isolate of *Isotricha prostoma* and *Isotricha intestinalis* from ovine rumen contents. *Comparative Biochemistry and Physiology* 74B: 357–364.

Yarlett, N., Lloyd, D. and Williams, A.G. 1982. Respiration of the rumen ciliate *Dasytricha ruminantium* Schuberg. *Biochemical Journal* 206: 259–266.

Yarlett, N. Lloyd, D. and Williams, A.G. 1985. Butyrate formation from glucose by the rumen protozoon *Dasytricha ruminantium*. *Biochemical Journal* 228: 187–192.

Yarlett, N., Orpin, C.G., Munn, E.A., Yarlett, N.C. and Greenwood, C.A. 1986. Hydrogenosomes in the rumen fungus *Neocallimastix patriciarum Biochemical Journal* 236: 729–739.

Yarlett, N., Scott, R.I., Williams, A.G. and Lloyd, D. 1983b. A note on the effects of oxygen on hydrogen production by the rumen protozoon *Dasytricha ruminantium* Schuberg. *Journal of Applied Bacteriology* 55: 359–361.

Yoder, R.C., Trenkle, A. and Burroughs, W. 1966. Influence of rumen protozoa and bacteria upon cellulose digestion *in vitro*. *Journal of Animal Science* 25: 609–612.

Yoshida, J., Nakamura, Y. and Nakamura, R. 1982. Effects of protozoal fraction and lactate on nitrate metabolism of microorganisms in sheep rumen. *Japanese Journal of Zootechnical Science* 53: 677–685.

Yoshida, M. and Katsuki, T. 1980. Ingestion and digestion of carbohydrates by rumen ciliate protozoa *in vitro*. *Bulletin of the Nippon Veterinary and Zootechnical College* 29: 63–67.

Youssef, F.G. and Allen, D.M. 1968. Part played by ciliate protozoa in rumen function. *Nature, London* 217: 777–778.

Zenbayashi, M., Kawashima, R. and Uesaka, S. 1969. Studies on importance of trace elements in farm animal feeding. XXXXI. Effect of molybdenum on carbohydrate fermentation by rumen protozoa. *Bulletin of the Research Institute for Food Science Kyoto University* 32: 32–38.

Zielyk, M.W. 1961. Ophryoscolecid fauna from the stomach of the white-tailed deer (*Odocoileus virginianus borealis*), and observations on the division of *Entodinium dubardi* Buisson 1923 (Ciliata Entodiniomorpha). *Journal of Protozoology* 8: 33–41.

Index

D

Brock/Springer Series in Contemporary Bioscience

(Continued from page ii)

John H. Andrews
COMPARATIVE ECOLOGY OF MICROORGANISMS AND MACROORGANISMS

Ryszard J. Chróst (Editor)
MICROBIAL ENZYMES IN AQUATIC ENVIRONMENTS

Alan G. Williams and Geoffrey S. Coleman
THE RUMEN PROTOZOA

John H. Andrews and Susan S. Hirano (Editors)
MICROBIAL ECOLOGY OF LEAVES

CPSIA information can be obtained at www.ICGtesting.com
Printed in the USA
LVOW04s2007181114

414349LV00003B/23/P

9 781461 276647